Vibrations of Continuous Systems

Vibrations of Continuous Systems

Arthur W. Leissa, Ph.D.
Mohamad S. Qatu, Ph.D.

New York Chicago San Francisco
Lisbon London Madrid Mexico City
Milan New Delhi San Juan
Seoul Singapore Sydney Toronto

The McGraw-Hill Companies

Cataloging-in-Publication Data is on file with the Library of Congress

1 2 3 4 5 6 7 8 9 0 DOC/DOC 1 7 6 5 4 3 2 1

ISBN 978-0-07-171479-2
MHID 0-07-171479-0

Sponsoring Editor
Michael Penn

Editing Supervisor
Stephen M. Smith

Production Supervisor
Richard C. Ruzycka

Acquisitions Coordinator
Michael Mulcahy

Project Manager
Aloysius Raj, Newgen Publishing and Data Services

Copy Editor
Paul Pineda

Proofreader
Sai Deepika

Indexer
Doreen McLaughlin

Art Director, Cover
Jeff Weeks

Composition
Newgen Publishing and Data Services

Printed and bound by RR Donnelley.

This book is printed on acid-free paper.

About the Authors

Arthur W. Leissa, Ph.D., is Professor Emeritus in the Mechanical Engineering Department at Ohio State University. A world-leading researcher in the vibrations of continuous systems, he has published more than 100 papers in this field. He is the author of two other books, *Vibration of Plates* and *Vibration of Shells,* which were reprinted in 1993 by the Acoustical Society of America as "classics in vibration," and have been cited by others in publications hundreds of times.

Dr. Leissa founded two biennial conferences: the Pan American Congress of Applied Mechanics (PACAM) in 1989 and the International Symposium on Vibrations of Continuous Systems (ISVCS) in 1997.

From 1987 to 1988 he was President of the American Academy of Mechanics, and from 1993 to 2008 he served as Editor-in-Chief of *Applied Mechanics Reviews,* the top international journal publishing review articles in applied mechanics. Dr. Leissa is a member of the editorial boards of *Journal of Sound and Vibration, International Journal of Mechanical Sciences, Composite Structures,* and *Journal of Vibration and Control.*

Mohamad S. Qatu, Ph.D., is a Professor of Mechanical Engineering at Mississippi State University. Prior to his academic career, he held consulting, senior research, and managerial positions at Ford Motor Company, Dana Corporation, Dresser Industries, and Honda North America. He is the author of *Vibration of Laminated Shells and Plates* and the co-author of two books on vehicle dynamics. Dr. Qatu has published more than 40 papers on the vibrations of continuous systems and a similar number in automotive engineering, and holds two patents.

He is the Founder and Editor-in-Chief of the *International Journal of Vehicle Noise and Vibration,* and is a member of the editorial boards of *Composite Structures, Journal of Vibration and Control,* and *SAE International Journal of Passenger Cars—Mechanical Systems.* He is a Fellow of both ASME and SAE.

Contents

Preface

Every structure or machine element in mechanical, civil, aerospace, marine, biomedical, automotive, or other engineering applications constitutes a continuous system. When subjected to an oscillating load, this system undergoes a vibratory behavior. Vibrations are an engineering concern in these applications because they may cause a catastrophic failure (complete collapse) of the machine or structure because of excessive stresses and amplitudes (resulting mainly from resonance) or because of material fatigue over a period of time. Documented examples are numerous. One of these is the collapse of the newly completed Tacoma Narrows Bridge in 1940, opened barely four months before, which swayed and collapsed in a 42-mile-per-hour wind undergoing a torsional mode resonance. In addition, vibrations can cause difficulties to users either because of excessive amplitudes or because of manifesting themselves into noise (particularly at higher frequencies). Other applications of vibrations of continuous systems can be found in sound recognitions and acoustical and music fields.

Vibrations of continuous systems is an extremely interesting subject. Discovering theoretically how strings, rods, beams, plates, shells, and other continuous bodies vibrate—particularly, in what shapes and at what frequencies they vibrate freely—is fascinating. And how they respond when subjected to fluctuating exciting forces and pressures is also interesting, and especially important in practical applications. Moreover, vibrations of continuous systems is an ideal subject to help understand the behaviors and meanings of partial differential equations and eigenvalue problems. The interplay between mathematics and physical understanding is emphasized throughout this book.

Although this work has been written as a textbook to be used in classes, it is also suitable for independent study. Read carefully, the paragraphs follow as they would in a lecture. Students (or readers) should have beforehand at least a basic understanding of *ordinary* differential equations and, preferably, some background in the vibrations of *discrete* systems. Otherwise, they will need to do supplementary reading in these subjects as they proceed. Some understanding of *partial* differential equations would also be beneficial.

Basic descriptions and explanations of vibrational concepts and phenomena are given in Chap. 1 (Introduction). This should be read carefully when beginning the book, and then read again as one progresses subsequently. Chapter 1 explains to the reader how the following chapters evolve as parts of a general development of the subject.

Each chapter after the first has problems at its end. Most of them were used as homework problems in the classes taught by the first author. They are chosen so as to develop *understanding* of the topic by the student. Most of them require significant thought and time spent (more than one hour each). For most of them, use of a computer should reduce the time required, and improve accuracy.

This book was initially written by the first author over a 30-year period. The second author wrote Secs 4.13 and 4.14 and Chap. 9, in addition to contributions to introductory sections of many chapters and his overall sponsorship and supervision of production of the *whole* manuscript. It is the result of the first author's 50 years of research in the field of vibrations of continuous systems, and having taught a graduate-level course of the same title at Ohio State University for 35 years. His research in the field resulted in the monograph *Vibration of Plates,* published in 1969, which presented results (theoretical and experimental) from approximately 500 research papers and reports. *Vibration of Shells,* published in 1973, summarized approximately 1000 references. Sources worldwide in *all* languages were used. The first monograph has been cited many hundreds of times by others in their publications, the second one almost as many. In addition, the first author supervised 40 Ph.D. dissertations and 20 M.Sc. theses, most of which dealt with the vibrations of continuous systems. The graduate student research, as well as collaboration with others, resulted in more than 100 published technical papers with the first author on vibrations of continuous systems. This book is also the result of 20 years of experience in this field, mostly in industry, by the second author. During that time he published approximately 40 technical papers on the subject (in addition to a similar number in the area of automotive noise and vibration). The second author taught the material for about 10 years at Oakland and Mississippi State universities. He also published a book on *Vibration of Laminated Shells and Plates* in 2004 which reviewed hundreds of papers in the field and is the first on the subject. The second author is also the co-author of two recent books on vehicle dynamics published in 2008 and 2009. He is supervising 10 graduate students (mostly Ph.D.s working on vibrations of continuous systems).

While the first author wishes to thank all his graduate students for their contributions through the years, the second author wishes to thank his current Ph.D. students M. Maleki, E. Asadi, W. Wang, and R. Wheeler for their help in preparing the manuscript.

Arthur W. Leissa, Ph.D.
Mohamad S. Qatu, Ph.D.

Vibrations of Continuous Systems

CHAPTER 1

Introduction

What do we mean by "vibrations of continuous systems"? "Vibrations" is a word generally understood by everyone. Webster's *New Collegiate Dictionary* gives its first (i.e., primary) definition of vibration as a "periodic motion of the particles of an elastic body or medium in alternately opposite directions from the position of equilibrium where that equilibrium has been disturbed." In the same source, "periodic" is defined as "occurring or recurring at regular intervals." These definitions are meant primarily for the layman, and they suit our technical needs reasonably well. For us the motions will be periodic in time. Rigorously periodic means the motion repeats itself exactly. In this book, we will also encounter "nearly periodic" motions as, for example, in the case of damped free vibrations.

A "continuous system" is not so obvious. In our present study, perhaps "continuous body" or "continuum" would be more immediately clear. But "continuous system" has been the terminology generally used for the past century or more for what we will deal with here, probably to contrast it with a "discrete system." Continuous systems will be described and discussed in some detail in the following section.

1.1 What Is a Continuous System?

Consider a bar (or rod) of elastic material which is fixed at the left end, as depicted in Fig. 1.1(a), and is completely free otherwise. The material of the bar is *continuous*. If its cross-sectional area (it may be circular, square, or otherwise) is A, and its total mass is M, then, the mass density at every point in the rod is $\rho = M/A\ell$, where ℓ is the bar length. This assumes that the material is *homogeneous*. Otherwise, ρ would not be a constant, but vary from point to point.

A *discrete* model of the bar is seen in Fig. 1.1(b). The mass has been "lumped" at five equally spaced points, including two at the ends and three in the interior. One could regard the bar as having been divided into three interior segments of length $\ell/4$, with the entire mass of each segment concentrated at the centers ($x = \ell/4$, $\ell/2$, $3\ell/4$); and two shorter segments, each of length $\ell/8$, to represent the ends.

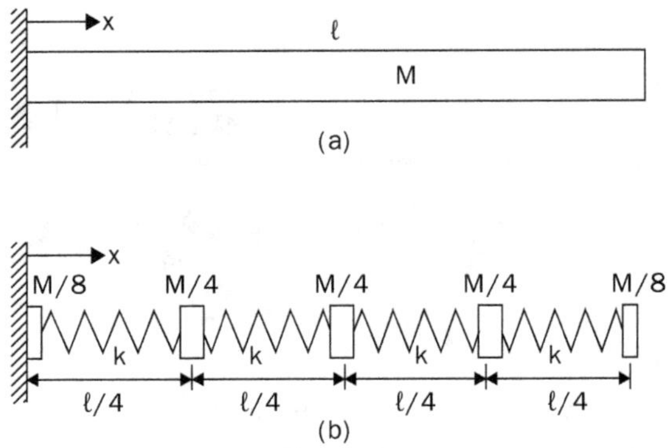

FIGURE 1.1 (a) A continuous bar of mass M; (b) a discrete model of the bar.

Thus, the interior discrete masses are each $M/4$, and the masses of the end segments are $M/8$, placed at the two ends of the bar.

The bar also has *stiffness* in the longitudinal (x) direction. In the continuous bar, the stiffness occurs uniformly along it. In the discrete model, it is represented by *massless* springs connecting the points of concentrated mass. If the bar is uniform (that is, having the same cross-section everywhere), then, the springs are each four times as stiff as the overall bar stiffness. Thus, each spring has a stiffness $k = 4\ AE/\ell$, where E is the modulus of elasticity (Young's modulus) of the material, assumed to be *linearly elastic* everywhere.

Now consider the displacements of points along the bar due to the forces acting longitudinally along it, either applied at its ends or distributed throughout it. The forces may be either static or dynamic. This one-dimensional (1D) continuous system represents the system accurately, especially if the bar is slender (i.e., ℓ is much greater than the average cross-sectional dimension). The discrete system is an approximation. That is, the continuous model can determine the displacements accurately, whereas the discrete model can only approximate them. As more points of concentrated mass are utilized, the approximation is improved. We will return to this example in more detail in Sec. 1.2.

Another example of a continuous system is a perfectly flexible string, fixed at its two ends and stretched with a tensile force (T). This is shown in Fig. 1.2(a). Ignoring gravity, the string would be stretched into a straight line by the tension. But due to transverse forces (static or dynamic) or vibratory motions, it is also shown in a typical deformed shape. The string has continuous mass all along, its total mass being M. Its transverse displacement (w) is a continuous function of the coordinate (x) which locates points along the string.

Figure 1.2(b) shows a possible discrete model of the continuous string in its transversely deformed shape. The distributed mass is

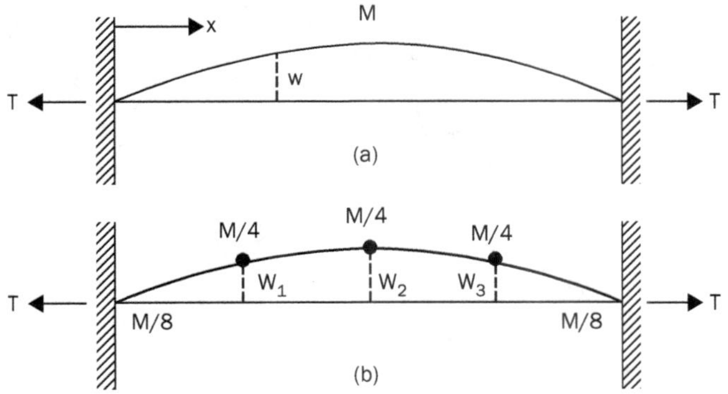

FIGURE 1.2 (a) A continuous string of mass M, displaced transversely; (b) a discrete model of the string.

replaced by three equally spaced particles of mass $M/4$ in its interior. The mass particles are interconnected by *massless* filaments, which are straight lines between the particles. The transverse displacement of the complete string is characterized now by the displacements of only the three mass points (w_1, w_2, w_3). The continuous system, which has infinite degrees of freedom (d.o.f.) in the transverse direction, has been replaced by a system with only three d.o.f.

Returning to the bar in Fig. 1.1, instead of longitudinal displacements, it could undergo *torsional* displacements, as measured by the angle (θ) by which each cross-section rotates about the axis of the bar. In this situation, the rod is frequently called a "shaft," as used in some mechanical equipment. Or, alternatively, the rod could undergo transverse displacements (w). In this latter situation the bar is typically called a "beam," which undergoes bending. A discrete representation for torsion would involve concentrated mass moments of inertia, with interconnecting massless torsional springs. The beam discrete model would be significantly more complicated, involving both translations and rotations of discrete masses, connected by translational and rotational springs.

The examples given above (string, bar, shaft, beam) are all *one-dimensional* problems. That is, the displacements of points along the body are functions of a single coordinate (x) along it. However, in each case the continuous system has infinite d.o.f., because the body has an infinite number of points of mass, each capable of moving differently than the others.

Still more examples of continuous systems are membranes, plates, and shells. Figure 1.3 shows a flat membrane of arbitrary shape, stretched in its plane by tensile force (T) around its boundary. Like the string, the membrane is assumed to be perfectly flexible. The body is two-dimensional (2D) because it takes two coordinates (e.g., x and y) to locate points in it. Static or dynamic displacements may

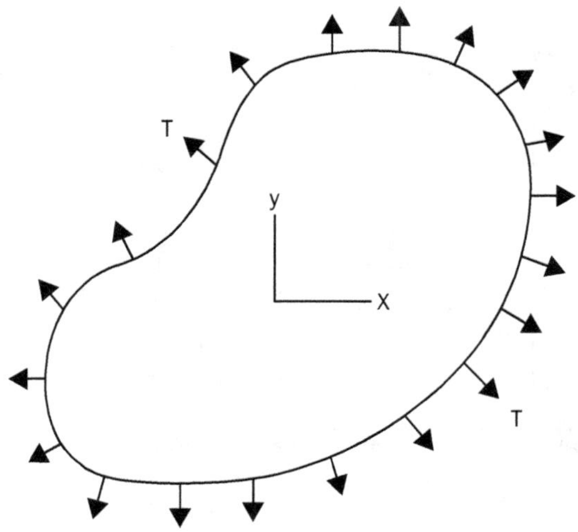

FIGURE 1.3 Flat membrane with tensile force (T) around its boundary.

occur either in the plane of the membrane, or out of its plane (transverse). Membranes are typically very thin, so that their bending stiffness is negligible. If significant bending (and/or twisting) stiffness is present, the body is considered to be a plate, and this stiffness is included in the analysis. If the body is not flat, but has curvature to form a surface, then it is a membrane shell (perfectly flexible) or a general shell (with bending and/or twisting stiffness). However, shells are also 2D because their displacements are determined by those of their middle surfaces (midway between the inner and out surfaces), and it takes only two coordinates to locate a point on a surface.

The foregoing 1D and 2D idealizations can often be made for structural elements. However, if they are not slender (or thin), then it may be necessary to carry out a 3D analysis of the displacements. Such an analysis is typically much more difficult than one which is 1D or 2D.

Typical structures are still more complicated. Examples of these are aircraft, buildings, automobiles, bridges, ships, and machinery. These may be regarded as assemblages of continuous systems (beams, plates, shells, etc.). In such cases, because of the geometrical complication involved, the structures are typically treated as discrete systems, with their components being approximated, using thousands of d.o.f. sometimes to represent the deformations of the entire structure. Nevertheless, understanding of the behavior of the relatively simple continuum models can often help greatly in the understanding of the more complicated structure, either a single part of it, or the entire body. For example, a submarine, an airplane wing, or a television transmission tower, each of them being a complicated structure, will typically have its most important lowest free vibration frequencies occurring in mode shapes which are similar to those encountered for bars and beams.

1.2 A Comparison of Frequencies Obtained from Continuous and Discrete Models

Let us return to the fixed-free bar described earlier in Fig. 1.1 and consider its longitudinal vibrations. Let its natural frequencies be written in *nondimensional form* by the parameter $\omega^* = \omega\sqrt{M\ell / AE}$. That is, for a given rod of mass (M), length (ℓ), cross-sectional area (A), and modulus of elasticity (E), knowing ω^* allows one to determine the frequency (ω). Moreover, as it can be easily shown (see Chap. 3) for bars of arbitrary M, ℓ, A, and E, the nondimensional frequencies for the continuous bar are exactly $\omega^* = \pi/2, 3\pi/2, 5\pi/2$, etc. ($\pi = 3.14159\ldots$).

Although Fig 1.1(b) shows a four d.o.f. discrete model of the rod, one could also model the rod by less (1, 2, 3) or more (> 4) vibrating, concentrated masses. Models having one, two, and three d.o.f. are shown in Fig. 1.4. Also shown is a discrete model having an *arbitrary* number (n) of d.o.f. The stiffness of the connecting springs are $k = n(AE/\ell)$ each.

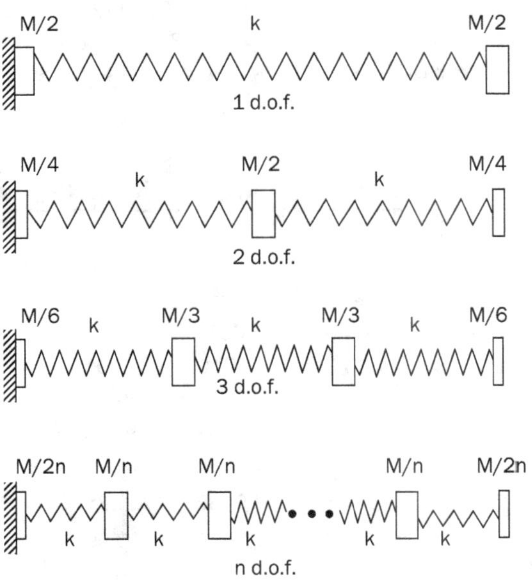

FIGURE 1.4 One, two, three, and multiple (n) degree-of-freedom discrete models of the fixed-free bar.

Drawing free body diagrams for each concentrated mass, and using either Newton's Laws or Lagrange's Equations (an energy formulation), one may obtain equations of motion for each of the systems. Assuming free (undamped) vibrations, the single d.o.f. system readily yields the frequency $\omega^* = \sqrt{2}$. For multiple (n) d.o.f. discrete representations, ω^* are found from the roots (eigenvalues) of the following determinant equation:

$$\begin{vmatrix} (2-\lambda) & 1 & 0 & 0 & 0 & - & - & - & 0 & 0 & 0 \\ -1 & (2-\lambda) & -1 & 0 & 0 & - & - & - & 0 & 0 & 0 \\ 0 & -1 & (2-\lambda) & -1 & 0 & - & - & - & 0 & 0 & 0 \\ - & - & - & - & - & - & - & - & - & - & - \\ 0 & 0 & 0 & 0 & 0 & - & - & - & -1 & (2-\lambda) & -1 \\ 0 & 0 & 0 & 0 & 0 & - & - & - & 0 & -1 & \frac{1}{2}(2-\lambda) \end{vmatrix} = 0 \quad (1.1)$$

where $\lambda = (\omega^*/n)^2$. For two d.o.f., one truncates the determinant, using only the top and bottom rows (note the 1/2 in the bottom row). For more d.o.f., one adds rows of the intermediate type shown, in between the top and bottom rows. Finding the roots of a determinant of order n of the type shown in (1.1) is a simple matter for a modern digital computer if n is not *extremely* large (say, $n > 1000$).

Table 1.1 lists the first (i.e., lowest) five values of ω^* for discrete models having $n = 1, 2, \ldots, 20$ d.o.f. It is seen that, as n increases, the frequencies approach the exact values of the continuous system. Moreover, the lowest frequency (called the "fundamental frequency") is obtained reasonably accurate ($\omega^* = 1.5643$) using only the five d.o.f. The resulting error of approximation is only 0.4 percent. However,

n	Mode 1	Mode 2	Mode 3	Mode 4	Mode 5
1	1.4142	-	-	-	-
2	1.5307	3.6955	-	-	-
3	1.5529	4.2426	5.7956	-	-
4	1.5607	4.4446	6.6518	7.8463	-
5	1.5643	4.5399	7.0711	8.9101	9.8769
7	1.5675	4.6239	7.4484	9.8995	11.8541
10	1.5692	4.6689	7.6537	10.4500	12.9890
15	1.5701	4.6930	7.7646	10.7510	13.6197
20	1.5704	4.7015	7.8036	10.8576	13.8447
∞ (exact)	1.5708	4.7124	7.8540	10.9956	14.1372

Table 1.1 Nondimensional Frequencies $\omega^* = \omega\sqrt{M\ell/AE}$ for n d.o.f. Discrete Models of Longitudinal Vibrations of a Fixed-Free Bar, as Described in Fig. 1.4

the errors for the second, third, fourth, and fifth frequencies are 3.7, 10.0, 26.6, and 54.2 percents, respectively.

1.3 A Preview of the Subsequent Chapters

Let us now have a preview of what will follow in this book, in the order in which it is presented. This should help significantly in the overall understanding of the subject, especially, for readers who have interests beyond the material specifically presented here.

Each chapter follows in the order of its mathematical complexity. This order is seen in Table 1.2. The transverse vibrations of strings are mathematically the most simple. Only one coordinate, measured along the string, is sufficient to define the classical problem of the transversely vibrating taut string, and the governing partial differential equation of motion is only of second order. The longitudinal or torsional vibrations of a straight bar are of the same mathematical complexity. Indeed, it will be shown that direct analogies exist between the string and rod problems, which allow the results from one to be applied to the other.

Analyzing the bending vibrations of a straight beam also requires only one coordinate, measured along the length. But the governing differential equation is then of fourth order, requiring satisfying two boundary conditions at each end of the beam, whereas the string and rod only require one. Thus, the differential equation and solutions of problems are somewhat more complicated.

The thin, flat, stretched membrane requires two coordinates to locate each point on it, and thus it is a 2D problem. Fortunately, like the string, the differential equation of motion for transverse vibrations only contains second derivatives. The inplane vibrations of such a membrane could also be analyzed but is not considered here. This would entail the solution of a plane elasticity problem, having a set of fourth-order partial differential equations. The

Chapter	Continuous systems	Dimensionality	Differential order
2	String	1	2
3	Bar	1	2
4	Beam	1	4
5	Membrane	2	2
6	Plate	2	4
7	Shell	2	8
8	Three dimensional	3	6

TABLE 1.2 Mathematical Complexity of Continuous Systems

resulting free vibration frequencies are typically of little interest, for they are usually at least one order of magnitude higher than those of transverse vibration.

A flat plate is typically thicker than a membrane, and has significant bending stiffness. Thus, like the beam, the equation of motion is now of fourth order. But, like the membrane, two coordinates are needed to locate each point on the midplane of the plate. Inplane vibrations would be the same as for the membrane, and seldom of interest.

Shells are like plates except that, instead of being flat, they have curvature. The curvature results in their being among the most efficient of all structural elements. Still, two coordinates can locate all points on their midsurfaces. But typical shells have both bending and stretching stiffness interacting with each other, resulting in eighth-order equations of motion. If the bending stiffness is negligible, the body is a membrane shell (e.g., a balloon), which may be examined as a special case of the shell vibration analysis.

Finally, the last category of continuous systems listed in Table 1.2 is denoted as "three-dimensional." This simply means that none of the simplifying kinematics assumptions used to develop equations of motion for the forging system are employed. The governing 3D equations of motion are not particularly complicated, but the necessity of using three coordinates causes greater difficulty in solving typical vibration problems.

In the subsequent chapters the governing equations of motion are first developed for each of the continuous system. This is done by making classical assumptions of structural mechanics about the material (linearly elastic, isotropic, homogeneous). The exact solutions of these partial differential equations are developed for free vibration, and boundary conditions (edge restraints) are applied. This results in mathematical eigenvalue problems, the solutions of which are the eigenvalues (nondimensional frequencies) and corresponding eigenfunctions (mode shapes). It is interesting to mention here that the term "eigenvalues," used in mathematics, comes from a partial translation of the German word "Eigenwert." A translation of "wert" is "value." A more complete, and perhaps better, translation would be "proper value."

The emphasis in this work is strongly on the free vibration problem—determining natural frequencies and corresponding mode shapes. Some attention is given in Chaps. 2 through 5 to strings, rods, beams, and membranes subjected to time-periodic exciting forces or displacements. Similar methods could be applied to the more complicated plate, shell, and 3D forced vibration problems. But in typical vibration studies it is most important to know the free vibration frequencies, where "resonance" (large displacement and stress amplitude) may occur. The magnitudes of such amplitudes require further calculation, and being able to quantify the damping presence (e.g., viscous, aerodynamic, dry friction, internal). Moreover, as it will be demonstrated in Chaps. 2 through 5, the standard forced vibration analysis of a continuous system is expressed in terms of the orthogonal eigenfunctions of the free vibration analysis.

Exact solutions of the partial differential equations of motion for continuous systems are only possible for a limited set of problems, depending on the necessary end conditions (mathematical boundary conditions) to be satisfied. For other cases, it is necessary to use an approximate method. The method should be able to generate free vibration frequencies and mode shapes which are sufficiently accurate, and require a reasonable amount of computational capability and time. The finite element method, in all its variations, is undoubtedly most widely used now for this purpose, especially for complex structures (e.g., aircraft, turbomachinery, buildings, bridges, naval vessels). In this book the approximate methods of Rayleigh and Ritz are used throughout for the relatively simple geometrical shapes analyzed. Like the finite element methods, they are based on energy principles (instead of differential equations) and, if used properly, will converge to exact frequencies and mode shapes if sufficient d.o.f. are made available. The Ritz method, in particular, has been used in hundreds of published research papers.

With one exception, the analysis carried out in the subsequent chapters is all *linear.* That is, assumptions are made such that the differential equations and boundary conditions utilized are all linear. The one exception is at the end of Chap. 2, where the nonlinear vibrations of the taut string are taken up. Generally, nonlinear effects became significant in all the continuous systems in this book when the vibratory displacements became "large." In the case of the string, it is seen that, unless the initial tension is extremely large, relatively small transverse vibration amplitudes can cause significant nonlinear effects. Nonlinearity may also affect the problem because of nonlinear material behavior. The solution of nonlinear vibration problems for continuous systems is an extremely complex and difficult subject, and therefore will be otherwise omitted here.

Some topics are looked into for the 1D configurations (strings, rods, beams) which are not taken up for the subsequent 2D and 3D situations, namely:

- Elastic supports, internally or at the boundaries
- Discontinuous bodies
- Buckling

These are not addressed in the latter chapters simply because the problems became more difficult and the solutions more lengthy. But, in principle, the same logic can be applied there as in the 1D problems. Particularly important are buckling aspects under certain static loading conditions. These are explained carefully for beams in

Sec. 4.11, but described only briefly for plates (Sec. 6.8), and mentioned for membranes (Example 5.3). In general, observing when any natural frequency approaches zero as static loading is increased is an excellent way to determine buckling loads of structural elements, both theoretically and experimentally.

CHAPTER 2

Transverse Vibrations of Strings

A string constitutes one of the most fundamental continuous systems. It is an important element in engineering and physical sciences. In engineering, strings can exist in many applications. These include belts in automotive systems and power transmission machines, cables in many structures and machines, electric power transmission lines, ropes in many devices, as well as other uses. In biomedical engineering, human cords are actually strings. Their vibrational characteristics are important in many applications including voice recognition.

In music and acoustics, strings constitute a major element in many musical instruments. Stringed instruments can be divided into different groups. There are ones in which the strings are supported by a neck and a bridge, for instance, a guitar or a violin. In other groups, the instruments have the strings contained within a frame or mounted on a body, such as a piano, cimbalom, or autoharp. The vibrational characteristics of these strings are the basic element of design in these instruments.

In physical sciences, strings may be used to study waves, their characteristics and propagation. The fundamental equations of wave propagation in strings have many analogies in physical sciences, including sound propagation.

The transverse vibration of strings will be studied in this chapter. Longitudinal vibrations of a string are also possible. For such motion, a string behaves the same as a bar, which will be taken up in Chap. 3. The fundamental differential equations of transverse vibration will be derived. Free vibrations will then be explored to determine the natural frequencies and mode shapes. Vibratory motion resulting from initial conditions will be investigated. Forced vibration with and without damping will also be treated. Approximate methods, particularly the Rayleigh and Ritz methods, will be explored. Various complicating effects like gravity, attached mass, and discontinuous strings will be covered. A section on nonlinear vibration of strings is also included.

2.1 Differential Equation of Motion

Figure 2.1 shows a completely flexible string (that is, having no bending stiffness) of length ℓ which is stretched to an initial tension (T_0). To be specific, the string is shown as being held at both ends by rigid walls, although the differential equation of motion to be derived does not depend on the end (or boundary) conditions. Indeed, as we shall see later, other physically meaningful boundary conditions can exist for the string.

The longitudinal coordinate (x) in Fig. 2.1 is taken in the direction of the initial, undeformed string. The coordinate origin is shown at the left wall, but this is arbitrary; it may be chosen anywhere. The string is shown in a representative, deformed shape that it has at a typical instant while undergoing vibration. Let the transverse direction be z, and w be the displacement of the string at any instant in the z direction, measured from the straight line, static equilibrium position (assuming that transverse gravity forces are not present). This notation for displacement components will be used consistently throughout this book. That is,

Coordinate	Associated displacement
x	u
y	v
z	w

A typical infinitesimal element of length ds, measured along the deformed string, is also shown in Fig. 2.1. The displacement w describes the motion of this typical element, and it depends on (i.e., is a function of) both its longitudinal location (x) and time (t). That is, $w = w(x,t)$.

In Fig. 2.2, the infinitesimal element is drawn enlarged and all forces acting on it are shown, yielding a free body diagram. The tension (T) is not necessarily constant, but may vary both with x (or s)

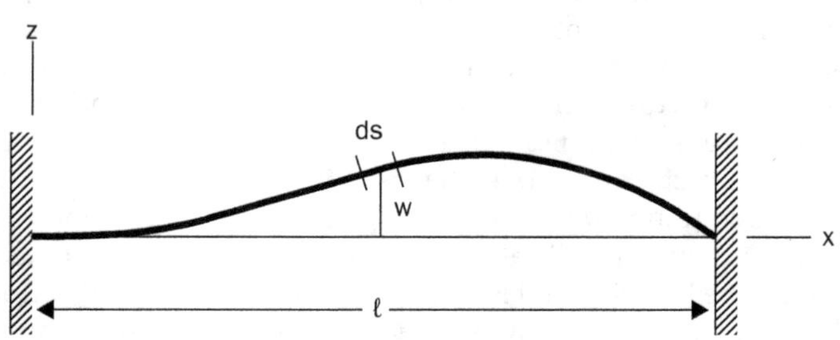

FIGURE 2.1 A flexible string of length ℓ.

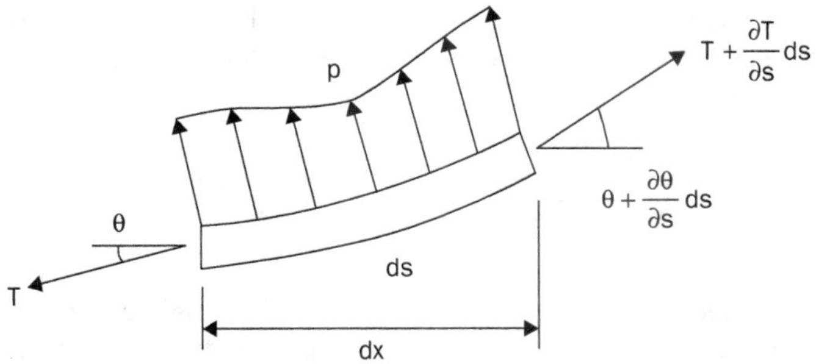

FIGURE 2.2 A free body diagram of a string infinitesimal element.

and t. The angle (θ) that the string makes with the horizontal axis also varies with x (or s) and t. In this chapter (except in Sec. 2.14), it will be assumed that the vibratory displacements are small (compared to ℓ), and that θ is also small everywhere. An external, distributed force (p, having dimensions of force per unit length) is also shown, acting normal (perpendicular) to the string.

Summing forces in the z direction gives

$$\sum F_z = -T\sin\theta + \left(T + \frac{\partial T}{\partial s} ds\right)\sin\left(\theta + \frac{\partial \theta}{\partial s} ds\right) + (p\,ds)\cos\theta = (\rho\,ds)\frac{\partial^2 w}{\partial t^2} \tag{2.1}$$

where $\rho = \rho(x)$ is the mass density *per unit length* of string. Using the trigonometric identity for the sine of the sum of two angles, and replacing $\sin(\alpha)$ by α, and $\cos(\alpha)$ by 1 for small angles, yields

$$\sin\left(\theta + \frac{\partial \theta}{\partial s} ds\right) = \sin\theta \bullet \cos\left(\frac{\partial \theta}{\partial s} ds\right) + \cos\theta \bullet \sin\left(\frac{\partial \theta}{\partial s} ds\right) = \sin\theta + \cos\theta \bullet \frac{\partial \theta}{\partial s} ds \tag{2.2}$$

Substituting (2.2) into (2.1), expanding the indicated product, canceling out the $T\sin(\theta)$ terms, one obtains

$$\left(T\cos\theta\frac{\partial \theta}{\partial s} + \frac{\partial T}{\partial s}\sin\theta + \frac{\partial T}{\partial s} ds \bullet \cos\theta\frac{\partial \theta}{\partial s} + p\cos\theta\right)ds = \rho\frac{\partial^2 w}{\partial t^2} ds \tag{2.3}$$

It is seen that all terms except one in (2.3) contain ds (i.e., they are first-order differentials). The remaining one contains $(ds)^2$, which is of higher order, and can therefore be discarded with no error. Doing so, and dividing through by the length ds, (2.3) may be rewritten as

$$\frac{\partial}{\partial s}(T\sin\theta)+p\cos\theta=\rho\frac{\partial^2 w}{\partial t^2} \tag{2.4}$$

As written, (2.4) is a nonlinear partial differential equation, with T depending on w. Indeed, if one wanted to consider large amplitude vibrations of the string, it could be useful. But we will avoid this difficult problem by making certain linearizing assumptions. First, assume that the slope of the displaced string ($\partial w/\partial x$) is small for all values of x and t. Then,

$$\sin\theta\approx\tan\theta=\frac{\partial w}{\partial x},\quad \cos\theta\approx 1,\quad \frac{\partial}{\partial s}=\frac{\partial}{\partial x} \tag{2.5}$$

This allows one to replace (2.4) by

$$\frac{\partial}{\partial x}\left(T\frac{\partial w}{\partial x}\right)+p=\rho\frac{\partial^2 w}{\partial t^2} \tag{2.6}$$

Second, assume that the initial tension T_0 is sufficiently large, and that the transverse displacement w is sufficiently small, so that T may be assumed *constant* during the motion (i.e., $T = T_0$). Then, (2.6) simplifies further to

$$T_0\frac{\partial^2 w}{\partial x^2}+p=\rho\frac{\partial^2 w}{\partial t^2} \tag{2.7}$$

which is the equation of motion governing forced vibration of the string. To simplify notation, we will not bother with the subscript on T. We will return to this equation in Sec. 2.9.

For the free vibration problem, $p = 0$, and (2.7) becomes

$$\boxed{T\frac{\partial^2 w}{\partial x^2}=\rho\frac{\partial^2 w}{\partial t^2}} \tag{2.8}$$

This is the well-known one-dimensional "wave equation" of physics. Why it should be called this will be seen in Sec. 2.5. Note that in all of the above equations, the string density may vary, i.e., $\rho = \rho(x)$. However, for ease in obtaining mathematical solutions, ρ will be taken as constant for the next several sections of this chapter.

2.2 Free Vibrations; Classical Solution

For the case ρ = constant, the classical method of separation of variables will now be employed to obtain a solution to (2.8). It must be noted, however, that this procedure *cannot* be used for solving all partial differential equations, as will be seen in Chap. 6.

Assume that the solution to (2.8) may be written in variables separable form; that is,

$$w(x,t) = X\phi \tag{2.9}$$

where $X = X(x)$ and $\phi = \phi(t)$ are each a function of one of the independent variables. Substituting (2.9) into (2.8) yields

$$TX''\phi = \rho X\phi'' \tag{2.10}$$

where the double primes indicate differentiation twice with respect to the arguments x and t of functions X and ϕ, respectively. Dividing through (2.10) by $X\phi$, and collecting constants on one side of the equation, allows us to rewrite (2.10) as

$$\frac{X''}{X} = \left(\frac{\rho}{T}\right)\frac{\phi''}{\phi} \tag{2.11}$$

Now the fundamental argument is made that since the left-hand side of (2.11) is a function only of x, and the right-hand side (R.H.S.) of (2.11) is a function only of t, then the equation can only be valid if each side equals a constant. For convenience, call this constant $-\alpha^2$. As a consequence, we may write

$$X'' + \alpha^2 X = 0 \tag{2.12a}$$

$$\phi'' + \omega^2\phi = 0 \tag{2.12b}$$

where

$$\omega^2 = \frac{T\alpha^2}{\rho} \tag{2.13}$$

Before considering solutions to (2.12) let us first write down the *boundary conditions* for the problem, assuming both ends of the string are fixed, as in Fig. 2.1. Because (2.8) is a second-order differential equation, we can expect one boundary condition per end, viz

$$w(0,t)=0 \tag{2.14a}$$

$$w(\ell,t)=0 \tag{2.14b}$$

These imply, from (2.9),

$$X(0)=0 \tag{2.15a}$$

$$x(\ell)=0 \tag{2.15b}$$

There are three possible types of solution to (2.12) depending on whether α^2 is negative, zero, or positive.

Case I: $\alpha^2 < 0$. Let $\beta^2 \equiv \alpha^2$. Then, (2.12a) has the solution in terms of hyperbolic functions

$$X = A\sinh\beta x + B\cosh\beta x \tag{2.16}$$

where A and B are constants to be determined. Substituting (2.16) into (2.15a), and recalling that sinh 0 = 0 and cosh 0 = 1, yields $B = 0$. Further, substituting (2.16) into (2.15b) yields A sinh $\beta\ell = 0$. This can be satisfied if either $A = 0$ or sinh $\beta\ell = 0$. If $A = B = 0$, then (2.16), together with (2.9), shows that no motion can exist. This trivial solution is the special case of static equilibrium with no displacement, but is of no interest to us. If sinh $\beta\ell = 0$, then, $\beta = 0$. But, $\beta = B = 0$ yields again the trivial solution according to (2.16).

Case II: $\alpha^2 = 0$. Then, (2.12a) has the solution

$$X = Ax + B \tag{2.17}$$

Applying boundary conditions, (2.15) requires $A = B = 0$, which again is a trivial solution.

Case III: $\alpha^2 > 0$. Then, (2.12a) has the solution in terms of trigonometric functions

$$X = A\sin\alpha x + B\cos\alpha x \tag{2.18}$$

Substituting (2.18) into (2.15a) requires that $B = 0$. Further, (2.15b) yields A sin $\alpha\ell = 0$. If $A = 0$, a trivial solution again results. Thus, the only nontrivial possibility is sin $\alpha\ell = 0$, which gives

$$\alpha\ell = m\pi \quad (m = 1, 2, \ldots, \infty) \tag{2.19}$$

The solution to (2.12b) is

$$\phi = C\sin\omega t + D\cos\omega t \tag{2.20}$$

where, from (2.13) and (2.19)

$$\omega = \frac{m\pi}{\ell}\sqrt{\frac{T}{\rho}} \quad (m = 1, 2, \ldots, \infty) \tag{2.21}$$

and C and D are constants of integration to be determined from the *initial conditions* for a particular problem. We recognize from (2.20) that ω is the *circular frequency* of free vibration (usually determined in radians/sec). The cyclic frequency (f) is related to ω by

$$f = \frac{\omega}{2\pi} \tag{2.22}$$

and has dimensions of cycles per second or hertz (Hz). Equation (2.21) tells us that the freely vibrating string has an *infinite* number of possible frequencies, and that the frequencies are integer multiples of the first frequency (which is called the *fundamental frequency*). However, as it will be seen in subsequent problems for continuous systems, this fortunate circumstance is most unusual. Equation (2.21) also shows that each frequency is increased as the tension in the string is increased, or as the length or density is decreased. This physical behavior is not unexpected, and is obvious enough to anyone who has played around with a guitar. While $\rho\ell$ is a measure of the total mass present in the system, T/ℓ is a measure of the stiffness. The period (τ_m) for the mth vibration frequency (ω_m) is the reciprocal of the cyclic frequency, or $2\pi/\omega_m$.

Equation (2.19) tells us that $\alpha\ell$ are the *eigenvalues* of the problem. That is, they are the "proper values" which, if chosen, permit us to obtain a nontrivial solution satisfying both the differential equation and boundary conditions, all of which are homogeneous. For this problem, the $\alpha\ell$ parameters are the *nondimensional frequencies*. That is, (2.19) may be rewritten as

$$\alpha\ell = \omega\ell\sqrt{\frac{\rho}{T}} = m\pi \quad (m = 1, 2, \ldots, \infty) \tag{2.23}$$

which, of course, is a rearrangement of (2.21). Continuing, the sin αx are the *eigenfunctions*. For each $\alpha\ell$, determined from (2.21), there exists one eigenfunction. This gives the displaced shape of the string vibrating in this mode and, in the usual terminology, is called the *mode shape*. While the constant B in (2.18) was found to be zero, the remaining constant A remains arbitrary, and is combined with C and D from (2.20) when rewriting (2.9). Thus, the amplitude of the eigenfunction is arbitrary at this stage and will ultimately be determined from the initial conditions.

Summarizing the above results, it may be said that a taut string is capable of executing free vibrations with a displacement function w given by

$$w_m(x,t) = \sin\alpha_m x(C_m \sin\omega_m t + D_m \cos\omega_m t) \tag{2.24}$$

The subscript m has been added to identify the mth mode shape (or, simply, mode), which vibrates with a frequency ω_m, and with an amplitude determined by C_m and D_m. The first four mode shapes (i.e., the four having the lowest frequencies) are depicted in Fig. 2.3. It is observed that $m = 1, 3, \ldots$ yields the symmetric modes, and that $m = 2, 4, \ldots$ furnishes the antisymmetric modes (with respect to the symmetry axis for the problem, which is at the center of the string). The points of zero displacement are called the "node points." The mode shapes are drawn in Fig. 2.3 as if they each have the same amplitude; but it is clear that the small slope assumptions (2.5) would be seriously violated for $m = 3$ and 4, as drawn.

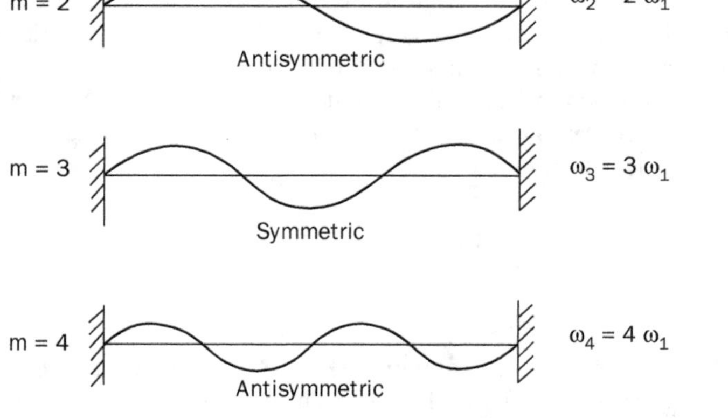

FIGURE 2.3 The first four mode shapes of a fixed string.

2.3 Initial Conditions

In a general case, a string could be set into motion by giving it both initial displacement and velocity. These conditions could be written as

$$w(x,0) = f(x) \tag{2.25a}$$

$$\frac{\partial w}{\partial t}(x,0) = g(x) \tag{2.25b}$$

where $f(x)$ and $g(x)$ are any functions continuous over the interval $0 \leq x \leq \ell$ which also satisfy the boundary conditions $f(0) = f(\ell) = g(0) = g(\ell) = 0$ (assuming that both ends of the string are fixed). In the special case when a string is plucked at one or more points, $g(x) = 0$. Another special case is where $f(x) = 0$. This is possible if the string is set into motion by an initial impact.

Because (2.24) is a solution to the linear differential equation (2.8) for every value of m, then a linear superposition of such solutions is also a solution. Thus, a general solution of (2.8) may be taken as

$$w(x,t) = \sum_{m=1}^{\infty} \sin\alpha_m x\left(C_m \sin\omega_m t + D_m \cos\omega_m t\right) \tag{2.26}$$

Thus, the free vibration of the string is assumed to be represented by the superposition of its free vibration modes, each having its own amplitude and its own frequency. Substituting (2.26) into (2.25a) yields

$$f(x) = \sum_{m=1}^{\infty} D_m \sin\alpha_m x \tag{2.27}$$

Multiplying both sides of (2.27) by $\sin\alpha_n x$, where $\alpha_n = n\pi/\ell$, and n is also an integer, and integrating both sides of the equation from 0 to ℓ, one finds that

$$\int_0^\ell \sin\frac{m\pi x}{\ell} \cdot \sin\frac{n\pi x}{\ell} dx = \begin{cases} 0 & \text{if } m \neq n \quad (2.28\text{a}) \\ \ell/2 & \text{if } m = n \quad (2.28\text{b}) \end{cases}$$

The first of these two equations is a statement of the *orthogonality* of the eigenfunctions. Because of this, all terms but one on the R.H.S. of 2.27 vanish. The sole remaining term, which exists when $m = n$, results from using (2.28b):

$$D_m = \frac{2}{\ell}\int_0^\ell f(x)\sin(\alpha_m x)dx \tag{2.29}$$

Thus, (2.27) is the Fourier series expansion for $f(x)$, and (2.29) is the well-known formula for calculating the Fourier coefficients.

Similarly, substituting (2.26) into (2.25b) and carrying out the same operations described above gives

$$C_m = \frac{2}{\omega_m \ell} \int_0^\ell g(x) \sin(\alpha_m x)\, dx \tag{2.30}$$

where attention must be called to ω_m in the denominator.

In the special case of initial displacement only, with no initial velocity, (2.30) gives $C_m = 0$, with the D_m being determined from (2.29). Conversely, for initial velocity only, $D_m = 0$, and C_m is given by (2.30).

Example 2.1 A taut string of length ℓ is plucked at its one-quarter point as shown in Fig. 2.4 and is released from rest. Determine the ensuing motion.

Solution

$$w(x,0) = f(x) = \sum_{m=1}^{\infty} D_m \sin \alpha_m x = \begin{cases} 4\delta \dfrac{x}{\ell} & \text{if } 0 \le x \le \dfrac{\ell}{4} \\ \dfrac{4}{3}\delta\left(1 - \dfrac{x}{\ell}\right) & \text{if } \dfrac{\ell}{4} \le x \le \ell \end{cases}$$

$$\begin{aligned} D_m &= \frac{2}{\ell} \int_0^\ell f(x) \sin \alpha_m x\, dx \\ &= \frac{2}{\ell} \int_0^{\ell/4} \left(4\frac{\delta}{\ell}\right) x \sin \alpha_m x\, dx + \frac{2}{\ell} \int_{\ell/4}^{\ell} \left(\frac{4}{3}\delta\right)\left(1 - \frac{x}{\ell}\right) \sin \alpha_m x\, dx \end{aligned}$$

Integrating by parts, where needed, gives

$$D_m = \frac{32}{3} \frac{\delta}{(m\pi)^2} \sin \frac{m\pi}{4}$$

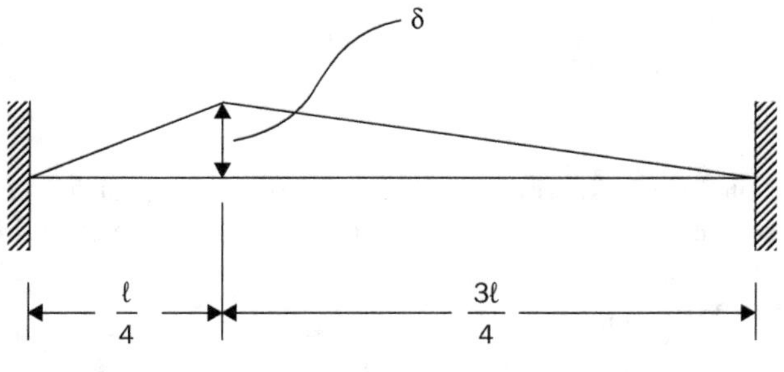

FIGURE 2.4 Initial shape of plucked string in Example 2.1.

whence

$$w(x,t)=\frac{32\delta}{3\pi^2}\sum_{m=1}^{\infty}\frac{1}{m^2}\sin\frac{m\pi}{4}\sin\alpha_m x\cos\omega_m t$$

where

$$\alpha_m=\frac{m\pi}{\ell},\qquad \omega_m=\frac{m\pi}{\ell}\sqrt{\frac{T}{\rho}}$$

In detail, the Fourier series expansion of the initial shape is

$$w(x,0)=\delta\left[0.7642\sin\frac{\pi x}{\ell}+0.2702\sin\frac{2\pi x}{\ell}+0.0849\sin\frac{3\pi x}{\ell}\right.$$
$$\left.+\,0-0.306\sin\frac{5\pi x}{\ell}+\cdots\right]$$

This shows the relative amplitudes with which the various modes are excited by this particular initial shape. The second and third modes have amplitudes that are 35 percent and 11 percent as large, respectively, as the first mode. Moreover, the 4th, 8th, 12th, ... modes are not excited at all by this initial shape. These modes have node points where the string is plucked. This is a first example of the more general observation that one cannot generate a vibration mode if the excitation occurs at the node point of the mode. Thus, the relative strengths of the overtones, compared with the fundamental tone, that one hears from the string of a musical instrument depend on what point it is plucked, and some overtones may not be excited at all.

A plot of the first three sums of the Fourier series for the displacement when $t = 0$ is shown in Fig. 2.5. Using only the first three (or four) terms of the series is seen to give a poor approximation to the exact shape in the vicinity of the maximum displacement.

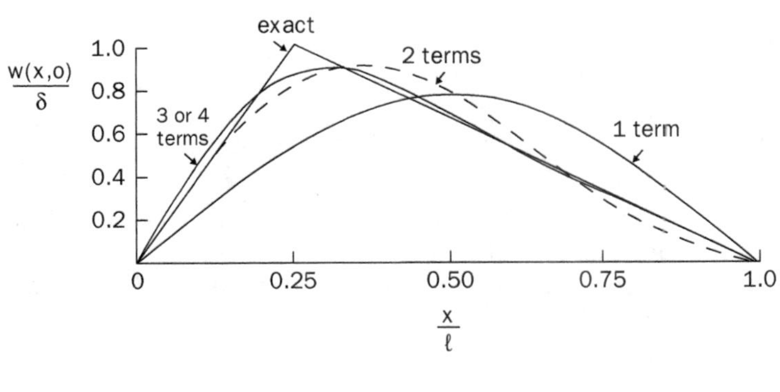

FIGURE 2.5 A plot of the first three sums of the Fourier series for the displacement when $t = 0$ in Example 2.1.

2.4 Consideration of Transverse Gravity

In the preceding sections, it was implicitly assumed that the free vibration takes place in the absence of gravitational forces. This could be possible if the system were in space, sufficiently far removed from significant gravitational effects, or if the forces were acting perpendicular to the plane of motion (i.e., in the y direction in Fig. 2.1).

Let us now suppose that gravity acts in the negative z direction in Fig. 2.1, and that the change in T due to it is negligible. The resulting infinitesimal force $p\,dx$ in (2.1) is then replaced by $-\rho g\,ds$, where g is the gravitational acceleration constant, and (2.8) becomes

$$T\frac{\partial^2 w}{\partial x^2} - \rho g = \rho\frac{\partial^2 w}{\partial t^2} \tag{2.31}$$

Equation (2.31) is a nonhomogeneous differential equation, but it is linear. Its solution $w(x,t)$ may therefore be regarded as the sum of two parts, i.e.,

$$w = w_c + w_p \tag{2.32}$$

where w_c is the complementary solution, obtained when ρg is ignored, and w_p is a suitable particular solution. The word "suitable" should be emphasized here, for various forms of particular solution may be possible, all differing from each other by terms of the complementary solution.

If we choose w_p to be the *static* displacement of the string due to the gravitational force, then w_p is the solution of

$$T\frac{\partial^2 w_p}{\partial x^2} = \rho g \tag{2.33}$$

which is

$$w_p = \left(\frac{\rho g}{2T}\right)x^2 + C_1 + C_2 x \tag{2.34}$$

where C_1 and C_2 are arbitrary constants of integration. If C_1 and C_2 are chosen so that w_p also satisfies the boundary conditions $w_p(0) = w_p(\ell) = 0$, then, $C_1 = 0$, $C_2 = -\rho g\ell/2T$ and the general solution (2.32) becomes

$$w = w_c - \left(\frac{\rho g}{2T}\right)(\ell x - x^2) \tag{2.35}$$

where w_c is the previous solution ignoring gravity, given by (2.26).

Thus, (2.35) may be regarded as a vibrational motion (w_c) superimposed on a static displacement (w_p) due to gravity. Furthermore, *the free vibration frequencies are unaffected by g*, inasmuch as the α_m are determined to be the same as without gravity in order to satisfy $w(\ell)= 0$. However, one must be careful not to extrapolate this conclusion to other problems. It applies to a *string* subjected to *transverse* gravity. For some other types of problems, gravity will be seen to have an effect on the vibration frequencies (but for others, it will not). However, this should not be a surprise, for in elementary texts on vibrations, it is quickly shown that even for single degree-of-freedom (d.o.f.) systems, gravity may or may not affect the frequency, depending on the problem considered.

2.5 Free Vibrations; Traveling Wave Solution

It was shown in Sec. 2.3 that if the vibration of a string is commenced by giving the string an initial displacement $f(x)$ and releasing it, the subsequent motion is given by

$$w(x,t)=\sum_{m=1}^{\infty} D_m \sin\alpha_m x \bullet \cos\omega_m t \tag{2.36}$$

But since, from (2.19) and (2.21), $\omega_m = \alpha_m c$, where $c=\sqrt{T/\rho}$, then (2.36) may be written as

$$w(x,t)=\sum_{m=1}^{\infty} D_m \sin\alpha_m x \bullet \cos\alpha_m ct \tag{2.37}$$

Using the trigonometric identity

$$\sin\alpha_m x \bullet \cos\alpha_m ct = \frac{1}{2}[\sin(\alpha_m x+\alpha_m ct)+\sin(\alpha_m x-\alpha_n ct)] \tag{2.38}$$

allows us to rewrite (2.37) as

$$w(x,t)=\frac{1}{2}\sum_{m=1}^{\infty} D_m[\sin\alpha_m(x+ct)] + \frac{1}{2}\sum_{m=1}^{\infty} D_m[\sin\alpha_m(x-ct)] \tag{2.39}$$

But, using (2.27), (2.39) may be simply stated as

$$w(x,t) = \frac{1}{2}f(x+ct) + \frac{1}{2}f(x-ct) \tag{2.40}$$

where, as before, $f(x)$ is the initial displacement function.

If one were to plot an arbitrary mathematical function $f(x)$, then, $f(x - ct)$ and $f(x + ct)$ would be the same function shifted forwards and backwards an amount ct, as shown in Fig. 2.6. Thus, (2.40) tells us that, as an alternative to (2.36), the displacement at any instant of time (t) after the beginning of motion ($t = 0$) may be regarded as the superposition of two functions. Each function has the same shape as the initial displacement, but only half the amplitude, shifted forwards and backwards an amount ct. Physically, this corresponds to two half-amplitude waves, one moving forwards and the other moving backwards along the string, with a wave velocity $c = \sqrt{T/\rho}$. Because (2.40) is also a solution to (2.8), the latter is often called the "wave equation."

Example 2.2 A taut string is initially deformed into the shape shown in Fig. 2.7 and released from rest. Use the traveling wave solution to determine its subsequent motion.

Solution

Figure 2.8 shows the string at seven equally spaced time intervals, beginning with $t = 0$ and ending with $t = \tau_1/2$ where $\tau_1 = 2\pi\ell 2_1 = 2\ell\sqrt{\rho/T}$ is the period of the vibration (i.e., the time when the initial shape reappears), as may be determined from (2.23). In the sketch, the string is drawn between the rigid walls $x = 0$ and $x = \ell$, but an additional length ℓ is also shown on each side of the actual string to permit us to keep track of the function (i.e., the traveling waves) beyond the limits of the boundaries. Actually, the complete function is a periodic one from $-\infty < x < \infty$.

At $t = 0$, the displaced shape is shown in $0 \le x \le \ell$ as a solid line, whereas the dashed line shows one of the functions of half-amplitude which will travel with increasing t. The function is antisymmetric with respect to $x = 0$

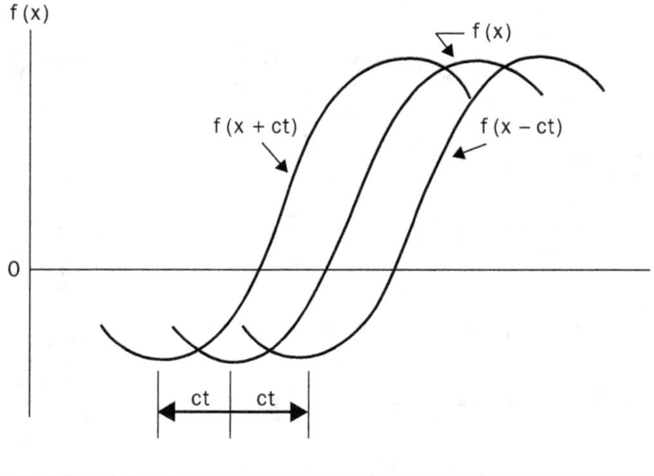

Figure 2.6 The function $f(x)$ shifted forwards and backwards an amount ct.

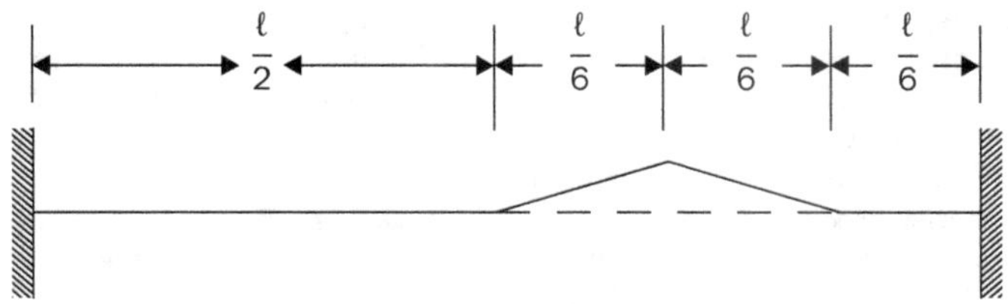

FIGURE 2.7 Initially deformed string of Example 2.2.

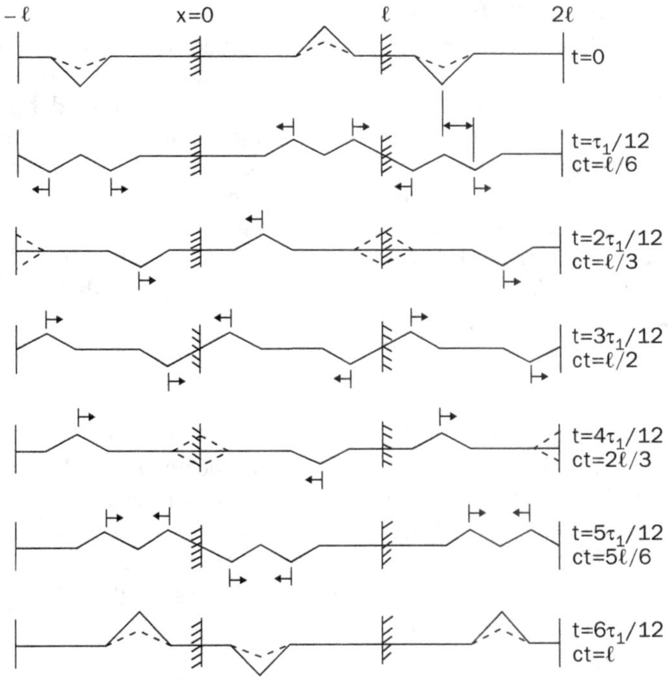

FIGURE 2.8 Travelling wave of Example 2.2 at various times.

[i.e., $f(x) = -f(-x)$] and it is periodic in x, with period 2ℓ [i.e., $f(x + 2\ell) = f(x)$], both properties being clear from (2.27). These properties are present for all t and readily permit one to draw $\frac{1}{2}f(x-ct)$ and $\frac{1}{2}f(x+ct)$ outside of the interval $0 \le x \le \ell$ once they are known within the interval.

At $t = \tau_1/12$, the two half-waves are completely separated, and the one traveling to the right is on the verge of infringing on the boundary at $x = \ell$. But at the same instant, another half-wave traveling to the left from the subsequent interval $\ell \le x \le 2\,\ell$ begins to make contact with $x = \ell$.

During the time $\tau_1/12 < t < \tau_1/6$, the two half-waves partially cancel each other and at $t = \tau_1/6$ they exactly cancel each other in the vicinity of the wall, $5\ell/6 < x < 7\ell/6$. The model of a *reflected* wave may also be used in place of the left-traveling wave to explain the cancellation. At every t, the two half-waves superimpose so that $w(\ell, t) = 0$ exactly.

The displaced shapes of the string for $t > \tau_1/2$ are obtained by proceeding *upwards* from one sketch to another in Fig 2.8, beginning with the one for $t = \tau_1/2$. That is, for example, $w(x, 7\tau_1/12) = w(x, 5\tau_1/12)$.

It is seen that the traveling wave solution permits one to determine the shape of the string at any instant by a relatively simple graphical procedure, compared with the large amount of numerical computation typically required by the classical eigenfunction superposition approach laid out in Sec. 2.3.

2.6 Other End Conditions

Only a string having both of its ends fixed has thus far been considered. Let us now take up a more general case, as shown in Fig. 2.9. There one sees a string that has one end ($x = 0$) fixed, but the other end is attached to a mass (M) that can move transversely without friction. Moreover, the mass is constrained by a spring of stiffness k. Gravity is ignored, for, as seen in Sec. 2.4, it has no effect on the free vibration frequencies or mode shapes.

The solution to (2.8) given by (2.9), (2.18), and (2.20) will again be used. The boundary condition $w(0, t) = 0$ again requires that $B = 0$, so that (2.18) again reduces to

$$X = A \sin \alpha x \tag{2.41}$$

where α is a constant yet to be determined.

The boundary condition at $x = \ell$ is a complicated one, which must be determined by drawing a free body diagram of the attached mass in a typical position, displaced $w(\ell, t)$ from the static equilibrium position. This is seen in Fig. 2.10. The three forces shown are the tension in the string (T), the normal force (N) of the constraining

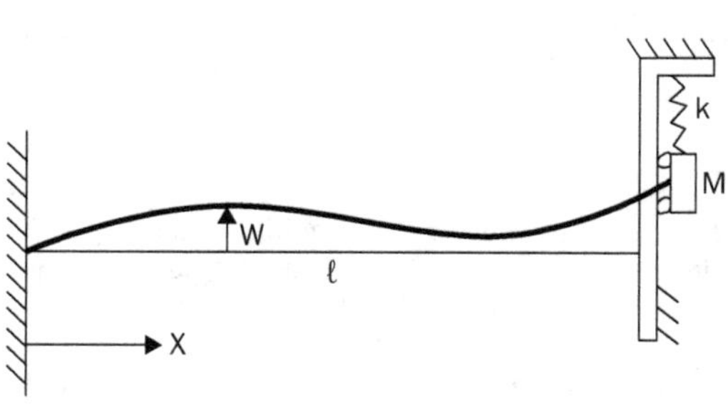

FIGURE 2.9 A more general end condition for a string.

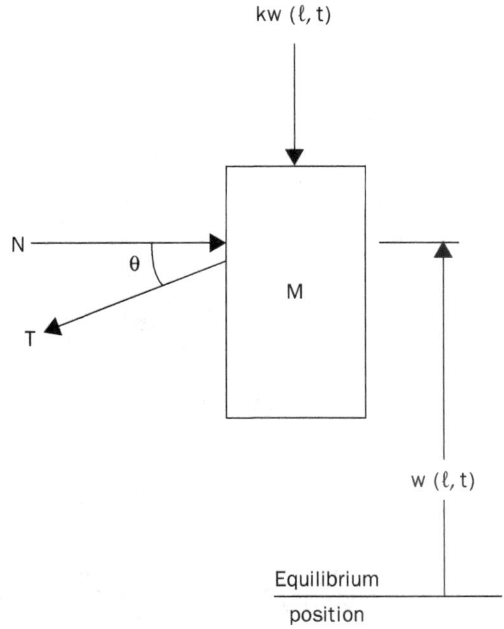

FIGURE 2.10 A free body diagram of a mass connected at a string end and attached to a spring.

boundary against the mass, and the restoring force in the constraining spring (kw), all drawn in their positive directions.

Summing forces vertically, we then obtain

$$-T\sin\theta - kw(\ell,t) = M\frac{\partial^2 w}{\partial t^2}(\ell,t) \tag{2.42}$$

But $\sin\theta = \partial w(\ell,t)/\partial x$ for small θ. Assuming $w(x,t) = X(x)\phi(t)$,

$$-TX'(\ell) - kX(\ell) = -M\omega^2 X(\ell) \tag{2.43}$$

or, using (2.41),

$$(M\omega^2 - k)A \bullet \sin\alpha\ell = TA\alpha \bullet \cos\alpha\ell \tag{2.44}$$

Dividing by A cos $\alpha\ell$ and then rearranging gives

$$\tan\alpha\ell = \frac{T\alpha}{M\omega^2 - k} \tag{2.45}$$

Substituting (2.13) and rewriting (2.45) in nondimensional form yields the frequency equation

$$\tan\beta = \frac{1}{M^*\beta - k^*/\beta} \tag{2.46}$$

where M^* and k^* are the nondimensional mass and stiffness ratios, respectively, defined by

$$M^* \equiv \frac{M}{\rho\ell}, \qquad k^* \equiv \frac{k}{T/\ell} \tag{2.47}$$

and $\beta \equiv \alpha\lambda$ is the nondimensional frequency:

$$\beta = \omega\ell\sqrt{\frac{\rho}{T}} \tag{2.48}$$

For selected values of M^* and k^*, the values of β which satisfy (2.46) are the eigenvalues for the problem. There are infinite numbers of β for each choice of M^* and k^*. The eigenfunctions (mode shapes) are then given by (2.41), which can be written in the more convenient form

$$X = A\sin\beta\epsilon \tag{2.49}$$

where $\epsilon \equiv x/\ell$.

To find the eigenvalues of the transcendental equation (2.46), one may rearrange it as

$$\tan\beta - \frac{1}{M^*\beta - k^*/\beta} = f(\beta) = 0 \tag{2.50}$$

Then, for fixed values of M^* and k^*, one may determine the roots of (2.50). This may be done by simply plotting $f(\beta)$ and determining its zero values; alternatively, various numerical techniques may be used. However, these procedures require using at least a programmable hand calculator, if not a small computer, if many eigenvalues are to be found.

If extensive numerical results are sought, it is easier to fix β in (2.46) and then solve for corresponding sets of M^* and k^* from the *linear* form

$$M^*\beta - \frac{k^*}{\beta} = \cot\beta \tag{2.51}$$

This procedure was used to make the plots shown in Fig. 2.11. For a chosen value of M^*, a number of values of the nondimensional

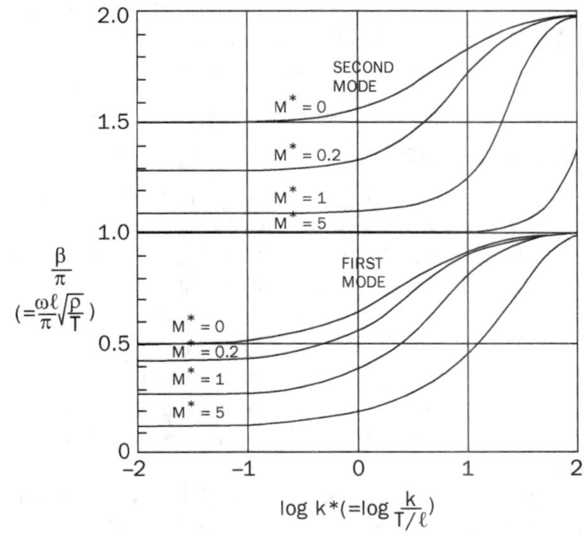

FIGURE 2.11 Frequency parameters for a string with an attached mass one end.

frequency parameter β/π were taken, and (2.51) was used to calculate *explicitly* the corresponding k^*, thereby achieving any single curve of Fig. 2.11.

Figure 2.11 shows how the frequency of the system varies as the stiffness ratio is changed. For good physical understanding, one should imagine a string having fixed values of ℓ, ρ, and T. Then the abscissa of Fig. 2.11 is obtained by varying the external spring stiffness (k), and the various curves are obtained by varying the amount of external mass (M). To see better the change in ω due to changing k, the common logarithm of k^* is used as the abscissa, giving the range $10^{-2} \le k^* \le 10^2$. Two sets of curves are drawn, with each set having four values of M^*, ranging from zero (no additional external mass) to five (large external mass). The first set, having the smaller values of β/π, yields the *fundamental* frequencies. The second set, having the larger values, yields the second mode frequencies.

For $M^* = 0$, β/π varies from 0.5 (for $k^* = 0$; i.e., $\log k^* = -\infty$) to 1.0 (for $k^* = \infty$) for the fundamental mode. The latter case is that of the rigid wall, which the curves approach for all values of M^* as k^* is increased. In the former case, one has the problem of a string having a *free* end—that is, constrained longitudinally, but not transversely—which would be difficult to achieve physically without having significant M^*. For the second mode, $\beta/\pi \to 2$ as $k^* \to \infty$, for all M^*, and $\beta/\pi \to 1.5$ as $k^* \to 0$ for $M^*= 0$. Moreover, the curves for the second mode change more rapidly than those for the fundamental mode. For very large M^* (say $M^* = 100$), the curves would change extremely rapidly from $\beta/\pi = 0$ (or 1) to $\beta/\pi = 1$ (or 2) as k^* is increased.

2.7 Discontinuous Strings

A discontinuity in a string may arise in various ways. One example is the string to which a concentrated mass (i.e., a particle) is attached at an intermediate point. Another example is a string having one density over part of its length, and another over the remaining part. A straightforward approach to such problems is to use a separate solution to the equation of motion (2.8) for each part of the string which is continuous, enforcing the necessary continuity or discontinuity conditions at the junction points, along with the boundary conditions. This will be illustrated below with an example of the first type mentioned above.

A taut string of density ρ has a particle of mass M attached to it at one-fourth its length, as shown in Fig. 2.12. We will investigate the free vibrations of this system.

The equations of motion for the two segments of string are

$$T\frac{\partial^2 w_1}{\partial x_1^2} = \rho\frac{\partial^2 w_1}{\partial t^2} \quad (0 < x_1 < \frac{\ell}{4}) \tag{2.52a}$$

$$T\frac{\partial^2 w_2}{\partial x_2^2} = \rho\frac{\partial^2 w_2}{\partial t^2} \quad (0 < x_2 < \frac{3\ell}{4}) \tag{2.52b}$$

Two separate coordinates x_1 and x_2 are used for the two segments, and are measured from the two rigid boundaries as shown in Fig. 2.12 to simplify the solution. The transverse displacements in the two segments are also separately denoted by w_1 and w_2.

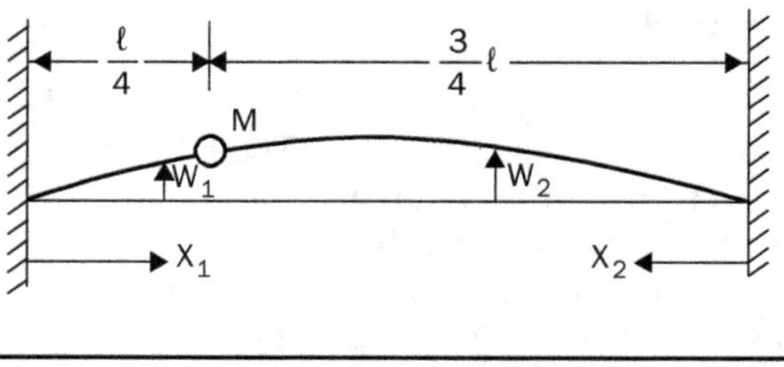

Figure 2.12 A taut string with a particle of mass *M* attached to it at one-fourth its length.

Following the procedure used in Sec. 2.2, solutions to (2.52) may be taken as

$$w_1 = (x_1, t) = X_1(x_1)\sin(\omega_1 t + \phi_1), \quad w_2 = (x_2, t) = X_2(x_2)\sin(\omega_2 t + \phi_2)$$

$$\begin{aligned} X_1(x_1) &= A_1 \sin\alpha_1 x_1 + B_1 \cos\alpha_1 x_1, \\ X_2(x_2) &= A_2 \sin\alpha_2 x_2 + B_2 \cos\alpha_2 x_2 \end{aligned} \tag{2.53}$$

where $\omega_1 = \alpha_1\sqrt{T/\rho}$ and $\omega_2 = \alpha_2\sqrt{T/\rho}$. Applying the boundary conditions:

$$w_1(0,t) = 0 \rightarrow X_1(0) = 0 \rightarrow B_1 = 0$$

$$w_2(0,t) = 0 \rightarrow X_2(0) = 0 \rightarrow B_2 = 0 \tag{2.54}$$

At the junction point, there must be continuity of displacements. That is,

$$w_1\left(\frac{\ell}{4}, t\right) = w_2\left(\frac{3\ell}{4}, t\right) \tag{2.55}$$

or

$$A_1 \sin\frac{\alpha_1\ell}{4} \bullet \sin(\omega_1 t + \phi_1) = A_2 \sin\frac{3\alpha_2\ell}{4} \bullet \sin(\omega_2 t + \phi_2) \tag{2.56}$$

For (2.56) to be satisfied for *all* t requires that $\omega_1 = \omega_2$ and that $\phi_1 = \phi_2$. Consequently from above, $\alpha_1 = \alpha_2$, and will simply be called α. Then, (2.56) reduces to

$$A_1 \sin\frac{\alpha\ell}{4} = A_2 \sin\frac{3\alpha\ell}{4} \tag{2.57}$$

The slopes are discontinuous at the junction. However, they are related to each other through the equation of motion for the particle. Using the free body diagram in Fig. 2.13, this equation is seen to be

$$-T\frac{\partial w_1}{\partial x_1}\left(\frac{\ell}{4}, t\right) - T\frac{\partial w_2}{\partial x_2}\left(\frac{3\ell}{4}, t\right) = M\frac{\partial^2 w_1}{\partial t^2}\left(\frac{\ell}{4}, t\right) \tag{2.58}$$

where, it is noted, the slopes in Fig. 2.13 have been drawn to be *positive* in accordance with the positive directions of x_1, x_2, w_1, and w_2. Using (2.53) and some of the previously determined results, (2.58) becomes

$$-TA_1\alpha\cos\frac{\alpha\ell}{4} - TA_2\alpha\cos\frac{3\alpha\ell}{4} = -M\omega^2 A_1 \sin\frac{\alpha\ell}{4} \tag{2.59}$$

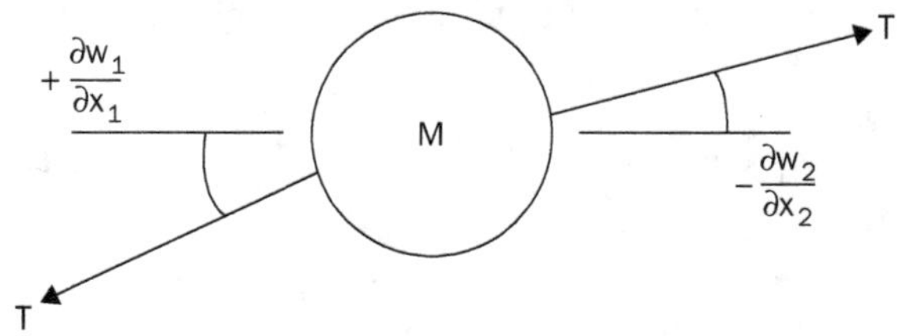

FIGURE 2.13 Free body diagram of a mass connecting two taut strings.

It is observed that the R.H.S. of (2.58) could have been equally well given as $M\partial^2 w_2(3\ell/4,t)/\partial t^2$, but then the R.H.S. of (2.59) could be obtained by substituting (2.57).

Equations (2.57) and (2.59) form a set of two homogeneous simultaneous equations in A_1 and A_2. They may be written in matrix form as

$$\begin{bmatrix} \sin\frac{\beta}{4} & -\sin\frac{3\beta}{4} \\ \left(M^*\beta\sin\frac{\beta}{4}-\cos\frac{\beta}{4}\right. & \left.-\cos\frac{3\beta}{4}\right) \end{bmatrix} \begin{bmatrix} A_1 \\ A_2 \end{bmatrix} = \begin{bmatrix} 0 \\ 0 \end{bmatrix} \tag{2.60}$$

where $\beta \equiv \alpha\ell$ and $M^* \equiv M/\rho\ell$. One solution of (2.60) is clearly $A_1 = A_2 = 0$, which is a trivial solution of no value to us. However, if we were to attempt to solve (2.60) *directly* for A_1, for example, by means of the well-known Cramer's Rule, the solution would be presented as the quotient of two determinants, as follows:

$$A_1 = \frac{\begin{vmatrix} 0 & -\sin\frac{3\beta}{4} \\ 0 & -\cos\frac{3\beta}{4} \end{vmatrix}}{\begin{vmatrix} \sin\frac{\beta}{4} & -\sin\frac{3\beta}{4} \\ \left(M^*\beta\sin\frac{\beta}{4}-\cos\frac{\beta}{4}\right. & \left.-\cos\frac{3\beta}{4}\right) \end{vmatrix}} \tag{2.61}$$

The numerator determinant is clearly zero. The only hope for a nonzero (also called "nontrivial") solution for A_1 is to have the denominator determinant also zero, which would put the quotient into the indeterminate form 0/0. Therefore, we try setting:

$$\begin{vmatrix} \sin\frac{\beta}{4} & -\sin\frac{3\beta}{4} \\ \left(M^*\beta\sin\frac{\beta}{4}-\cos\frac{\beta}{4}\right) & -\cos\frac{3\beta}{4} \end{vmatrix} = 0 \tag{2.62}$$

Expanding (2.62) yields

$$-\sin\frac{\beta}{4}\bullet\cos\frac{3\beta}{4}+\sin\frac{3\beta}{4}\bullet\left(M^*\beta\sin\frac{\beta}{4}-\cos\frac{\beta}{4}\right)=0 \tag{2.63}$$

This can be put into a more simple form by using the trigonometric identity for the sine of the sum of the angles $\beta/4$ and $3\beta/4$. Then, (2.63) becomes

$$\sin\beta - M^*\beta\bullet\sin\frac{\beta}{4}\bullet\sin\frac{3\beta}{4}=0 \tag{2.64}$$

Or, dividing through (2.63) by $\sin\beta/4 \bullet \sin 3\beta/4$ yields an alternative form, which has a nice "symmetry" and could be even more useful:

$$\cot\frac{\beta}{4}+\cot\frac{3\beta}{4}=M^*\beta \tag{2.65}$$

As we will soon see, Eqs. (2.64) and (2.65) do have nontrivial solutions for β. Either one may be called the "frequency equation" (or the "characteristic equation") for the problem, for the $\beta=\omega\ell\sqrt{\rho/T}$ satisfying them are the eigenvalues, or nondimensional frequencies. Correspondingly, the determinant displayed in (2.62) is called the "frequency determinant" (or "characteristic determinant"). If any β so determined is substituted back into *either* of the two equations (2.57) or (2.59) that generated the frequency determinant, then the corresponding *eigenvector* A_2/A_1 can be determined. Clearly, (2.57) is the simpler form to use in the present case, yielding

$$\frac{A_2}{A_1}=\frac{\sin(\beta/4)}{\sin(3\beta/4)} \tag{2.66}$$

The eigenfunctions for the problem are then, using (2.53) and (2.54),

$$X_1(\xi_1)=A_1\sin\beta\xi_1 \quad (0\le\xi_1\le 0.25)$$

$$X_2(\xi_2)=A_1\left[\frac{\sin(\beta/4)}{\sin(3\beta/4)}\right]\sin\beta\xi_2 \quad (0\le\xi_2\le 0.75) \tag{2.67}$$

where $\xi_1 = x_1/\ell$ and $\xi_2 = x_2/\ell$ are nondimensional coordinates. The amplitude A_1 is arbitrary, and may be set equal to unity for simplicity.

Choosing $M^* = 0.5$, thereby adding considerable, but not overwhelming, concentrated mass to the string, (2.65) yields the eigenvalues, the first six of which are listed in the first column of Table 2.1. [If a more comprehensive parametric study were desired for a wide range of M^*, one could assume values of β and solve (2.65) explicitly for the corresponding M^*.] For comparison, the values for $M^* = 0$ (no added mass) are also included. It is seen that each of the first six frequencies are significantly reduced by the added mass, except the fourth one ($\beta = 4\pi = 12.566$). Because for this mode the particle lies at a node point of the fourth mode of the string with no added mass, it does not affect this frequency. Similarly, it would not affect the frequencies of the higher modes having multiples of four half-waves.

The third column of Table 2.1 shows the results of a possible "engineering approximation" to the problem when $M^* = 0.5$. Here the added mass is "smeared out"—that is, the mass is uniformly distributed over the length of the string—yielding a string of uniform mass density 1.5 ρ. Numerical data are then obtained for Table 2.1 by replacing ρ by 1.5ρ, which results in the values of column 2 being divided by $\sqrt{1.5}$ to yield column 3. It is seen that the representation provides a reasonable approximation for the first frequency (6.3 percent error), but a poor one for the others. Usually, "smearing out" a concentrated mass uniformly will result in fundamental frequencies higher than the exact values. However, this might not be the case if the particle were located close to a wall.

Let us now determine the mode shapes for the first two frequencies when $M^* = 0.5$. Substituting the values $\beta = 2.4137$ and

Exact solutions		"Smeared out" mass ($M^* = 0.5$)
$M^* = 0.5$	$M^* = 0$	
2.4137	3.1416	2.5651
4.8033	6.2832	5.1302
8.6413	9.4248	7.6953
12.566	12.566	10.260
13.336	15.708	12.854
16.922	18.850	15.391

TABLE 2.1 Frequency Parameters $\beta = \omega\ell\sqrt{\rho/T}$ for a String Having a Particle Attached at Its One-Quarter Length.

4.8033 from Table 2.1 into (2.66) yields $A_2/A_1 = 0.584$ and -2.097, respectively. Setting $A_1 = 1$ and plotting (2.67) yield the first two mode shapes, depicted in Fig. 2.14. In these sketches the eigenfunctions are also plotted as dashed lines beyond their limits of application to show their shapes more clearly.

2.8 Damped Free Vibrations

If a string is given an initial displacement and released from rest, the motion will not continue unabated, as assumed in Sec. 2.3, but in reality will diminish as time increases due to the inevitable presence of damping. Damping of various types can exist, independently or concurrently. Viscous damping occurs if the string vibrates in a viscous fluid, and the damping force can then be assumed proportional to the velocity. Aerodynamic damping may also occur, and then the damping force is taken to be proportional to the *square* of the velocity. Dry friction (or Coulomb) damping could exist if the

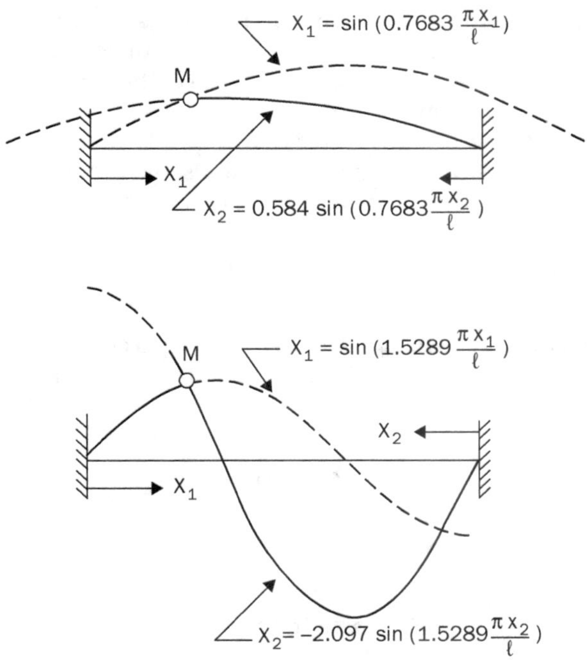

Figure 2.14 The mode shapes for the lowest two frequencies of a string with an attached mass $M^* = 0.5$ at $x = \lambda/4$.

string were supported by a horizontal plane having friction, while the string undergoes horizontal motion. Or these types of damping could also enter through the boundary conditions. Another important type is inherent material damping (also called "hysteretic" or "structural" damping). The present section will concern itself only with viscous damping, so as to provide an introduction to the topic of damped free vibrations.

We begin by returning to the equation of motion. For the case of viscous damping, the externally applied force per unit length (p) in (2.1) is taken to be $-c(\partial w/\partial t)$, where c is a constant which establishes the magnitude of the damping force. The negative sign is required because p was assumed to be in the positive direction of w (Fig. 2.2), whereas the damping force must be in opposition to a positive velocity. Making this substitution in (2.7) result to

$$T\frac{\partial^2 w}{\partial x^2} = \rho\frac{\partial^2 w}{\partial t^2} + c\frac{\partial w}{\partial t} \tag{2.68}$$

for the equation of motion for free, damped vibrations of a string.

In addition to the viscous shearing force, it should be noted that the inertia of the surrounding medium may also be transmitted to the string by normal force. This is usually accommodated in (2.68) by replacing ρ by a larger "effective mass" coefficient. We will not consider this complication here.

One could solve (2.68) by separation of variables, but let us simply assume that the string can vibrate in the same shape as it would without damping. That is, if the string has both ends fixed,

$$w_m(x,t) = \Phi_m(t)\sin\alpha_m x \quad (m = 1, 2, \ldots) \tag{2.69}$$

for the mth mode, where $\alpha_m = m\pi/\ell$ as before. This clearly satisfies the boundary conditions at both ends of the string. Substituting (2.69) into (2.68) gives

$$(\rho\phi_m'' + c\phi_m' + T\alpha_m^2\phi_m)\sin\alpha_m x = 0 \tag{2.70}$$

For this to be satisfied for all x, then

$$\rho\phi_m'' + c\phi_m' + T\alpha_m^2\phi_m = 0 \tag{2.71}$$

which is the classical equation of motion for the free, damped vibration of a single d.o.f. system. In this case, the d.o.f. is modal, rather than the displacement of a single point.

The solution of (2.71) is obtained by direct analogy with the equation for the single d.o.f. system, as given in Appendix A, replacing m by ρ, k by $T\alpha_m^2$, and retaining c (although it has different dimensions in the two problems).

Consequently, in the case of overdamped motion the mth mode responds as

$$\phi_m(t) = \mathrm{e}^{-\zeta_m \omega_m t}\left(C_m \mathrm{e}^{\sqrt{\zeta_m^2 - 1}\,\omega_m t} + D_m \mathrm{e}^{-\sqrt{\zeta_m^2 - 1}\,\omega_m t}\right) \tag{2.72}$$

where ω_m is the *undamped* natural frequency, given by (2.21), $\zeta_m \equiv c/c_{cm}$ is the damping ratio, and where

$$c_{cm} = 2\sqrt{T\alpha_m^2 \rho} = 2\frac{m\pi}{\ell}\sqrt{T\rho} = m c_{c1} \tag{2.73}$$

is the *critical* damping coefficient for the mth *mode*. Thus, as for the single d.o.f. system, overdamped motion requires that $\zeta_m > 1$. Since c_{cm} is different for each mode, then so is ζ_m. And because c_{cm} increases with m—indeed, (2.73) shows that c_{cm} is an integer multiple of its value for the first mode—then the possibility exists for a string to have its lower modes overdamped ($\zeta_m > 1$) while its higher modes are underdamped ($\zeta_m < 1$). Indeed, from (2.73), $\zeta_m = c/mc_{c1} = \zeta_1/m$. Thus, for an initial displacement shape which is arbitrary, all modes will be present, some being overdamped, other underdamped. However, if the first mode is underdamped, then (2.73) tells us that *all* modes are underdamped.

In the case of *underdamped* motion, the solution of (2.71) is

$$\phi_m(t) = \mathrm{e}^{-\zeta_m \omega_m t}\left(C_m \sin\sqrt{1-\zeta_m^2}\;\omega_m t + D_m \cos\sqrt{1-\zeta_m^2}\;\omega_m t\right) \tag{2.74}$$

with ω_m, ζ_m, and c_{cm} again defined as above.

The two types of motion described by (2.72) and (2.74) are both exponentially decaying in amplitude: the first one being nonoscillatory (and in no sense a vibration) and the second one being oscillatory. In the second case, the "damped natural frequency" of the oscillatory motion is $\sqrt{1-\zeta_m^2}\,\omega_m$, which clearly decreases as the damping coefficient (c) increases. In both cases, the coefficients C_m and D_m are determined by the initial conditions.

For *arbitrary* initial conditions, more than a single mode will be involved. Then,

$$w(x,t) = \sum_{m=1}^{\infty} \phi_m(t) \sin \alpha_m x \tag{2.75}$$

where ϕ_m is given either by (2.72) or (2.74). For example, if the string m is released from rest such that $w(x,0) = f(x)$, and if the modes are all underdamped, then at $t = 0$, (2.74) and (2.75) yield

$$\sum_{m=1}^{\infty} D_m \sin\alpha_m x = f(x) \tag{2.76}$$

Thus, the D_m are obtained by (2.29). Then, applying $\partial\omega/\partial t(x,0) = 0$ yields

$$C_m = \frac{\zeta_m}{\sqrt{1-\zeta_m^2}} D_m \quad (m = 1, 2, \ldots, \infty) \tag{2.77}$$

and the motion is completely determined.

Finally, let us consider in more detail the rate of decay of the motion, which is governed by the exponential coefficient $\zeta_m\omega_m$ in both (2.72) and (2.74). Using (2.21) and (2.73), rewriting this yields

$$\frac{c}{2\rho} = \zeta_m\omega_m = \left(\frac{c}{c_{cm}}\right)\omega_m = \left(\frac{c}{mc_{c1}}\right)m\omega_1 = \left(\frac{c}{c_{c1}}\right)\omega_1 \tag{2.78}$$

This shows us that not only $\zeta_m\omega_m$ can be expressed in terms of the parameters c_{c1} and w_1 of the first mode but, more importantly, that it is the same for all modes. Thus, the decay rates are the same for all modes of a simple string on which viscous damping is acting.

2.9 Forced Vibrations; Eigenfunction Superposition Method

Suppose now that a string is subjected to a transverse distributed force, $p(x,t)$, which has the dimensions of force per unit length. Physically, one may regard this as being a pressure which excites motion, except that the pressure has been multiplied by a proper cross-sectional dimension of the string, so that p has the required units. In general, p may vary independently with both x and t. It is taken to be positive in the direction of positive z (or w), as shown in Fig. 2.15.

If a viscous damping force is present, then considering (2.7) and (2.68), the equation of motion for the system can be written as

$$T\frac{\partial^2 w}{\partial x^2} + p(x,t) = \rho\frac{\partial^2 w}{\partial t^2} + c\frac{\partial w}{\partial t} \tag{2.79}$$

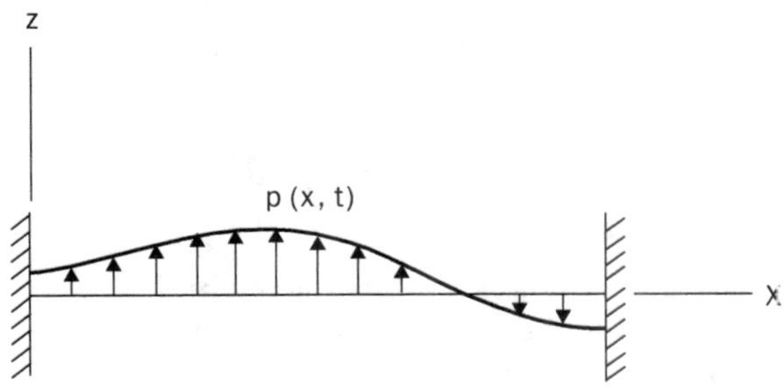

FIGURE 2.15 A forcing function for a string.

The displacement response $w(x,t)$ may be regarded as the sum of two parts. The first part is the complementary solution of (2.79), obtained from the equation itself by setting $p(x,t) = 0$. This solution has already been presented in Sec. 2.8. The second part is the particular solution of (2.79). In this section, we will concern ourselves only with the particular solution. In the case of periodic exciting force, the particular solution yields the steady-state response of the system, whereas the complementary solution yields the transient response. The latter decays with time and eventually vanishes, for all practical purposes, but the steady-state response continues for all time.

Let us assume once again that both ends of the string are fixed. One procedure to determine the steady-state solution of (2.79) is to assume that p may be expressed as

$$p(x,t) = \sum_{m=1}^{\infty} p_m(t) \sin\frac{m\pi x}{\ell} \tag{2.80}$$

That is, the pressure is assumed to be made up of components, that each component has the same shape as an eigenfunction of the free, undamped vibration problem, and that each have the possibility of independent variation over time. One may then further assume that displacement response can be expressed in terms of the responses of the individual eigenfunctions (or mode shapes), i.e.,

$$w(x,t) = \sum_{m=1}^{\infty} \phi_m(t) \sin\frac{m\pi x}{\ell} \tag{2.81}$$

Substituting (2.80) and (2.81) into (2.79), and recognizing that the latter must be satisfied for all points along the string (i.e., arbitrary choice of x), the following ordinary differential equation results:

$$\rho\phi_m'' + c\phi_m' + (T\alpha_m^2)\phi_m = p_m(t) \quad (m = 1, 2, \ldots, \infty) \tag{2.82}$$

where $\alpha_m = m\pi/\ell$ as before. Let us assume further that the exciting pressure varies sinusoidally with time. Then, (2.80) can be expressed as

$$p(x,t) = P(x)\sin\Omega t = \sin\Omega t \sum_{m=1}^{\infty} P_m \sin\alpha_m x \tag{2.83}$$

where Ω is the forcing (or exciting) frequency, and P_m is the Fourier coefficient representing the amplitude of each component of exciting pressure. Following the same procedure which was used to obtain (2.29), the P_m are found from

$$P_m = \frac{2}{\ell}\int_0^{\ell} P(x)\sin\alpha_m x\, dx \tag{2.84}$$

Thus, for sinusoidal exciting pressure, (2.82) becomes

$$\rho\phi_m'' + c\phi_m' + (T\alpha_m^2)\phi_m = P_m \sin\Omega t \quad (m = 1, 2, \ldots, \infty) \tag{2.85}$$

Equation (2.85) determines the response of each mode to the corresponding pressure component. It is of the same form as the equation of motion for a single d.o.f., discrete system (mass, dashpot, spring, and exciting force). However, in this case, it represents the response of a single *modal* d.o.f. of the system. Using the summary of results for a forced vibration of a single d.o.f. system in Appendix A, replacing m by ρ, k by $T\alpha_m^2$, and F_0 by P_m, the solution of (2.85) may be set down as

$$\phi_m = A_m \sin\Omega t - B_m \cos\Omega t \tag{2.86}$$

where

$$\frac{A_m}{(\delta_{\text{st}})_m} = \frac{1-(\Omega/\omega_m)^2}{\Delta_m} \tag{2.87a}$$

$$\frac{B_m}{(\delta_{\text{st}})_m} = \frac{2\zeta_m(\Omega/\omega_m)^2}{\Delta_m} \tag{2.87b}$$

$$\Delta_m = \left[1-\left(\frac{\Omega}{\omega_m}\right)^2\right]^2 + \left[2\zeta_m\left(\frac{\Omega}{\omega_m}\right)\right]^2 \tag{2.87c}$$

In the above expressions, $(\delta_{\text{st}})_m = P_m/T\alpha_m^2$, which is the amplitude of the *static* displacement of the mth mode if the string were subjected to a static pressure $P_m \sin\alpha_m x$, and ζ_m is the damping ratio used in

Sec. 2.8; i.e., $\zeta_m = c/c_{cm}$, $c_{cm} = 2\sqrt{T\alpha_m^2\rho} = mc_{c1}$. Equation (2.86) has a $\sin\Omega t$ term, which represents the component of the motion in-phase with the exciting force, and a $\cos\Omega t$ term, which represents the out-of-phase component. The latter always lags the in-phase response; hence, the term is written in (2.86) with a minus sign preceding it, so that both A_m and B_m are positive coefficients.

Alternatively, the response may be expressed in terms of an amplitude (C_m) and a phase angle (ϕ_m):

$$\phi_m = C_m \sin(\Omega t - \phi_m) \tag{2.88}$$

Because the in- and out-of-phase components may be regarded as orthogonal components of a rotating vector, it follows that C_m is the magnitude of the resultant vector, i.e., $C_m = \sqrt{A_m^2 + B_m^2}$. By simple manipulation, using a trigonometric identity, as is done in introductory vibration texts, it can be shown that

$$C_m = \frac{(\delta_{\text{st}})_m}{\sqrt{\Delta_m}} \tag{2.89a}$$

$$\tan\phi_m = \frac{B_m}{A_m} = \frac{2\zeta_m(\Omega/\omega_m)}{1-(\Omega/\omega_m)^2} \tag{2.89b}$$

In the case of no damping, $B_m = 0$ in (2.86), and $\phi_m = 0$ in (2.88). Then, by (2.87a) or (2.89a) and (2.87c), the amplitude of the motion becomes

$$C_m = A_m = \frac{(\delta_{\text{st}})_m}{1-(\Omega/\omega_m)^2} \tag{2.90}$$

It is worthwhile to take the time to review the plots of $C_m/(\delta_{\text{st}})_m$ and ϕ_m, which are the same as the curves of C/δ_{st} and ϕ for the discrete, single d.o.f. system shown in Appendix A. Most importantly, it is seen that as $\Omega \to \omega_m$, if no damping is present ($\zeta_m = 0$), the amplitude of the mth *mode* becomes infinite, and that if damping is small, the amplitude becomes large.

If one were to excite a string with pressure $p(x, t) = P(x) \sin \Omega t$, where $P(x)$ is some rather general function, beginning with $\Omega = 0$ and slowly increasing Ω, (2.87c) together with (2.89a) show that at every natural frequency (ω_m), the mth mode would be strongly excited if damping were small. This large amplitude response, which would occur at each natural frequency, is called a *resonance*. Thus, as in any continuous system, an infinite number of resonances would exist. However, if some damping is present (as in the case of any real system), then, the amplitudes at the resonant frequencies will typically decrease as m is increased. To show this, let us rewrite (2.89a) in the case when the exciting frequency exactly equals one of the natural frequencies. Then $\Delta_m = (2\zeta_m)^2$, and

$$(C_m)_{\Omega=\omega_m} = \frac{(P_m/T\alpha_m^2)}{2\zeta_m} = \frac{P_m\ell^2}{2\pi^2 T}\left(\frac{c_{c1}}{c}\right)\frac{1}{m} \tag{2.91}$$

Equation (2.91) shows that the amplitude of the mth mode varies inversely with m in an explicit manner. Moreover, the magnitudes of the Fourier coefficients (P_m) of the loading representation typically decrease as m increases, which also reduces C_m as m is increased. P_m need not decrease at the beginning of the series. For example, if a loading function $P(x)$ should resemble the fifth mode, then P_5 would be the largest magnitude; but thereafter, as m increases, P_m would decrease.

It should also be said that, as for a discrete system, if absolutely no damping were present, and if the exciting frequency were exactly equal to one of the natural frequencies, the amplitude of the motion would not instantly become very large. Rather, it would take time for the amplitude to grow. As it is shown in the introductory vibration texts, (2.86) and (2.88) are not valid solutions of (2.85) when Ω is the same as a natural frequency, and no damping is present. Instead, if $\Omega = \omega_m$ with $c = 0$, then, the particular solution of (2.85) is

$$\phi_m(t) = -\frac{1}{2}(\delta_{\text{st}})_m \Omega t \cos\Omega t \tag{2.92}$$

which may be regarded as a sinusoidal function, shifted $\pi/2$ in time, having an amplitude which increases linearly with time. That it should take time for the amplitude to grow to an unbounded value is physically clear, for as the amplitude of the motion increases, so does the energy (kinetic plus potential) of the system, and work must be done by the external pressure $p(x, t)$ over a finite displacement, and consequently over a finite length of time, to cause the energy to increase.

Ultimately, one is usually interested in knowing the behavior of a particular point, say $x = x_0$, on the string. Most importantly, one wishes to know the *total amplitude* of the displacement of the point. Equation (2.89a) supplies only the amplitude of a single mode. These amplitudes *cannot* be straightforwardly added to obtain the total amplitude if damping is present, because the phase angles ϕ_m are different for each mode. Returning to (2.81) and (2.86), the total displacement of the point $x = x_0$ is

$$\begin{aligned} w(x_0,t) &= \sum_{m=1}^{\infty}(A_m \sin\Omega t - B_m \cos\Omega t)\sin\alpha_m x_0 \\ &= \left(\sum A_m \sin\alpha_m x_0\right)\sin\Omega t - \left(\sum B_m \sin\alpha_m x_0\right)\cos\Omega t \\ &= W\sin(\Omega t - \Phi) \end{aligned} \tag{2.93}$$

where W and Φ are defined by

$$W = \sqrt{\left(\sum A_m \sin\alpha_m x_0\right)^2 + \left(\sum B_m \sin\alpha_m x_0\right)^2} \tag{2.94a}$$

$$\phi = \tan^{-1}\frac{\sum B_m \sin\alpha_m x_0}{\sum A_m \sin\alpha_m x_0} \tag{2.94b}$$

the summations, of course, extending for all m. Thus, the total amplitude W may be regarded as the *vector* sum of the amplitudes of the separate displacement components at point $x = x_0$. Equation (2.94a) shows that the in-phase components $A_m \sin\alpha_m x_0$ are added up separately, and so are the out-of-phase components $B_m \sin\alpha_m x_0$ and the total resultant amplitude is found by the same procedure used to obtain the magnitude of any vector. Equation (2.94b) suggests that the phase angle of the motion varies along the string, that is, $\phi = \phi(x)$.

To generalize the analysis, let us now consider the response of a string subjected to an exciting pressure which is *periodic in time,* as depicted by the representative function of Fig. 2.16. Then $P_m(t)$ in (2.80) and (2.82) is such a periodic function, having a period, τ. Further, expand $P_m(t)$ into its Fourier components in time, i.e.,

$$p_m(t) = a_{m0} + \sum_{m=1}^{\infty}(a_{mn}\cos n\Omega t + b_{mn}\sin n\Omega t) \tag{2.95}$$

where Ω is the frequency of the periodic forcing function, $\Omega = 2\pi/\tau$. Then, by a procedure similar to that followed in Sec. 2.3, the Fourier coefficients are determined as

$$a_{m0} = \frac{1}{\tau}\int_0^{\tau} p_m(t)\,dt \tag{2.96a}$$

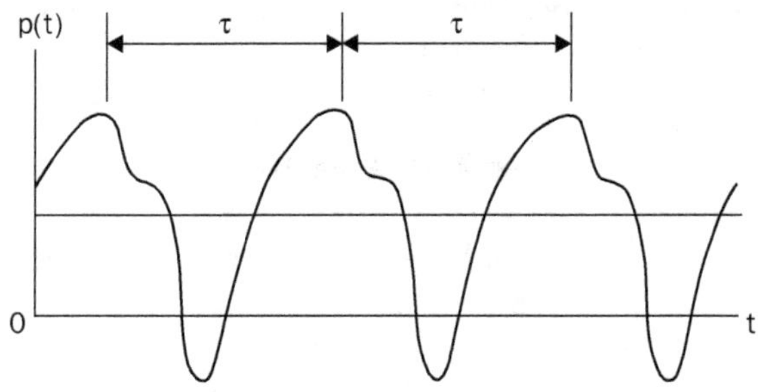

FIGURE 2.16 A representative periodic function.

$$a_{mn} = \frac{2}{\tau}\int_0^{\tau} p_m(t)\cos n\Omega t\, dt \tag{2.96b}$$

$$b_{mn} = \frac{2}{\tau}\int_0^{\tau} p_m(t)\sin n\Omega t\, dt \tag{2.96c}$$

From (2.96a), it can be seen that a_{m0} represents the average value of the forcing function over the period. This may be regarded as a constant, static pressure acting along the string, causing a static deflection, whereas the a_{mn} and b_{mn} cause vibratory displacements about that static position. Solving (2.82), the static displacement (ϕ_{m0}) of the mth mode due to a_{m0} is $\phi_{m0} = a_{m0}/T\alpha_m^2$. The dynamic displacement is obtained by substituting the time-varying terms of (2.95) on the R.H.S. of (2.82) and solving the ordinary differential equation for the response $\phi_m(t)$ of the mth mode. The behavior of the mth mode then is of the form

$$\phi_m = \phi_{m0} + \sum_{n=1}^{\infty}(A_{mn}^* \sin n\Omega t + B_{mn}^* \cos n\Omega t) \tag{2.97}$$

where A_{mn}^* and B_{mn}^* are constants.

Example 2.3 A stretched string of length ℓ with two fixed ends is subjected to a uniformly distributed, sinusoidal forcing pressure

$$p(x,t) = q_0 \sin\Omega t$$

where q_0 and Ω are constants, with transverse, viscous damping present. Investigate the steady-state, dynamic response of a point located at the middle of the string.

Solution

Locating the coordinate origin at the left end of the string, and expressing the transverse loading in terms of the eigenfunctions,

$$p(x,t) = \sin \Omega t \sum_{m=1}^{\infty} P_m \sin \alpha_m x$$

$$P_m = \frac{2}{\ell}\int_0^{\ell} q_0 \sin \alpha_m x \, dx = \frac{4q_0}{m\pi} \quad (m = 1,3,5,\ldots)$$

The motion is expressed by (2.81), with terms defined by (2.86) and (2.87). To use these, we must evaluate:

$$(\delta_{st})_m = \frac{P_m}{T\alpha_m^2} = \frac{4q_0}{m\pi} \bullet \frac{\ell^2}{T(m\pi)^2} = \frac{4q_0\ell^2}{m^3\pi^3 T} \quad (m = 1,3,5,\ldots)$$

Thus, the static displacement (for $\Omega = 0$) can be written as

$$\delta(x) = \frac{4q_0\ell^2}{\pi^3 T} \sum_{m=1,3}^{\infty} \frac{1}{m^3} \sin \frac{m\pi x}{\ell}$$

and at the middle ($x = \ell/2$), it is

$$\delta\left(\frac{\ell}{2}\right) \equiv \delta = \frac{4q_0\ell^2}{\pi^3 T} \sum_{m=1,3}^{\infty} \frac{1}{m^3} \sin \frac{m\pi}{2}$$

This may also be obtained from the static equilibrium equation, which is the specialized form of (2.79):

$$T\frac{\partial^2 W}{\partial x^2} + q_0 = 0$$

Solving it, and evaluating the two constants of integration from the boundary conditions,

$$W(x) = \frac{q_0}{2T} x(\ell - x)$$

Therefore, at the middle,

$$\delta = \frac{q_0 \ell^2}{8T}$$

and the nondimensional displacement may be written as

$$\frac{w(x,t)}{\delta} = \sum_{m=1,3}^{\infty} \frac{(\delta_{st})_m}{\delta}\left[\frac{A_m}{(\delta_{st})_m}\sin \Omega t - \frac{B_m}{(\delta_{st})_m}\cos \Omega t\right]\sin \frac{m\pi x}{\ell}$$

From the above work, it can be seen that $(\delta_{st})_m/\delta = 32/m^3\pi^3$, so that at the middle,

$$\frac{w(\ell/2,t)}{\delta} = \frac{32}{\pi^3}\sum_{m=1,3}^{\infty} \frac{1}{m^3}\left[\frac{A_m}{(\delta_{st})_m}\sin \Omega t - \frac{B_m}{(\delta_{st})_m}\cos \Omega t\right]\sin \frac{m\pi}{2}$$

with $A_m/(\delta_{st})_m$ and $B_m/(\delta_{st})_m$ given by (2.87). For these expressions,

$$\zeta_m = \frac{c}{c_{cm}} = \frac{c}{2\sqrt{T\rho}}\left(\frac{\ell}{m\pi}\right)$$

$$\frac{\Omega}{\omega_m} = \frac{\Omega}{\omega_1}\left(\frac{\omega_1}{\omega_m}\right) = \frac{1}{m}\frac{\Omega}{\omega_1}$$

where $\omega_1 = (\pi/\ell)\sqrt{T/\rho}$.

Let us look first at the response of the individual modes in the case of *no* damping. From above, their nondimensional amplitudes are determined to be

$$\frac{W_m}{\delta} = \frac{32}{m^3\ell^3}\frac{A_m}{(\delta_{st})_m}\sin\frac{m\pi}{2} = \frac{32}{m^3\pi^3}\cdot\frac{(-1)^{(m-1)/2}}{1-(1/m^2)(\Omega/\omega_1)^2} \qquad (m = 1, 3, \ldots)$$

where W_m is the amplitude of the mth mode displacement. A plot of W_m/δ as a function of the frequency ratio $\Omega/\omega_1 = 0$ is shown in Fig. 2.17 for the first three modes excited (m = 1, 3, 5). For the first mode $W_1/\delta = 1.032$ at $\Omega/\omega_1 = 0$. This component of the static deflection exceeds unity. But when other components are added ($W_m/\delta = -0.038$, $W_5/\delta = 0.008$, etc), the *total* static deflection is unity. We now study the changes in the amplitudes of the modes (W_m) as the forcing frequency (Ω) is slowly increased, by observing how W_m/δ changes with Ω/ω_1. For a *given system* (given ℓ, T, q_0), δ and ω_1 are constants. As Ω is increased from $\Omega = 0$, the vibration amplitude of the first mode increases slowly at first, but soon it grows rapidly. At $\Omega/\omega_1 = 0.5$, $W_1/\delta = 1.33$. But, by $\Omega/\omega_1 = 0.8$, it is becoming large ($W_1/\delta = 2.87$), and it continues to increase at an increasing rate as $\Omega/\omega_1 \to 1$ (for example, at $\Omega/\omega_1 = 0.9$, $W_1/\delta = 5.43$). As Ω/ω_1 passes through unity, there is a phase reversal (ϕ_1 changes from 0° to 180°), which causes the amplitude to change suddenly from ∞ to $-\infty$. We shall see below that this is the behavior of the mathematical solution of this highly idealized problem (namely, absolutely no damping), and that this flip-flop cannot really occur in physical reality. For $\Omega/\omega_1 > 1$, W_1/δ decreases rapidly in magnitude at first, and has small values at the subsequent resonance intervals ($W_1/\delta = -0.13$ at $\Omega/\omega_1 = 3$ and $W_1/\delta = -0.04$ at $\Omega/\omega_1 = 5$).

For the second excited mode (W_3/δ), the amplitude is quite small for all $\Omega/\omega_1 < 3$, except for exciting frequencies close to resonance ($W_3/\delta = -0.07$ at $\Omega/\omega_1 = 2$, $W_3/\delta = -0.12$ at $\Omega/\omega_1 = 2.5$), where it increases extremely rapidly ($W_3/\delta = -1.92$ at $\Omega/\omega_1 = 2.97$). At resonance ($\Omega/\omega_1 = 3$), W_3/δ flip-flops from $-\infty$ to ∞, and decreases very rapidly thereafter. Thus, the second excited mode ($m = 3$) has a much sharper resonance behavior than the first mode. For the third excited mode ($m = 5$), the aforementioned characteristics become still more exaggerated. That is, the initial increase of the curve is still slower ($W_5/\delta = 0.009$ at $\Omega/\omega_1 = 1$, $W_5/\delta = 0.013$ at $\Omega/\omega_1 = 3$), and no attempt is made to show these small values in Fig. 2.17 until resonance is approached. There, the amplitude climbs extremely rapidly ($W_5/\delta = 0.42$ at $\Omega/\omega_1 = 4.95$ and $W_5/\delta = 2.07$ at $\Omega/\omega_1 = 4.99$), before another flip-flop occurs.

Now, let us consider the *total* response of the system—that is, the superposition of the modal responses—and include the effects of damping. This is obtained

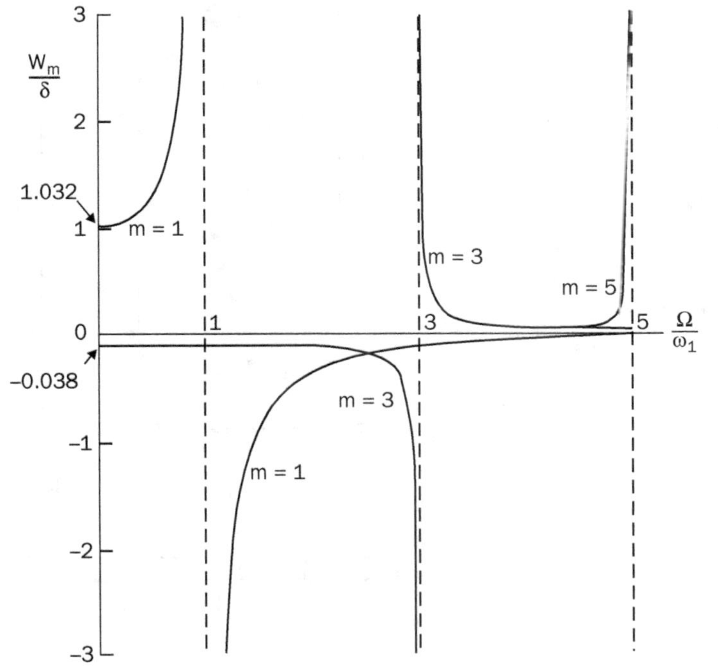

FIGURE 2.17 A plot of W_m/δ as a function of the frequency ratio $\Omega/\omega_l = 0$ for the first three modes excited in Example 2.3.

by (2.94a) and is plotted in Fig. 2.18. There, nondimensional amplitude (W/δ) is plotted versus Ω/ω_1 over a frequency range which includes the first three resonances. The six curves shown correspond to various nondimensional damping ratios c/c_{c1}, where $c_{c1} = (2\pi/\ell)\sqrt{T\rho}$ [see (2.73)] is the critical damping coefficient of the first mode. Thus, for a given system having prescribed ℓ, T, and ρ, c/c_{c1} is a measure of the amount of damping present. For no damping ($c/c_{c1} = 0$), the curve shown is a superposition of the curves for the separate modes (Fig. 2.17) discussed above. It is shown with all positive values, instead of flip-flopping from $+\infty$ to $-\infty$ at $\Omega/\omega_1 = 1$. To explain this, we could say that we are plotting the absolute value of W/δ. But a more simple explanation, which is also physically more appealing, is to say that the curve shown is actually for a very small amount of damping present (say, $c/c_{c1} < 10^{-6}$).

The curve for "small" damping ($c/c_{c1} = 0.1$) in Fig. 2.18 peaks at $W/\delta = 5.18$ for $\Omega/\omega_1 = 0.99$. It should be mentioned here that while $c/c_{c1} = 0.1$ seems to be a small number, it represents a rather *large* amount of viscous damping in a physical system. Peaks occur again at $W/\delta = 0.59$ for $\Omega/\omega_1 = 2.98$, and $W/\delta = 0.21$ for $\Omega/\omega_1 = 5.01$. Thus, with any damping present, the uniformly distributed exciting pressure causes a much larger response in the vicinity of the first natural frequency than it does for the subsequent ones. The positions of peak amplitude are shifted to the *left* of $\Omega/\omega_1 = 1$ and $\Omega/\omega_1 = 3$ with damping present, as occurs in a single d.o.f. system. (The vicinity of $\Omega/\omega_1 = 3$ is seen better in the

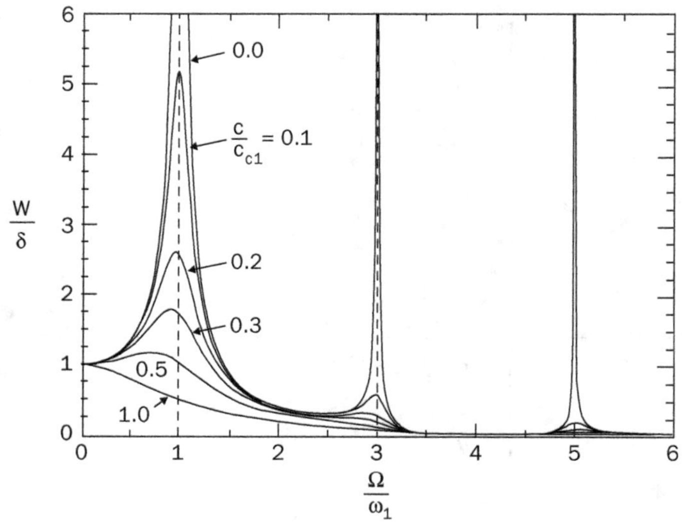

Figure 2.18 The total response of the system, including damping, of Example 2.3.

magnified view of Fig. 2.19.) However, the peak amplitude in the vicinity of $\Omega/\omega_1 = 5$ is shifted to the right. Between $3.6 < \Omega/\omega_1 < 4.5$, the vibratory motion virtually ceases when small damping is present, as can be seen in Fig. 2.20. Remarkably, with no damping, the amplitude is exactly zero for $\Omega/\omega_1 = 4$.

All the curves in Fig. 2.18 begin with $W/\delta = 1$ for $\Omega/\omega_1 = 0$. When damping increases, the peak amplitudes are reduced, with peak values being shifted further to the left or to the right of resonance values. For $c/c_{c1} = 1$ the peak has been shifted to the origin $\Omega/\omega_1 = 0$, where it remains for larger c/c_{c1}.

Figure 2.21 is a plot showing how the phase angle ϕ, as determined by (2.94b), changes with the frequency ratio Ω/ω_1 for a point taken at the center of the string. For $0 < \Omega/\omega_1 < 1.5$, ϕ is seen to vary in a manner quite similar to a one d.o.f. system; however, the curves have slightly different values of ϕ (approximately 90°) at $\Omega/\omega_1 = 1$. In the range $2 < \Omega/\omega_1 < 4$, ϕ is seen to vary between 180° and 360° for small or moderate damping. If the plot were extended to larger Ω/ω_1, one would find that the curves for the intervals $1 < \Omega/\omega_1 < 5$, $5 < \Omega/\omega_1 < 9$, $9 < \Omega/\omega_1 < 13$, and so on, would look quite similar for each interval. That is, the plot of the phase angle is almost periodic, with a period of $\Omega/\omega_1 = 4$. One also finds that for any fixed values of c/c_{c1} and Ω/ω_1, the phase angle varies with x. That is, different points along the string have different phase angles.

2.10 Forced Vibrations; Closed Form Exact Solutions

Now let us consider another approach to the problem of forced vibration of a string. We will limit ourselves to forcing functions which are sinusoidal in time.

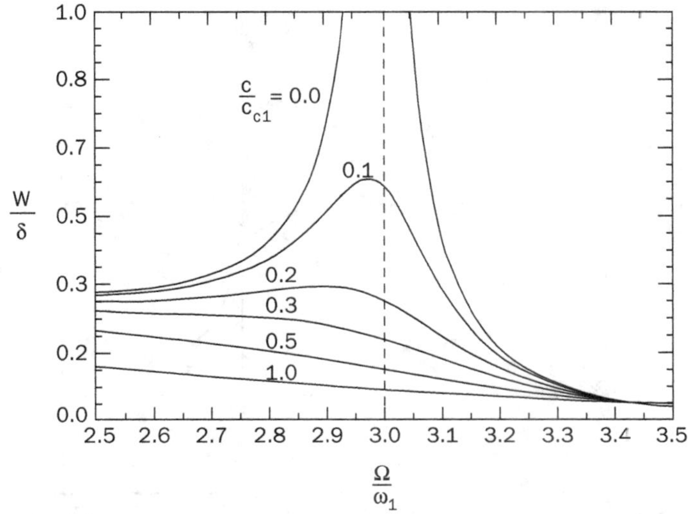

FIGURE 2.19 Magnified vibratory amplitudes in the vicinity of $\Omega/\omega_1 = 3$.

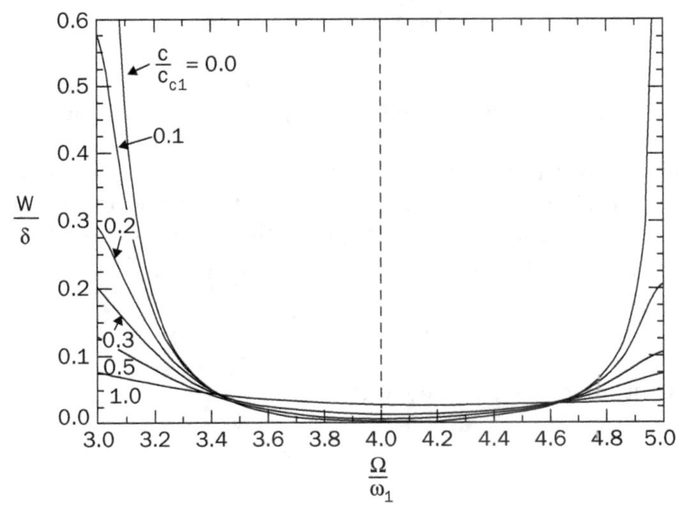

FIGURE 2.20 Magnified vibratory amplitudes in the range $3.0 < \Omega/\omega_1 < 5.0$.

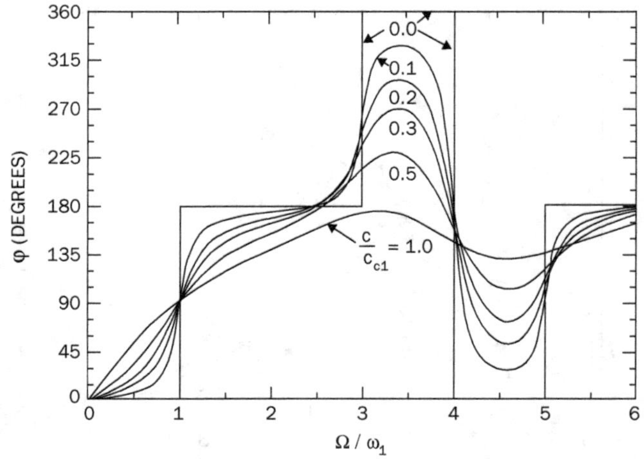

FIGURE 2.21 The phase angle ξ as a function of the frequency ratio Ω/ω_1 for a point taken at the center of the string.

Consider first the case of *no damping*, which simplifies matters considerably. The equation of motion (2.79) then becomes

$$T\frac{\partial^2 w}{\partial x^2} + P(x)\sin\Omega t = \rho\frac{\partial^2 w}{\partial t^2} \tag{2.98}$$

It would seem reasonable to assume that, if an elastic system were to be excited by a sinusoidal force, the response would also be sinusoidal in time and that, if no damping were present, the motion would be in-phase with the exciting force (or perhaps 180° out-of-phase, which is obtained by changing the sign of the response amplitude). Therefore, we can assume that

$$w(x,t) = X(x)\sin\Omega t \tag{2.99}$$

Substituting this into (2.98) yields

$$TX'' + \rho\Omega^2 X = -P(x) \tag{2.100}$$

which has the solution

$$X(x) = C_1\sin\beta x + C_2\cos\beta x + X_p \tag{2.101}$$

where X_p is a suitable particular solution, and where

$$\beta^2 = \frac{\rho\Omega^2}{T} = \frac{\rho\omega_1^2}{T} \times \left(\frac{\Omega}{\omega_1}\right)^2 \tag{2.102}$$

In (2.102), ω_1 is the first natural frequency of the string. But since $\rho\omega_1^2/T = (\pi/\ell)^2$

$$\beta = \frac{\pi}{\ell}\left(\frac{\Omega}{\omega_1}\right) \tag{2.103}$$

Equation (2.101) is fascinating. One must first find a particular solution X_p for the loading condition at hand, and evaluate C_1 and C_2 from the boundary conditions, before (2.101) is completely determined. But, after these are known, (2.101) allows one to calculate in *closed form* the displacement of a point $x = x_0$ along the string, instead of having to sum an infinite series [i.e., (2.93) with $B_m = 0$]. Even more interestingly, the arguments of the trigonometric functions each change as Ω is varied. Thus, (2.101) is quite a simple expression to describe the amplitude and shape of the string as Ω is varied. Certainly, it is easier to evaluate at any point $x = x_0$ than to follow the eigenfunction superposition method laid out in the preceding section.

Example 2.4 Use the closed-form solution method to evaluate the displacement at the middle of the string in Example 2.3 when no damping is present.

Solution

Setting $P(x)$ equal to the constant q_0 in (2.100), a particular solution is easily found, so that the complete solution for the displacement is given by

$$X(x) = C_1 \sin\beta x + C_2 \cos\beta x - \frac{q_0}{\rho\Omega^2}$$

Applying the boundary conditions:

$$X(0) = 0 \rightarrow C_2 = \frac{q_0}{\rho\Omega^2}$$

$$X(\ell) = 0 \rightarrow C_1 = \frac{q_0}{\rho\Omega^2}(\csc\beta\ell - \cot\beta\ell)$$

Furthermore,

$$\frac{q_0}{\rho\Omega^2} = \frac{q_0\ell^2}{8T} \bullet \frac{8T}{\rho\Omega^2\ell^2} = \delta \bullet \frac{8T}{\omega_1^2\ell^2\rho}\left(\frac{\omega_1}{\Omega}\right)^2 = \frac{8\delta}{\pi^2}\left(\frac{\omega_1}{\Omega}\right)^2$$

where δ is the *static* displacement of the string at its middle, as determined in Example 2.3. Substituting these, the displacement at the middle of the string is given by

$$X\left(\frac{\ell}{2}\right)=\frac{8\delta}{\pi^2}\left(\frac{\omega_1}{\Omega}\right)^2\left[(\csc\beta\ell-\cot\beta\ell)\sin\frac{\beta\ell}{2}+\cos\frac{\beta\ell}{2}-1\right]$$

with $\beta\ell=\pi(\Omega/\omega_1)$ from (2.103). By means of trigonometric identities, this result may be simplified, and finally written in nondimensional form as

$$\frac{X(\ell/2)}{\delta}=\frac{8}{\lambda^2}\left(\sec\frac{\lambda}{2}-1\right)$$

where $\lambda = \beta\ell = \pi(\Omega/\omega_1)$. It is observed that $\lim_{\lambda\to 0}\{[X(\ell/2)]/2\}=1$, as expected.

When *damping is present*, solution (2.99) would not be expected to fit, for one would expect a phase angle between the applied force and the displacement response. An attempt to substitute (2.99) into (2.79), with $p(x,t) = P(x)\sin(\Omega t)$, substantiates this. It is therefore necessary to generalize the assumed solution to

$$w(x,t)=X_1(x)\sin\Omega t-X_2(x)\cos\Omega t \tag{2.104}$$

The minus sign is used in (2.104) because it is expected, from previous experience with forced vibration problems, that the motion will *lag* the excitation.

Substituting (2.104) into (2.79) yields

$$\begin{aligned}&[TX_1''+\rho\Omega^2X_1-c\Omega X_2+P(x)]\sin\Omega t\\&\quad-[TX_2''+\rho\Omega^2X_2+c\Omega X_1]\cos\Omega t=0\end{aligned} \tag{2.105}$$

For (2.105) to be satisfied for all time (t), because of the orthogonality of $\sin\Omega t$ and $\cos\Omega t$, the bracketed coefficients of $\sin\Omega t$ and $\cos\Omega t$ in (2.105) must independently be zero. That is

$$TX_1''+\rho\Omega^2X_1-c\Omega X_2=-P(x)$$

$$TX_2''+\rho\Omega^2X_2+c\Omega X_1=0 \tag{2.106}$$

Equations (2.106) are a fourth-order system of ordinary differential equations having constant coefficients, which are coupled together through the damping terms. Before solving them, let us put them into nondimensional form. First, define $\xi = x/\ell$, where ℓ is the length of the string. Then, $X''(x) = X''(\xi)/\ell^2$, and (2.106) becomes

$$X_1'' + \left(\frac{\rho\Omega^2\ell^2}{T}\right)X_1 - \left(\frac{c\Omega\ell^2}{T}\right)X_2 = -\frac{P(\xi)\ell^2}{T} \tag{2.107a}$$

$$X_2'' + \left(\frac{\rho\Omega^2\ell^2}{T}\right)X_2 - \left(\frac{c\Omega\ell^2}{T}\right)X_1 = 0 \tag{2.107b}$$

Utilizing (2.102) and (2.103)

$$\frac{\rho\Omega^2\ell^2}{T} = (\beta\ell)^2 = \pi^2\left(\frac{\Omega}{\omega_1}\right)^2 \equiv \lambda^2 \tag{2.108}$$

Further, utilizing the damping ratio $\zeta_m = c/c_{cm}$ and (2.73),

$$\begin{aligned}\frac{c\Omega\ell^2}{T} &= \frac{c}{c_{c1}} \cdot \frac{2\pi}{\ell}\sqrt{T\rho} \cdot \frac{\Omega}{\omega_1} \cdot \frac{\pi}{\ell}\sqrt{\frac{T}{\rho}} \cdot \frac{\ell^2}{T} \\ &= 2\pi\lambda\zeta_1\end{aligned} \tag{2.109}$$

where $\zeta_1 = c/c_{c1}$. Finally, define $P(\xi) \cdot \ell^2/T = Q(\xi)$. Then, (2.107) may be rewritten in nondimensional form as

$$X_1'' + \lambda^2 X_1 - (2\pi\lambda\zeta_1)X_2 = -Q(\xi) \tag{2.110a}$$

$$X_2'' + \lambda^2 X_2 - (2\pi\lambda\zeta_1)X_1 = 0 \tag{2.110b}$$

Solving (2.110b) for X_1 and substituting into (2.110a) gives

$$X_2^{IV} + 2\lambda^2 X_2'' + (\lambda^4 + 4\pi^2\lambda^2\zeta_1^2)X_2 = 2\pi\lambda\zeta_1 Q \tag{2.111}$$

The solution to (2.111) consists of the sum of complementary (X_{2c}) and particular (X_{2p}) solutions; that is

$$X_2 = X_{2c} + X_{2p} \tag{2.112}$$

To determine the complementary solution, first rewrite the homogeneous equation as

$$[D^4 + 2\lambda^2 D^2 + (\lambda^4 + \gamma^4)]X_{2c} = 0 \tag{2.113}$$

where

$$\gamma \equiv 2\pi\lambda\zeta_1 = \frac{c\Omega\ell^2}{T}, \quad D \equiv \frac{d}{d\xi} \tag{2.114}$$

Assuming

$$X_{2c} = e^{p\xi} \tag{2.115}$$

then, (2.113) yields

$$\begin{aligned} p^4 + 2\lambda^2 p^2 + (\lambda^4 + \gamma^4) &= 0 \\ (p^2 + \lambda^2)^2 &= -\gamma^4 \\ p^2 + \lambda^2 &= \pm i\gamma^2, \quad i = \sqrt{-1} \\ p^2 &= -(\lambda^2 \pm i\gamma^2) \end{aligned}$$

Let

$$\tan\theta \equiv \frac{\gamma^2}{\lambda^2}, \quad R = \sqrt{\lambda^4 + \gamma^4} \tag{2.116}$$

Then,

$$\begin{aligned} \lambda^2 + i\gamma^2 = \mathrm{Re}^{i\theta}, &\quad \lambda^2 - i\gamma^2 = \mathrm{Re}^{-i\theta} \\ p^2 &= -\mathrm{Re}^{i\theta}, -\mathrm{Re}^{-i\theta} \\ p &= \pm i\sqrt{R}\, e^{i\theta/2}, \pm i\sqrt{R}\, e^{-i\theta/2} \end{aligned}$$

Therefore, the four roots for p may be written as

$$P_{1,2} = \pm i\sqrt{R}\left(\cos\frac{\theta}{2} + i\sin\frac{\theta}{2}\right) = -a + ib,\ a - ib$$

$$P_{3,4} = \pm i\sqrt{R}\left(\cos\frac{\theta}{2} - i\sin\frac{\theta}{2}\right) = a + ib,\ -a - ib$$

where

$$a \equiv \sqrt{R}\ \sin\frac{\theta}{2} = \sqrt{\frac{R(1-\cos\theta)}{2}} = \sqrt{\frac{R(1-\lambda^2/R)}{2}} = \sqrt{\frac{R-\lambda^2}{2}} \qquad (2.117a)$$

$$b \equiv \sqrt{R}\ \cos\frac{\theta}{2} = \sqrt{\frac{R(1+\cos\theta)}{2}} = \sqrt{\frac{R(1+\lambda^2/R)}{2}} = \sqrt{\frac{R+\lambda^2}{2}} \qquad (2.117b)$$

Since, by (2.116), $R > \lambda^2$ for nonzero damping, constants a and b defined by (2.117) must always be real numbers. The complementary solution to (2.111) is therefore

$$\begin{aligned} X_{2c} &= C_1' e^{(a+ib)\xi} + C_2' e^{(a-ib)\xi} + C_3' e^{(-a+ib)\xi} + C_4' e^{(-a-ib)\xi} \\ &= e^{a\xi}(C_1^* \sin b\xi + C_2^* \cos b\xi) + e^{-a\xi}(C_3^* \sin b\xi + C_4^* \cos b\xi) \\ &= C_1 \sinh a\xi \bullet \sin b\xi + C_2 \cosh a\xi \bullet \sin b\xi \\ &\qquad + C_3 \sinh a\xi \bullet \cos b\xi + C_4 \cosh a\xi \bullet \cos b\xi \end{aligned} \qquad (2.118)$$

where $C'_1, \ldots, C'_4$, $C_1^*, \ldots, C_4^*$, and $C_1, \ldots, C_4$ are each sets of four constants of integration. Of the three solution forms shown in (2.118), the last one is ordinarily the most useful. To verify that it is, indeed, a complementary solution to (2.111), one may substitute it back into (2.113), using (2.117) for a and b, and (2.116) for R, as needed. A particular solution to (2.111) depends on $Q = Q(\xi)$, and may be found straightforwardly.

Substituting (2.112) and (2.118) back into (2.110b) permits direct determination of X_1. It is found to be

$$\begin{aligned} X_1 &= C_4 \sinh a\xi \bullet \sin b\xi + C_3 \cosh a\xi \bullet \sin b\xi \\ &\qquad - C_2 \sinh a\xi \bullet \cos b\xi - C_1 \cosh a\xi \bullet C_2 \cos b\xi + X_{1p} \end{aligned} \qquad (2.119)$$

where X_{1p} arises from the substitution of X_{2p} into (2.110b), and $C_1, \ldots, C_4$ are the same constants of integration as in (2.118).

To evaluate $C_1, \ldots, C_4$, one must apply the boundary conditions. For example, suppose both ends of the string ($\xi = 0, 1$) are fixed. Then the boundary conditions are:

$$w(0,t) = 0 \Rightarrow X_1(0) = 0, \text{ and } X_2(0)$$

$$w(1,t) = 0 \Rightarrow X_1(1) = 0, \text{ and } X_2(1) \qquad (2.120)$$

This yields four simultaneous equations to determine the four unknowns $C_1,\ldots,C_4$. If the loading function $P(\xi)$ and boundary conditions are both symmetric with respect to the middle of the string, then only the *even* functions of (2.118) and (2.112) need be retained (i.e., $C_2 = C_3 = 0$), and C_1 and C_4 are determined from only two simultaneous equations. These arise from the boundary conditions:

$$w\left(\pm\frac{1}{2},t\right)=0\rightarrow X_1\left(\frac{1}{2}\right)=0,\quad X_2\left(\frac{1}{2}\right)=0 \tag{2.121}$$

The determination and evaluation of closed-form, exact solutions for the forced response of a string with viscous damping present will ordinarily be an easier procedure to follow than that using eigenfunction superposition (Sec. 2.9).

Example 2.5 Generalize the solution of Example 2.4 to include the effects of viscous damping.

Solution

Setting $P(\xi)$ equal to the constant q_0 in (2.107), Q in (2.110a) and (2.111) is given by

$$Q=\frac{q_0\ell^2}{T}\equiv Q_0$$

A particular solution to (2.111) is

$$X_{2p}=\frac{2\pi\lambda\zeta_1 Q_0}{\lambda^2+4\pi^2\lambda^2\zeta_1^2}=\frac{\gamma^2 Q_0}{\lambda^4+\gamma^4}$$

From (2.110b), X_{1p} is found to be

$$X_{1p}=-\frac{\lambda^2 Q_0}{\lambda^4+\gamma^4}$$

which is inserted in (2.119). The boundary conditions (2.121) will then yield

$$\begin{bmatrix}\cosh\frac{a}{2}\bullet\cos\frac{b}{2} & -\sinh\frac{a}{2}\bullet\sin\frac{b}{2}\\ \sinh\frac{a}{2}\bullet\sin\frac{b}{2} & \cosh\frac{a}{2}\bullet\cos\frac{b}{2}\end{bmatrix}\begin{bmatrix}C_1\\ C_2\end{bmatrix}=\begin{bmatrix}-\lambda^2\\ \gamma^2\end{bmatrix}\frac{Q_0}{\lambda^4+\gamma^4}$$

Inverting to solve for C_1 and C_4,

$$\begin{bmatrix}C_1\\ C_4\end{bmatrix}=\frac{Q_0}{(\lambda^4+\gamma^4)\Delta}\begin{bmatrix}\cosh\frac{a}{2}\bullet\cos\frac{b}{2} & \sinh\frac{a}{2}\bullet\sin\frac{b}{2}\\ -\sinh\frac{a}{2}\bullet\sin\frac{b}{2} & \cosh\frac{a}{2}\bullet\cos\frac{b}{2}\end{bmatrix}\begin{bmatrix}-\lambda^2\\ \gamma^2\end{bmatrix}$$

where Δ is the determinant of the coefficient matrix, given by

$$\Delta = \sinh^2 \frac{a}{2} \bullet \sin^2 \frac{b}{2} + \cosh^2 \frac{a}{2} \bullet \cos^2 \frac{b}{2}$$

The displacement of an arbitrary point is therefore given by

$$w(\xi, t) = X_1(\xi) \sin \Omega t - X_2(\xi) \cos \Omega t$$

where

$$X_1(\xi) = C_4 \sinh a\xi \bullet \sin b\xi - C_1 \cosh a\xi \bullet C_2 \cos b\xi - \frac{\gamma^2}{\lambda^4 + \gamma^4} Q_0$$

$$X_2(\xi) = C_1 \sinh a\xi \bullet \sin b\xi + C_4 \cosh a\xi \bullet C_2 \cos b\xi + \frac{\gamma^2}{\lambda^4 + \gamma^4} Q_0$$

and λ, γ, R, and a and b are determined from (2.108), (2.114), (2.116), and (2.117), respectively. The displacement may also be normalized with respect to the static displacement at the center of the string (δ) by dividing through X_1 and X_2 by

$$\delta = \frac{1}{8} \frac{q_0 \ell^2}{T} = \frac{1}{8} Q_0$$

The displacement at any value of ξ may also be expressed as

$$w(\xi, t) = \sqrt{X_1^2 + X_2^2} \sin(\Omega t - \phi)$$

where the phase angle ϕ by which the motion lags the exciting force is

$$\phi(\xi) = \tan^{-1} \frac{X_2}{X_1}$$

2.11 Energy Functionals for a String

A string undergoing free, undamped vibrations will at a typical instant of time possess both potential and kinetic energy. Its potential energy will be greatest when it is in its maximum displacement state, measured from the equilibrium position. Its kinetic energy will be greatest as it passes through the equilibrium position with its maximum velocity. As in any system undergoing free, undamped vibrations, potential energy decreases as kinetic energy increases, and vice versa, the sum of the two quantities being constant at all times.

As we shall see in the next two sections, useful methods exist for analyzing the free vibrations of a string which utilize the potential and kinetic energies, rather than its differential equation of motion. Let us therefore formulate expressions for these quantities.

Consider an infinitesimal element of string which has the length dx in its straight, equilibrium position. As before, it has an initial tensile force (T) in it, which is sufficiently large to be considered constant during subsequent transverse displacements. Returning to Fig. 2.2, which shows the element in a typical, displaced position, we see that it has stretched to the length ds, with $dx = ds \cos\theta$, where $\tan\theta$ is the slope of the string at the instant. The increase in potential energy of the string is measured by the increase in its internal strain energy. For the element, this is the product of the internal force (T), which remains unchanged during deformation (in contrast with other physical systems, where it begins at zero and increases) and the elongation of the element. Thus, the change in potential energy of the element is

$$d(PE) = T(ds - dx) = T(ds - ds\cos\theta) \tag{2.122}$$

But

$$\cos(\theta) = 1 - \frac{\theta^2}{2!} + \frac{\theta^4}{4!} - \cdots \tag{2.123}$$

Therefore, for small θ

$$d(PE) = \frac{T\theta^2}{2}ds = \frac{1}{2}T\left(\frac{\partial w}{dx}\right)^2 dx \tag{2.124}$$

and the potential energy (PE) of the entire string is

$$PE = \frac{1}{2}\int_0^{\ell} T\left(\frac{\partial w}{\partial x}\right)^2 dx \tag{2.125}$$

T is left inside the integrand for generality, for in certain types of problems (e.g., a string hanging freely by its weight, undergoing transverse oscillations), T may be a function of x, even though it does not depend on w or t.

The kinetic energy (KE) of the element shown in Fig. 2.2 at any instant is

$$d(KE) = \frac{1}{2}(\rho ds)\left(\frac{\partial w}{\partial t}\right)^2 \tag{2.126}$$

where again ρ is mass per unit length.

The kinetic energy of the entire string is then

$$KE = \frac{1}{2}\int_0^{\ell} \rho \left(\frac{\partial w}{\partial t}\right)^2 dx \tag{2.127}$$

where ρ is left inside the integrand to provide for strings that have variable density, $\rho = \rho(x)$.

The quantities PE and KE given by (2.125) and (2.127), respectively, are functions of the displacement (w). However, w is in turn a function of both x and t. Therefore, these "functions of functions" are termed "functionals."

2.12 Rayleigh Method

If a string is vibrating with a frequency ω in one of its mode shapes of free, undamped vibration, the maximum potential energy (PE_{max}) occurs when the displacements of all points along the string are maximum, and their velocities are zero. Thus, PE_{max} occurs when $KE = 0$. Conversely, the maximum kinetic energy (KE_{max}) occurs when the velocities of all points are maximum, which occurs while the string passes through the equilibrium position (i.e., $w = 0$). Thus, KE_{max} exists when $PE = 0$. Because the total energy of the system must be conserved (i.e., $PE + KE$ = constant), then

$$PE_{max} = KE_{max} \tag{2.128}$$

Regardless of the initial conditions, it is always possible to begin the time coordinate for the problem so that the displacement in the vibrating mode may be expressed as

$$w(x,t) = X(x)\sin \omega t \tag{2.129}$$

so that the more general form, $\sin(\omega t + \phi)$ is not needed. Then, substituting (2.129) into (2.125), PE_{max} is observed to occur where $\sin^2 \omega t = 1$, whence

$$PE_{max} = \frac{1}{2}\int_0^{\ell} T(X')^2\, dx \tag{2.130}$$

Substituting (2.129) into (2.127), KE_{max} is observed to occur where $\cos^2 \omega t = 1$, whence

$$KE_{max} = \omega^2 KE^*_{max} \tag{2.131}$$

where

$$KE^{*}_{max} = \frac{1}{2}\int_0^{\ell} \rho X^2\, dx \tag{2.132}$$

Substituting (2.131) into (2.128) yields the frequency formula

$$\omega^2 = \frac{PE_{max}}{KE^{*}_{max}} \tag{2.133}$$

If one takes $X(x) = \sin m\pi x/\ell$, which is the *exact* shape of the mth mode of a uniform string fixed at both ends as described by (2.24), and substitutes this into (2.133), using (2.130) and (2.132), the exact value of the mth frequency is found, which was given in (2.21).

In his classic book *Theory of Sound* (Sections 88 and 89), first published in 1877, Lord Rayleigh [1] made an important, practical extension of the use of (2.133) when the exact mode shape is not known. For this reason, the R.H.S. of (2.133) for vibrating systems, in general, is called "Rayleigh's Quotient." Rayleigh's extension is to *assume* a mode shape, if it is not known exactly, and use (2.133), together with (2.130) and (2.132) in the case of a string, to determine an approximate value of the frequency. The displacement function selected should clearly satisfy the geometric boundary conditions present (i.e., zero displacement for the string); that is, it must be an *admissible function*. If the function chosen approximates closely the exact eigenfunction, then, the frequency determined by (2.133) will closely approximate the exact frequency. Finally, because the approximating function does not give the string the freedom it needs to vibrate in the shape it desires, constraint is added to the system. The effect of adding constraint to a vibrating system is to increase the frequencies of the modes constrained. This is called "Rayleigh's Principle." Thus, if (2.133) gives an approximate frequency, it will be an upper bound to the true fundamental frequency. If several different approximate, admissible functions are chosen for $X(x)$ and used in (2.133), they will all yield upper bounds to the exact ω, and the lowest will be the most accurate (the *least* upper bound). The Rayleigh method has been used by many in the published literature. Most notable is the short monograph by Temple and Bickley [2].

Example 2.6 Let the fundamental mode shape of a vibrating string be approximated by a simple polynomial, and use Rayleigh's method to obtain an approximate frequency.

Solution

With the coordinates shown in Fig. 2.1, the most simple polynomial which satisfies the boundary conditions is

$$X(x) = (x - 0)(x - \ell) = x(x - \ell)$$

From (2.130) and (2.132), we obtain

$$PE_{max} = \frac{T\ell^3}{6}, \qquad KE^*_{max} = \frac{\rho\ell^5}{60}$$

whence (2.133) yields

$$\omega^2 = 10\frac{T}{\rho\ell^2}$$

or

$$\omega\ell\sqrt{\frac{\rho}{T}} = \sqrt{10} = 3.1623$$

Comparing this with the exact fundamental frequency of $\omega\ell\sqrt{\rho/T} = \pi = 3.1416$, obtained from (2.21), one finds the upper bound approximate frequency to be quite close—within 0.66 percent of the exact value.

A slightly more complicated function could have been chosen as the cubic polynomial

$$X(x) = x^2\,(x - \ell)$$

for this, too, is an admissible function. However, this yields $\omega\ell\sqrt{\rho/T} = \sqrt{14} = 3.7417$ which is 19.1 percent too high. This large error is caused by the extremely poor representation of the first mode shape by the cubic polynomial. The cubic function forces zero slope (as well as zero displacement) at $x = 0$, and a skewed mode shape upon the string. In contrast, the previous second-degree function, which is a parabola symmetric about the midpoint of the string, fits the half-sine wave exact eigenfunction very closely (Fig. 2.22).

2.13 Ritz Method

Because Rayleigh's Quotient (2.133) yields an upper bound on the exact frequency, one would like to choose admissible functions so as to minimize the value of ω^2 determined by it. This may be done by trial and error, choosing various $X(x)$, and comparing results.

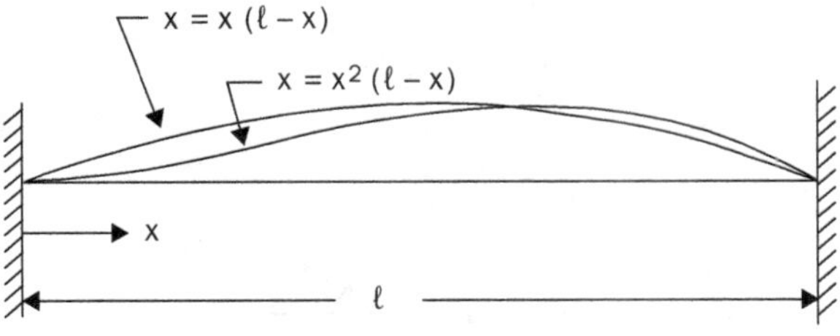

Figure 2.22 Possible polynomials for representing the first mode.

However, it would be preferable to have a procedure which would straightforwardly improve on the previous value of ω^2 and, even more importantly, would converge to the exact value if followed sufficiently long.

Such a method was presented by Ritz [3] in 1908, and was subsequently used by him on a much more complicated problem of plate vibrations. However, the method is applicable to any free vibration problem.

The Ritz method depends on choosing a set of admissible functions $\phi_i(x)$, each of which satisfies at least the *geometric* boundary conditions for the problem, and letting the desired mode shape be approximated by a sum of such functions; i.e.,

$$X(x) = \sum_{i=1}^{I} C_i \phi_i(x) \tag{2.134}$$

where the C_i are arbitrary coefficients which will be determined from minimization of the frequency. Equation (2.134) may be substituted into (2.133). After the integrations required by (2.130) and (2.132) are made, ω^2 is a function of the I coefficients C_i, which may be regarded as variables to be chosen so as to minimize ω^2. The direct method of minimizing a function of several variables is to take partial derivatives of the function with respect to each of the variables, in turn, and to solve the resulting set of simultaneous equations for the values of the variables corresponding to the minimum.

Before carrying out the above procedure in detail for any specific problem, it is desirable to apply it first in general to (2.133) to obtain an algebraically more simple form of the simultaneous minimizing equations. The minimizing equations are

$$\frac{\partial(\omega^2)}{\partial C_i} = 0 \quad (i = 1, 2, \dots, I) \tag{2.135}$$

Substituting (2.133) into (2.135),

$$\frac{\partial}{\partial C_i}\left(\frac{PE_{max}}{KE^*_{max}}\right) = \frac{KE^*_{max}\left(\dfrac{\partial PE_{max}}{\partial C_i}\right) - PE_{max}\left(\dfrac{\partial KE^*_{max}}{\partial C_i}\right)}{(KE^*_{max})^2} = 0$$

or

$$KE^*_{max}\frac{\partial PE_{max}}{\partial C_i} - PE_{max}\frac{\partial KE^*_{max}}{\partial C_i} = 0$$

Substituting $PE_{max} = \omega^2 KE^*_{max}$ from (2.133), we obtain

$$KE^*_{max}\left(\frac{\partial PE_{max}}{\partial C_i} - \omega^2 \frac{\partial KE^*_{max}}{\partial C_i}\right) = 0$$

Because KE^*_{max} is never zero except in a trivial solution, the foregoing may be divided through by it to yield the following more useful set of minimizing equations:

$$\frac{\partial}{\partial C_i}(PE_{max} - \omega^2 KE^*_{max}) = 0 \quad (i = 1,2,\ldots,I) \tag{2.136}$$

This is a set of I simultaneous, linear, algebraic equations in the unknown C_i. However, the equations are homogeneous (zero right-hand-sides). For a nontrivial solution, the determinant of the coefficient matrix is set equal to zero. The roots of the determinant are the I values of ω^2. The lowest value of ω^2 is an upper bound approximation to the fundamental frequency, and the higher values are also upper bound approximations (albeit, usually less accurate) to higher frequencies for the string. Substituting any value of ω^2 so obtained back into the I simultaneous equations in C_i, and ignoring any one of the equations, the remaining set of I-1 simultaneous equations may be made nonhomogeneous by dividing through by one of the C_i, and they may be solved for the eigenvectors corresponding to the ω^2 used.

In the published technical literature, one sometimes finds the Ritz method called the "Rayleigh–Ritz method." But, as we have seen above, the Rayleigh and Ritz methods are two different procedures. This is further clarified in [4].

Example 2.7 Determine the fundamental frequency of a uniform string of length ℓ hanging freely in a vertical position, loaded by its own weight.

Solution

Choose coordinates as shown in Fig. 2.23. The tension in any section of the string is then

$$T = \rho g(\ell - x)$$

where ρ is mass per unit length, and g is the gravitational acceleration constant. The differential equation of motion (2.6) is

$$\frac{\partial}{\partial x}\left[\rho g(\ell - x)\frac{\partial w}{\partial x}\right] = \rho\frac{\partial^2 w}{\partial t^2}$$

or

$$g(\ell - x)\frac{\partial^2 w}{\partial x^2} - g\frac{\partial w}{\partial x} = \frac{\partial^2 w}{\partial t^2}$$

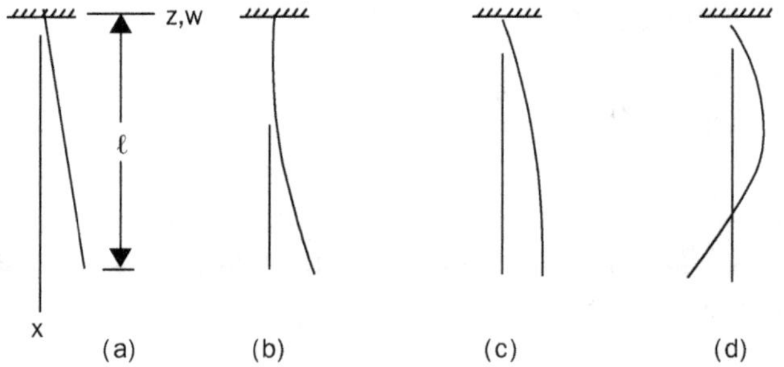

FIGURE 2.23 Possible admissible functions for hanging string vibrations.

This variable coefficient differential equation may be solved by the method of Frobenius, or it may be transformed into a form of Bessel's equation, but let us get relatively simple, straightforward solutions by Rayleigh and Ritz methods.

One simple admissible function is the straight line, depicted in Fig. 2.23(a):

$$X(x) = x$$

Substituting this into (2.130) and (2.132) yields $PE_{max} = \rho g \ell^2/4$ and $KE^*_{max} = \rho \ell^3/6$, whence (2.133) gives $\omega = \sqrt{3g/2\ell} = 1.225\sqrt{g/\ell}$. This result may be compared with that for the classical, single d.o.f. pendulum, where all mass is concentrated at the free end, and is connected to a hinge at the fixed end by means of a rigid, massless rod. This highly idealized configuration has the well-known frequency, $\omega = \sqrt{g/\ell}$. Another well-known result is the frequency of a rigid bar pendulum having uniformly distributed mass, $\omega = \sqrt{3g/2\ell}$. This frequency is exactly the same as that found above for the string having a straight line mode shape, for it is exactly the same problem—uniformly distributed mass and weight, and all points constrained to move in a straight line. Since it seems unlikely that a perfectly flexible string should behave the same as a rigid bar, the result of introducing curvature into the problem [Fig. 2.23(b) and (c)] should be investigated.

Consider next the positive curvature shown in Fig. 2.23(b). This would seem to be a reasonable mode shape, because the fundamental mode shape of a double pendulum is similar. This mode shape could be represented by the simple parabola

$$X(x) = x^2$$

Substituting this function into (2.130)–(2.132) gives $\omega = \sqrt{5g/3\ell} = 1.291\sqrt{g/\ell}$. Because both approximate solutions are upper bounds on the exact frequency, the parabolic shape is clearly a worse approximation than the straight-line shape. This could mean either that the curvature should be opposite as in Fig. 2.23(c), or that the parabolic function is locally inaccurate for the string (i.e., the curvature should be positive, but different).

The parabolic function of the shape shown in Fig. 2.23(c) is

$$X(x) = 2\ell x - x^2$$

Using this with the Rayleigh method gives $\omega = 1.369\sqrt{g/\ell}$, which is the worst result thus far!

From the above information, it would seem that a two-function approximation consisting of a linear function, plus a second-degree correction function, could be used with the Ritz method to improve on the linear function approximation. That is, choose

$$X(x) = C_1 x + C_2 x^2$$

Adding the x^2 term cannot yield a result worse than that from the linear function alone, for the Ritz method could yield $C_2 = 0$ if the x^2 term gave no improvement. Substituting this two term polynomial into (2.130), (2.132), and (2.136) yields two simultaneous equations, which may be written in matrix form as

$$\begin{bmatrix} \frac{g}{\ell} - \frac{2}{3}\omega^2 & \frac{2}{3}\frac{g}{\ell} - \frac{1}{2}\omega^2 \\ \frac{2}{3}\frac{g}{\ell} - \frac{1}{2}\omega^2 & \frac{2}{3}\frac{g}{\ell} - \frac{2}{5}\omega^2 \end{bmatrix} \begin{bmatrix} C_1 \\ C_2\ell \end{bmatrix} = \begin{bmatrix} 0 \\ 0 \end{bmatrix}$$

It should be mentioned that the derivatives of (2.136) should be taken before the integrations are carried out. This reduces the labor significantly. The coefficient matrix given above has the typical characteristics of one generated by the Ritz method; that is,

1. It is fully populated (no zero elements).
2. It has ω^2 in each element.
3. It is symmetric.

Setting its determinant equal to zero, and expanding it, yields the following frequency equation:

$$3\lambda^2 - 32\lambda + 40 = 0$$

where $\lambda = \omega^2\ell/\gamma$ is the square of the nondimensional frequency. The roots of the frequency equation are

$$\lambda_1 = \frac{16 - 2\sqrt{34}}{3} = 1.4460, \qquad \lambda_2 = \frac{16 + 2\sqrt{34}}{3} = 9.2206$$

The corresponding frequencies are

$$\omega_1 = 1.2025\sqrt{g/\ell}, \qquad \omega_2 = 3.0365\sqrt{g/\ell}$$

Exact values of the first two frequency parameters are obtained from a solution involving Bessel functions ([5], p. 107). They are (to five significant figures):

$$\omega_1 = 1.2024\sqrt{g/\ell}, \qquad \omega_2 = 2.7600\sqrt{g/\ell}$$

Thus, one sees that the Ritz approximation to the first frequency is extremely close (0.008 percent error), whereas the second frequency is much worse (10 percent error). This is also typical of the Ritz method. That is, both

Exact	Rayleigh method			Ritz method	Simple pendulum
	$X = x$	$X = x^2$	$X = 2\ell x - x^2$		
1.2024	1.2247	1.2910	1.3693	1.2025	1.0000

TABLE 2.2 Values of $\omega\sqrt{g/\ell}$ for the Fundamental Frequency of a Freely Hanging String, by Various Methods

frequencies obtained by the Ritz method are upper bounds on the first two frequencies; however, the lowest one is usually a closer upper bound.

The eigenvectors for the two modes are found by returning to the matrix equation in C_1 and C_2, and substituting the eigenvalues (λ) into either one of them. This yields the eigenvectors

$$\left(\frac{C_2}{C_1}\right)_1 = \frac{0.638}{\ell}, \qquad \left(\frac{C_2}{C_1}\right)_2 = -\frac{1.305}{\ell}$$

Consequently, the first two approximate mode shapes are

$$X_1(\xi) = \xi + 0.638\,\xi^2, \qquad X_2(\xi) = \xi - 1.305\,\xi^2$$

where $\xi = x/\ell$. It is clear that both terms of the originally assumed displacement contribute significantly to each mode shape, although the linear term is most important for the first mode and the second degree term is most important for the second mode. Thus, the fundamental mode shape resembles Fig. 2.23(b), except that it has significant slope at its attachment point ($\partial y/\partial x \neq 0$ at $x = 0$). The second mode shape is depicted in Fig. 2.23(d).

Fundamental frequencies obtained by the various approximate displacement functions are summarized in Table 2.2.

2.14 Large Amplitude Vibrations

In deriving the equation of motion (2.8) for the free vibrations of a string, two assumptions were made in order that it resulted in being linear:

1. The tension (T) in the static equilibrium position is large and the transverse displacement is small so that the tension may be treated as a constant during the motion.
2. The slope of the string is everywhere small during the motion, so that $\sin\theta \approx \tan\theta = \partial w/\partial x$ and $\cos\theta \approx 1$, where θ is the angle the string makes with the straight line equilibrium position, as shown in Fig. 2.2.

Of these two assumptions, the first is the most important, for as the tension in the string increases from its equilibrium value during finite amplitude, transverse displacement, the frequencies are increased. In the extreme case of a string which is slack (no tension when in equilibrium), the linear theory gives zero frequencies, which is clearly wrong.

A generalization of the equation of motion to account for changing tension was presented by Kirchhoff [6] more than a century ago. It can be written as

$$\left[T_0 + \frac{EA}{2\ell}\int_0^\ell \left(\frac{\partial w}{\partial x}\right)^2 dx\right]\frac{\partial^2 w}{\partial x^2} = \rho\frac{\partial^2 w}{\partial t^2} \tag{2.137}$$

where T_0 now represents the tensile force in the equilibrium position and the second term in the brackets is the tension increase due to transverse displacement averaged over the length of the string. It will be recalled from Sec. 2.11 that, for small slopes, the change in length of an infinitesimal portion due to w is

$$ds - dx = \frac{1}{2}\left(\frac{\partial w}{\partial x}\right)^2 dx \tag{2.138}$$

Therefore, the average strain in the string is

$$\bar{\epsilon} = \frac{1}{2L}\int_0^\ell \left(\frac{\partial w}{\partial x}\right)^2 dx \tag{2.139}$$

Assuming that the string stretches linearly in accordance with Hooke's Law, then the average value of tension increase is

$$T_1 = EA\bar{\epsilon} \tag{2.140}$$

where E is Young's modulus and A is the cross-sectional area of the string (assumed to be constant in this analysis). Thus, T_1 is the second term in the brackets of (2.137). The mass density per unit length is ρ, as before, also assumed to be constant with respect to time.

The integro-differential equation (2.137) is nonlinear, and appears formidable in comparison with (2.8). However, a variables separable form of solution is possible, and it is exact. Assume that

$$w(x,t) = \phi(t)\sin\frac{m\pi x}{\ell} \tag{2.141}$$

That is, it is assumed that the mode shapes are the same as those of the linear analysis. Clearly, the boundary conditions $w(0,t) = w(\ell,t) = 0$ are thereby satisfied.

Substituting (2.141) into (2.137) yields

$$\phi'' + \frac{T_0}{\rho}\left(\frac{m\pi}{\ell}\right)^2 \phi + \frac{EA}{4}\left(\frac{m\pi}{\ell}\right)^4 \phi^3 = 0 \tag{2.142}$$

This nonlinear, ordinary differential equation in ϕ is the well-known Duffing equation of motion for a single d.o.f. system having a "hard" spring. It has an exact solution in terms of elliptic integrals (cf. [7], pp. 312–316). The motion is periodic in time, although not simple harmonic (which the linear solution is, for single mode excitation). It should be noted that, while the linear analysis permits superposition of responses of the individual modes to permit dealing with arbitrary initial conditions, the nonlinear analysis does not, for (2.137) would not be satisfied by the superimposed solution.

If slopes become significant (in comparison with unity), then, the more general equation of motion

$$\frac{\partial}{\partial x}(T \sin\theta) = \rho \frac{\partial^2 w}{\partial t^2} \tag{2.143}$$

may be used, obtained by summing forces in the transverse direction as in Sec. 2.1. For large slopes, longitudinal motion (in the x-direction in Fig. 2.1) also becomes significant, and enters the problem through the equation of motion

$$\frac{\partial}{\partial x}(T \cos\theta) = \rho \frac{\partial^2 u}{\partial t^2} \tag{2.144}$$

where u is the longitudinal component of displacement. Assuming a linearly elastic material, it may be shown that T is a function of both w and u, given by [8]

$$T = EA\left[(1+\epsilon_0)\sqrt{\left(1+\frac{\partial u}{\partial x}\right)^2 + \left(\frac{\partial w}{\partial x}\right)^2} - 1\right] \tag{2.145}$$

where $\epsilon_0 = T_0/EA$ is the equilibrium strain. Substituting (2.145) into (2.143) and (2.144) results in a set of highly nonlinear, coupled equations in w and u. An exact solution of these equations is intractable; however, accurate approximate solutions were obtained [8] by means of a form of the Galerkin method using incremental time steps, and also by finite differences. It is found that, if the string is displaced into a single half-sine wave mode shape initially and released from rest, the resulting motion is not periodic, although for moderate amplitudes ($\delta/\ell = 0.1$, where δ is the displacement amplitude), it is nearly so.

Table 2.3 presents the ratio of the fundamental frequency obtained from the two types of nonlinear solutions discussed above to that from linear analysis for a variety of nondimensional amplitudes ($\delta/\ell = 0.01, 0.05, 0.1, 0.2, 0.4$) and of equilibrium strains ($\epsilon_0 = 10^{-1}, 10^{-2}, 10^{-3}, 10^{-4}, 10^{-5}$). From these results, it is clear that the linear solution is reasonably accurate only under very restricted conditions. For example, for a moderate equilibrium strain ($\epsilon_0 = 10^{-3}$) and a very small transverse displacement ($\delta/\ell = 0.01$), the linear frequency is too small by 8.9 percent. If ϵ_0 is reduced to 10^{-4}, the frequency of the nonlinear model becomes 67 percent higher than that of the linear analysis. If instead the initial displacement becomes visually obvious ($\delta/\ell = 0.1$), then, the frequency is seen to be 4.298 times as great as the linear one. It is also clear that for a string which is nearly slack in its equilibrium position ($\epsilon_0 < 10^{-4}$), the linear theory gives extremely inaccurate results.

For metallic strings (i.e., wires), one can tolerate only small equilibrium strains ($\epsilon_0 < 10^{-3}$) so that the yield stress of the material is not exceeded; otherwise, the linear stress–strain assumption of the theory is not valid. Thus, for example, a steel wire undergoing $\epsilon_0 = 10^{-3}$, having a Young's modulus $E = 30 \times 10^6$ psi, has a tensile stress of 300,000 psi if the linear relationship holds, which is in the vicinity of the yield stress for high-strength, small-diameter wire which can be found in a piano. Transverse displacement in the shape of a half-sine wave causes an additional average strain of

$$\epsilon_1 = \frac{\pi^2}{4}\left(\frac{\delta}{L}\right)^2 \tag{2.146}$$

Thus, a very small displacement of $\delta_1/L = 0.01$ would result in an additional average strain of 0.25×10^{-3}, which would cause the yield stress (and perhaps the rupture stress) of the wire having $\epsilon_0 = 10^{-3}$ to be exceeded. Similarly, if $\delta/\ell = 0.10$, (2.146) yields $\epsilon_1 = 2.5 \times 10^{-3}$, which would cause failure, no matter how small ϵ_0 is.

However, for some nonmetallic strings, the full ranges of δ/ℓ and ϵ_0 used in Table 2.3 may be practical.

Note that the frequencies listed in Table 2.3 determined from Kirchhoff's relatively simple equation (2.137) agree remarkably well with those from the coupled, large-slope analysis for most of the ranges of δ/ℓ and ϵ_0. Significant disagreement occurs, as expected, for $\delta/\ell = 0.40$, because of the large slopes involved.

2.15 Some Concluding Remarks

Although many aspects of the vibrating string were considered in this chapter, a few were not taken up which have received some attention in the published literature.

δ/L	ϵ_0	Coupled, large-slope equations	Kirchhoff equation (2.137)
0.01	10^{-1}	1.050	1.001
	10^{-2}	1.014	1.009
	10^{-3}	1.089	1.088
	10^{-4}	1.673	1.673
	10^{-5}	4.324	4.331
0.05	10^{-1}	1.073	1.023
	10^{-2}	1.213	1.207
	10^{-3}	2.335	2.338
	10^{-4}	6.718	6.732
	10^{-5}	21.02	21.07
0.10	10^{-1}	1.142	1.148
	10^{-2}	1.674	1.673
	10^{-3}	4.298	4.331
	10^{-4}	13.23	13.35
	10^{-5}	41.73	42.10
0.20	10^{-1}	1.366	1.314
	10^{-2}	2.785	2.850
	10^{-3}	8.208	8.479
	10^{-4}	25.76	26.64
	10^{-5}	81.40	84.17
0.40	10^{-1}	1.896	1.966
	10^{-2}	4.844	5.420
	10^{-3}	14.87	16.86
	10^{-4}	46.76	53.24
	10^{-5}	*	168.3

*Results not available due to numerical instability.

TABLE 2.3 Ratio (ω/ω_L) of Nonlinear to Linear Fundamental Frequencies

Variable coefficient, linear differential equations arise in the case of a string hanging by its own weight, $T = T(x)$, or having variable density, $\rho = \rho(x)$. It is recommended that such problems be dealt with by the Rayleigh and Ritz methods as discussed in Secs. 2.12 and 2.13. In *some* cases, these differential equations may be solved in terms of Bessel functions, if the proper transformation can be found, but in general, they cannot be. If $T(x)$ or $\rho(x)$ can be expressed as Cx^n, where C is a constant and n is an integer, then an exact solution of the governing differential equation is possible in terms of infinite power series of x (the Frobenius method). Bessel's solution is one example of this.

References

1. J. W. Strutt (Lord Rayleigh), *Theory of Sound*, vol. 1, The MacMillan Co., 1877; reprint, Dover Publications, 1945.
2. G. Temple and W. G. Bickley, *Rayleigh's Principle and Its Applications to Engineering*, Oxford University Press, 1933; reprint, Dover Publications, 1956.
3. W. Ritz, "Über eine neue Methode zur Lösung gewisser Variationsprobleme der mathematischen Physik," *Journal für die Reine und Angewandte Mathematik*, 1908.
4. A. W. Leissa, "The historical bases of the Rayleigh and Ritz methods," *J. Sound Vib.*, 207 (2005): 961–78.
5. N. Y. McLachlan, *Bessel Functions for Engineers*, 2nd ed., Oxford University Press, 1955.
6. G. Kirchhoff, *Vorlesungen über Mathematische Physik: Mechanik*, Leipzig, 1876.
7. H. M. Hansen and P. F. Chenea, *Mechanics of Vibration*, John Wiley and Sons, 1952.
8. A. W. Leissa and A. M. Saad, "Large amplitude vibrations of strings," *J. Appl. Mech.*, 61 (1994)(2): 296–301.

Problems

1 A string of length ℓ has both ends fixed. At time $t = 0$, its initial shape is described by

$$w(x, 0) = \begin{cases} \sin(2\pi x/\ell) & \text{if } 0 \le x \le \ell/2 \\ 0 & \text{if } \ell/2 \le x \le \ell \end{cases}$$

and is released from rest. Use the modal superposition method to determine its displacement at instant $t = \frac{1}{8}\tau_0$ where τ_0 is the fundamental period of the string. Plot this function to show the shape of the string at this instant.

2 Solve Problem 1 utilizing the traveling wave method. Show the displaced shape at $t/\tau_0 = 0$, 1/16, 1/8, 1/4, 3/8, and 1/2.

3 Solve Example 2.1 by the traveling wave method. Show the displaced shape of at $t/\tau_0 = 0$, 0.1, 0.2, 0.3, 0.4, 0.5.

4 In the case when $k^* = 1$, plot the mode shapes corresponding to the frequencies found on Fig. 2.11. Draw the four curves (for $M^* = 0, 0.2, 1, 5$) for the first mode together on a single graph, and those for the second mode together on a second graph. From physical considerations, add the curves for $M^* = \infty$, and verify them from the limiting cases of the frequency and mode shape equations.

5 A. Set up a frequency determinant for the example of the string having a particle attached at its one-quarter point (Sec. 2.7). Use two displacement functions, w_1 and w_2 and measure x from the left boundary for both functions, so that w_1 and w_2 are valid for $0 < x < \ell/4$ and $\ell/4 < x < \ell$, respectively.

B. Expand the determinant and, by means of trigonometric identities, try to reduce the resulting equation to the form of (2.63). If unsuccessful, verify that the determinant is *probably* correct by showing that $\beta = 2.4136$ is a root of it when $M^* = 0.5$, as it is for (2.63).

6 A composite string of total length ℓ consists of two segments $a\ell$ and $b\ell$ (Fig. 2.24), each having uniform, but different mass densities $c\rho$ and $d\rho$, respectively, as shown below (a, b, c, and d are constants), and ρ is here the *average* density of the entire string.

A. Derive the frequency equation for the system. Put it into a form containing $\beta = \omega\ell\sqrt{\rho/T}$, a/b, and c/d as nondimensional parameters. (Hint: To simplify the algebra, use *two* coordinate origins x_1 and x_2 for the two portions of the string, with origins at the two fixed ends.)

B. Let $a/b = 1$ and $c/d = 5$. Solve for the first four, nondimensional frequencies. Compare these with the ones for a *uniform* string having the same total mass.

C. Plot the first two mode shapes for Part B.

D. Choose appropriate values of a, b, c, and d to permit an approximate representation of the concentrated mass problem of Sec. 2.7, determine the first four eigenvalues, and compare them with those found in the first column of Table 2.1.

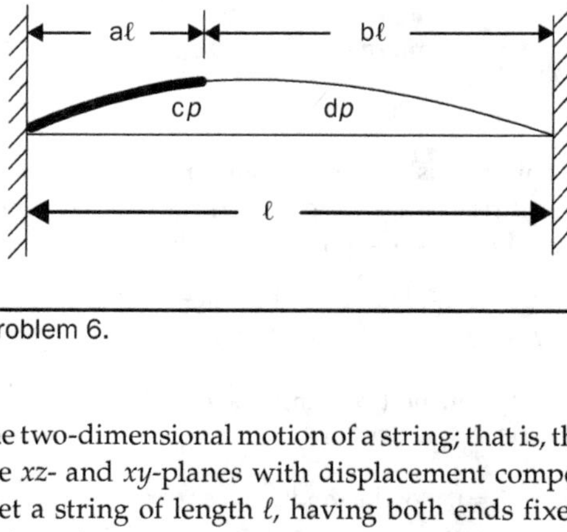

FIGURE 2.24 Problem 6.

7 Consider the two-dimensional motion of a string; that is, the simultaneous vibration in the xz- and xy-planes with displacement components w and v, respectively. Let a string of length ℓ, having both ends fixed, be given the following initial conditions:

$$w(x,0) = w_0 \sin\frac{\pi x}{\ell}, \qquad \frac{\partial w}{\partial t}(x,0) = 0$$

$$v(x,0) = 0, \qquad \frac{\partial v}{\partial t}(x,0) = v_0 \sin\frac{\pi x}{\ell}$$

where w_0 and v_0 are constants indicating the magnitudes of the initial displacement and velocity components.

A. Suppose that there is no damping. Show that the subsequent motion of the string is a whirling one, and find the path in the yz-plane of a typical point located at $x = x_0$. Determine the condition under which the path is circular.

B. Suppose now that the string is immersed in a viscous medium, with a damping constant c effective for every point on the string. Assuming that initially the string is whirling with all points executing circular paths, determine the subsequent motion of a typical point when the string is released from the initial path. Assume the system to be underdamped.

C. Let $\zeta = c/c_c = 0.2$ for the system of Part B. For a typical point on the string, make a plot of one displacement component (v) versus the other (w) over the length of time $0 \le t \le 2\tau$, where $\tau = 2\pi/\omega_1 = 2\ell\sqrt{\rho/T}$.

8 Consider again the free vibration of the string plucked at its quarter point and released from rest, as in Example 2.1. However, now let the string be immersed in a viscous medium with a damping resistance such that the damping ratio for the first vibration mode is $\zeta_1 = 0.3$.

A. Show that the first mode completes one cycle of decaying motion in the time $t = \tau = 2\pi/\sqrt{1-\zeta_1^2}\,\omega_1$ where ω_1 is the undamped frequency of the first mode.

B. If the maximum initial displacement is δ, make plots of the shape of the string at $t = \tau/2$ and $t = \tau$.

9 The string shown (Fig. 2.25) is subjected to a distributed pressure given by

$$p(x,t) = q_{\max}\frac{x}{\ell}\sin\Omega t$$

Assuming no damping, determine the responses of the first three modes at the point $x = 2\ell/3$. Plot them in nondimensional form on a single graph having Ω/ω_1 as the abscissa. If any modes are not excited, explain physically why this should be.

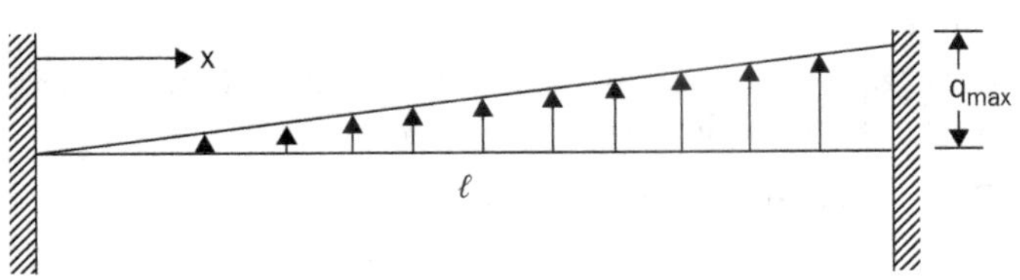

Figure 2.25 Problem 9.

10 Use the eigenfunction superposition method to determine the total response of the point $x = 2\ell/3$ in Problem 9 in the case of small damping ($c/c_{c1} = 0.1$). Make a plot similar to the curve for $c/c_{c1} = 0.1$ in Fig. 2.18, and explain any significant differences.

11 Do Problem 10 utilizing the closed form solution method.

A. For $c/c_{c1} = 0$

B. For $c/c_{c1} = 0.1$

12 A string of length ℓ (Fig. 2.26) is immersed in a viscous medium and is subjected to a transverse pressure given by

$$p(x,t) = \sin\frac{\pi x}{\ell}\Psi(t)$$

where $\Psi(t)$ is a function periodic in time, with period τ, as shown.

A. Determine the steady state response of the string.

B. Let the damping vanish. Plot the steady state displacement of the middle of the string ($x = \lambda/2$) as a function of time for $\tau = 0.9\ \tau_0$, $1.5\ \tau_0$, and $2\ \tau_0$ where τ_0 is the period of the fundamental free vibration mode.

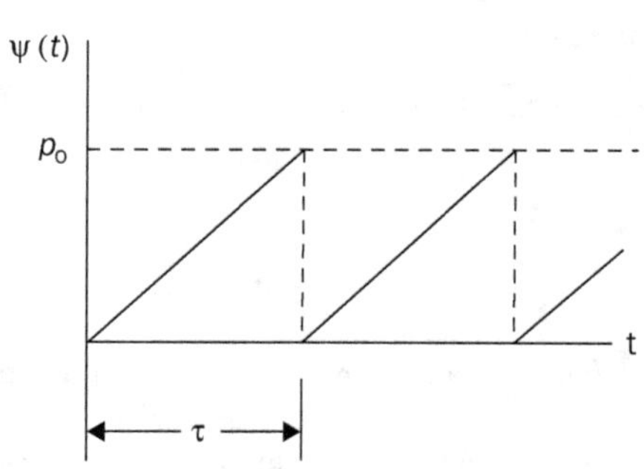

Figure 2.26 Problem 12.

13 Do Problem 12 when the transverse pressure is uniformly distributed along the length of the string, with intensity p_0.

14 A. Prove that the orthogonality relationship

$$c\int_0^{a\ell} X_{1m}X_{1n}\,dx_1 + d\int_0^{b\ell} X_{2m}X_{2n}\,dx_2 = 0 \quad (m \neq n)$$

is valid for the composite string of Problem 6, where x_1 and x_2 are the two coordinates suggested in Part A of Problem 6, and where X_m and X_n are eigenfunctions defined, for example, by

$$X_m = \begin{cases} X_{1m}(x_1) & \text{for } 0 \le x_1 \le a\ell \\ X_{2m}(x_2) & \text{for } 0 \le x_2 \le b\ell \end{cases}$$

If you cannot prove the relationship in general, then let $a/b = 1$ and $c/d = 5$, and use the eigenfunctions of Problem 6 to verify the relationship computationally for all combinations of $m,n = 1,2,3,4$. But be sure to do this with sufficient digits to preserve accuracy.

B. Let the composite string of Problem 6B be given an initial displacement in the shape of half-sine wave, i.e.,

$$f(x) = A\sin\frac{\pi x}{\ell}$$

where A is a constant. The string is then released from rest. There is no damping. Use the above orthogonality relationship to determine the subsequent motion $w(x,t)$. On a single graph, plot the shape of the string at $t = 0$, $\tau_1/6$, $\tau_1/3$, $\tau_1/2$, $2\tau_1/3$, $5\tau_1/6$, and τ_1 where τ_1 is the period of free vibration of the *fundamental* (lowest frequency) mode.

15 A string of length λ and total mass M has a circular cross section, the radius of which varies linearly from R at one end to $2R$ at the other. The string is horizontal, and both ends are fixed. Use the Rayleigh method with an assumed mode shape in the form of the static deflection of the string to find the fundamental frequency. Plot the approximate mode shape.

16 Solve Problem 15 by the Ritz method, using the sum of two admissible functions to represent the mode shape. Choose your functions carefully so that a result better than that of Problem 15 will be found.

CHAPTER 3

Longitudinal and Torsional Vibrations of Bars

Among the fundamental components of continuous systems are bars. Bars are components that have one dimension (length) considerably larger than the other two. In that, they share the same definition as a beam with only one distinction: the loading. Loading in beams is in the transverse direction. Their motion is mainly in a direction perpendicular to their longitudinal axis. Beams will be studied in the next chapter. Bars, on the other hand, take axial or torsional loads and can deform longitudinally and rotationally around their longitudinal axis.

Bars are often referred to as shafts, rods, or columns. In static analysis, we often refer to them as bars when they take axial tensile load. If they are under compressive loads, we call them columns. If subjected to torsion, we refer to them as shafts or rods. In a dynamic analysis, if the motion is rotational around their longitudinal axis, they are usually denoted as shafts. If their motion is in the longitudinal direction, they are simply called bars or rods.

Longitudinal and torsional vibrations of bars are typically higher in frequencies than in their beam-like transverse bending modes. They could, however, be vulnerable for excitation in certain applications causing potential engineering challenges. Truss members, hydraulic cylinders, as well as other components can be subjected to axial forces that excite mainly their longitudinal frequencies. Shafts in automotive and general power transmission applications may be subjected to torsional loading from the engine or from electric or other motors exciting their torsional frequencies.

The longitudinal and torsional vibrations of bars will be studied in this chapter. The fundamental equations will first be derived. Free longitudinal and torsional vibration is treated. A striking similarity with the wave equation of the transverse vibration of a string is noted, leading to analogy in many solutions. Forced vibration is studied with emphasis on internal material damping. Energy functionals are then generated and used in numerical analyses by the Rayleigh and Ritz methods.

3.1 Equation of Motion for Longitudinal Vibrations

A bar (or rod) of length ℓ and cross-sectional area A is depicted in Fig. 3.1. The area need not be constant; however, cross-sectional shapes at all values of x will be assumed similar in order to avoid coupling between longitudinal and torsional displacements. The material of the bar is elastic; however, it need not be homogeneous. Its material properties may vary as a function of x. Consider the longitudinal (or extensional) vibration of the bar (i.e., motion in the x-direction).

A differential element of length dx is taken at a typical coordinate location x. Its free body diagram is shown in Fig. 3.2. The force P is the resultant of the longitudinal stress σ_x acting internally on A, where σ_x is assumed to be uniform over the cross-section. P varies along the length, and is also a function of time, i.e., $P = P(x,t)$. In addition, a distributed force (p) is shown, having dimensions of force per unit length of bar, which results from external sources, either internally or externally applied. Examples of this would be magnetic or surface traction forces.

Summing forces in the x-direction:

$$-P + \left(P + \frac{\partial P}{\partial x} dx \right) + p dx = \rho A \frac{\partial^2 u}{\partial t^2} dx \tag{3.1}$$

where ρ is now mass per unit volume, and u is the displacement in the x-direction. Substituting $P = \sigma_x A$ and simplifying gives

$$\frac{\partial}{\partial x}(\sigma_x A) + p = \rho A \frac{\partial^2 u}{\partial t^2} \tag{3.2}$$

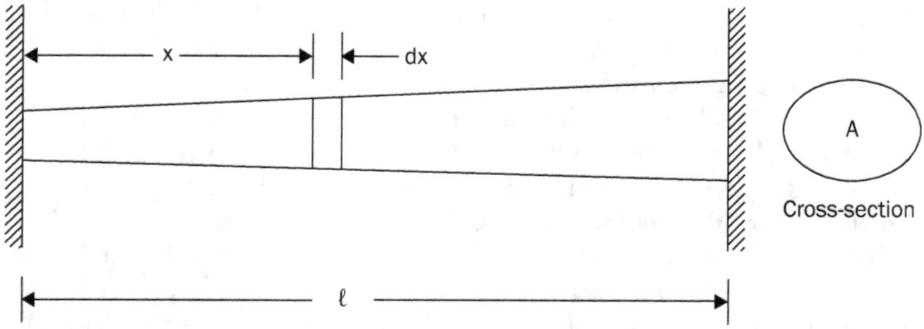

FIGURE 3.1 A bar (or rod) of length ℓ and cross-sectional area A.

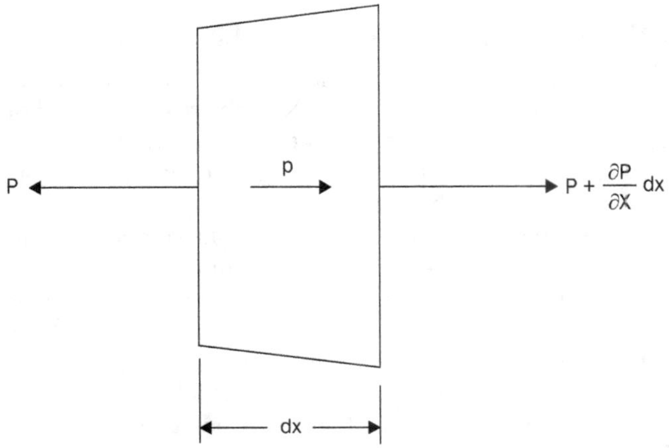

FIGURE 3.2 Free body diagram of a differential element of length dx subjected to longitudinal forces.

For a linearly elastic material subjected to uniaxial stress, $\sigma_x = E\epsilon_x$, where E is Young's modulus and ϵ_x is the longitudinal strain. The latter is related to the displacement by $\epsilon_x = \partial u/\partial x$. Therefore, (3.2) becomes

$$\frac{\partial}{\partial x}\left(AE\frac{\partial u}{\partial x}\right) + p = \rho A\frac{\partial^2 u}{\partial t^2} \tag{3.3}$$

In (3.3), A, E, and ρ may all be functions of x, whereas p may be a function of both x and t. This is the equation of motion for a rather general class of problems.

If A and E are both constant, then (3.3) becomes

$$AE\frac{\partial^2 u}{\partial x^2} + p = \rho A\frac{\partial^2 u}{\partial t^2} \tag{3.4}$$

which is identical to (2.7), except that T has been replaced by AE, ρ by ρA, and w by u.

For free vibrations, $p = 0$, and (3.4) reduces to

$$\boxed{E\frac{\partial^2 u}{\partial x^2} = \rho\frac{\partial^2 u}{\partial t^2}} \tag{3.5}$$

which is analogous to (2.8).

3.2 Equation of Motion for Torsional Vibrations

Let us consider now the torsional vibrations of the same bar, shown in Fig. 3.1. This consists of the rotation of each cross-section about the longitudinal axis which passes through the centroids of the cross-sections. We will restrict ourselves to cross-sections having at least two symmetry axes (such as the ellipse seen in Fig. 3.1) to avoid coupling between twisting and bending displacements.

A typical element of length dx, determined by parallel planes located at x and $x + dx$, is again chosen. A free body diagram of it is drawn in Fig. 3.3. A twisting moment M_t is shown acting on the cross-section taken at the x-plane. This moment is the resultant of the internal shear stresses τ_{xy} and τ_{xz} (Fig. 3.4), which exist on the cross-section and vary as functions of the transverse coordinates y and z (as well as with x and t). The twisting moment is related to the shear stresses by (see Fig. 3.4).

$$M_t = \iint_A (\tau_{xz} y - \tau_{xy} z) dydz \tag{3.6}$$

M_t is therefore a function of x and t. The possibility of an externally applied twisting moment m_t, having the dimensions of moment per unit length is also shown in Fig. 3.3.

Summing moments about the x-axis:

$$-M_t + \left(M_t + \frac{\partial M_t}{\partial x} dx \right) + m_t dx = dI_c \frac{\partial^2 \theta}{\partial t^2} \tag{3.7}$$

where dI_c is the *mass* moment of inertia of the infinitesimal element about the x-axis, and θ is the rotation angle (in radians) of the cross-section; $\theta = \theta(x,t)$. Looking in detail at dI_c:

$$dI_c = \iint_A r^2 dm = \rho dx \iint r^2 dA = \rho J dx \tag{3.8}$$

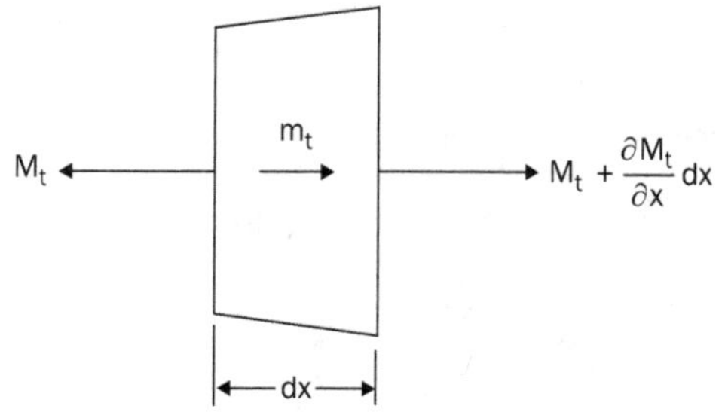

Figure 3.3 Free body diagram of a typical element of length *dx* subjected to torsional moments.

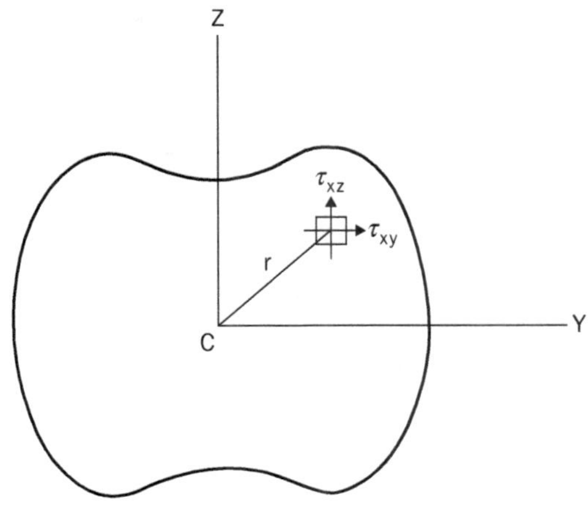

FIGURE 3.4 Internal shear stresses τ_{xy} and τ_{xz} that result in a twisting moment M_t.

where r is the polar coordinate of a typical point in the cross-section (see Fig. 3.4), ρ is mass per unit volume (assumed now to be constant throughout the cross-section), and J is the "polar *area* moment of inertia" of the cross-section (more properly, the polar second moment of the area), defined by

$$J = \iint_A r^2 dydz \tag{3.9}$$

For example, $J = \pi R^4/2$ for a circle with radius R, and $J = a^4/6$ for a square with side a. Substituting (3.8) into (3.7) and simplifying,

$$\frac{\partial M_t}{\partial x} + m_t = \rho J \frac{\partial^2 \theta}{\partial t^2} \tag{3.10}$$

The twisting moment is related to the angle of twist by a linear relationship of the form

$$M_t = k_\theta G \frac{\partial \theta}{\partial x} \tag{3.11}$$

where k_θ is the torsional stiffness coefficient for the cross-section, which may be evaluated by the St. Venant formulation of classical elasticity theory (cf. [1], pp. 291–315), and G is the shear modulus for the material. A partial listing of J and k_θ for various cross-sectional shapes is given in Table 3.1. Numerous data for k_θ are available for other shapes (cf. [2], pp. 194–199).

Shape		J	k_θ	k_θ/J
Circle	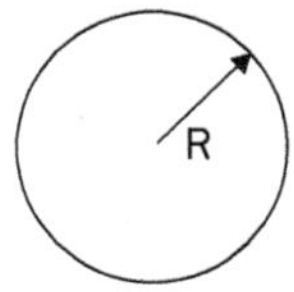	$\frac{mR^4}{2}$	$\frac{mR^4}{2}$	1
Hollow Circle	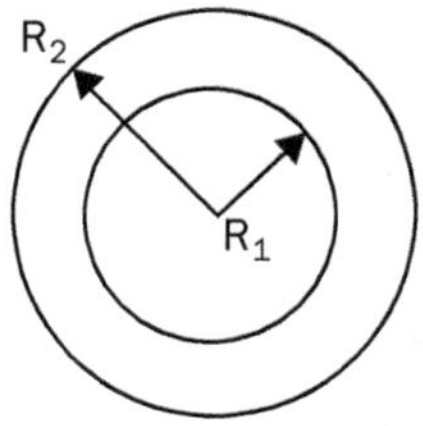	$\frac{\pi}{2}(R_2^4-R_1^4)$	$\frac{\pi}{2}(R_2^4-R_1^4)$	1
Ellipse	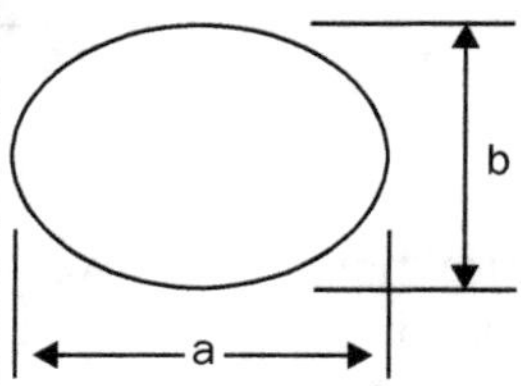	$\frac{\pi ab(a^2+b^2)}{64}$	$\frac{\pi a^3b^3}{16(a^2+b^2)}$	$\frac{4}{\left(\frac{a}{b}+\frac{b}{a}\right)^2}$
Square	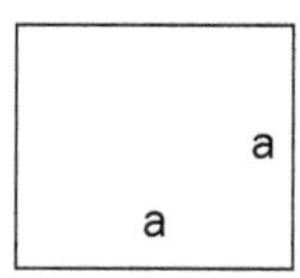	$\frac{a^4}{6}$	$0.1406a^4$ ([1], p. 313)	0.8436
2×1 Rectangle	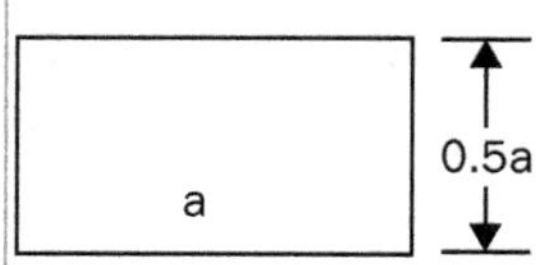	$\frac{5}{96}a^4$	$0.0286a^4$ ([1], p. 312)	0.549
Equilateral Triangle	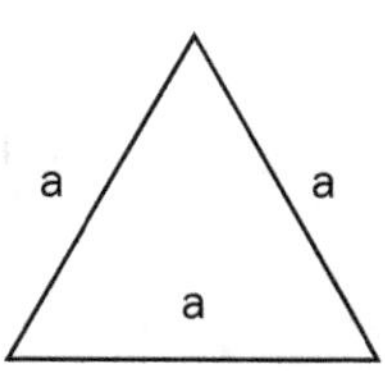	$\frac{\sqrt{3}a^4}{48}$	$\frac{\sqrt{3}a^4}{80}$	0.600

Table 3.1 J and k_θ for Various Cross-Sectional Shapes

It should also be mentioned that the classical elasticity theory analysis described above assumes that cross-sections are free to warp out of their planes during torsional displacements. All but circular cross-sections will typically warp. If one or both ends of a bar are rigidly fixed, so that an end cannot warp, and if the bar is not slender (so that the end effects are small), then the additional stiffness due to warping constraint must be considered.

Substituting (3.11) into (3.10) yields

$$\frac{\partial}{\partial x}\left(k_\theta G \frac{\partial \theta}{\partial x}\right) + m_t = \rho J \frac{\partial^2 \theta}{\partial t^2} \tag{3.12}$$

In (3.12) k_θ, G, ρ, and J may all be functions of x, whereas m_t may be a function of both x and t. If k_θ, G, ρ, and J are constants, and if free vibrations are of interest, then (3.12) may be written as

$$\boxed{k_\theta G \frac{\partial^2 \theta}{\partial x^2} = \rho J \frac{\partial^2 \theta}{\partial t^2}} \tag{3.13}$$

Equation (3.13) is analogous to (2.8) and (3.5).

3.3 Free Vibration of Bars

It has been shown in the preceding two sections that the equations of longitudinal or torsional motion, (3.5) or (3.13), for uniform rods are of the same form as that for a uniform string (2.8). Therefore, all the free vibration results for strings obtained in Chap. 2 are also directly applicable to the other two problems simply by replacing T and ρ by AE and ρA, respectively, for longitudinal vibrations, and by $k_\theta G$ and ρJ, respectively, for torsional vibrations. If the mathematical boundary conditions applied to either u or θ on the bar are the same as those for w on the string, then the problems are entirely analogous.

Thus, for the rod of length ℓ having both ends fixed, (2.23) yields the frequencies of longitudinal vibrations

$$\omega_m \ell \sqrt{\frac{\rho}{E}} = m\pi \qquad (m=1,2,\ldots,\infty) \tag{3.14}$$

and the frequencies of torsional vibrations

$$\omega_m \ell \sqrt{\frac{\rho J}{k_\theta G}} = m\pi \qquad (m=1,2,\ldots,\infty) \tag{3.15}$$

It should be remembered that ρ for the string is mass per unit length, whereas for both types of rod vibrations, it is mass per unit volume.

It is seen from (3.14) and (3.15) that neither longitudinal nor torsional frequencies are affected by the cross-sectional size of the bar. Longitudinal frequencies also do not depend on the cross-sectional shape. However, torsional frequencies are proportional to $\sqrt{k_\theta / J}$, which does depend on the shape. Values of k_θ/J for various shapes are listed in Table 3.1. It is seen, for example, that the torsional frequencies of bars of elliptical cross-section having $a/b = 2$ are 20 percent less than those of circular bars, whereas bars of rectangular cross-section with $a/b = 2$ have frequencies 26 percent less.

From (3.14) and (3.15) it is seen that the ratio of the longitudinal frequencies to the torsional frequencies of a bar is $\sqrt{(E/G)(J/k_\theta)}$. Because $E/G > 1$ and $J/k_\theta \geq 1$, the longitudinal frequencies are greater than the torsional ones for all cross-sectional shapes.

Adapting further the results from Chap. 2, longitudinal and torsional mode shapes (eigenfunctions) for bars having both ends fixed are

$$X_m(x) = \sin \alpha_m x \tag{3.16}$$

where $a_m = m\pi/\ell$ and $m = 1, 2, \ldots, \infty$. While bars, like strings, can vibrate in a single mode if the motion is begun properly (e.g., initial displacement in the shape of the eigenfunction, and no initial velocity), arbitrary initial conditions require a superposition of modes to represent the motion, with $u(x,t)$ or $\theta(x,t)$ given by (2.26), and the Fourier coefficients given by (2.29) and (2.30) as before. The traveling wave solution described by (2.40) is applicable to longitudinal and torsional rod vibrations by using wave velocities $c = \sqrt{E/\rho}$ and $c = \sqrt{k_\theta G/\rho J}$, respectively.

Other boundary conditions exist for longitudinal and torsional vibrations of bars which are physically possible and important. Particularly important are the fixed-free and free-free boundaries. For example, the boundary conditions for the fixed-free bar in the case of longitudinal vibrations are

$$u(0,t)=0 \rightarrow X(0)=0 \tag{3.17a}$$

$$\frac{\partial u}{\partial x}(\ell,t)=0 \rightarrow X'(\ell)=0 \tag{3.17b}$$

Condition (3.17b) is derived from

$$P(\ell,t) = AE\epsilon_x(\ell,t) = AE\frac{\partial u}{\partial x}(\ell,t) = 0 \tag{3.18}$$

Applying (3.17a) to the differential equation solution (2.18) once again requires $B = 0$, so that the eigenfunction is

$$X(x) = \sin \alpha x \tag{3.19}$$

Condition (3.17b) yields

$$\alpha \cdot \cos \alpha \ell = 0 \tag{3.20}$$

whence,

$$\alpha \ell = \frac{(2m-1)\pi}{2} \quad (m=1,2,\ldots,\infty) \tag{3.21}$$

Since $\omega^2 = E\alpha^2/\rho$

$$\omega_m \ell \sqrt{\frac{\rho}{E}} = \frac{(2m-1)\pi}{2} \quad (m=1,2,\ldots,\infty) \tag{3.22}$$

is the infinite set of nondimensional frequencies. The first four mode shapes may be seen in Fig. 3.5. Similar results are applicable to the torsion problem by analogy. The corresponding boundary conditions for a fixed-free string are mathematically possible, but physically they are unrealistic. The free end of the string would have to be attached to a *massless* bead, which must slide freely in the transverse direction *without friction,* and at the same time apply the necessary string tension!

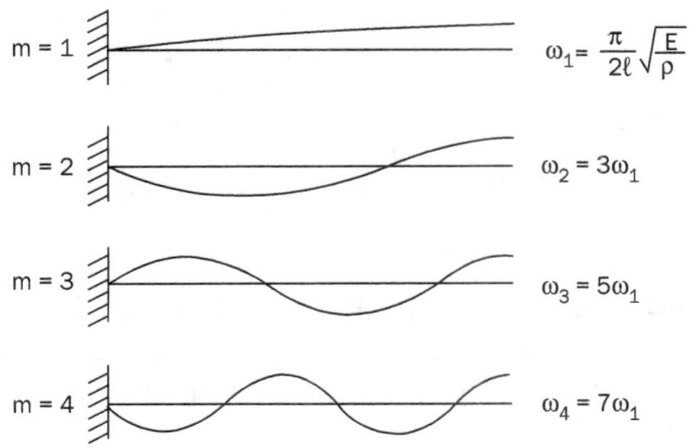

FIGURE 3.5 The first four mode shapes for the longitudinal vibration of a bar with one end fixed and other free.

3.4 Other Solutions by Analogy

Discontinuities may arise in bars for the same reasons as in strings (see Sec. 2.7), and the previous solutions obtained for strings in such cases are directly applicable to the longitudinal and torsional vibrations of equivalent bars by analogy. Thus, for example, the solution in Sec. 2.7 for a string having an intermediate mass M attached, given by (2.65) and (2.67), may be used for the longitudinal vibrations of a bar also having such a mass. To do so, $\alpha = \beta/\ell$ is simply redefined as $\omega\sqrt{\rho/E}$, and M^* as $M/\rho A\ell$.

For forced vibrations of bars with viscous damping, (3.4) yields

$$AE\frac{\partial^2 u}{\partial x^2} + p(x,t) = \rho A\frac{\partial^2 u}{\partial t^2} + c\frac{\partial u}{\partial t} \tag{3.23}$$

for longitudinal motions. The damping force term $c\partial u/\partial t$ would typically arise due to drag forces acting along the sides of the bar in opposition to the motion. Similarly, for torsional vibrations (3.12) gives

$$k_\theta G\frac{\partial^2 \theta}{\partial x^2} + m_t(x,t) = \rho J\frac{\partial^2 \theta}{\partial t^2} + c\frac{\partial \theta}{\partial t} \tag{3.24}$$

where m_t is typically an external exciting torque distributed along the sides of the bar. The damping constant c has different dimensions in (3.23) and (3.24).

For *free vibrations with damping*, p and m_t are omitted from (3.23) and (3.24). The resulting equations are seen to be in the same form as (2.68). Thus, the solutions presented for the string in Sec. 2.8 may be used for longitudinal or torsional motions of bars by replacing T and ρ either by AE and ρA, or by $k_\theta G$ and ρJ, as appropriate.

The same comment applies to *forced vibration* problems. That is, comparing (3.23) and (3.24) with (2.79), it is seen that the same substitutions permit one to use the results of Secs. 2.9 and 2.10 straightforwardly for bar vibrations in the case of viscous damping.

3.5 Free Vibrations of Bars with Variable Cross-Section

If a bar is made of a material which is homogeneous, then E, p, and G are constants in (3.3) and (3.12). The resulting equations of motion for free vibration are

$$\frac{\partial}{\partial x}\left(A\frac{\partial u}{\partial x}\right) = \frac{\rho A}{E}\frac{\partial^2 u}{\partial t^2} \tag{3.25}$$

$$\frac{\partial}{\partial x}\left(k_\theta \frac{\partial \theta}{\partial x}\right) = \frac{\rho J}{G}\frac{\partial^2 \theta}{\partial t^2} \tag{3.26}$$

where A, k_θ, and J are functions of x if the cross-section is varying in a *continuous* manner (i.e., a step change in cross-section would be represented by a discontinuous bar, as discussed in Sec. 3.4).

Consider first the longitudinal vibrations. Assuming sinusoidal time response in the case of free vibrations,

$$u(x,t) = X(x)\sin(\omega t + \phi) \tag{3.27}$$

(3.25) becomes

$$\frac{d^2X}{dx^2} + \frac{1}{A}\frac{dA}{dx}\frac{dX}{dx} + \left(\frac{\rho\omega^2}{E}\right)X = 0 \tag{3.28}$$

Suppose the area varies as

$$A = A_0 x^n \tag{3.29}$$

where A_0 is a constant. Then (3.28) becomes:

$$X'' + \frac{n}{x}X' + \frac{\rho\omega^2}{E}X = 0 \tag{3.30}$$

where $X' = dX/dx$, etc. Define ν and k by the following equations:

$$\nu = \frac{n-1}{2}, \quad k^2 = \frac{\rho\omega^2}{E} \tag{3.31}$$

and (3.30) may be rewritten as

$$X'' + \frac{2\nu+1}{x}X' + k^2X = 0 \tag{3.32}$$

Now make a substitution of dependent variables such that

$$X(x) = x^{-\nu}U(x) \tag{3.33}$$

which converts (3.32) to

$$x^2U'' + xU' + \left(k^2x^2 - \nu^2\right)U = 0 \tag{3.34}$$

This is the classical form of Bessel's equation. Its solution may be written as

$$U = C_1 J_v(kx) + C_2 Y_v(kx) \tag{3.35}$$

where J_v and Y_v are Bessel functions of the first and second kind, respectively. The "order" of the Bessel functions is v, which may be an integer or not.

A brief summary of some of the important properties of Bessel functions may be found in Appendix B. A bibliography of useful references on Bessel functions is also given at the end of Appendix B.

Substituting (3.33) and (3.35) into (3.27) yields

$$u(x,t) = \left[C_1 J_v(kx) + C_2 Y_v(kx)\right] x^{-v} \sin(\omega t + \phi) \tag{3.36}$$

where v and k are defined by (3.31). The frequencies and mode shapes of free vibration are determined by applying the boundary conditions.

Example 3.1

A. Determine the frequency equation for the longitudinal vibrations of a bar of circular cross-section and length ℓ, fixed at both ends, where the radius varies linearly as

$$r = r_0 \frac{x}{a}$$

and where the radius at the small end is r_0 as shown in Fig. 3.6.

B. Evaluate the nondimensional frequencies and mode shapes for the case when $a = \ell$. Compare these with those of a uniform thickness bar having the same *average* radius ($3r_0/2$).

Solution

Part A. The cross-sectional area varies with x according to

$$A = \pi r^2 = \pi \left(\frac{r_0}{a}\right)^2 x^2$$

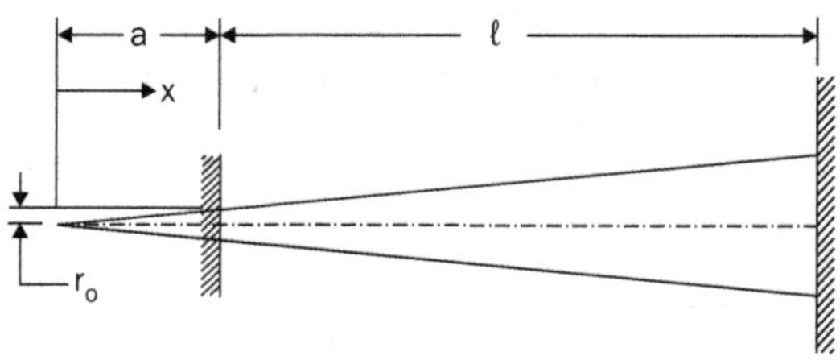

FIGURE 3.6 A bar of circular cross-section with a linearly varying radius fixed at both ends.

Thus, $A_0 = \pi (r_0/a)^2$ and $n = 2$ in (3.29). Referring to (3.31) and (3.35), (3.33) becomes

$$X(x) = x^{-1/2}[C_1 J_{1/2}(kx) + C_2 Y_{1/2}(kx)]$$

Bessel functions of order 1/2 have the particular relationship $Y_{1/2}(z) = -J_{-1/2}(z)$, and are also related to trigonometric functions by

$$J_{1/2}(z) = \sqrt{\frac{2}{\pi z}} \sin z$$

$$J_{-1/2}(z) = \sqrt{\frac{2}{\pi z}} \cos z$$

Therefore, the solution above may be rewritten as

$$X = \frac{1}{x}(C_1 \sin kx + C_2 \cos kx)$$

where $\sqrt{2/\pi k}$ and a minus sign have been incorporated into the arbitrary coefficients C_1 and C_2.

For the particular case when $A = A_0 x^2$, an exact solution to the equation of motion may be found without using Bessel functions. Returning to (3.3), setting $p = 0$, and substituting the area variation,

$$E\frac{\partial}{\partial x}\left(x^2 \frac{\partial u}{\partial x}\right) = \rho x^2 \frac{\partial^2 u}{\partial t^2}$$

or

$$x\frac{\partial^2 u}{\partial x^2} + 2\frac{\partial u}{\partial x} = \left(\frac{\rho}{E}\right) x \frac{\partial^2 u}{\partial t^2}$$

which may be rewritten as

$$\frac{\partial^2}{\partial x^2}(ux) = \left(\frac{\rho}{E}\right)\frac{\partial^2}{\partial t^2}(ux)$$

This is once again our wave equation (2.8), except that the dependent variable is replaced by the product ux. The solution is therefore

$$ux = (C_1 \sin kx + C_2 \cos kx) \cdot \sin(\omega t - \phi)$$

with $k^2 = \rho\omega^2/E$, which agrees with the result obtained above using Bessel functions.

Applying the boundary conditions at $x = a$ and $x = a + \ell = b$,

$$U(a) = 0 \rightarrow C_1 \sin ka + C_2 \cos ka = 0$$

$$U(b) = 0 \rightarrow C_1 \sin kb + C_2 \cos kb = 0$$

For a nontrivial solution,

$$\begin{vmatrix} \sin ka & \cos ka \\ \sin kb & \cos kb \end{vmatrix} = 0$$

whence, $\sin ka \bullet \cos kb - \sin kb \bullet \cos ka = 0$

or, using the well-known trigonometric identity,

$$\sin k(a-b) = -\sin k\ell = 0 \qquad \text{(frequency equation)}$$

Part B. The roots of the frequency equation are clearly

$$k\ell = m\pi \quad (m = 1, 2, \ldots, \infty)$$

and those are seen from (3.31) to be the nondimensional frequencies

$$\omega\ell\sqrt{\frac{\rho}{E}} = m\pi \quad (m = 1,2,\ldots,\infty)$$

The frequencies for the bar with linear taper are seen to be the same as those for the untapered (constant radius) bar. This unusual result applies only when $n = 2$ in (3.29), with both ends fixed. For other n, or other end conditions, it would not apply. The frequencies do not depend on the actual cross-sectional size. Moreover, they do not depend on the actual shape, but rather on how the area varies. Thus, a bar of rectangular cross-section having linear taper in both dimensions, so that $A = A_0 x^2$, would have the same longitudinal frequencies as the circular one in this example.

To determine the mode shapes, we substitute the eigenvalues $k\ell$ back into one of the two boundary condition equations, say the first, whence the mth eigenvector is

$$\frac{C_2}{C_1} = -\tan ka = -\tan\frac{m\pi a}{\ell} = -\tan m\pi = 0$$

and the mth eigenfunction (mode shape) is

$$X_m = \frac{C_1}{x}\left(\sin\frac{m\pi x}{\ell} - \tan\frac{m\pi a}{\ell} \bullet \cos\frac{m\pi x}{\ell}\right)$$

where C_1 is an arbitrary amplitude coefficient determined by the initial conditions. The mode shape is significantly different than that for the bar having uniform cross-section (3.16).

The torsional vibrations of variable cross-section bars are complicated by the existence of two parameters, k_θ and J, in (3.26), both depending on x. Whereas the area (A) varies with the second power of the cross-sectional dimension, both k_θ and J vary with the fourth power. This is clear for the configurations shown in Table 3.1. For an elliptical cross-section, for example, one may write

$$J = \frac{\pi a^4}{64}\left(\frac{b}{a}\right)\left[1+\left(\frac{b}{a}\right)^2\right] \tag{3.37}$$

$$k_\theta = \frac{\pi a^4}{16(a/b)[(a/b)^2+1]} \tag{3.38}$$

so that both parameters depend on the aspect ratio (a/b), but vary with the fourth power of the cross-sectional dimension (a). Assuming sinusoidal motion (3.27), then (3.26) becomes

$$\frac{d^2X}{dx^2} + \frac{1}{k_\theta}\frac{dk_\theta}{dx}\frac{dX}{dx} + \left(\frac{\rho J\omega^2}{Gk_\theta}\right)X = 0 \tag{3.39}$$

From (3.37) and (3.38), it is seen that $\rho J\omega^2/Gk_\theta$ is a constant. If k_θ is taken to vary as

$$k_\theta = k_0 x^n \tag{3.40}$$

where k_0 is a constant, substitution of (3.40) into (3.39) results in

$$X'' + \frac{n}{x}X' + k^2 X = 0 \tag{3.41}$$

where $k^2 = \rho J\omega^2/Gk_\theta$. This equation is of the same form as (3.30), so a solution in Bessel functions is again possible. However, since the area varies only with the square of the cross-sectional dimension, while k_θ varies with the fourth power, the Bessel functions required for the torsional vibration analysis are of different order than those for the longitudinal vibrations.

3.6 Forced Vibrations of Bars; Material Damping

In Sec. 3.4 it was shown that forced vibrations for the longitudinal or torsional vibrations of bars subjected to distributed viscous damping could be analyzed, by analogy, with the same methods as for the transverse vibrations of strings. However, another type of damping is inherent in the material from which bars are made. This is called material damping (also known in the technical literature as "hysteretic" or "structural" damping). It does not enter the linearized analysis for the string.

Suppose a completely free rod is placed in a vacuum chamber, being held in the horizontal position by two threads from which it is hanging. Suppose further that the air is evacuated, and that one or more of the longitudinal free vibration modes are excited, perhaps by impact. As the bar executes its free vibrations, it will be seen that the amplitude decreases, even though viscous or aerodynamic damping forces are not present. The decay in amplitude is due to material damping.

Material damping may be incorporated into the problem by regarding the elastic moduli as complex quantities, that is,

$$E^* = E(1+i\eta), \qquad G^* = G(1+i\eta) \tag{3.42}$$

where $i = \sqrt{-1}$, and η is the *loss factor* (also known as "damping factor") for the material (the values of η may be somewhat different for E^* and G^*, but for simplicity will be taken the same in this introductory study). Values of η are typically very small for metals ($10^{-6} < \eta < 10^{-3}$), but can be quite large for other materials such as rubber or plastic ($10^{-2} < \eta < 1$). For more understanding of material damping, including the concept of complex modulus, the excellent monograph by Snowdon [3] is recommended.

With material damping the equations of motion for longitudinal and torsional vibrations are written as

$$AE(1+i\eta)\frac{\partial^2 u}{\partial x^2} + p(x,t) = \rho A \frac{\partial^2 u}{\partial x^2} \tag{3.43a}$$

$$k_\theta E(1+i\eta)\frac{\partial^2 \theta}{\partial x^2} + m_t(x,t) = \rho J \frac{\partial^2 \theta}{\partial x^2} \tag{3.43b}$$

For excitation which is sinusoidal in time, it is convenient to express the distributed longitudinal force or twisting moment as

$$p(x,t) = P(x)e^{i\Omega t}, \quad m_t(x,t) = M_t(x)e^{i\Omega t} \tag{3.44}$$

where Ω is the exciting frequency. The responses $u(x,t)$ and $\theta(x,t)$ are then also taken in complex form.

Consider the longitudinal vibration problem. A solution for the displacement may be assumed as

$$u(x,t) = [U_1(x) - iU_2(x)]e^{i\Omega t} \tag{3.45}$$

where the real part (U_1) is in phase with the exciting force, and the imaginary part ($-iU_2$) lags the exciting force by 90 degrees. Substituting (3.44) and (3.45) into (3.43a) gives

$$AE(1+i\eta)\left(U_1'' - iU_2''\right)e^{i\Omega t} + Pe^{i\Omega t} + \rho A\Omega^2(U_1 - iU_2)e^{i\Omega t} = 0 \tag{3.46}$$

Since this must be satisfied at all times, separating the coefficient of $e^{i\Omega t}$ into its real and imaginary parts yields

$$U_1'' + \left(\frac{\rho\Omega^2\ell^2}{E}\right)U_1 + \eta U_2'' = -\frac{P\ell^2}{AE} \tag{3.47a}$$

$$U_2'' + \left(\frac{\rho\Omega^2\ell^2}{E}\right)U_2 - \eta U_1'' = 0 \tag{3.47b}$$

where the nondimensional variable $\xi = x/\ell$ is now being used. These may be written as

$$U_1'' + \lambda^2 U_1 + \eta U_2'' = -Q(\xi) \tag{3.48}$$

$$U_2'' + \lambda^2 U_2 - \eta U_1'' = 0 \tag{3.49}$$

where $\lambda^2 = \dfrac{\rho\Omega^2\ell^2}{E}, \quad Q(\xi) = \dfrac{P(\xi)\ell^2}{AE}$

Equations (3.48) and (3.49) are seen to be of a different form than (2.110), which arose from viscous damping. Most notably, the coupling terms involve the second derivatives in the present case, rather than the functions themselves. Using the following linear transformation of variables

$$X_1 = U_1 + \eta U_2 \tag{3.50a}$$

$$X_2 = -\eta U_1 + U_2 \tag{3.50b}$$

or

$$U_1 = \frac{1}{1+\eta^2}(X_1 - \eta X_2) \tag{3.51a}$$

$$U_2 = \frac{1}{1+\eta^2}(\eta X_1 + X_2) \tag{3.51b}$$

then (3.48) and (3.49) become

$$X_1'' + \frac{\lambda^2}{1+\eta^2}(X_1 - \eta X_2) = -Q(\xi) \tag{3.52a}$$

$$X_2'' + \frac{\lambda^2}{1+\eta^2}(X_2 + \eta X_1) = 0 \tag{3.52b}$$

Equations (3.52) are seen to have the same form as (2.110), except that $\lambda^2/(1+\eta^2)$ is in place of λ^2, and $\lambda^2\eta/(1+\eta^2)$ is in place of $2\pi\lambda\zeta_1$. Using these replacements, the solutions (2.118) and (2.119) for viscously damped strings and rods are seen to be directly applicable to the rod having material damping as well.

Example 3.2 A rod has one end fixed, while the other end is subjected to an exciting force $P_0 \sin \Omega t$, as shown in Fig. 3.7. The material has a damping loss factor η. Determine the amplitude of the motion of the free end, U_0, as a function of P_0, Ω, and η.

Solution

For this problem, the R.H.S. of (3.52a) is zero; therefore, only the complementary solution of (3.52) is needed. The exciting force will enter the problem through a boundary condition.

Defining $\beta^2 = \dfrac{\lambda^2}{1+\eta^2}$ and $\gamma^2 = \dfrac{\eta\lambda^2}{1+\eta^2} = \eta\beta^2$, (3.52) becomes

$$X_1'' + \beta^2 X_1 - \gamma^2 X_2 = 0$$

$$X_2'' + \beta^2 X_1 + \gamma^2 X_1 = 0$$

Using the solutions (2.118) and (2.119) found previously for (2.110),

$$X_1 = C_4 \sinh a\xi \cdot \sin b\xi + C_3 \cosh a\xi \cdot \sin b\xi - C_2 \sinh a\xi \cdot \cos b\xi - C_1 \cosh a\xi \cdot \cos b\xi$$

$$X_2 = C_1 \sinh a\xi \cdot \sin b\xi + C_2 \cosh a\xi \cdot \sin b\xi + C_3 \sinh a\xi \cdot \cos b\xi + C_4 \cosh a\xi \cdot \cos b\xi$$

where $a = \sqrt{\dfrac{R-\beta^2}{2}}$, $b = \sqrt{\dfrac{R+\beta^2}{2}}$, $R = \sqrt{\beta^4 + \gamma^4} = \dfrac{\lambda^2}{\sqrt{1+\eta^2}}$

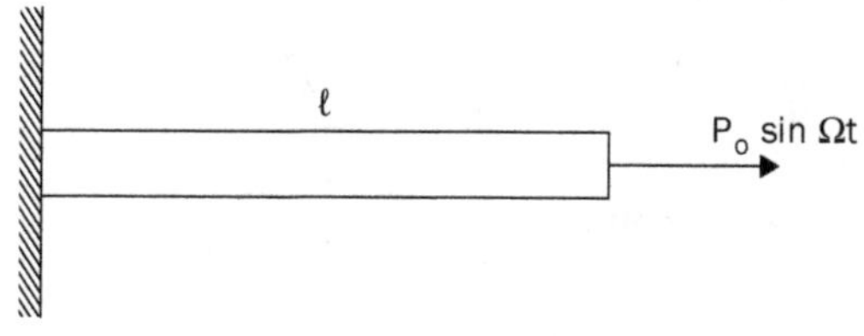

FIGURE 3.7 A rod has one end fixed and the other subjected to a longitudinal exciting force $P_0 \sin\Omega t$.

B.C.1 $u(0,t)=0 \rightarrow U_1(0)=0, \quad U_2(0)=0.$

$$\left.\begin{aligned} U_1(0)=0 \rightarrow X_1(0)-\eta X_2(0)=0 \rightarrow -C_1-\eta C_4=0 \\ U_2(0)=0 \rightarrow \eta X_1(0)+X_2(0)=0 \rightarrow -\eta C_1+C_4=0 \end{aligned}\right\} \rightarrow C_1=C_4=0$$

The solutions for U_1 and U_2, therefore, reduce to

$$U_1=\frac{1}{1+\eta^2}\left[(C_3-\eta C_2)\cosh a\xi \cdot \sin b\xi-(C_2+\eta C_3)\sinh a\xi \cdot \cos b\xi\right]$$

$$U_2=\frac{1}{1+\eta^2}\left[(\eta C_3+C_2)\cosh a\xi \cdot \sin b\xi+(-\eta C_2+C_3)\sinh a\xi \cdot \cos b\xi\right]$$

Defining new constants as $A_1=\dfrac{C_3-\eta C_2}{1+\eta^2}, \quad A_2=\dfrac{C_2+\eta C_3}{1+\eta^2}$, then

$$U_1=A_1\cosh a\xi \cdot \sin b\xi - A_2\sinh a\xi \cdot \cos b\xi$$

$$U_2=A_2\cosh a\xi \cdot \sin b\xi + A_1\sinh a\xi \cdot \cos b\xi$$

B.C.2. $AE\dfrac{\partial u}{\partial x}(\ell,t)=P_o\sin\Omega t \rightarrow U_1'(1)=\dfrac{P_o\ell}{AE}=\delta_{st}, \quad U_2'(1)=0$

where δ_{st} is the displacement of the end of a rod loaded *statically*. The last two equations allow one to solve for A_1 and A_2 as

$$A_1=\frac{c}{c^2+d^2}\delta_{st}, \qquad A_2=\frac{-d}{c^2+d^2}\delta_{st}$$

where $c = a\sinh a \cdot \sin b + b\cosh a \cdot \cos b$

$d = a\cosh a \cdot \cos b - b\sinh a \cdot \sin b$

$c^2 + d^2 = (a^2 + b^2)(\sinh^2 a \cdot \sin^2 b + \cosh^2 a \cdot \cos^2 b)$

Then the amplitude of the motion of the free end is given by

$$U_0=\sqrt{U_1^2(1)+U_2^2(1)}$$

Substituting U_1 and U_2 from above, then A_1 and A_2, expanding, and collecting terms finally yields the relatively simple closed form:

$$\frac{U_0}{\delta_{st}}=\sqrt{\frac{\sinh^2 a \cdot \cos^2 b+\cosh^2 a \cdot \sin^2 b}{(a^2+b^2)(\sinh^2 a \cdot \sin^2 b+\cosh^2 a \cdot \cos^2 b)}}$$

In evaluating U_0/δ_{st}, it is useful to rewrite λ as

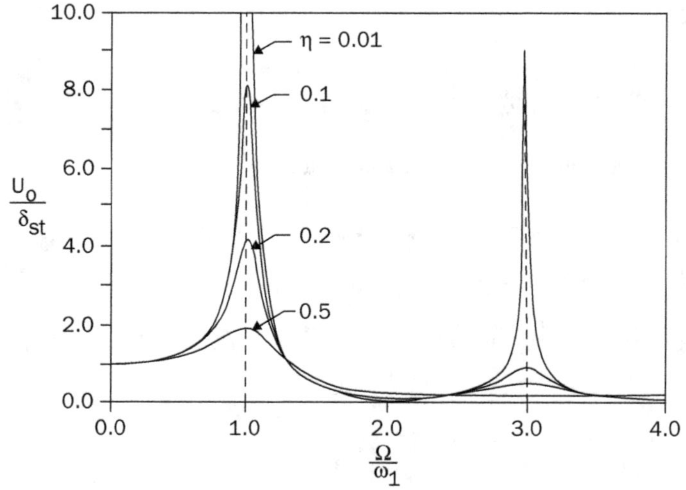

FIGURE 3.8 A plot of U_0 / δ_{st} versus Ω / ω_1 for Example 3.2.

$$\lambda = \Omega\ell\sqrt{\frac{\rho}{E}} = \left(\frac{\Omega}{\omega_1}\right)\left(\omega_1\ell\sqrt{\frac{\rho}{E}}\right) = \frac{\pi}{2}\left(\frac{\Omega}{\omega_1}\right)$$

This is used to calculate a and b.

A plot of U_0 / δ_{st} versus Ω/ω_1 over the frequency range $0 \leq \Omega/\omega_1 \leq 4$ is shown in Fig. 3.8. Resonances are seen to occur in the vicinities of $\Omega/\omega_1 = 1$ and 3, as expected. However, unlike a single degree-of-freedom system with material damping, (cf. [4]), the maximum amplitudes do not occur exactly at the natural frequencies.

3.7 Energy Functionals and Rayleigh and Ritz Methods

In order to employ Rayleigh and Ritz methods with free vibration problems for bars, it is necessary to have the proper energy functionals.

For *longitudinal vibrations*, the potential energy of the system consists of the strain energy due to deformation. Consider again the infinitesimal volume of the bar having cross-sectional area A and length dx, shown in Fig. 3.2. The infinitesimal quantity of strain energy in this volume is

$$d(PE) = \frac{1}{2} P\, de \tag{3.53}$$

where de is the infinitesimal elongation of the bar element during vibratory motion. The "1/2" is required because, during the elongation, the longitudinal force begins at zero as elongation begins. For our linearly elastic material, the force increases proportionally with the elongation. Thus, $P/2$ represents the average force applied during the elongation. Since $P = \sigma_x A$ and $de = \epsilon_x dx$, where σ_x is the longitudinal stress and ϵ_x is the longitudinal strain,

$$d(PE) = \frac{1}{2}\sigma_x A \epsilon_x dx \tag{3.54}$$

Further, $\sigma_x = E\epsilon_x$ and $\epsilon_x = \partial u/\partial x$. Substituting these into (3.54) and integrating the strain energy over the entire bar, we obtain

$$PE = \frac{1}{2}\int_0^\ell AE\left(\frac{\partial u}{\partial x}\right)^2 dx \tag{3.55}$$

If A and E are not functions of x, they may be brought in front of the integral sign. Otherwise, (3.55) is capable of dealing with variable cross-section and/or nonhomogeneous (E varying with x) bars.

The kinetic energy of the volume of mass depicted in Fig. 3.2 is clearly

$$d(KE) = \frac{1}{2}(\rho A dx)\left(\frac{\partial u}{\partial t}\right)^2 \tag{3.56}$$

Thus, the kinetic energy of longitudinal vibrations for the entire bar is

$$KE = \frac{1}{2}\int_0^\ell \rho A\left(\frac{\partial u}{\partial t}\right)^2 dx \tag{3.57}$$

For *torsional vibrations*, the strain energy stored in the infinitesimal volume shown in Fig. 3.3 is

$$d(PE) = \frac{1}{2}M_t\frac{\partial \theta}{\partial x}dx \tag{3.58}$$

where $(\partial\theta/\partial x)dx$ is the rotation of one end with respect to the other. Substituting (3.11) into (3.58) and integrating over the length, the potential energy in the entire bar is then

$$PE = \frac{1}{2}\int_0^\ell k_\theta G\left(\frac{\partial \theta}{\partial x}\right)^2 dx \tag{3.59}$$

The kinetic energy of the bar element undergoing torsional rotation about its centroidal axis is

$$d(KE) = \frac{1}{2} dI_c \left(\frac{\partial \theta}{\partial t} \right)^2 \tag{3.60}$$

Substituting (3.8) into (3.60) and integrating over the length yields the kinetic energy for the entire bar:

$$KE = \frac{1}{2} \int_0^{\ell} \rho J \left(\frac{\partial \theta}{\partial t} \right)^2 dx \tag{3.61}$$

The energy functionals may also be obtained from those of the vibrating string by use of the analogy described at the beginning of Sec. 3.3. Thus, (3.55) or (3.59) could be obtained from (2.125) by replacing T by AE or by $k_\theta G$, respectively. Similarly, (3.57) and (3.61) are also found from (2.127) by replacing ρ by ρA and ρJ, respectively.

Rayleigh and Ritz methods are applied to the longitudinal and torsional vibration of bars as was discussed for the string in Secs. 2.12 and 2.13. Using the Rayleigh method, one describes either $u(x,t)$ or $\theta(x,t)$ as in (2.129), where again $X(x)$ must satisfy the geometric constraints on the boundaries. An approximate natural frequency is then calculated by use of Rayleigh's Quotient (2.133), with the maximum potential energy obtained by replacing u and θ by X in (3.55) and (3.59), and the maximum kinetic energy by replacing $\partial u/\partial t$ and $\partial \theta/\partial t$ by ωX in (3.57) and (3.61). If the mode shape $X(x)$ is chosen with reasonable accuracy, then the corresponding frequency calculated from Rayleigh's Quotient will be reasonably accurate. More accurate results may be obtained with the Ritz method by choosing a set of admissible functions as in (2.134) and writing the minimizing equations (2.136), where

$$KE_{max} = \omega^2 KE^*_{max} \tag{3.62}$$

References

1. S. P. Timoshenko and J. N. Goodier, *Theory of Elasticity*, 3rd ed., McGraw-Hill, 1970.
2. R. J. Roark, *Formulas for Stress and Strain*, 4th ed., McGraw-Hill, 1965.
3. J. C. Snowdon, *Vibration and Shock in Damped Mechanical Systems*, John Wiley and Sons, Inc., 1968.
4. J. E. Ruzicka and T. E. Derby, *Influence of Damping in Vibration Isolation*, Shock and Vibration Information Center, Washington, D.C., 1971.

Problems

1 A. Determine the torsional vibration frequencies of a bar of length l having both ends free.

B. Plot the first four mode shapes and locate the node points.

C. Show that the antisymmetric modes and frequencies are the same as that of a fixed-free bar of length $\ell/2$.

D. Show that $\omega = 0$ is not only a solution to the problem, but that a nontrivial motion is described by it.

2 Consider the longitudinal and torsional vibrations of a rod made of aluminum ($G/E = 0.4$).

A. Determine the ratio of the frequencies of longitudinal vibration (ω_L) to those of torsion (ω_T) for rods having both ends fixed, and cross-sections which are:

1. Circular
2. Square
3. A 4 × 1 rectangle

B. Prove whether or not the ratios found in Part A are valid for rods having *all possible* end conditions of the same type.

3 Return to Example 2.1 in Chap. 2. Show carefully to what longitudinal and torsional vibration problems for rods the results correspond. Draw sketches of the rods in their initially deformed positions.

4 A bar of length ℓ has one end fixed and the other free. The free end is subjected to a static, longitudinal force P, causing the end to displace an amount $\delta = P\ell/AE$. Suddenly the force P is removed. Determine the subsequent motion $u(x,t)$ of the bar as a summation of the responses of the free vibration mode shapes (see Sec. 2.3).

5 Solve Problem 4 using a traveling wave solution (see Sec. 2.5).

A. Determine $u(x,t)$.

B. Plot the displacement of points along the bar at $t/\tau_1 = 1/16, 1/4, 5/16$.

C. Plot the velocity of points along the bar at $t/\tau_1 = 1/16, 1/4, 5/16$.

D. Plot the stress (σ_x) along the bar at $t/\tau_1 = 1/16, 1/4, 5/16$.

6 A bar is hanging freely with its own weight $\rho g A \ell$. The longitudinal axis is vertical. Prove, whether or not, for all boundary conditions, the frequencies and mode shapes of longitudinal motion are the same as those of the bar when gravity forces are ignored.

7 A rod of length ℓ is fixed at one end, and the other end is attached to a rigid mass M, which is free to move.

A. Obtain the transcendental equation for the longitudinal vibration frequencies.

B. Evaluate the first four frequencies $\omega\ell\sqrt{\rho/E}$ for $M/\rho A\ell = 0.2$, 1, and 5, where A is the cross-sectional area of the rod. Tabulate your results.

C. On a single graph, plot the fundamental mode shape for $M/\rho Al = 0$, 0.2, 1, 5, and ∞.

8 A shaft of noncircular cross-section and length l has both ends fixed. A mass M having polar mass moment of inertia I_M about the shaft axis is attached at an intermediate location $x = \eta\ell(0 < \eta < 1)$ measured from one end.

A. Apply boundary conditions to determine the frequency equation for free vibration frequencies of *torsional* motion.

B. Let $\eta = 3/4$ in the result of Part A. Compare this with (2.65). Explain any similarities or differences between the two equations.

9 A. Find the natural frequencies of longitudinal vibration for a tapered circular bar of length ℓ which comes to a point (and is free) at one end, and is fixed at the other (Fig. 3.9). Its radius varies as $r = r_0 x/\ell$. Note: the *stress* at the free end is not zero.

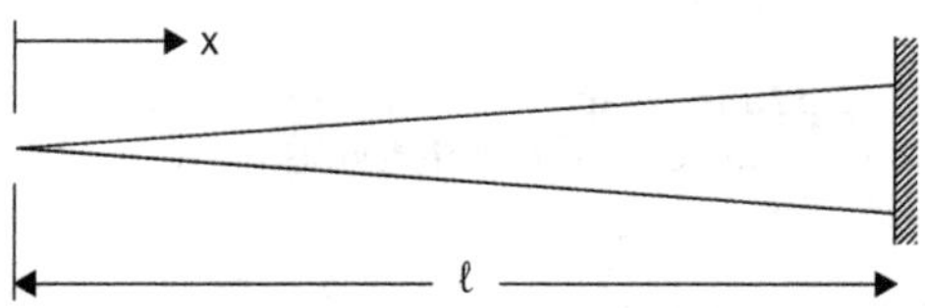

FIGURE 3.9 Problem 9.

B. Determine the stress within the bar at a typical point as it vibrates in its fundamental mode. Express this in terms of displacement amplitude (δ) at the free end. Evaluate the stress at the free tip ($x = 0$).

10 Determine the fundamental longitudinal frequency and mode shape of a bar of length ℓ which has a circular cross-section, with an *area* that varies linearly between A_1 at one end and A_2 at the other, when both ends are *free*. Compare this with the results for a bar having a uniform diameter. If the result depends on A_2/A_1, let this ratio be 4.

11 Repeat Example 3.1 for the problem of torsional vibrations.

12 A bar has one end fixed ($x = 0$) and the other free ($x = \ell$). The fixed end is excited longitudinally with a displacement $\delta \sin\Omega t$. Assuming a small amount or viscous damping is present, determine the steady state response $u(x,t)$.

13 Use the results of Problem 4 or 5 for $u(x,t)$ to determine the potential and kinetic energies within the bar at any instant of time. Plot these on the same graph for $0 \le t \le \tau_0$, where τ_0 is the fundamental period of the bar. Plot also on the same graph the total energy in the system as a function of time.

14 A. Use the Rayleigh method with an assumed displacement in the form of an algebraic polynomial to obtain an approximate value of the fundamental frequency for the bar of Example 3.1 (with $a = \ell$).

B. Add a term to the polynomial of Part A, and use the Ritz method to find an improved approximate frequency.

C. Compare the frequencies from above with the exact ones.

D. On a single graph, make a plot of the two approximations for the first mode shape from above, as well as the exact one. Normalize the mode shape curves so that each has a maximum value of unity.

15 A. Use the Rayleigh method to obtain a reasonably accurate approximate value of the fundamental frequency for Problem 8B. Explain why you think your solution is "reasonably accurate."

B. Improve the solution of Part A by using the Ritz method, with the displacement taken as the sum of *two* admissible functions, each one applicable over the entire length of the shaft.

CHAPTER 4

Beam Vibrations

A beam is typically described as a structural element having one dimension (length) which is many times greater than its other dimensions (width and depth). It may be straight or curved.

Beams are one of the most fundamental structural and machine components. Almost every structure or machine one can think of has one or more beam components. Buildings, steel framed structures, and bridges are examples of beam applications in civil engineering. In these applications, beams exist as structural elements or components supporting the whole structure. In addition, the whole structure can be modeled at a preliminary level as a beam. For example, a high-rise building can be modeled as a cantilever beam, or a bridge modeled as a simply supported beam. In mechanical engineering, rotating shafts carrying pulleys and gears are examples of beams. In addition, frames in machines (e.g., a truck) are beams. Robotic arms in manufacturing are modeled as beams as well. In aerospace engineering, beams (curved and straight) are found in many areas of the aircraft or space vehicle. In addition, the whole wing of a plane is often modeled as a beam for some preliminary analysis. Innumerable other examples of beams exist.

In many of these applications, beams are subjected to dynamic loads. Imbalance in driveline shafts, combustion in crank shaft applications, wind or earthquake on a bridge or a structure, and impact load when a vehicle goes over a bump are all examples of possible dynamic loadings that beam structures can be exposed to. All of these loads and others can excite the vibration of the beam structure. This can cause durability concerns (because of potentially excessive dynamic stresses) or discomfort because of the resulting noise and vibration.

This chapter will be dedicated to the study of transverse vibrations of beams. The equations of motion will first be derived. Solutions are then found and discussed for the natural frequencies and mode shapes of various boundary conditions using exact methods. Beams with added masses and springs are also treated. Forced vibration is then discussed when the beam is subjected to transverse periodic loading. Beams that have internal supports or discontinuities are investigated. Energy functionals are developed for possible use in approximate methods such as those of Rayleigh and Ritz as well as in finite element methods. Beams subjected to axial forces while undergoing transverse vibrations are also studied, leading to a physical understanding of buckling. Moderately thick beams, where both shear deformation and rotary inertia must be included, are also studied in this chapter.

4.1 Equations of Motion for Transverse Vibrations

Figure 4.1 shows a beam of length ℓ which is clamped at its left end. Its right end is supported by knife-edges which are capable of applying transverse forces, but no bending moment. This is one example of a simply supported end. Vibrational motion is assumed to take place strictly in the z-direction, with a displacement component $w = w(x,t)$. For the sake of some increased generality, the beam shown has a cross-section of varying size, although it is assumed that all cross-sections are similar in geometrical shape (to avoid coupling with motions which are either torsional or in the y-direction). Cross-sections are assumed to have at least one axis of symmetry (the z-axis), also to avoid coupling with torsional or y-directional motions. For cross-sections having no symmetry, a more general theory, including coupling, would need to be used. It will also be assumed for the present that the length is at least 20 times the average depth (z-coordinate dimension). This restriction will be relaxed in Sec. 4.12.

The beam described above is very much like the rod shown previously in Fig. 3.1. The *essential* difference between the rod and the beam is in their displacement characteristics—the former may undergo either longitudinal or torsional vibrations, while the latter vibrates transversely (in the z-direction) only.

The usual strength of materials assumptions will be made for the transverse displacements of a beam. This makes the problem both mathematically and physically 1 dimensional; that is, w depends on x, but not on y or z. Thus, the motion of the beam is completely determined by the motion of its centerline (the line joining all centroids of cross-sections), and w is defined more precisely to be the

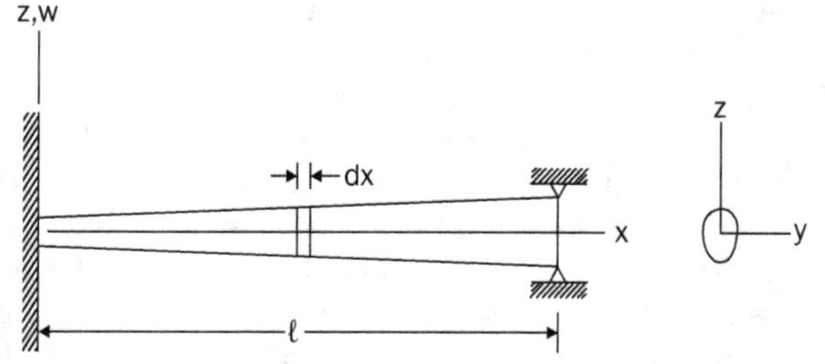

FIGURE 4.1 A differential element of a beam.

displacement of the centerline away from its straight equilibrium position (transverse gravitational effects may be ignored in the vibration problem). The displaced centerline may be envisioned like that of the string seen previously in Fig. 2.1.

Consider the differential beam element of length dx shown in Fig. 4.1. Figure 4.2 shows it in a typical displaced position with its centerline unstretched ($ds = dx$). One observes a transverse shearing force (V) and a bending moment (M) acting on the left face. Both quantities vary along the length of the beam and with time, $V = V(x,t)$ and $M = M(x,t)$, whether measured in the original, horizontal direction (x), or in the curvilinear direction (s). The bending moment is caused by stresses acting normal to the faces, which are zero at the unstretched centerline (the "neutral axis" of the beam) and vary linearly through the depth. These are the results of the classical Euler–Bernoulli beam theory which assumes that plane cross-sections normal to the centerline remain plane and normal during deformation. Also shown in Fig. 4.2 is an external, distributed force $p = p(x,t)$, having dimensions of force per unit length.

Summing forces in the direction normal to the centerline at the center of the element yields

$$V - \left(V + \frac{\partial V}{\partial s} ds\right) + pds = (\rho A ds)\frac{\partial^2 w}{\partial t^2} \tag{4.1}$$

where it has been assumed that the beam is undergoing small amplitude vibrations, so that the slope of the centerline ($\partial w/\partial x$) is everywhere small, and where ρ is mass per unit *volume* and A is the cross-sectional area. Cancelling terms in (4.1), and dividing by ds, it becomes

$$-\frac{\partial V}{\partial x} + p = \rho A \frac{\partial^2 w}{\partial t^2} \tag{4.2}$$

where $\partial/\partial x = \partial/\partial s$ for small slopes.

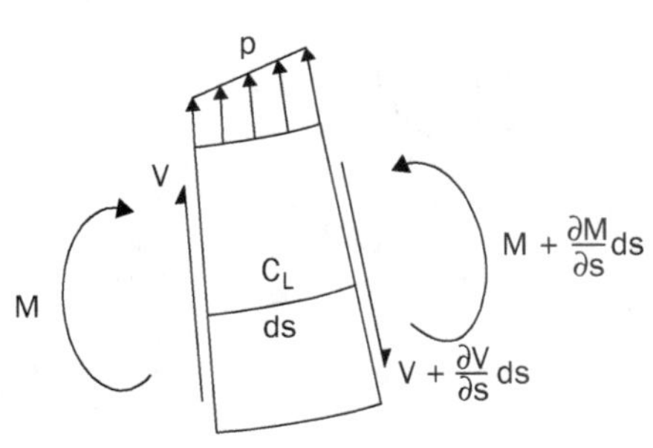

FIGURE 4.2 Free body diagram of a differential beam element.

Summing moments about the center of the element and omitting some quantities which are higher order differentials, yields

$$-M+\left(M+\frac{\partial M}{\partial s}ds\right)-V ds=0 \tag{4.3}$$

Here, setting the right-hand-side (R.H.S.) of (4.3) equal to zero implies that the rotary inertia of the element is being neglected (it will be included in Sec. 4.12). Cancelling terms, (4.3) becomes:

$$V=\frac{\partial M}{\partial s}=\frac{\partial M}{\partial x} \tag{4.4}$$

Returning to the classical, Euler-Bernoulli theory, as it is shown in standard strength of materials textbooks, the curvature at any location x along the beam is proportional to the bending moment there. In detail

$$M=EI\frac{\partial^2 w}{\partial x^2} \tag{4.5}$$

where $\partial^2 w/\partial x^2$ is the linearized (i.e., small slope) curvature, and EI is the flexural rigidity (E is Young's modulus of the material, and I is the second moment of the cross-sectional area with respect to the neutral axis, or "area moment of inertia").

Substituting (4.4) and (4.5) into (4.2) results in

$$\frac{\partial^2}{\partial x^2}\left(EI\frac{\partial^2 w}{\partial x^2}\right)+\rho A\frac{\partial^2 w}{\partial t^2}=p \tag{4.6}$$

In this form the equation of motion is applicable to variable cross-section beams, with $A = A(x)$ and $I = I(x)$, and even nonhomogeneous beams, with $E = E(x)$ and $\rho = \rho(x)$. Of course, the standard Euler–Bernoulli beam theory limits one to linearly elastic materials.

If the beams studied are of homogeneous material and constant cross-section, (4.6) can be simplified to

$$EI\frac{\partial^4 w}{\partial x^4}+\rho A\frac{\partial^2 w}{\partial t^2}=p \tag{4.7}$$

which is the equation governing forced vibration of the most simple type of beam, the forcing pressure being $p = p(x,t)$.

For *free vibration*, $p = 0$, and (4.7) becomes

$$EI\frac{\partial^4 w}{\partial x^4} + \rho A\frac{\partial^2 w}{\partial t^2} = 0 \tag{4.8}$$

4.2 Solution of the Differential Equation for Free Vibrations

Following the same separation of variables procedure used previously for the string in Sec. 2.2, we assume that the solution to (4.8) may be written as

$$w(x,t) = X\Phi \tag{4.9}$$

where $X = X(x)$ and $\Phi = \Phi(t)$. Substituting (4.9) into (4.8) yields

$$EIX^{IV}\Phi + \rho AX\Phi'' = 0 \tag{4.10}$$

where $X^{IV} = d^4X/dx^4$. Dividing through (4.10) by $X\Phi$ and collecting constants in one term, (4.10) may be rewritten as

$$\frac{X^{IV}}{X} = -\left(\frac{\rho A}{EI}\right)\frac{\Phi''}{\Phi} \tag{4.11}$$

Thus, the variables have been separated, and each side of (4.11) must equal a constant, say, α^4. We therefore may write

$$X^{IV} - \alpha^4 X = 0 \tag{4.12a}$$

$$\Phi'' + \omega^2\Phi = 0 \tag{4.12b}$$

where $\omega^2 = \dfrac{EI\alpha^4}{\rho A}$ (4.13)

To solve (4.12a), rewrite it first in operator form as

$$(D^4 - \alpha^4)X = 0 \tag{4.14}$$

where $D = d/dx$, $D^2 = d^2/dx^2$, etc. Equation (4.14) may be factored into

$$(D^2 + \alpha^2)(D^2 - \alpha^2)X = 0 \tag{4.15}$$

Because (4.15) is a linear differential equation, its solution is the sum of the solutions obtained from the separate equations

$$(D^2 + \alpha^2)X = 0 \tag{4.16a}$$

$$(D^2 - \alpha^2)X = 0 \tag{4.16b}$$

Assuming α^2 to be positive, (4.16a) obviously has a solution in terms of trigonometric functions, and the solution to (4.16b) may be written in terms of either exponential or hyperbolic functions. The latter are preferred, for they are either even or odd, and where symmetry (or antisymmetry) of motion is present in a problem, the solution is simplified. Thus, the solution to (4.12a) becomes

$$X(x) = C_1 \sin \alpha x + C_2 \cos \alpha x + C_3 \sinh \alpha x + C_4 \cosh \alpha x \tag{4.17}$$

where $C_1, \ldots, C_4$ are constants of integration, and where sinh and cosh are the hyperbolic sine and cosine functions, respectively. Some useful information concerning the hyperbolic functions is given in Appendix C. Equation (4.15) shows that the solution for negative α^2 is the same as (4.17). Solutions for $\alpha^2 = 0$ will be trivial, which may be shown by applying any set of boundary conditions.

The solution to (4.12b) is the sinusoidal behavior in time expected in a free, undamped vibration problem:

$$\Phi(t) = D_1 \sin \omega t + D_2 \cos \omega t \tag{4.18}$$

where D_1 and D_2 are additional constants of integration. If (4.17) is multiplied by (4.18) to obtain $w(x,t)$, the resulting form will have eight combined constants of integration. These are all determined from the boundary conditions and initial conditions for a particular problem, as will be seen later.

4.3 Classical Boundary Conditions—Frequencies and Mode Shares

The physical conditions which exist at the two ends of a beam must be expressed mathematically as boundary conditions, which are then applied to the solution (4.17). Two types of useful end conditions often found on beams have already been illustrated in Fig. 4.1—clamped and simply supported. The free end is another commonly encountered situation. The corresponding boundary conditions for an end, say $x = \ell$, are listed below:

$$\text{Clamped: } w = 0, \frac{\partial w}{\partial x} = 0 \tag{4.19}$$

$$\text{Simply supported: } w = 0,\ M = 0 \rightarrow \frac{\partial^2 w}{\partial x^2} = 0 \tag{4.20}$$

$$\text{Free: } M = 0 \rightarrow \frac{\partial^2 w}{\partial x^2} = 0,\ V = 0 \rightarrow \frac{\partial^3 w}{\partial x^3} = 0 \tag{4.21}$$

Expression of the moment and shear boundary conditions in terms of w as given above is accomplished with the use of (4.5) and (4.4), respectively.

Considering all "simple" boundary conditions which are mathematically possible, they may be written as

$$w = 0 \quad \text{or} \quad V = 0 \tag{4.22a}$$

$$\frac{\partial w}{\partial x} = 0 \quad \text{or} \quad M = 0 \tag{4.22b}$$

Taking all combinations of (4.22), it is clear that there are four. Three have already been described in (4.19), (4.20), and (4.21). The fourth possibility is:

$$\frac{\partial w}{\partial x} = 0, \quad V = 0 \tag{4.23}$$

This condition is almost never considered in the published literature, because it is seldom encountered in structural applications. It could be found with a beam having an end cut normal to its neutral axis against a lubricated, rigid wall. Otherwise, it is virtually impossible to impose a physical constraint which prevents rotation at an end without causing significant shearing force during the vibratory motion, because of the translational inertia generated by the restraining device.

Vibration frequencies and mode shapes are determined by substituting (4.17) into four boundary conditions, two at each end of the beam. This yields a set of homogeneous simultaneous equations in the four unknowns $C_1, \ldots, C_4$. For a nontrivial solution, the determinant of the coefficient matrix is set equal to zero. The roots of this determinant are the eigenvalues, which are typically $\alpha\ell$. From (4.13), the corresponding nondimensional frequencies are found:

$$\beta^2 = (\alpha\ell)^2 = \omega\ell^2\sqrt{\frac{\rho A}{EI}} \tag{4.24}$$

There are infinite sets of eigenvalues and nondimensional frequencies for each problem. Substituting any eigenvalue into any three of the four homogeneous boundary condition equations permits one to solve for the corresponding amplitude ratios (or eigenvectors) C_2/C_1, C_3/C_1, and C_4/C_1. These determine the corresponding eigenfunction, which is (4.17) rewritten as

$$X(x) = C_1\left(\sin\alpha x + \frac{C_2}{C_1}\cos\alpha x + \frac{C_3}{C_1}\sinh\alpha x + \frac{C_4}{C_1}\cosh\alpha x\right) \qquad (4.25)$$

The mode shape X/C_1 may then be plotted. The determination of C_1 depends on the initial conditions (see Sec. 4.6).

Two example problems will now be solved to demonstrate the application of the procedure described above.

Example 4.1 Determine the free vibration frequencies and mode shapes for a beam of length ℓ which is simply supported at both ends.

Solution

From (4.20) the boundary conditions may be stated as

$$w(0,t) = 0 \rightarrow X(0) = 0$$

$$\frac{\partial^2 w}{\partial x^2}(0,t) = 0 \rightarrow X''(0) = 0$$

$$w(\ell,t) = 0 \rightarrow X(\ell) = 0$$

$$\frac{\partial^2 w}{\partial x^2}(\ell,t) = 0 \rightarrow X''(\ell) = 0$$

Substituting (4.17) into the first two boundary conditions results in $C_2 = C_4 = 0$. The last two conditions will result in

$$\begin{bmatrix} \sin\alpha\ell & \sinh\alpha\ell \\ -(\alpha\ell)^2\sin\alpha\ell & (\alpha\ell)^2\sinh\alpha\ell \end{bmatrix}\begin{bmatrix} C_1 \\ C_3 \end{bmatrix} = \begin{bmatrix} 0 \\ 0 \end{bmatrix}$$

Setting the determinant of the coefficient matrix equal to zero, and expanding it, yields:

$$2(\alpha\ell)^2 \sin\alpha\ell \cdot \sinh\alpha\ell = 0$$

Both $\alpha\ell = 0$ and $\sinh\alpha\ell = 0$ (which implies $\alpha = 0$) give trivial solutions ($\omega = 0$). The nontrivial solution is

$$\sin\alpha\ell = 0 \rightarrow \alpha\ell = m\pi \qquad (m = 1, 2, \ldots, \infty)$$

Substituting this result into either of the two equations arising from the second two boundary conditions gives $C_3 = 0$. Thus, this problem yields the same, simple, sinusoidal mode shapes as were seen previously for a string having both ends fixed (Sec. 2.2). The nondimensional frequencies are

$$\omega\ell^2\sqrt{\frac{\rho A}{EI}} = \pi^2,\ 4\pi^2,\ 9\pi^2, \ldots$$

Example 4.2 Repeat Example 4.1 for a beam having both ends free.

Solution

The preceding example was fortuitous in that two boundary conditions immediately eliminated two constants, resulting in a determinant of only second-order generated by the remaining two conditions. If the same procedure were followed for the present problem, a fourth-order determinant would result, which would be considerably more complicated to evaluate. But the determinant orders may always be reduced if symmetry is present and one takes advantage of it. Then all vibration modes will be either symmetric or antisymmetric. To do so, one locates the origin of the coordinate system at the *center* of the beam.

Symmetric modes: $X(x) = C_2 \cos \alpha x + C_4 \cosh \alpha x$

$$X''\left(\pm\frac{\ell}{2}\right) = 0 = -C_2\alpha^2 \bullet \cos\frac{\alpha\ell}{2} + C_4\alpha^2 \bullet \cosh\frac{\alpha\ell}{2}$$

$$X'''\left(\pm\frac{\ell}{2}\right) = 0 = C_2\alpha^3 \bullet \sin\frac{\alpha\ell}{2} + C_4\alpha^3 \bullet \sinh\frac{\alpha\ell}{2}$$

Expanding the frequency determinant, we have

$$\sin\frac{\beta}{2} \bullet \cosh\frac{\beta}{2} + \cos\frac{\beta}{2} \bullet \sinh\frac{\beta}{2} = 0 \qquad (\beta = \alpha\ell)$$

or

$$\tan\frac{\beta}{2} - \tanh\frac{\beta}{2} = 0$$

This has roots $\beta = 0, 4.730, 10.996 \ldots$. The zero root in this case corresponds to a rigid body translational motion. From the boundary conditions the amplitude ratio is found to be

$$\frac{C_4}{C_2} = -\frac{\sin\beta/2}{\sinh\beta/2} = \frac{\cos\beta/2}{\cosh\beta/2}$$

which determines the mode shape completely for each β (except $\beta = 0$).

Antisymmetric modes: $X(x) = C_1 \sin \alpha x + C_3 \sinh \alpha x$

$$X''\left(\pm\frac{\ell}{2}\right) = 0 = -C_1\alpha^2 \bullet \sin\frac{\alpha\ell}{2} + C_3\alpha^2 \bullet \sinh\frac{\alpha\ell}{2}$$

$$X'''\left(\pm\frac{\ell}{2}\right) = 0 = -C_1\alpha^3 \bullet \cos\frac{\alpha\ell}{2} + C_3\alpha^3 \bullet \cosh\frac{\alpha\ell}{2}$$

Frequency equation: $\sin\frac{\beta}{2} \bullet \cosh\frac{\beta}{2} - \cos\frac{\beta}{2} \bullet \sinh\frac{\beta}{2} = 0$

or

$$\tan\frac{\beta}{2} - \tanh\frac{\beta}{2} = 0$$

Roots: $\beta = 0$ (rigid body rotation), 7.853, 14.137,...

Amplitude ratios: $\dfrac{C_3}{C_1} = \dfrac{\sin\beta/2}{\sinh\beta/2} = \dfrac{\cos\beta/2}{\cosh\beta/2}$ (except for $\beta = 0$)

If one did not consider symmetry, but took the coordinate origin at one end of the beam, and expanded the resulting fourth-order determinant, after some algebra and use of trigonometric identities, the following remarkably simple frequency equation would result:

$$\cos\beta \bullet \cosh\beta = 1 \qquad (\beta = \alpha\ell)$$

Using identities for cos (β/2) and cosh (β/2) (see Appendix C), one may show that this equation may be expanded into

$$\tan^2\frac{\beta}{2} - \tanh^2\frac{\beta}{2} = 0$$

or

$$\left(\tan\frac{\beta}{2} + \tanh\frac{\beta}{2}\right)\left(\tan\frac{\beta}{2} - \tanh\frac{\beta}{2}\right) = 0$$

which contains both frequency equations obtained previously for the symmetric and antisymmetric modes taken separately.

In Example 4.2, two mode shapes which corresponded to $\omega = 0$ were found. These were "rigid body" modes. Although they are not what we would usually call free vibration modes, they may be required in a forced vibration analysis. Therefore, a clear determination of them will now be made.

If $\omega = 0$, then $\alpha = 0$, and (4.14) becomes

$$\frac{d^4X}{dx^4} = 0 \tag{4.26}$$

The solution to (4.26) is

$$X(x) = C_1 + C_2x + C_3x^2 + C_4x^3 \tag{4.27}$$

For the beam with both ends free, and the coordinate origin chosen at the center, for the symmetric modes, $X(x) = C_1 + C_3x^2$, the condition $X'''(\pm\ell/2) = 0$ is assured, and

$$X''\left(\pm\frac{\ell}{2}\right) = 0 \rightarrow \; C_3 = 0, \quad \therefore \; X(x) = C \tag{4.28}$$

For the antisymmetric modes, $X(x) = C_2x + C_4x^3$

$$\left.\begin{aligned} X''\left(\pm\frac{\ell}{2}\right) &= 0 \\ X'''\left(\pm\frac{\ell}{2}\right) &= 0 \end{aligned}\right\} \rightarrow C_4 = 0, \quad \therefore \; X(x) = C_2x \tag{4.29}$$

Thus, $X = C_1$ corresponds to rigid body translation, and $X = C_2x$ corresponds to rigid body rotation (of small amplitude).

In a similar way, it may be shown that the simply supported-free (SS–F) beam has one rigid body mode (rotation).

A summary of the nondimensional free vibration frequencies for all six combinations of classical end conditions for beams is given in Table 4.1. It is remarkable that the SS–F and F–F beams have

m	C–C	C–SS	C–F	SS–SS	SS–F	F–F
1	22.373	15.418	3.5160	9.8696	0	0
2	61.673	49.965	22.034	39.478	15.418	0
3	120.903	104.248	61.697	88.826	49.965	22.373
4	199.859	178.270	120.902	157.914	104.248	61.673
5	298.556	272.031	199.860	246.740	178.270	120.903
>5	$(2m+1)^2\pi^2/4$	$(4m+1)^2\pi^2/16$	$(2m-1)^2\pi^2/4$	$m^2\pi^2$	$(4m-3)^2\pi^2/16$	$(2m-3)^2\pi^2/4$

TABLE 4.1 Frequency Parameters $\beta^2 = \omega \ell^2 \sqrt{\rho A / EI}$ for Beams

the same frequencies as the C–SS and C–C beams, respectively, except for the one and two zero frequencies corresponding to rigid body modes. However, their mode shapes are considerably different. In Example 4.1, it was shown for SS–SS beams that the eigenvalues $\beta = \alpha\ell$, which by (4.24) are the square roots of the frequency parameters, are all separated by π. Table 4.1 shows that the eigenvalues for all end conditions also become separated by π as higher frequencies are considered. As written, the eigenfunction in each case has its coordinate origin at the first-mentioned end of the beam.

Frequency equations and eigenfunctions for each of the six cases are summarized below.

Clamped–clamped:

$$\cos\beta \cdot \cosh\beta = 1 \tag{4.30a}$$

$$X = (\cosh\beta\xi - \cos\beta\xi) - \gamma(\sinh\beta\xi - \sin\beta\xi) \tag{4.30b}$$

$\gamma = 0.98250,\ 1.00078,\ 0.99997,\ 1.00000, \ldots$

Free–free:

$$\cos\beta \cdot \cosh\beta = 1 \tag{4.31a}$$

$$X = (\cosh\beta\xi + \cos\beta\xi) - \gamma(\sinh\beta\xi + \sin\beta\xi) \tag{4.31b}$$

$\gamma =$ same as clamped-clamped

Clamped–SS:

$$\tan\beta = \tanh\beta \tag{4.32a}$$

$$X = (\cosh\beta\xi - \cos\beta\xi) - \gamma(\sinh\beta\xi - \sin\beta\xi) \tag{4.32b}$$

$\gamma = 1.00078, 1.00000, \ldots$

Free–SS:

$$\tan\beta = \tanh\beta \tag{4.33a}$$

$$X = (\cosh\beta\xi + \cos\beta\xi) - \gamma(\sinh\beta\xi + \sin\beta\xi) \tag{4.33b}$$

$\gamma =$ same as clamped-SS

Clamped–free:

$$\cos\beta \cdot \cosh\beta = -1 \tag{4.34a}$$

$$X = (\cosh\beta\xi - \cos\beta\xi) - \gamma(\sinh\beta\xi - \sin\beta\xi) \tag{4.34b}$$

$\gamma = 0.73410,\ 1.01847,\ 0.99922,\ 1.00003,\ 1.00000, \ldots$

SS–SS:

$$\sin\beta = 0 \tag{4.35a}$$

$$X = \sin\beta\xi \tag{4.35b}$$

In the above equations, $\xi = x/\ell$ is measured in each case from the left end of the beam. The values of β are the square roots of the frequency parameters listed in Table 4.1. More accurate values of β and γ are available in the classical study of Young and Felgar [1].

Figure 4.3 depicts the first four mode shapes for each of the six cases. The rigid body modes of the SS–F and F–F beams are included. For each of the six cases, the first mode has no node point, the second mode has one node point, and each higher mode has one additional node point. For the higher modes of the SS–F beam, conservation of angular momentum about the hinged left end must be preserved at all times. Similarly, the higher modes of the F–F beam must conserve translational momentum, as well as rotational momentum about the center point of the equilibrium line for the beam. In Fig. 4.3 it is also observed that all mode shapes for the C–C, SS–SS, and F–F beams alternate between being symmetric and being antisymmetric.

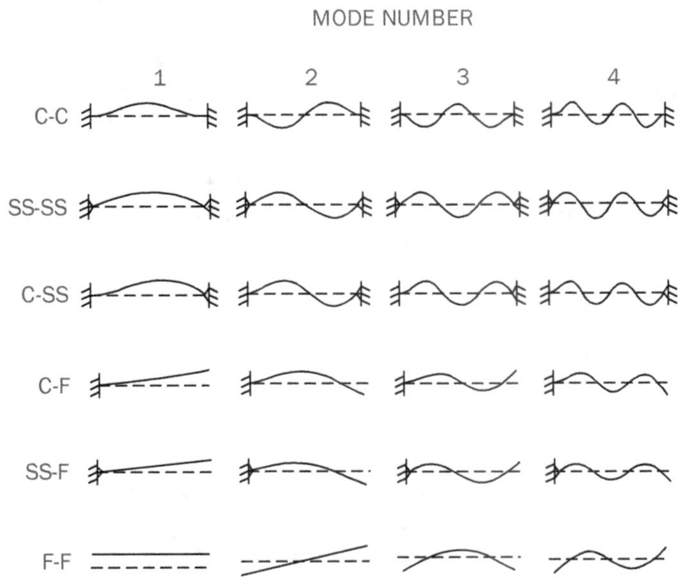

Figure 4.3 The first four mode shapes for beams with different boundaries.

4.4 Other Boundary Conditions—Added Masses or Springs

The mathematical boundary conditions at a clamped end of a beam as used in Sec. 4.3 were zero displacement and zero rotation. However, in reality this never occurs. Even a steel beam perfectly welded to an infinitely large constraint block (i.e., an infinite half-space) would undergo rotation at this "clamped" end during vibration. This may cause significant decreases in the natural frequencies, especially the lower ones, as was shown in an interesting study by MacBain and Genin [2] for a cantilever. In real structures the rigid constraints acting on beams must often be replaced by elastic constraints. Then, the translational and rotational stiffnesses of the structure at the attachment end must be known from other analyses. Moreover, the mass of the attached structure may significantly affect the beam vibration frequencies and mode shapes.

Figure 4.4 shows a beam of length ℓ having rigid masses M_1 and M_2 attached to its ends. The masses are supported by translational springs at each end, having stiffnesses k_1 and k_2. Rotational springs with stiffnesses K_1 and K_2 are also shown attached to the masses. The origin of the x-coordinate system is placed off the beam to emphasize that it may be chosen anywhere.

Free vibration frequencies and mode shapes for the beam of Fig. 4.4 may be straightforwardly determined by using the solution (4.17) of the equation of motion and applying the proper boundary conditions at the ends. To determine the proper boundary conditions, it is essential that free body diagrams be shown for the two end masses. This is done in Fig. 4.5. The masses are shown having positive displacements (w) and positive rotations ($\partial w/\partial x$). These cause forces and moments within the external springs which act on the masses as shown. An incremental length of beam is shown attached at the proper side of each mass, and the internal shearing force (V) and bending moment (M) in each is also indicated, drawn in the positive directions as previously used in Fig. 4.2.

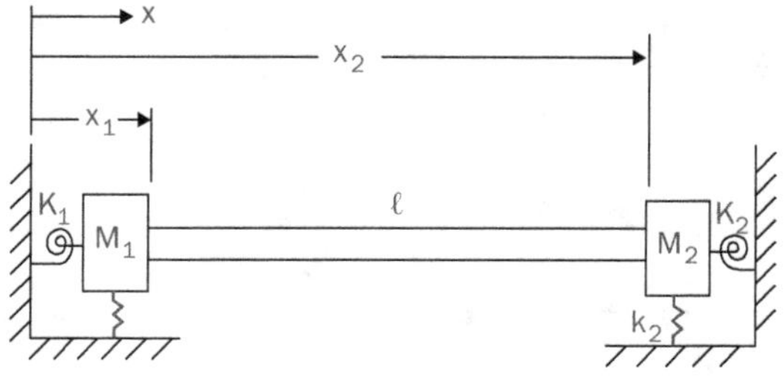

Figure 4.4 A beam with springs and rigid masses and springs at each end.

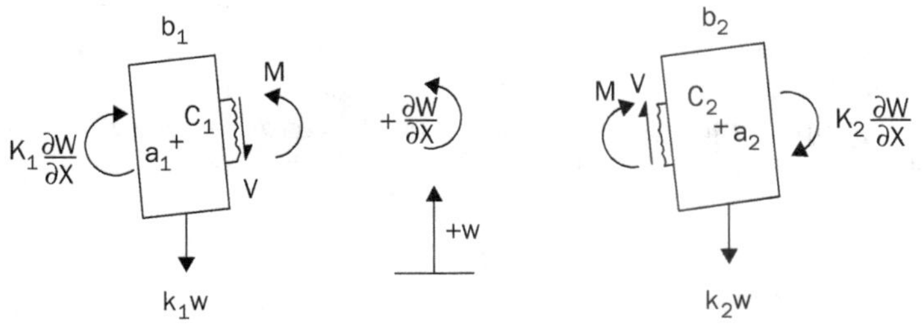

FIGURE 4.5 Free body diagrams for the two end masses.

For mass M_1, summing forces in the z-direction (assuming small slopes, as usual):

$$-V - k_1 w = M_1 \frac{\partial^2 w}{\partial t^2} \quad (\text{at } x = x_1) \tag{4.36a}$$

Summing moments about the center of mass C_1, we obtain

$$M - K_1 \frac{\partial w}{\partial x} - V \bullet \left(\frac{b_1}{2}\right) = I_{M_1} \frac{\partial^2}{\partial t^2}\left(\frac{\partial w}{\partial x}\right) \quad (\text{at } x = x_1) \tag{4.36b}$$

where I_{M_1} is the mass moment of inertia of M_1 about C_1. The term on the R.H.S. of (4.36b) is the rotary inertia of M_1, which could be large.

Similarly, for mass M_2 one obtains the following equations of motion:

$$V - k_2 w = M_2 \frac{\partial^2 w}{\partial t^2} \quad (\text{at } x = x_2) \tag{4.37a}$$

$$-M - K_2 \frac{\partial w}{\partial x} - V \bullet \left(\frac{b_2}{2}\right) = I_{M_2} \frac{\partial^2}{\partial t^2}\left(\frac{\partial w}{\partial x}\right) \quad (\text{at } x = x_2) \tag{4.37b}$$

It is important to note that in writing (4.36) and (4.37):

1. Positive directions for force and moment summations were chosen in the directions of positive displacement and rotation, so that the inertia terms may be left with positive signs.
2. The signs of V and M change from (4.36) to (4.37).

Errors in signs due to incorrect reasoning in either of the above are easily made. A single sign error will typically yield worthless results for free vibration frequencies and mode shapes.

Assuming that $w(x,t) = X(x) \sin \omega t$ or $X(x) \cos \omega t$ for free vibrations, and using (4.4) and (4.5), then (4.36) and (4.37) become:

$$\left.\begin{aligned} &EIX''' + (k_1 - M_1\omega^2)X = 0 \\ &EIX'' - \left(\frac{EIb_1}{2}\right)X''' - (K_1 - I_{M_1}\omega^2)X' = 0 \end{aligned}\right\} \text{ at } x = x_1 \tag{4.38}$$

$$\left.\begin{aligned} &EIX''' + (k_2 - M_2\omega^2)X = 0 \\ &EIX'' - \left(\frac{EIb_2}{2}\right)X''' - (K_2 - I_{M_2}\omega^2)X' = 0 \end{aligned}\right\} \text{ at } x = x_2 \tag{4.39}$$

Substituting (4.17) into (4.38) and (4.39) yields four homogeneous equations in the unknown constants $C_1, \ldots, C_4$. The fourth-order frequency determinant is written and eigenvalues αl are determined by plotting the determinant versus αl and finding its zeros.

The situation described above is extremely general, and many important problems may be solved as special cases. One example is the simply supported beam having rotational springs at both ends. In that case, $k_1 = k_2 = \infty$ and $M_1 = M_2 = 0$ in Fig. 4.4 and in (4.38) and (4.39). Another interesting example which is a special case will now be taken up.

Example 4.3 A beam of length ℓ has one end completely free. The other end is constrained by a translational spring and a rotational spring, as shown in Fig. 4.6.

A. Derive the characteristic equation for free vibration frequencies.
B. Show that the characteristic equations for free–free, hinged–free, and clamped–free beams arise as special cases.
C. Let $k_1\ell^3/EI = K_1\ell/EI = K$. Determine K so that the fundamental frequency is one-half that of a clamped–free (i.e., cantilever) beam.

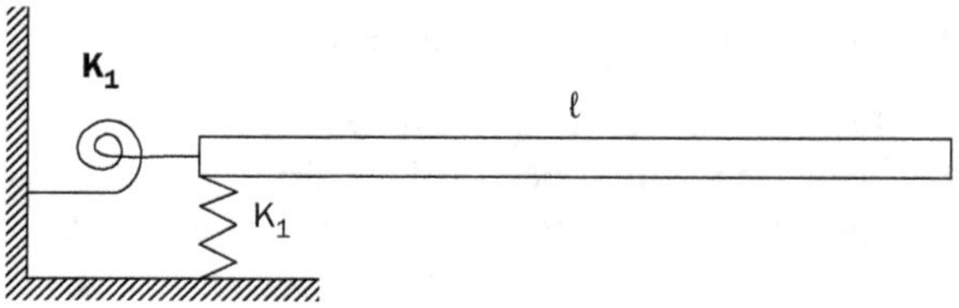

Figure 4.6 A beam with one end constrained by springs and the other end completely free.

D. For the value of K determined in Part C, calculate the first four nondimensional frequencies of the beam. Compare these with the corresponding ones for F–F and C–F beams.
E. Plot the mode shape of the fundamental frequency in Part D.

Solution

Part A: Choosing the coordinate origin at the left end and substituting (4.17) into the boundary conditions at $x = 0$ and $x = \ell$:

$$\begin{bmatrix} -\beta^3 & K_W & \beta^3 & K_W \\ -K_\theta & -\beta & -K_\theta & \beta \\ -\sin\beta & -\cos\beta & \sinh\beta & \cosh\beta \\ -\cos\beta & \sin\beta & \cosh\beta & \sinh\beta \end{bmatrix} \begin{bmatrix} C_1 \\ C_2 \\ C_3 \\ C_4 \end{bmatrix} = \begin{bmatrix} 0 \\ 0 \\ 0 \\ 0 \end{bmatrix}$$

where P, K_W, and K_θ are the nondimensional parameters:

$$\beta = \alpha\ell, \quad K_W = \frac{k_1\ell^3}{EI}, \quad K_\theta = \frac{K_1\ell}{EI}$$

Expanding the frequency determinant and using $\cos^2\beta + \sin^2\beta = 1$ and $\cosh^2\beta - \sinh^2\beta = 1$, it yields the characteristic equation:

$$K_W K_\theta(\cos\beta \cdot \cosh\beta + 1) - K_W\beta(\sin\beta \cdot \cosh\beta - \cos\beta \cdot \sinh\beta)$$
$$- K_\theta\beta^3(\sin\beta \cdot \cosh\beta + \cos\beta \cdot \sinh\beta) - \beta^4(\cos\beta \cdot \cosh\beta - 1) = 0$$

Part B: For an F–F beam, $K_W = K_\theta = 0$, yielding (4.31a). For an SS–F beam, divide through the above equation by K_W; then set $K_W \to \infty$, $K_\theta = 0$, which yields (4.33a). For a C–F beam, divide through by $K_W \cdot K_\theta$; then set $K_W \to \infty$, $K_\theta \to \infty$, which yields (4.34a). In the above steps we are careful to divide by infinity, but not by zero.

Part C: The fundamental frequency of a cantilever beam is $\omega = 3.5160\sqrt{EI/\rho A\ell^4}$ (see Table 4.1). Therefore, take $\beta = \sqrt{3.5160/2} = 1.3259$. Using this, and setting $K_W = K_\theta = K$, the characteristic equation becomes

$$1.4887K^2 - 7.5772K + 1.5803 = 0$$

This has *two* roots: $K = 0.2179$ and 4.8719. However, substituting $K = 0.2179$ into the characteristic equation and solving for β yields $\beta = 0.6423, 1.3259, 4.7742, 7.8806$, and the value 1.3259 corresponds to the *second* frequency for this beam having small end stiffness, so this root is inappropriate.

Part D: Using the second root ($K = 4.8719$) yields $\beta = 1.3259, 2.3690, 5.1972, 8.2177$. Squaring these values permits comparison of nondimensional frequencies with those of the F–F and C–F beams, as seen below:

Mode number	$\omega\ell^2\sqrt{\rho A/EI}$		
	K = 0 (F–F)	**K = 4.8719**	**K= ∞ (C–F)**
1	0	1.7580	3.5160
2	0	5.6126	22.034
3	22.373	27.011	61.697
4	61.673	67.531	120.902

As expected, the frequencies of the elastically constrained beam fall between those of the F–F and C–F beams.

Part E: To plot mode shapes it is first necessary to determine the eigenfunctions. For the desired values of $K_W = K_\theta = K = 4.8719$ and $\beta = 1.3259$, one may solve any three of the four simultaneous equations shown in Part A for the ratios C_2/C_1, C_3/C_1, and C_4/C_1. This part of the problem would be simplified if the coordinate origin had been chosen at the *right* end, rather than the left. Then the free end conditions would require that $C_1 = C_3$ and $C_2 = C_4$, and (4.17) could be written as:

$$X(\xi) = (\sin \beta\xi + \sinh \beta\xi) + \frac{C_2}{C_1}(\cos \beta\xi + \cosh \beta\xi)$$

where $\xi = x/\ell$. Then C_2/C_1 is determined from either of the two boundary conditions at the left end as

$$\frac{C_2}{C_1} = \frac{\beta^3(\cos\beta - \cosh\beta) + K_W(\sin\beta + \sinh\beta)}{\beta^3(\sin\beta + \sinh\beta) - K_W(\cos\beta + \cosh\beta)}$$

$$= -\frac{\beta(\sin\beta - \sinh\beta) - K_\theta(\cos\beta + \cosh\beta)}{\beta(\cos\beta - \cosh\beta) + K_\theta(\sin\beta - \sinh\beta)}$$

The fundamental mode shape for $K_W = K\theta = K = 4.8719$ is shown in Fig. 4.7. It is seen that the elastically constrained left end translates and rotates while the beam undergoes flexure.

4.5 Orthogonality of the Eigenfunctions

The eigenfunctions of free, undamped vibrations of beams form an orthogonal set of functions. The property of orthogonality will be found useful in dealing with initial conditions (Sec. 4.6) and with forced vibrations (Sec. 4.8). In order to take advantage of this property

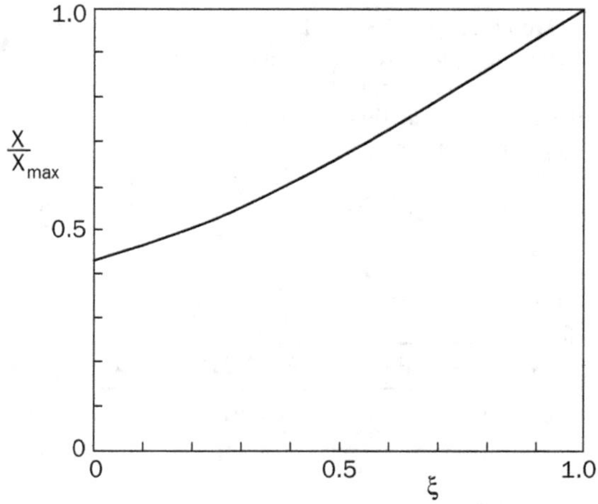

FIGURE 4.7 The fundamental mode shape for a beam with elastic supports at one end and the other free. $K_W = K_\theta = K = 4.8719$.

later, the orthogonality property will be proven in this section, first for the classical type of boundary conditions seen in Sec. 4.3, and secondly for a representative problem of Sec. 4.4.

Consider the mth free vibration mode of a beam. In order for it to satisfy the equation of motion, its eigenfunction must be a solution of (4.12a). That is,

$$X_m^{IV} - \alpha_m^4 X_m = 0 \tag{4.40}$$

where α_m is the constant obtained from the mth eigenvalue, $\beta_m = \alpha_m \ell$, as described in Sec. 4.3 and 4.4.

Consider further another mode, designated by the subscript n. Then its eigenfunction must satisfy

$$X_n^{IV} - \alpha_n^4 X_n = 0 \tag{4.41}$$

Multiply (4.40) and (4.41) by X_n and X_m, respectively, and subtract the two equations, yielding

$$X_n X_m^{IV} - X_m X_n^{IV} - (\alpha_m^4 - \alpha_n^4) X_m X_n = 0 \tag{4.42}$$

Integrating over the length of the beam yields

$$(\alpha_m^4 - \alpha_n^4)\int_0^\ell X_m X_n dx = \int_0^\ell (X_n X_m^{IV} - X_m X_n^{IV})\, dx \tag{4.43}$$

Let us integrate by parts a typical term of the R.H.S.:

$$\begin{aligned}\int_0^\ell X_n X_m^{IV} dx &= [X_n \ X_m''']_0^\ell - \int_0^\ell X_n' X_m''' \ dx \\ &= [X_n \ X_m''']_0^\ell - [X_n' \ X_m'']_0^\ell + \int_0^\ell X_n'' X_m'' dx\end{aligned} \tag{4.44}$$

Similarly,

$$\int_0^\ell X_m X_n^{IV} dx = [X_m \ X_n''']_0^\ell - [X_m' \ X_n'']_0^\ell + \int_0^\ell X_m'' X_n'' \, dx \tag{4.45}$$

Substituting (4.44) and (4.45) into (4.43) gives

$$\begin{aligned}(\alpha_m^4 - \alpha_n^4)\int_0^\ell X_m X_n dx &= [X_n \ X_m''']_0^\ell - [X_n' \ X_m'']_0^\ell \\ &\quad - [X_m \ X_n''']_0^\ell + [X_m' \ X_n'']_0^\ell\end{aligned} \tag{4.46}$$

For the classical boundary conditions of Sec. 4.3, (4.22) requires that either $X_m = 0$ or $X_m''' = 0$, and that either $X_m' = 0$ or $X_m'' = 0$. Similarly, either $X_n = 0$ or $X_n''' = 0$, and either $X_n' = 0$ or $X_n'' = 0$. Thus, all bracketed terms on the R.H.S. of (4.46), when evaluated at the boundaries $x = 0$ and $x = \ell$, are zero. Thus, (4.46) becomes

$$(\alpha_m^4 - \alpha_n^4)\int_0^\ell X_m X_n dx = 0 \tag{4.47}$$

If the modes have different frequencies, which is typically the case, then $\alpha_m \neq \alpha_n$, and

$$\int_0^\ell X_m X_n dx = 0 \qquad (m \neq n) \tag{4.48}$$

which is the orthogonality property. Using this, from (4.40) one can also show that

$$\int_0^\ell X_n X_m^{IV} dx = 0 \qquad (m \neq n) \tag{4.49}$$

and from (4.44)

$$\int_0^\ell X_n' X_m''' dx = 0 \qquad (m \neq n) \tag{4.50a}$$

$$\int_0^\ell X_n'' X_m'' dx = 0 \qquad (m \neq n) \tag{4.50b}$$

To deal with orthogonality for the more general boundary conditions described in Sec. 4.4, consider again the beam having elastic translational and rotational constraints at the left end ($x = 0$), and a free right end ($x = \ell$), studied in Example 4.3. For the reasons given above, the bracketed quantities on the R.H.S. of (4.46) will all be zero for $x = \ell$. At $x = 0$, the boundary conditions become:

$$EIX''' + k_1 X = 0 \tag{4.51a}$$

$$EIX'' - K_1 X' = 0 \tag{4.51b}$$

for both X_m and X_n. Solving (4.51a) for X_m and X_n in terms of X_m''' and X_n''', respectively, and (4.51b) for X_m' and X_n' in terms of X_m'' and X_n'', respectively, and substituting into the R.H.S. of (4.46), it is found that all terms cancel, yielding again (4.47).

Of course, the eigenfunctions could be proven orthogonal by substituting them in detail into (4.48) and evaluating the integral directly. But this would typically be much more complicated than the general procedures laid out above.

4.6 Initial Conditions

From the results of Sec. 4.2, a general solution for the free, undamped, vibration displacement may be expressed as

$$w(x,t) = \sum_{m=1}^{\infty} X_m(x)(A_m \sin \omega_m t + B_m \cos \omega_m t) \tag{4.52}$$

where X_m and ω_m are the eigenfunction and natural frequency of the mth mode, respectively. Thus, a general motion may be represented by the superposition of free vibration modes, each vibrating at its own frequency.

Suppose the beam is given an initial displacement and velocity of arbitrary forms, $f(x)$ and $g(x)$, respectively, which are consistent with the boundary conditions at the ends, i.e.,

$$w(x,0) = f(x) \tag{4.53a}$$

$$\frac{\partial w}{\partial t}(x,0) = g(x) \tag{4.53b}$$

But, from (4.52),

$$w(x,0) = \sum_m B_m X_m \tag{4.54a}$$

$$\frac{\partial w}{\partial t}(x,0) = \sum_m A_m \omega_m X_m \tag{4.54b}$$

Therefore,

$$f(x) = \sum_m B_m X_m \tag{4.55a}$$

$$g(x) = \sum_m A_m \omega_m X_m \tag{4.55b}$$

Multiply both sides of (4.55a) by X_n, where X_n represents the nth eigenfunction, and integrate over the length of the beam; i.e.,

$$\int_0^{\ell} f X_n dx = \int_0^{\ell} \sum_m B_m X_m X_n dx \tag{4.56}$$

Interchanging the order of summation and integration on the R.H.S. of (4.56), and considering the orthogonality of X_m and X_n as expressed by (4.48), it is seen that all terms of the infinite sum on the R.H.S. of (4.56) vanish, except when $m = n$. Thus, (4.56) yields

$$B_m = \frac{\int_0^\ell f X_m dx}{\int_0^\ell X_m^2 dx} \tag{4.57}$$

Following the same procedure, one obtains from (4.55b)

$$A_m = \frac{1/\omega_m \int_0^\ell g X_m dx}{\int_0^\ell X_m^2 dx} \tag{4.58}$$

Equations (4.57) and (4.58) are formulas which may be straightforwardly used to determine the amplitude components A_m and B_m of each mode as the beam responds to initially imposed displacements and/or velocities. In the special case when the beam is given an initial displacement shape and released from rest, then $g = 0$ and (4.58) tells us that $A_m = 0$. Conversely, if the beam were in its straight, equilibrium position and subjected to an impulsive loading of short duration (i.e., an impact), the subsequent free vibration motion would begin with $f = 0$, whence $B_m = 0$.

The eigenfunctions $X(x)$ given previously in (4.30–4.35) for five of the six cases of combinations of simple edge conditions have already been normalized, that is, the dominator integrals needed for (4.57) and (4.58) evaluate to ℓ. The simple sine function given by (4.35b) is not normalized. The integral of its square is $\ell/2$.

Example 4.4 A cantilever beam is initially deformed by a statically applied end moment M_0, as depicted in Fig. 4.8. Suddenly the moment is released.

Make a plot of $\delta(t)/\delta_0$ versus t/τ_1 for $0 < t/\tau_1 < 6$, where $\delta(t)$ and δ_0 are the vibratory and static displacements of the free end, respectively, and where τ_1 is the first natural period of free vibration. Is $\delta(t)$ a *periodic* function?

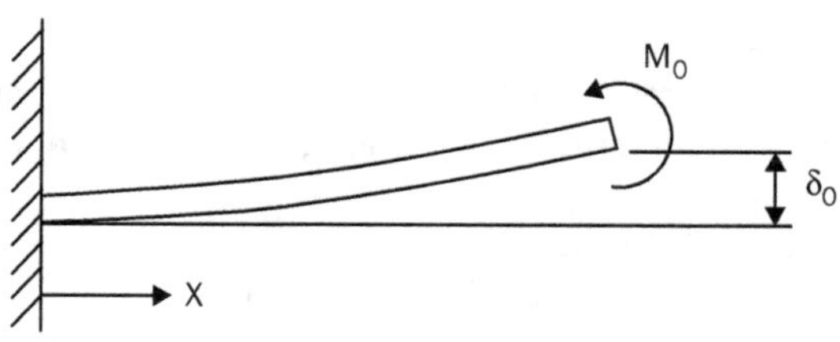

FIGURE 4.8 A cantilever beam deformed by a statically applied end moment M_0.

Solution

It is first necessary to determine the static displacement shape of the beam, from which it is released from rest. This is found by integrating the equation of motion (4.8) in the special case of equilibrium, i.e.,

$$EI\frac{d^4w_0}{dx^4}=0$$

which yields $w_0 = C_0 + C_1x + C_2x^2 + C_3x^3$. Applying the two boundary conditions at each end, the four constants are determined, and the initial, static displacement is found to be

$$w_0(x) = f(x) = \frac{M_0x^2}{2EI}, \quad \text{and} \quad w_0(\ell) = \delta_0 = \frac{M_0\ell^2}{2EI}$$

Thus, $A_m = 0$, and the B_m are determined by means of (4.57), where $X_m(x)$ for the cantilever beam is given by (4.34b).

Evaluation of integrals needed for the numerator (4.57) is rather complicated. It may be done exactly, wherein considerable integration by parts is required for the numerator integrals. (As mentioned above, the denominator of (4.57) evaluates as ℓ.) Alternatively, the integration may be done numerically. In either case, the constants β and γ in the eigenfunction (4.34b) must be determined with precision if accurate values of B_m are to be obtained. The first five B_m are

$B_1 = 0.44539$
$B_2 = -0.03937$
$B_3 = 0.00825$
$B_4 = -0.00301$
$B_5 = 0.00142$

It is seen that B_1 is by far the largest coefficient, and so the response of the first mode strongly dominates the ensuing motion. This is as expected, because the parabolic initial shape approximates the first vibration mode shape reasonably well. If the beam were initially displaced in *exactly* the first mode shape, it would vibrate subsequently in that single mode—no other modes would be involved at all.

The graph shown in Fig. 4.9 is a plot of $\delta(t)/\delta_0$ versus t/τ_1, where $\tau_1 = 2\pi/\omega_1$ is the fundamental period, for six cycles of the response of the first mode. It is seen that, while the first mode dominates the response, the higher frequency modes make small contributions to the motion which disturb its smoothness.

Moreover, the motion is *not periodic* in time. If only two modes were superimposed the motion would be almost, but not exactly, periodic. For two superimposed modes to result in periodic motion, the ratio of the frequencies (ω_m/ω_n) must be a rational number. If more modes are involved, then *all* the frequency ratios must be rational numbers to have periodic motion. For the most simple string problems (see Secs. 2.2 and 2.3) and for beams having both ends simply supported this is the situation, but in general, it is not.

For example, consider two modes having a frequency ratio of exactly 1.4 = 14/10 = 7/5. Then the superimposed motion would be periodic, with a period of $5\tau_1$, or $7\tau_2$, where τ_1 and τ_2 are the periods of the two modes. If the ratio were exactly 1.41 = 141/100, then the superimposed period would be $100\tau_1 = 141\tau_2$. But if $\omega_2/\omega_1 = \sqrt{2} = 1.41421356\ldots$, then the superimposed motion would never repeat.

Figure 4.10 shows the shape of the beam at various times taken during the period of the first mode (τ_1). The beam passes through the equilibrium position

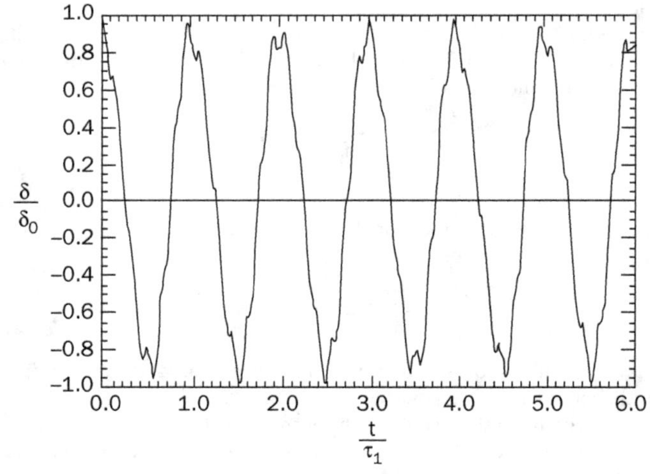

FIGURE 4.9 Tip displacement of a cantilever beam subjected to a bending moment at the end which is then released.

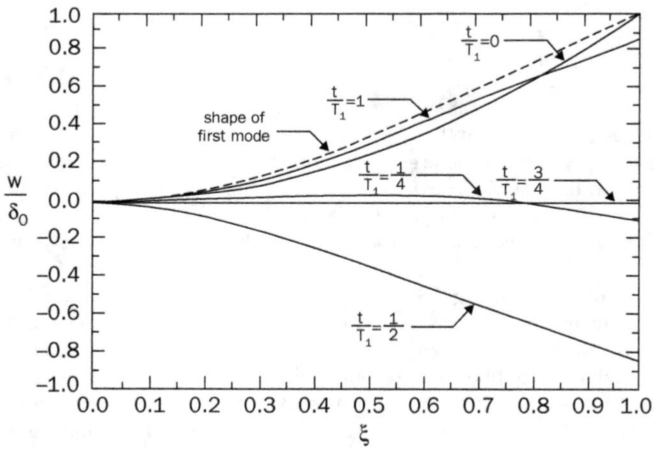

FIGURE 4.10 The shape of the cantilever beam with a released end moment at various times.

after a quarter cycle of motion with a shape that has significant curvature. At the end of a half cycle, the shape is not the mirror image of the initial shape, as it would be if only a single mode were excited.

The nonperiodic dynamic behavior of a cantilever beam subjected to various types of initial static loadings, and then released, is discussed in more detail in Ref. [3].

4.7 Continuous and Discontinuous Beams

Figure 4.11 shows a beam of length ℓ supported at its ends (A and C) and at an intermediate point (B). Such a beam is often called "continuous" or "two-span." The span lengths are ℓ_1 and ℓ_2, as shown. It is assumed that the amplitude of free vibrations is sufficiently small so that the weight of the beam will keep it in contact with all three supports at all times. If not, then knife-edge supports are added to the top side of the beam at the same locations. If contact were permitted to break at any of the points, the problem would become much more complicated.

One straightforward approach to solving the free vibration problem for the continuous beam depicted in Fig. 4.11 is to divide the beam into two parts of length ℓ_1 and ℓ_2, each having its displacement w_1 and w_2, respectively. Each displacement must satisfy the equation of motion (4.8) for its own region of applicability, as well as the proper continuity conditions at point B. Boundary conditions at A and C must also be enforced.

In detail, for the beam of Fig. 4.11, the equations of motion are

$$EI\frac{\partial^4 w_1}{\partial {x_1}^4}+\rho A\frac{\partial^2 w_1}{\partial t^2}=0 \qquad (0 \le x_1 \le \ell_1) \tag{4.59a}$$

$$EI\frac{\partial^4 w_2}{\partial {x_2}^4}+\rho A\frac{\partial^2 w_2}{\partial t^2}=0 \qquad (0 \le x_2 \le l_2) \tag{4.59b}$$

Solutions to each equation may be found in the forms of (4.9), (4.17), and (4.18). Applying the boundary conditions at both ends ($x_1 = 0$ and $x_2 = 0$) reduces them to the following.

$$w_1(x,t)=(A_1\sin\alpha_1 x_1 + B_1\sinh\alpha_1 x_1)\sin\omega t \tag{4.60a}$$

$$w_2(x,t)=(A_2\sin\alpha_2 x_2 + B_2\sinh\alpha_2 x_2)\sin\omega t \tag{4.60b}$$

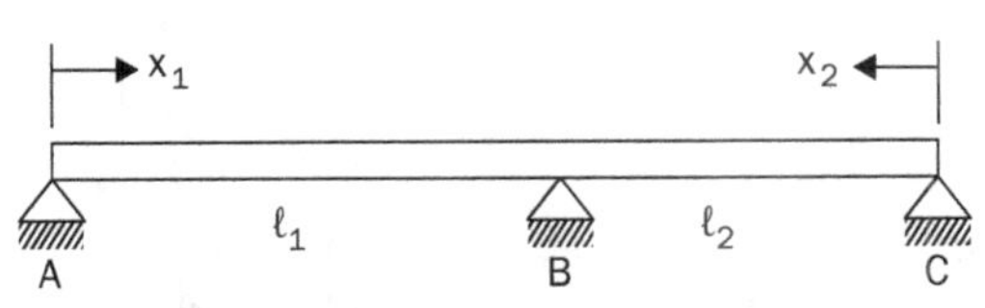

FIGURE 4.11 A simply supported continuous beam.

The frequency ω must be the same in both parts of the beam to permit subsequent satisfaction of the continuity conditions at B. In accordance with (4.13), it is related to α_1 and α_2 by

$$\alpha_1^4 = \frac{\omega^2 \rho A}{EI} = \alpha_2^4 \equiv \alpha^4 \tag{4.61}$$

Since we will only concern ourselves with finding frequencies and mode shapes, and not in applying initial conditions, the sin (ωt) form in (4.60) is adequate.

The conditions at the common support point B are

$$w_1(\ell_1,t)=0 \tag{4.62a}$$

$$w_2(\ell_2,t)=0 \tag{4.62b}$$

$$\frac{\partial w_1}{\partial x_1}(\ell_1,t) = -\frac{\partial w_2}{\partial x_2}(\ell_2,t) \tag{4.62c}$$

$$M(\ell_1,t) = M(\ell_2,t) \tag{4.62d}$$

where careful note of the sign differences in the slopes of (4.62c) should be made. The shearing force across B is, of course, not continuous, and an external reaction (R) is supplied by the support. This reaction may be found later, in terms of a unit amplitude of vibratory displacement, after the problem is solved. The *downward* force R_B required for positive w_1 and w_2 is found from an infinitesimal length of beam taken across point B, and is

$$R_B = V_1(\ell_1,t) + V_2(\ell_2,t) = EI\frac{\partial^3 w_1}{\partial x_1{}^3}(\ell_1,t) + EI\frac{\partial^3 w_2}{\partial x_2{}^3}(\ell_2,t) \tag{4.63}$$

The reactions at points A and C may be found in a similar manner.

Substituting (4.60) into the four conditions (4.62) yields four homogeneous equations in A_1, B_1, A_2, and B_2, which may be written in matrix form as

$$\begin{bmatrix} \sin\beta & \sinh\beta & 0 & 0 \\ 0 & 0 & \sin r\beta & \sinh r\beta \\ \cos\beta & \cosh\beta & \cos r\beta & -\cosh r\beta \\ \sin\beta & -\sinh\beta & \sin r\beta & \sinh r\beta \end{bmatrix} \begin{bmatrix} A_1 \\ B_1 \\ A_2 \\ B_2 \end{bmatrix} = \begin{bmatrix} 0 \\ 0 \\ 0 \\ 0 \end{bmatrix} \tag{4.64}$$

where $\beta = \alpha\ell_1$ and $r = \ell_2/\ell_1$. The nondimensional frequencies β^2 are found from the roots of the fourth-order determinant arising from (4.64). The eigenvector components B_1/A_1, A_2/A_1, and B_2/A_1 are then found from any three of the equations in (4.64), which then determines the mode shapes according to (4.60).

It should be noted that in the special case when $r = 1$ ($\ell_2 = \ell_1$), the problem simplifies greatly. In that case the beam is symmetrically supported and all mode shapes are either symmetric or antisymmetric. The symmetric mode shapes and frequencies are the same as those of an S–C beam of length ℓ_1, and the antisymmetric ones are those of an S–S beam. Thus, for $r = 1$, the fundamental mode is antisymmetric.

Problems involving multiple spans may be solved in a similar manner. However, the amount of work involved increases considerably as the number of spans is increased. For general end conditions a frequency determinant of order $4N$ must be generated, where N is the number of spans. The problem of Fig. 4.11 required only a fourth-order determinant because the choice of coordinate origins at the two simply supported ends to more easily satisfy the four end conditions is readily seen.

An example of a discontinuous beam is seen in Fig. 4.12. In this case the depth of the beam is discontinuous, which changes the values of A and I in the equation of motion (4.8) in passing from one part of the beam to another. Similarly, a discontinuity in material could occur, causing abrupt changes in E and ρ. Still another type of discontinuity would be the imposition of a mass or a spring at a point along a beam. The vibration frequencies and mode shapes of any such discontinuous beams may be found straightforwardly by the method described earlier in this section. That is, separate solutions to (4.8) are taken. Two boundary conditions at each end and four conditions at the discontinuity point yield the needed characteristic determinant. For the beam of Fig. 4.12 the four conditions at point B are continuity of w, $\partial\omega/\partial x$, M, and V.

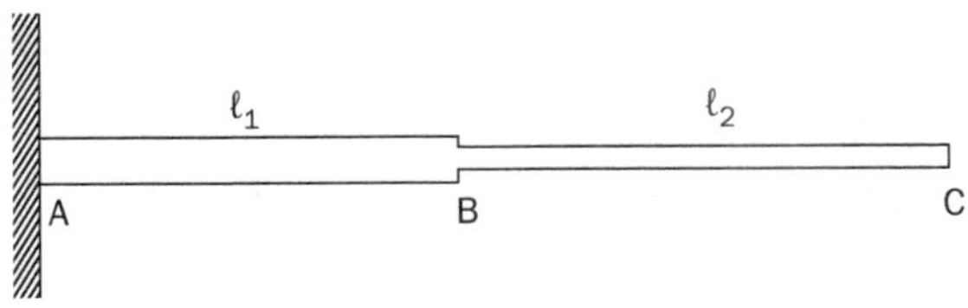

FIGURE 4.12 An example of a discontinuous beam.

4.8 Forced Vibrations

Generalizing the equation of motion (4.7) for the forced vibration of uniform cross-section beams, it becomes

$$EI\frac{\partial^4 w}{\partial x^4}+\rho A\frac{\partial^2 w}{\partial t^2}+c\frac{\partial w}{\partial t}=p \tag{4.65}$$

where $p = p(x,t)$ is a distributed force (with units of force/length) and c is a viscous damping coefficient [with units of force • time/(length)^2] depending on such parameters as the viscosity of the surrounding medium and the cross-sectional shape of the beam. The "effective mass" of the surrounding medium (see Sec. 2.8) is neglected here. If the material damping within the beam is considered, then E is replaced by $E(1+i\eta)$, as for bars (see Sec. 3.6).

In the previous chapters two exact methods were developed for the analysis of forced vibrations of strings and bars—eigenfunction superposition and closed form. Both are applicable to beam vibration analysis.

Following the *eigenfunction superposition* method, it is assumed that the loading may be expressed as

$$p(x,t)=\sum_{m=1}^{\infty}p_m(t)X_m(x) \tag{4.66}$$

where X_m is the eigenfunction of free, undamped vibration of the mth mode of the beam in question. Eigenfunctions for the six classical sets of boundary conditions were given by (4.30b)–(4.35b). Multiplying both sides of (4.66) by a typical eigenfunction X_n, integrating over the length of the beam, and taking advantage of the orthogonality relationship developed in (4.48), a formula for determining the $p_m(t)$ is obtained:

$$p_m(t)=\frac{\int_0^{\ell}p(x,t)X_m dx}{\int_0^{\ell}X_m^2 dx} \tag{4.67}$$

Further a solution to (4.65) is assumed as

$$w(x,t)=\sum_{m=1}^{\infty}\Phi_m(t)X_m(x) \tag{4.68}$$

Substituting (4.66) and (4.68) into (4.65) gives:

$$EI\sum_m \Phi_m X_m^{IV} + \rho A\sum_m \Phi_m'' X_m + c\sum_m \Phi_m' X_m = \sum_m p_m X_m \tag{4.69}$$

However, the X_m are solutions of the free vibration equation

$$EIX_m^{IV} - \rho A\omega_m^2 X_m = 0 \tag{4.70}$$

Substituting $EIX_m^{IV} = \rho A\omega_m^2 X_m$ into (4.69), multiplying both sides by X_n, integrating over the length of the beam, and making use of the orthogonality property of the eigenfunctions, there results

$$\rho A\Phi_m'' + c\Phi_m' + \rho A\omega_m^2\Phi_m = p_m(t) \qquad (m = 1,2,\ldots,\infty) \tag{4.71}$$

Or, in terms of the nondimensional frequency parameter β (see Sec. 4.3)

$$\beta_m^2 = \omega_m \ell^2\sqrt{\frac{\rho A}{EI}} \tag{4.72}$$

which was used in (4.30)–(4.35) and Table 4.1,

$$\rho A\Phi_m'' + c\Phi_m' + \beta_m^4\left(\frac{EI}{\ell^4}\right)\Phi_m = p_m(t) \qquad (m = 1,2,\ldots,\infty) \tag{4.73}$$

Equation (4.73) is the one d.o.f., forced vibration equation for the mth mode. It is in the same form as (2.85) which arose in the forced vibration of strings. For forcing loads which are sinusoidal in time, its solution is therefore expressed by (2.86) and (2.87), where ρ is replaced by ρA, and $T\alpha_m^2$ by $EI\beta_m^4/\ell^4$. The response $w(x_0,t)$ of a point on the beam ($x = x_0$) is then determined by (2.93) and (2.94). For more general, periodic forcing functions as described by (2.95) and (2.96), the response is given by (2.97).

If the *closed-form* method is followed, for a sinusoidal forcing function

$$p(x,t) = P(x)\sin\Omega t \tag{4.74}$$

a solution to (4.65) is assumed to be

$$w(x,t) = X_1(x)\sin\Omega t - X_2(x)\cos\Omega t \tag{4.75}$$

Substituting (4.74) and (4.75) into (4.65) and requiring the sin Ωt and cos Ωt components to satisfy the equation independently yields

$$X_1^{IV} - k^4 X_1 + \gamma^4 X_2 = Q(\xi) \tag{4.76a}$$

$$X_2^{IV} - k^4 X_2 + \gamma^4 X_1 = 0 \tag{4.76b}$$

where derivatives are taken with respect to the nondimensional coordinate $\xi = x/\ell$, and where

$$k^4 = \frac{\Omega^2 \rho A \ell^4}{EI}, \quad \gamma^4 = \frac{c\Omega \ell^4}{EI}, \quad Q(\xi) = \frac{P(\xi)\ell^4}{EI} \tag{4.77}$$

Equations (4.76) are similar in form to (2.110). That is, they are both coupled sets of ordinary differential equations, with the coupling terms existing due to the damping. However (2.110) is of fourth order, which yielded four independent constants of integration, whereas (4.76) is of eighth order, and will consequently have eight independent constants of integration. Nevertheless, the complementary solution may be found straightforwardly [4], and a suitable particular solution to yield the desired $Q(\xi)$ may be found. The eight constants of integration are then determined from the boundary conditions. If material damping is present, the closed-form solution may again be determined straightforwardly [4].

In the case of no damping, the motion is in phase (or 180° out of phase) with the exciting force, and the closed-form solution may be taken as

$$w(x,t) = X(x)\sin\Omega t \tag{4.78}$$

Substituting this into the equation of motion

$$EI\frac{\partial^4 w}{\partial x^4} + \rho A\frac{\partial^2 w}{\partial t^2} = P(x)\sin\Omega t \tag{4.79}$$

and using the nondimensional coordinates $\xi = x/\ell$, one obtains

$$X^{IV} - k^4 X = Q(\xi) \tag{4.80}$$

with k and Q again being defined by (4.77). The complementary solution to this equation is obtained by factoring the homogeneous equation into the operator form

$$(D^2 + k^2)(D^2 - k^2)X_C = 0 \tag{4.81}$$

where $D^2 = d^2/d\xi^2$. The solution to (4.81) is

$$X_C = C_1 \sin k\xi + C_2 \cos k\xi + C_3 \sinh k\xi + C_4 \cosh k\xi \tag{4.82}$$

Adding a particular solution which yields the proper R.H.S. of (4.80) yields the complete solution. The four constants $C_1, \ldots, C_4$ are obtained from the boundary conditions.

The forced vibration behavior of beams will now be illustrated by two example problems.

Example 4.5 A cantilever beam of length ℓ is excited by a uniform, distributed pressure $P(x,t) = q_0 \sin \Omega t$, where q_0 is a constant. It vibrates in a viscous fluid. Neglecting the mass of any fluid being moved, find the response of the free end of the beam.

Solution

The eigenfunction superposition method is used, with the eigenfunctions given by (4.34b). Figure 4.13 shows the nondimensional displacement, W/δ, at the free end (where δ is the *static*, displacement) as a function of Ω/ω_1, with Ω/ω_1 being varied over a wide enough range to include the first two resonances. Curves are shown for two values of the damping ratio, c/c_{c1}, where c_{c1} is the value of the critical damping for the first mode. That is,

$$\delta = \frac{q_0 \ell^4}{8EI}, \qquad \omega_1 = \frac{3.5160}{\ell^2}\sqrt{\frac{EI}{\rho A}}$$

$$c_{C1} = \frac{2\beta_1^2}{\ell^2}\sqrt{EI\rho A} = \frac{7.032}{\ell^2}\sqrt{EI\rho A}$$

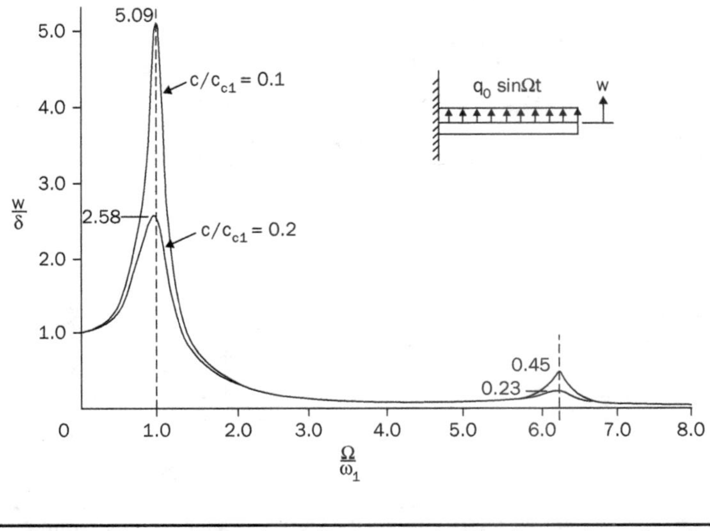

FIGURE 4.13 The nondimensional displacement of the free end as a function of Ω/ω_1.

The curves are similar to those seen previously for the string subjected to a uniformly distributed exciting pressure (Fig. 2.18). That is, the first peak value of W/δ occurs close to $\Omega/\omega_1 = 1$ and is much greater than for higher resonances.

Example 4.6 A uniformly distributed exciting force $p(x,t) = q_0 \sin \Omega t$ is applied to a beam with both ends clamped, as shown in Fig. 4.14. The effects of damping may be neglected. Determine the displacement, $w(x,t)$, and evaluate it at the middle of the beam.

Solution

The closed-form solution method is used. For a uniformly distributed loading, $Q = q_0\ell^4/EI$ in (4.80). A suitable particular solution to (4.80) is

$$X_p = -\frac{Q}{k^4} = -\frac{q_0}{\rho A \Omega^2}$$

Because there is symmetry in both the loading and in the boundary conditions, the coordinate origin is chosen at the center of the beam, and the odd functions of ξ in the solution are discarded by setting $C_1 = C_3 = 0$ in (4.82). Thus the complete solution is

$$X(\xi) = C_2 \cos k\xi + C_4 \cosh k\xi + X_p$$

Applying the boundary conditions

$$X\left(\frac{1}{2}\right) = X'\left(\frac{1}{2}\right) = 0$$

yields $\quad C_2 = -\dfrac{X_p}{\Delta}\sinh\dfrac{k}{2}, \quad C_4 = -\dfrac{X_p}{\Delta}\sin\dfrac{k}{2}$

where $\Delta = \sin\dfrac{k}{2} \cdot \cosh\dfrac{k}{2} + \cos\dfrac{k}{2} \cdot \sinh\dfrac{k}{2}$

Using Table 4.1, k in (4.77) may be rewritten as

$$k^2 = \frac{\Omega}{\omega_1}\left(\omega_1 \ell^2 \sqrt{\frac{\rho A}{EI}}\right) = 22.373\frac{\Omega}{\omega_1}$$

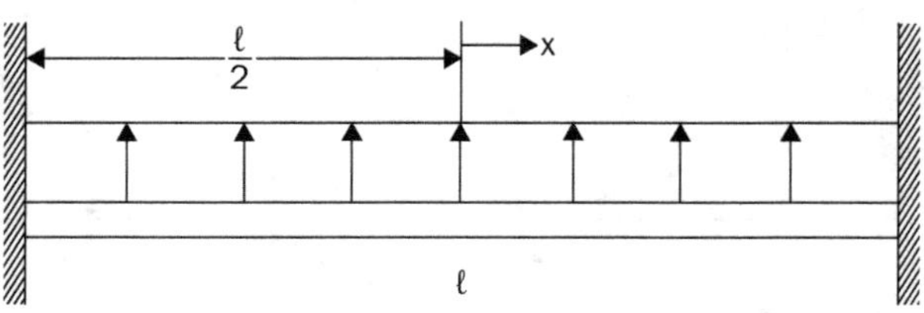

FIGURE 4.14 A uniformly distributed exciting force applied to a beam with both ends clamped.

If $\Omega = \omega_i$, where ω_i is any of the free undamped vibration frequencies of the beam, $\Delta = 0$ (which is the characteristic equation one solves to find the ω_i for the symmetric modes), and C_2 and C_4 both become infinite, as expected.

Finally, the complete solution may be expressed as

$$w(\xi,t) = \frac{q_0\ell^4}{EIk^4 \bullet \Delta}\left(\sinh\frac{k}{2}\cos k\xi + \sin\frac{k}{2}\cosh k\xi - \Delta\right)\sin\Omega t$$

At the middle of the beam,

$$w(0,t) = \frac{q_0\ell^4}{EIk^4 \bullet \Delta}\left(\sinh\frac{k}{2} + \sin\frac{k}{2} - \Delta\right)\sin\Omega t$$

which is a simple, closed-form expression to evaluate and plot.

4.9 Energy Functionals—Rayleigh Method

It was shown in Chaps. 2 and 3 that the Rayleigh and Ritz methods are very useful for determining the natural frequencies of strings and bars, particularly when no simple, exact solutions exist. The same may be said about beams. To use these methods it is necessary to have at one's disposal the expressions for the maximum potential and kinetic energies of the beam as it vibrates.

Potential energy is stored in the beam in the form of strain energy as it undergoes flexure. To evaluate it we consider an element of the beam having differential length dx, as shown in Fig. 4.15(a). This element may also be seen in Fig. 4.1. Figure 4.15(b) shows a view of the cross-section, looking along the x-axis. As stated in Sec. 4.1, the analysis is restricted to beams having at least one symmetry axis. A differential element of area dA is shown in the cross-section acting at a distance z from the neutral axis (N.A.) of the beam. Thus, the shaded rectangles shown in the two views represent the differential volume of the beam $dA \bullet dx$.

The strain energy in the differential volume is

$$d(PE) = \frac{1}{2}\sigma_x dA \bullet \epsilon_x dx \tag{4.83}$$

where σ_x is the bending stress at that location, and ϵ_x is the corresponding strain. The total strain energy in the beam is then

$$PE = \frac{1}{2}\int_{vol}\sigma_x\epsilon_x dA\,dx \tag{4.84}$$

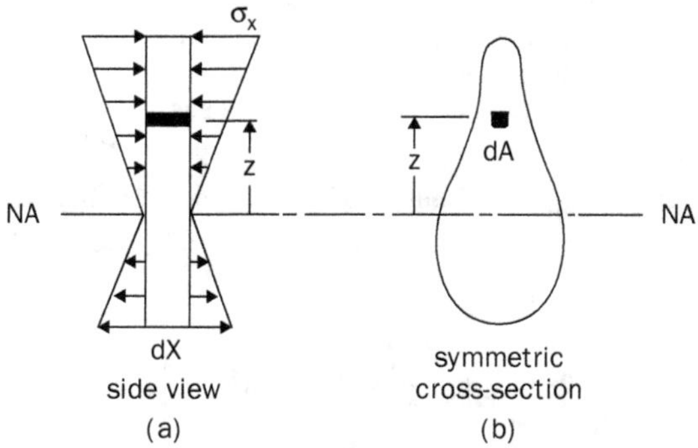

FIGURE 4.15 A beam with differential length *dx* and a cross-section with at least one symmetry axis.

The stress at a distance z from the neutral axis is well known from strength of materials as

$$\sigma_x = -\frac{Mz}{I} \tag{4.85}$$

Using this, along with the 1-dimensional stress–strain relationship, $\epsilon_x = \sigma_x/E$, (4.84) becomes

$$PE = \frac{1}{2}\int_0^\ell \left[\int_A \frac{1}{E}\left(\frac{Mz}{I}\right)^2 dA\right] dx \tag{4.86}$$

where the volume integration has been separated into two parts, along the length of the beam and throughout the area. The area moment of inertia may be a function of x, but is constant for a given cross-section. Moreover, assume that the material is homogeneous throughout the cross-section (but not necessarily along the length). Then

$$PE = \frac{1}{2}\int_0^\ell \frac{M^2}{EI^2}\left(\int_A z^2 dA\right) dx \tag{4.87}$$

The definition of I is

$$I = \int_A z^2 dA \tag{4.88}$$

Substituting this and (4.5) into (4.87) gives

$$PE = \frac{1}{2}\int_0^\ell EI\left(\frac{\partial^2 w}{\partial x^2}\right)^2 dx \tag{4.89}$$

Equation (4.89) is the total strain energy in the beam at any instant of time, since $w = w(x,t)$. With E and I retained in the integrand, (4.89) is capable of representing nonhomogeneous materials, $E = E(x)$, or variable cross-sections, $I = I(x)$.

The translational kinetic energy of the element of beam volume shown shaded in Fig. 4.15 is

$$d(KE) = \frac{1}{2}\rho(dA \bullet dx)\left(\frac{\partial w}{\partial t}\right)^2 \tag{4.90}$$

Assuming that the mass density (ρ) may vary along the length, $\rho = \rho(x)$, but not within a cross-section, then the total kinetic energy of the beam at any instant is

$$KE = \frac{1}{2}\int_0^\ell \rho A\left(\frac{\partial w}{\partial t}\right)^2 dx \tag{4.91}$$

To use the Rayleigh method, one sets

$$PE_{\max} = KE_{\max} \tag{4.92}$$

(see Sec. 2.12). Assuming sinusoidal motion for the beam,

$$w(x,t) = X(x)\sin\omega t \tag{4.93}$$

then

$$PE_{\max} = \frac{1}{2}\int_0^\ell EI(X'')^2\, dx \tag{4.94}$$

and

$$KE_{\max} = \frac{\omega^2}{2}\int_0^\ell \rho A X^2 dx \tag{4.95}$$

Thus, one determines a natural frequency with the Rayleigh method by setting $PE_{\max} = KE_{\max}$ and evaluating the quotient

$$\omega^2 = \frac{PE_{\max}}{KE^*_{\max}} \tag{4.96}$$

where $$KE^*_{max} = \frac{1}{2}\int_0^\ell \rho A X^2 dx \tag{4.97}$$

To use the Rayleigh method, one assumes a displacement function $X(x)$ which reasonably approximates the shape of the desired free vibration mode. The closer it resembles the correct mode shape, the more accurate the frequency approximation will be. The geometric boundary conditions or other constraints (e.g., an intermediate point support on a continuous beam) must be satisfied exactly. However, the generalized force types of boundary conditions (i.e., bending moment or transverse shearing force) need not be satisfied. Indeed, satisfying the latter requires additional effort, but does not necessarily give better results.

Example 4.7 Use the Rayleigh method to obtain an approximate value of the fundamental frequency of a cantilever beam.

Solution

Various functions will be tried as assumed mode shapes. The one that yields the lowest frequency will be the best, for all frequencies obtained will be upper bounds to the exact value. For this problem we happen to know the exact value as well, which permits calculations of errors.

First trial function: If we did not already have the exact first mode shape from earlier analysis (which is depicted in Fig. 4.3), some physical observation of the vibrations of a knife blade held flat to a table with one hand and excited with the other would suggest a parabolic mode shape. That is, in terms of the coordinate origin shown in Fig. 4.8,

$$X = x^2$$

This simple function is the lowest degree polynomial which satisfies the geometric boundary conditions $X(0) = X'(0) = 0$. Substituting it into (4.94) and (4.97), and then using (4.96) yields $\omega\ell^2\sqrt{\rho A/EI} = \sqrt{20} = 4.472$. This is not very accurate, being 27.2 percent higher than the exact value of 3.516 seen in Table 4.1.

Second trial function: Another function which resembles the expected mode shape and satisfies the geometric boundary conditions is

$$X = 1 - \cos\frac{\pi x}{2\ell}$$

Using this with (4.96) yields $\omega\ell^2\sqrt{\rho A/EI} = 3.664$, which is much better, as it has an error of only 4.2 percent.

Third trial function: Neither of the two preceding trial functions paid any attention to the boundary conditions of zero moment and shear at the free end. One wonders how much improvement would result if a function were used that satisfied both of them. For this purpose, choose a polynomial of higher degree, $x^2 + C_3x^3 + C_4x^4$, which satisfies $X(0) = X'(0) = 0$, and determine C_3 and C_4 so that $X''(\ell) = X'''(\ell) = 0$ as well. Doing so gives

$$X = \xi^2 - \frac{2}{3}\xi^3 + \frac{1}{6}\xi^4 \qquad \left(\xi = \frac{x}{\ell}\right)$$

Substituting this into (4.96) results to $\omega\ell^2\sqrt{\rho A/EI} = \sqrt{162/13} = 3.530$, which is a very good result, being only 0.4 percent higher.

Fourth trial function: The error in the fourth-degree polynomial assumption may be seen if one considers the *static equilibrium* equation

$$EIX^{IV} = p$$

where p is a distributed static loading along the beam. This is known from elementary strength of materials, and may be obtained from (4.7) as well. In a free vibration situation, p may be regarded as "inertia loading," $p = -\rho A\omega^2 X$, due to the acceleration. If a beam were to deflect into a parabolic shape during vibration, $X^{IV} = 0$, corresponding to zero inertia loading, which is completely unreasonable. Similarly, the fourth-degree function yields inertia loading which is constant along the beam. This is not a good representation, but at least it permits the inertia loading to be present in an *average* sense. Therefore, let us select a trial function which, when differentiated four times, would give us a second-degree inertia loading variation. Clearly, it would be a polynomial of *sixth* degree. If we simply choose

$$X = x^6$$

then the B.C. at $x = 0$ are satisfied as well. Rayleigh's Quotient (4.96) then yields the terrible result $\omega\ell^2\sqrt{\rho A/EI} = \sqrt{1300} = 36.061$! A sketch of the function shows that it varies much too rapidly as it approaches $x = \ell$. Furthermore, it causes very large residual values of bending moment and shear at $x = \ell$.

Fifth trial function: Choose $X = C_2x^2 + C_3x^3 + x^6$. This function makes X^{IV} second degree, regardless of the choice of C_2 and C_3, and satisfies $X(0) = X'(0) = 0$. The coefficients C_2 and C_3 are chosen so that $X'(\ell) = X'''(\ell) = 0$. This gives

$$X = 45\xi^2 - 20\xi^3 + \xi^6 \qquad \left(\xi = \frac{x}{\ell}\right)$$

Equation (4.96) gives $\omega\ell^2\sqrt{\rho A/EI} = 3.5164$ compared with the exact solution of 3.5160! Thus, this trial function gives an *extremely* accurate frequency, only 0.01 percent above the exact value.

Before leaving this problem, we should reflect a bit more on *why* the various trial functions gave results of such different accuracies. Let us first look at the functions themselves, which are plotted in Fig. 4.16, normalized so as to all have unit values of displacement at the beam tip ($\xi = x/\ell = 1$). The curves are numbered corresponding to the earlier discussion, and "0" identifies the exact solution eigenfunction. The fifth trial function falls virtually on the exact curve, and is so drawn (the largest difference is 0.0046, in the vicinity of $\xi = 0.5$). The third trial function used (fourth-degree polynomial) is also reasonably accurate, and is seen to lie close to the exact curve. The fourth function ($X = \xi^6$) gave very poor results, and lies far away from the other curves. Because KE^*_{max} depends on the integrals of the squares of these functions, one may visualize how the errors affect the denominator of Rayleigh's Quotient (4.96).

However, to understand better the error in PE_{max} one should look at the linearized *curvatures* (X'') of the exact and trial functions. These are shown in Fig. 4.17, and are proportional to the bending moments, according to (4.5). As one would expect, errors in second derivatives are usually worse than the errors in the functions themselves. The fifth trial function again falls nearly on

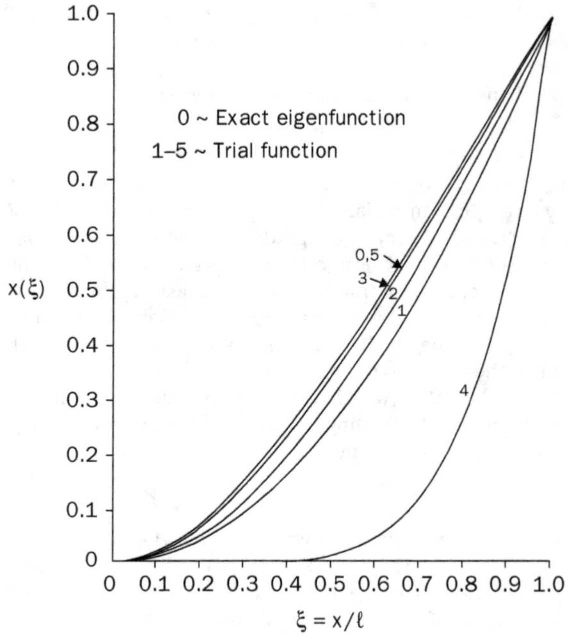

FIGURE 4.16 Functions used for analysis of a cantilever beam using Rayleigh's method.

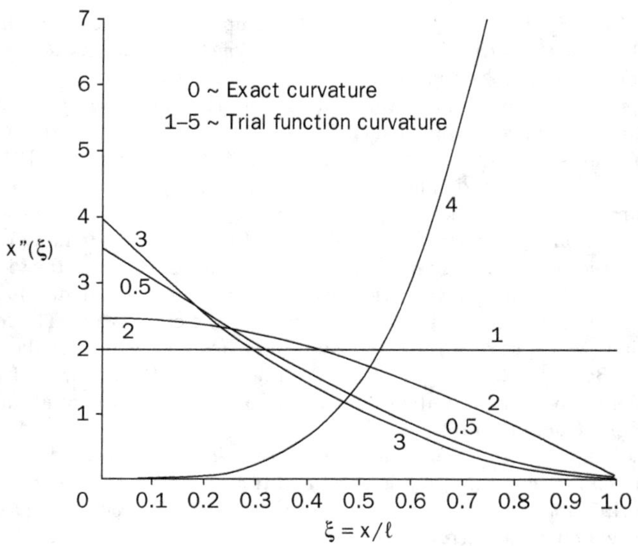

FIGURE 4.17 Exact and trial function curvatures for a cantilever beam.

the exact curve (the largest difference is 0.054, which is 1.5 percent, occurring at $\xi = 0$). The first and fourth functions do not yield the needed maximum X'' at the wall and zero X'' at the free end. Indeed $X = \xi^6$ yields zero displacement, slope, moment, and shearing force values at $\xi = 0$, a very strange situation indeed! The function $1 - \cos(\pi\xi/2)$ does not give zero shear force at the free end (the slope of X'' is not zero).

To summarize the preceding example, in solving problems by the Rayleigh method, the following considerations are involved in the selection of approximate trial functions which are to be sufficiently accurate:

1. Geometric boundary conditions *must* be satisfied.
2. It is usually better to satisfy all the generalized force boundary conditions as well.
3. The function to be used should be plotted (or at least sketched). Does it reasonably represent the mode shape expected?
4. Is the "inertia loading" obtained from taking higher derivatives (fourth order in the case of the beam) reasonably similar in shape to that of the displacement?

Accomplishing all of the above may result in a very complicated function. In that case, if necessary, differentiations and integrations may be performed numerically.

4.10 Ritz Method

To use the Ritz method a set of trial functions capable of representing the free vibration mode shapes is chosen as

$$X(x) = \sum_{i=1}^{I} C_i \phi_i(x) \tag{4.98}$$

where C_i are constants to be determined and the trial functions φ_i must each satisfy the geometric boundary conditions. The C_i are determined so that ω^2, as given by (4.96), is minimized.

The necessary minimizing equations are the same as those for the string (2.135). Following the procedure laid out in Sec. 2.13, these are cast into the form (2.136), which is more convenient for computations:

$$\frac{\partial}{\partial C_i}(PE_{\max} - \omega^2 KE^*_{\max}) = 0 \qquad (i = 1, 2, \dots, I) \tag{4.99}$$

where now $PE_{\max}$ and $KE^*_{\max}$ are the expressions (4.94) and (4.97) relevant to the beam.

Example 4.8 Use the Ritz method to obtain a reasonably accurate value of the fundamental frequency of a cantilever beam, and compare with the results found previously in Example 4.7, where the Rayleigh method was demonstrated.

Solution

Any algebraic polynomial of the form

$$X = \sum_{i=2}^{I} C_i x^i$$

satisfies the geometric boundary conditions $X(0) = X'(0) = 0$ exactly, where it must be carefully noted that the series begins with $i = 2$.

Choosing the two term polynomial $X = C_2x^2 + C_3x^3$ gives

$$PE_{\max} - \omega^2 KE^*_{\max} = \frac{EI}{2}\int_0^\ell (2C_2 + 6C_3x)^2 dx - \frac{\rho A\omega^2}{2}\int_0^\ell (C_2x^2 + C_3x^3)^2 dx$$

In applying (4.99), it is more efficient to carry out the minimizing differentiations $\partial/\partial C_i$ first, instead of first squaring the integrands and integrating. The computational savings become more readily apparent if three or four term polynomials are employed for X. Equations (4.99) then generates two equations in C_2 and C_3 which may be written in matrix form as

$$\begin{bmatrix} \left(4-\frac{\lambda}{5}\right) & \left(6-\frac{\lambda}{6}\right) \\ \left(6-\frac{\lambda}{6}\right) & \left(12-\frac{\lambda}{7}\right) \end{bmatrix} \begin{bmatrix} C_2 \\ C_3 \end{bmatrix} = \begin{bmatrix} 0 \\ 0 \end{bmatrix} \qquad \lambda = \frac{\omega^2\ell^4\rho A}{EI}$$

It is observed that the coefficient matrix above is symmetric. This is to be expected when using the Ritz method, and is a necessary check on the correctness of calculations. Finding least common denominators for each of the above equations to eliminate the fractions, and expanding the frequency determinant, results in

$$\lambda^2 - 1224\lambda + 15{,}120 = 0$$

This has two roots, $\lambda_{1,2} = 612 \mp (1/2)\sqrt{1{,}437{,}696}$, which yields the two non-dimensional frequencies $\omega\ell^2\sqrt{\rho A/EI} = 3.5327$ and 34.807. The lower one is a reasonable upper bound approximation (0.48 percent error) to the exact value of 3.5160. The higher one is an upper bound approximation, albeit a very poor one, to the second nondimensional frequency of 22.034. The corresponding approximate mode shapes are obtained by substituting the roots for λ, one at a time, into either of the two equations in C_2 and C_3 given above, and solving for C_3/C_2. They are $C_3/C_2 = -0.384/\ell$ for the first mode, and $C_3/C_2 = -1.216/\ell$ for the second.

Comparing the approximate fundamental frequency found above with those determined in Example 4.6, it is seen that the cubic polynomial used here, with the optimum choice of C_3/C_2 being determined by the Ritz method, is about equally accurate as the fourth-degree polynomial used with the Rayleigh method (third trial function). Moreover, the amount of computational work and time required is about the same in both solutions.

Further improvement in the accuracy of the first two frequencies, as well as upper bound approximations to higher frequencies, may be achieved by taking additional terms in the algebraic polynomial representing the displacement. A summary of the results obtained using 2, 3, 6, and 10 polynomial terms is made in Table 4.2. Percentages by which the Ritz solutions differ from the exact solution are shown in parentheses in Table 4.2. Thus, using a three-term polynomial,

Mode number	Number of polynomial terms				Exact values
	2	3	6	10	
1	3.53273 (0.48)	3.51707 (0.03)	3.51601 (0.00)	3.51601 (0.00)	3.51601
2	34.8068 (57.97)	22.2334 (0.90)	22.0348 (0.00)	22.0345 (0.00)	22.0345
3		118.1444 (91.49)	61.7163 (0.03)	61.6972 (0.00)	61.6972
4			128.389 (6.19)	120.904 (0.00)	120.902
5			223.551 (11.85)	199.886 (0.01)	199.860
6			1006.013 (236.96)	303.162 (1.54)	298.556
7				429.999 (3.12)	416.991
8				773.874 (39.40)	555.165
9				1082.564 (51.82)	713.079
10				5527.655 (520.57)	890.732

*Value in parentheses is percent error in Ritz solution (two decimal places)

Table 4.2 Nondimensional Frequencies $\omega\ell^2\sqrt{\rho A/EI}$ for a Cantilever Beam by the Ritz Method, Using Algebraic Polynomials

the fundamental frequency is obtained very accurately (0.03 percent error), and a reasonable approximation for the second frequency is found; however, the third frequency, as expected is very poor. Using six polynomial terms, the fundamental frequency is seen to converge to six significant figures. Higher frequencies also converge to their exact values as additional terms are taken, which demonstrates that the polynomials form a set of functions which are mathematically complete as sufficient terms are employed, and that numerical roundoff error is no significant problem with ten terms. Using more than ten terms, one quickly encounters round off error difficulties ("ill conditioning"), which may be circumvented by using *orthogonal* polynomials.

Comparing further the Rayleigh and Ritz methods, one may say, in general, that the former can obtain reasonably accurate results, at least for fundamental frequencies, without a great deal of computational effort. However, some ingenuity and physical understanding of the vibrational behavior is required. Reasonably accurate estimates of *higher* frequencies are usually difficult to achieve. The Ritz method proceeds straightforwardly and requires little physical understanding. Exact frequencies, fundamental and higher ones, will be approached as closely as desired as additional terms in the trial function are taken, provided that:

1. The functions satisfy the geometric boundary conditions exactly.
2. The functions form a mathematically *complete* set (i.e., capable of representing any possible deflected shape of the beam) as the number of terms used increases.
3. Computational round off errors are not excessive.

As the last statement implies, truly accurate solutions with the Ritz method typically require computer programming. For 1-dimensional problems such as the beam, very accurate results can usually be obtained with single precision (i.e., eight significant figures) arithmetic.

4.11 Effects of Axial Forces

A beam having one end clamped and the other end simply supported, subjected to an axial force T at each end, is shown in its static equilibrium position in Fig. 4.18. For the sketch shown, T is constant everywhere along the beam. However, in general, T may vary along the length—that is, $T = T(x)$—when gravitational or other distributed forces act in the axial direction.

Suppose now the beam vibrates in the transverse direction. The equation of motion is obtained from a free body diagram of a differential element of length dx. This element is shown previously in Fig. 4.2, with the axial force superimposed as it was seen for the string in Fig. 2.2. To the previous equation of motion (4.1) for the beam must

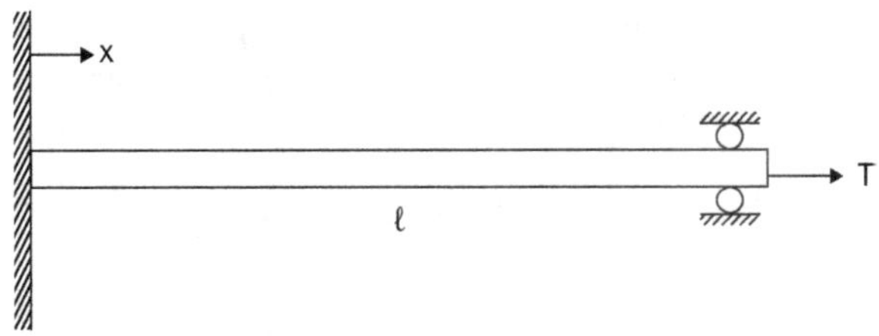

Figure 4.18 A beam subjected to an axial force T at each end.

be added the effect of the axial forces, as they were in (2.1). Making the same assumptions as for both problems previously—that is, the transverse displacements and slopes are small, and T does not depend on w or t, (4.2) becomes

$$\frac{\partial}{\partial x}\left(T\frac{\partial w}{\partial x}\right)-\frac{\partial V}{\partial x}+p=\rho A\frac{\partial^2 w}{\partial t^2} \tag{4.100}$$

Summing moments about the center of mass of the element, it is found that the terms added to (4.3) are of higher differential order. Thus, T has no effect on the moment equation (4.4). Substituting (4.4) and the beam stiffness relationship (4.5) into (4.100) results in

$$\frac{\partial^2}{\partial x^2}\left(EI\frac{\partial^2 w}{\partial x^2}\right)+\rho A\frac{\partial^2 w}{\partial t^2}=p+\frac{\partial}{\partial x}\left(T\frac{\partial w}{\partial x}\right) \tag{4.101}$$

This is the equation of motion for forced vibrations (with damping omitted).

In the special case when the flexural rigidity (EI) vanishes, (4.101) reduces to (2.6) for a string. The only inconsistency is that, in the present problem, mass per unit length is ρA, whereas in (2.6) it was ρ. The other limiting case is when $T = 0$, resulting in our previous equation (4.6) for the beam. Thus, (4.101) is a generalization of both (2.6) and (4.6). However, in the present problem, T may be negative (compressive) as well as positive (tensile), whereas for the string it may only be positive. For free, undamped vibrations, $p = 0$. The class of problems one should study first is when EI, ρA, and T all are constant. Then the equation of motion reduces to

$$EI\frac{\partial^4 w}{\partial x^4}+\rho A\frac{\partial^2 w}{\partial t^2}=T\frac{\partial^2 w}{\partial x^2} \tag{4.102}$$

which is seen to be a generalization of both (2.8) and (4.8).

The effects of axial forces on vibration frequencies may be most readily seen for beams having both ends simply supported. Assuming sinusoidal time response,

$$w(x,t) = X(x)\sin\omega t \tag{4.103}$$

(4.102) becomes

$$EIX^{IV} - TX'' - \rho A\omega^2 X = 0 \tag{4.104}$$

Because (4.104) contains only even derivatives of X, it is clear that

$$X(x) = \sin\frac{m\pi x}{\ell} \qquad (m = 1, 2, \ldots) \tag{4.105}$$

is capable of satisfying (4.104) as well as the conditions $X(0) = X''(0) = X(\lambda) = X''(\lambda) = 0$ required at simply supported boundaries. Substituting (4.105) into (4.104) yields a simple, closed-form formula for the square of the nondimensional frequency parameter:

$$\frac{\omega^2\ell^4\rho A}{EI} = m^2\pi^2\left(m^2\pi^2 + \frac{T\ell^2}{EI}\right) \tag{4.106}$$

Figure 4.19 is a plot of (4.106) which shows the variation of the first two frequencies (squared) with the tensile force. It is seen that, for a given beam (fixed values of E, I, ℓ, ρ, and A), the frequency squared increases with increasing T in a linear manner. The rate of increase (i.e., the slope) of ω^2 is four times as great for the second mode as for the first mode. For negative values of T, the frequencies are decreased. Thus, *tensile forces increase the frequencies, while compressive forces decrease them.* It is important to observe that (4.106) gives $\omega = 0$ for $m = 1$ when $T = -\pi^2 EI/\ell^2$. This is the Euler critical buckling load for a column having both ends simply supported. Physically, this means that as one increases the compressive force on a beam, the fundamental frequency is lowered. As the load approaches the buckling value, the frequency approaches zero (the period of oscillation becomes very long). This is an excellent way of determining a buckling load *experimentally,* without destroying the test specimen. That is, a curve of frequency versus axial force may be plotted and, as the frequency becomes small, the curve may be *extrapolated* to zero to determine the buckling load. However, one must be careful in approaching zero frequency, for the frequency itself (rather than its

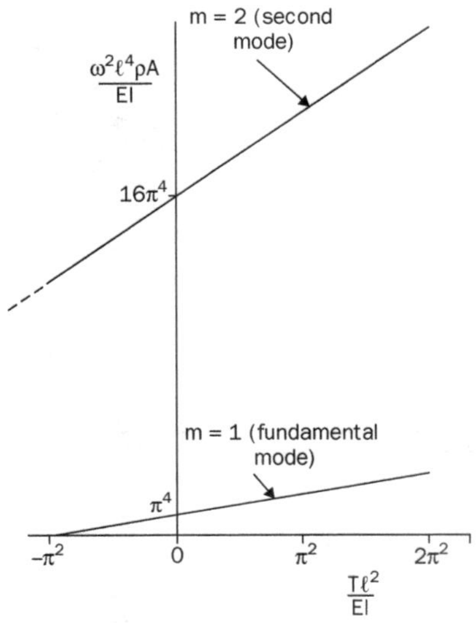

Figure 4.19 Variation of the first two frequencies (squared) of a simply supported beam with the tensile force.

square) changes rapidly as the critical load $T\ell^2/EI = -\pi^2$ is approached as a parabola (Fig. 4.20).

Figure 4.19 shows only the two lowest frequency curves. There are, of course, an infinite set of them. Correspondingly, there are an infinite set of buckling loads. However, as soon as the smallest one (in absolute value) is reached (the *critical* buckling load), it must be assumed that the beam will buckle, so that the higher frequencies have no physical meaning for larger compressive loads.

For beams having other end conditions, a more general solution to (4.104) must be derived. Rewriting (4.104) in operator form:

$$\left[D^4 - \left(\frac{T}{EI}\right)D^2 - \left(\frac{\omega^2\rho A}{EI}\right)\right]X = 0 \tag{4.107}$$

where $D \equiv d/dx$ This has an auxiliary equation:

$$m^4 - \left(\frac{T}{EI}\right)m^2 - \frac{\omega^2\rho A}{EI} = 0 \tag{4.108}$$

With roots
$$m^2 = \frac{T}{2EI} \pm \sqrt{\left(\frac{T}{2EI}\right)^2 + \frac{\rho A\omega^2}{EI}} \tag{4.109}$$

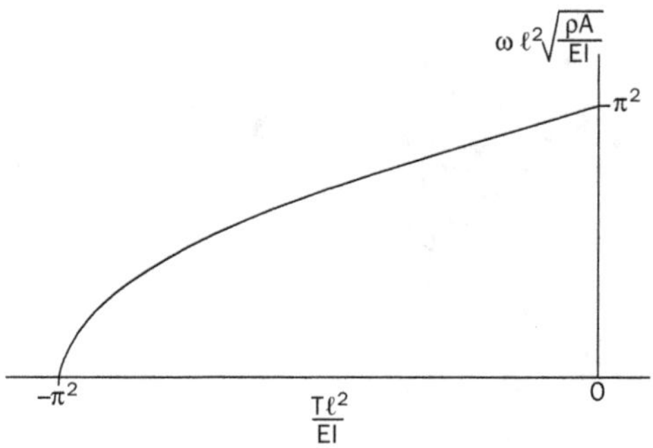

Figure 4.20 The frequency changes rapidly as the critical load parameter approaches the value ($-\pi^2$).

Thus (4.107) may be factored into

$$\left(D^2+\alpha_1^2\right)\left(D^2-\alpha_2^2\right)X=0 \tag{4.110}$$

where
$$\alpha_{1,2}^2=\mp\frac{T}{2EI}+\sqrt{\left(\frac{T}{2EI}\right)^2+\frac{\rho A\omega^2}{EI}} \tag{4.111}$$

In this form, α_1^2 and α_2^2 are both always positive for $\omega^2>0$. The solution of (4.107) is therefore

$$X(x)=C_1\sin\alpha_1 x+C_2\cos\alpha_1 x+C_3\sinh\alpha_2 x+C_4\cosh\alpha_2 x \tag{4.112}$$

with α_1 and α_2 given by (4.111), and $C_1,\ldots,C_4$ are constants of integration. In the more useful nondimensional form,

$$X(\xi)=C_1\sin\beta_1\xi+C_2\cos\beta_1\xi+C_3\sinh\beta_2\xi+C_4\cosh\beta_2\xi \tag{4.113}$$

where
$$\beta_{1,2}=\alpha_{1,2}\ell=\sqrt{\mp\frac{T\ell^2}{2EI}+\sqrt{\left(\frac{T\ell^2}{2EI}\right)^2+\frac{\rho A\omega^2\ell^4}{EI}}} \tag{4.114}$$

and $\xi=x/\ell$.

The free vibration problem is solved in the general case by applying two B.C. at each end of the beam to yield a fourth-order frequency determinant.

If T is not constant, then the free vibration equation is

$$EI\frac{\partial^4 w}{\partial x^4}+\rho A\frac{\partial^2 w}{\partial t^2}=T\frac{\partial^2 w}{\partial x^2}+\frac{\partial T}{\partial x}\frac{\partial w}{\partial x} \tag{4.115}$$

which is a differential equation with variable coefficients. If T is a simple algebraic polynomial (such as a linear function, for gravity loading, or a second-degree function, for centrifugal loading), then, exact solutions may be found [after time is taken out of the problem, using (4.103)] by the well-known method of Frobenius (cf. [5], p. 252), which yields $X(x)$ in the form of a power series. A suitable substitution of variables may also transform the differential equation into a recognizable form of Bessel's equation, whereupon $X(x)$ may be expressed in terms of tabulated Bessel functions (which are also power series, as seen in Appendix B).

However, for a general form of axial loading, $T = T(x)$, an exact solution is impossible. In such cases the Rayleigh and Ritz methods are very useful. Then the potential energy is composed of two parts—strain energy and load potential. The strain energy for a beam is as given by (4.89). The load potential was derived previously for a string, (2.130). Thus, the maximum potential energy during a cycle of free vibration is

$$PE_{\max} = \frac{1}{2}\int_0^\ell EI(X'')^2\,dx + \frac{1}{2}\int_0^\ell T(X')^2\,dx \tag{4.116}$$

and the maximum kinetic energy remains as in (4.95).

Example 4.9 A beam (or column) of length ℓ has one end clamped and the other one free, as shown in Fig. 4.21. The member is subjected to a body force per unit volume ρg, due to gravity or due to vertical acceleration of its base, causing internal compressive force.

Determine the fundamental frequency parameter $\omega\ell^2\sqrt{\rho A/EI}$ as a function of a suitable nondimensional load parameter containing g. Compare results with other known ones in special cases, where possible.

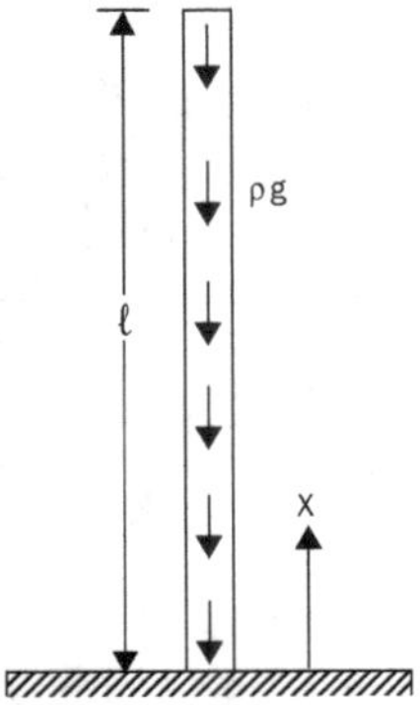

Figure 4.21 A beam (or column) subjected to a body force (e.g., gravity).

Solution

The compressive force at any cross-section is $\rho g A(\ell - x)$. Substituting $T = -\rho g A(\ell - x)$ into (4.115), and assuming sinusoidal motion (4.103), gives

$$EIX^{IV} - \rho g A(l-x)X'' - \rho g A X' - \rho A \omega^2 X = 0$$

One could solve this equation by the method of Frobenius, but much more simple solutions may be found using the Rayleigh method.

From our experience with the unloaded cantilever beam in Example 4.7, it would seem that a good choice for $X(x)$ would be the fourth-degree polynomial that satisfies all four boundary conditions:

$$X(x) = x^4 - 4\ell x^3 + 6\ell^2 x^2$$

Using this in (4.97) and (4.116), Rayleigh's Quotient (4.96) gives

$$\lambda = \frac{\omega^2 \ell^4 \rho A}{EI} = \frac{162}{13}\left(1 - \frac{1}{8}\gamma\right) = 12.46 - 1.558\gamma$$

where $\gamma = \rho g A \ell^3 / EI$ is a loading parameter which arises naturally in the solution of the equations. Increasing mass density (ρ), weight density (ρg), or axial acceleration (g, which would be the sum of downward gravitational acceleration and upward base acceleration) all cause proportional increases in γ. However, T should be regarded primarily as a measure of g, because ρ also appears in λ. The frequency formula may be compared with known results in three special cases:

1. Unloaded beam ($g = 0$, whence $\gamma = 0$). The exact result in this case, $\omega \ell^2 \sqrt{\rho A / EI} = 3.5160$ is known from Table 4.1.
2. Hanging string ($EI = 0$, with $g \rightarrow -g$ and $\gamma \rightarrow -\gamma$). The exact result $\omega \sqrt{\ell / g} = 1.2024$, is obtained from a solution using Bessel functions, as pointed out in Example 2.7.
3. Static buckling ($\omega = 0$). This problem also has an exact solution in terms of Bessel functions (cf. Timoshenko and Gere [6], p.103). The exact value of the critical loading parameter is $\gamma_{cr} = 7.837$.

For the present solution using the fourth-degree polynomial, results for the three special cases are found from the frequency formula to be:

1. $g = 0 \;\rightarrow\; \omega \ell^2 \sqrt{\rho A / EI} = \sqrt{12.46} = 3.530$ (0.4 percent error, as seen in Example 4.7
2. $EI = 0 \;\rightarrow\; \omega \sqrt{\ell / g} = \sqrt{1.558} = 1.248$ (3.8 percent error)
3. $\omega = 0 \;\rightarrow\; \gamma_{cr} = 12.46 / 1.558 = 8$ (2.1 percent error)

The fourth-degree trial function therefore yields very accurate vibration frequencies for small γ, but for tensile loading together with small EI it is less accurate.

The second trial function used in Example 4.7 was

$$X(x) = 1 - \cos\frac{\pi x}{2\ell}$$

This gave a reasonable estimate of the frequency when $g = 0$; $\omega \ell^2 \sqrt{\rho A / EI}$ = 3.664 (4.2 percent error). Using it in the present problem yields the frequency formula

$$\frac{\omega^2 \ell^4 \rho A}{EI} = 13.42 - 1.618\gamma$$

The other two special cases of the formula are

$$EI = 0 \quad \rightarrow \quad \omega\sqrt{\ell / g} = \sqrt{1.618} = 1.272 \quad \text{(5.8 percent error)}$$

$$\omega = 0 \quad \rightarrow \quad \gamma_{cr} = 13.42 / 1.618 = 8.294 \quad \text{(5.8 percent error)}$$

Therefore, this simple function appears to be worse than the preceding one in all respects.

Finally, let us use the very accurate sixth-degree polynomial of Example 4.7, which may be written as

$$X(x) = x^6 - 20\ell^3 x^3 + 45\ell^4 x^2$$

This gave an extremely accurate frequency when $g = 0$; $\omega\ell^2\sqrt{\rho A / EI} = 3.5164$ (0.01 percent error). Using it in the present problem results in

$$\frac{\omega^2 \ell^4 \rho A}{EI} = 12.36 - 1.573\gamma$$

The other two special case results are then

$$EI = 0 \quad \rightarrow \quad \omega\sqrt{\ell / g} = \sqrt{1.573} = 1.254 \quad \text{(4.3 percent error)}$$

$$\omega = 0 \quad \rightarrow \quad \gamma_{cr} = 12.36 / 1.573 = 7.858 \quad \text{(0.3 percent error)}$$

The three special cases are all upper bounds to the exact solution for each trial function used. The last function used gives the smallest values of ω for all positive γ; therefore, it is the most accurate for compressive loading. However, the first function is most accurate for large negative γ (such as the hanging string). None of the functions used can be very accurate for $EI = 0$ because all of them require $X'(0) = 0$. A better function would relax this constraint as $EI \rightarrow 0$.

4.12 Shear Deformation and Rotary Inertia

In deriving the equation of motion for a beam at the beginning of this chapter, the translational inertia of a beam element was taken into account, but the rotary inertia was neglected. More than a century ago, Rayleigh ([7], p. 293) showed that "rotatory inertia" (in the language of his day) could have a significant effect on beam frequencies if the beam is not very slender. Nearly a half-century later, Timoshenko [8,9] demonstrated that if rotary inertia effects are to be included, then the shear flexibility should be considered in addition to the bending flexibility of a beam, for it is typically at least as important as rotary inertia. In honor of these classical papers, beams having both shear deformation and rotary inertia considered in their theoretical representations are often called "Timoshenko beams." Less often, one finds the description "Rayleigh beam" for one where rotary inertia alone is added.

Let us modify our previously derived equations of motion, based on the classical, Euler–Bernoulli beam theory, to include both shear deformation and rotary inertia effects. Returning first to the moment equation (4.3), rotary inertia may be straightforwardly added to its R.H.S. giving

$$-M+(M+\frac{\partial M}{\partial s})ds - Vds = dJ_y \frac{\partial^2 \psi}{\partial t^2} \tag{4.117}$$

where, ψ is the rotation angle of the cross-section and dJ_y is the *mass* moment of inertia of the element in Fig. 4.2 about an axis through its center of mass and parallel to the y-axis. For a beam which is of homogeneous material throughout its cross-section, even though it may be nonhomogeneous along its length, that is, $\rho = \rho\,(x)$, dJ_y may be expressed in terms of the *area* moment of inertia (I_y, or simply I) by

$$dJ_y = \rho I ds \tag{4.118}$$

Substituting this into (4.117) and simplifying its results, we obtain

$$\frac{\partial M}{\partial s} - V = \rho I \frac{\partial^2 \psi}{\partial t^2} \tag{4.119}$$

in place of (4.4). For small rotations, $\partial M/\partial s$ may be replaced by $\partial M/\partial x$.

Consider next the shear deformation. This may perhaps be most easily understood for a cantilever beam subjected to an upwardly directed, static force P applied to its free end, as depicted in Fig. 4.22. The force P causes a uniform shearing force $V = -P$ along the entire length of the beam, and a uniform shear strain, $\gamma = \delta/\ell$, along the length, where δ is the tip deflection. All cross-sections move parallel to each other, without rotating. The shearing force is the integral of the shear stress over the cross-sectional area:

$$V = \int_A \tau dA = \tau_{avg} A \tag{4.120}$$

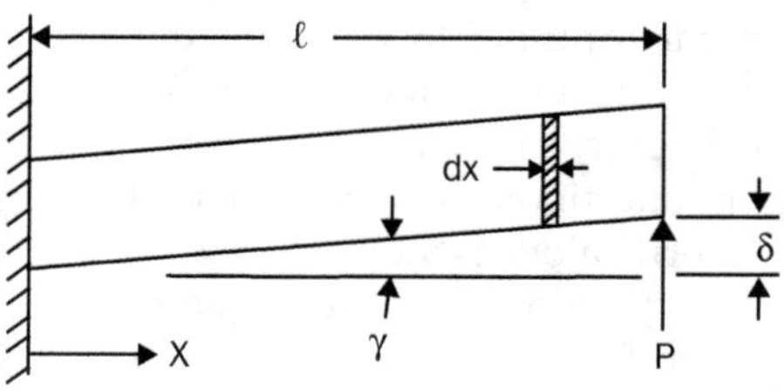

FIGURE 4.22 Shear deformation of a cantilever beam subjected to an upwardly directed, static force P applied to its free end.

where τ_{avg} is the *average* shear stress on a cross-section. If γ defines the shear strain at the neutral axis, then the stress–strain relationship is

$$\tau_{avg} = -kG\gamma \tag{4.121}$$

where k is a constant called the "shear correction factor," depending on the cross-sectional shape ($k < 1$). For a rectangular cross-section, for example, $k = 2/3 = 0.667$. For a circular cross-section, $k = 3/4 = 0.750$. The minus sign is needed in (4.121) because positive shearing force V causes negative shear strain, according to our sign convention.

The slope of the neutral axis, when both bending and shear effects are considered, is the sum of the slopes arising from both types of flexibility:

$$\frac{\partial w}{\partial x} = \psi + \gamma \tag{4.122}$$

The moment–curvature relationship (4.5) must be written in terms of bending slope only, i.e.,

$$M = EI\frac{\partial \psi}{\partial x} \tag{4.123}$$

With (4.121) and (4.122), (4.120) becomes

$$V = -kGA\left(\frac{\partial w}{\partial x} - \psi\right) \tag{4.124}$$

Substituting (4.124) into the force equation of motion (4.2), with kGA assumed to be constant,

$$kGA\left(\frac{\partial^2 w}{\partial x^2} - \frac{\partial \psi}{\partial x}\right) + p = \rho A\frac{\partial^2 w}{\partial t^2} \tag{4.125}$$

Substituting (4.123) and (4.124) into the moment equation of motion (4.119), we obtain

$$\frac{\partial}{\partial x}\left(EI\frac{\partial \psi}{\partial x}\right) + kGA\left(\frac{\partial W}{\partial x} - \psi\right) = \rho I\frac{\partial^2 \psi}{\partial t^2} \tag{4.126}$$

Equations (4.125) and (4.126) are a fourth-order set of coupled differential equations, expressed in terms of the two dependent variables $w(x,t)$ and $\psi\,(x,t)$. Alternatively, they could be expressed in terms of w and γ. For the free vibrations ($p = 0$) and uniform beams (EI = constant), (4.125) and (4.126) become

$$kGA\left(\frac{\partial^2 w}{\partial x^2} - \frac{\partial \psi}{\partial x}\right) = \rho A \frac{\partial^2 w}{\partial t^2} \tag{4.127a}$$

$$EI\frac{\partial^2 \psi}{\partial x^2} + kGA\left(\frac{\partial w}{\partial x} - \psi\right) = \rho I \frac{\partial^2 \psi}{\partial t^2} \tag{4.127b}$$

One can get a single, fourth-order equation by solving (4.127a) for $\partial\Psi/\partial\xi$ and substituting it into (4.127b):

$$EI\frac{\partial^4 w}{\partial x^4} + \rho A\frac{\partial^2 w}{\partial t^2} - \left(\rho I + \frac{EI\rho}{kG}\right)\frac{\partial^4 w}{\partial x^2 \partial t^2} + \frac{\rho^2 I}{kG}\frac{\partial^4 w}{\partial t^4} = 0 \tag{4.128}$$

In this form one sees the first two terms, representing classical beam theory, plus three additional terms. The last one is strange indeed, for it involves the fourth derivative of time. Rotary inertia is eliminated by setting terms containing ρI equal to zero (but not $EI\rho$). Shear flexibility is eliminated by letting $G \to \infty$. Thus, the last term in (4.128) is a coupling term which exists only if *both* effects are present. However, while (4.128) is interesting because it shows how the classical beam theory equation of motion is generalized to account for shear deformation and/or rotary inertia, it is not useful for solving most problems, for the boundary conditions typically involve both w and ψ.

In recent decades, considerable study of the shear coefficient k has been made by researchers. Different values of k have been proposed so that the shear deformation beam theory more accurately represents different aspects of more complete theories, including 3D theories in some cases. A summary of some of the various methods developed for selecting k may be found in the article by Cowper [10]. He also developed a procedure for determining values of k for low frequency vibrations which are quite consistent with the static, 3-dimensional theory of elasticity. Some of his results are summarized in Table 4.3. More complicated formulas for calculating k for structural sections are also available in Ref. [10].

The most useful boundary conditions are listed below.

$$\text{Clamped: } w = 0, \quad \psi = 0 \tag{4.129a}$$

$$\text{Simply supported: } \quad w = 0, \quad M = EI\frac{\partial \Psi}{\partial x} = 0 \tag{4.129b}$$

Shape		K
Rectangle	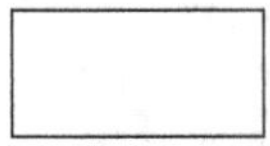	$\frac{10(1+\nu)}{12+11\nu}$
Circle	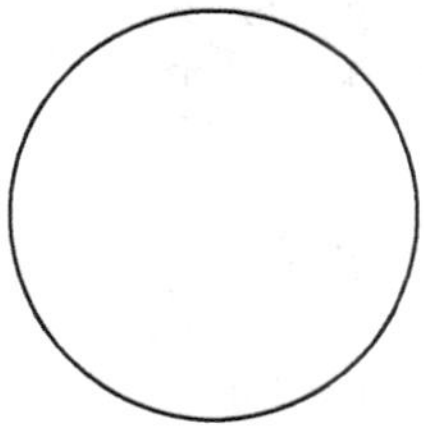	$\frac{6(1+\nu)}{7+6\nu}$
Hollow circle	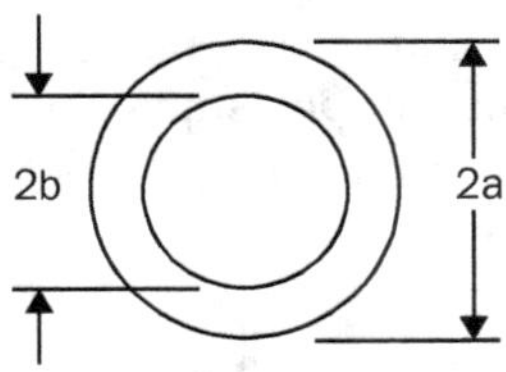 	$\frac{6(1+\nu)(1+m^2)^2}{(7+6\nu)(1+m^2)^2+(20+12\nu)m^2}$ where $m = b/a$
Ellipse	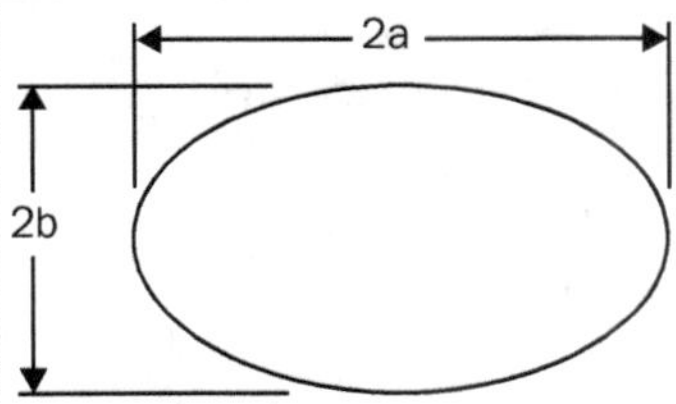 	$\frac{12(1+\nu)n^2(3n^2+1)^2}{(40+37\nu)n^4+(16+10\nu)n^2+\nu}$ where $n = a/b$
Semicircle	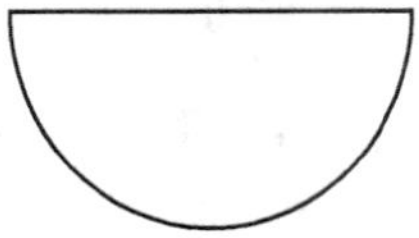	$\frac{(1+\nu)}{1.305+1.273\nu}$
Thin-walled circular tube	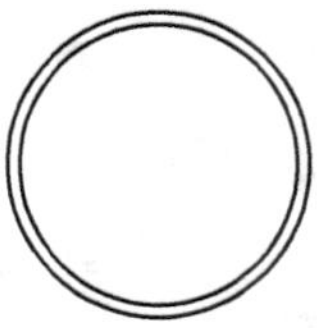	$\frac{2(1+\nu)}{4+3\nu}$
Thin-walled square tube		$\frac{20(1+\nu)}{48+39\nu}$

*v is Poisson's ratio.

Table 4.3 Shear Stress Correction Factors k according to Cowper [10]

Free: $$M = EI\frac{\partial \Psi}{\partial x} = 0, \quad V = -kGA\left(\frac{\partial w}{\partial x} - \Psi\right) = 0 \tag{4.129c}$$

As in previous beam vibration problems, the most simple solution for frequencies and mode shapes is found when both ends are simply supported. Translation and bending rotations are assumed as

$$w(x,t) = C\sin\alpha x \bullet \sin\omega t \tag{4.130a}$$

$$\psi(x,t) = D\cos\alpha x \bullet \sin\omega t \tag{4.130b}$$

where $\alpha = m\pi/\ell$, $(m = 1,2,...)$, and C and D are arbitrary constants. It is observed that (4.130) satisfy the simply supported boundary conditions exactly at $x = 0$ and $x = \ell$. Substituting (4.130) into (4.127) yields

$$\left[kGA(-C\alpha^2 + D\alpha) + \rho A\omega^2 C\right]\sin\alpha x \bullet \sin\omega t = 0$$

$$\left[-EI\alpha^2 D + kGA(C\alpha - D) + \rho I\omega^2 D\right]\cos\alpha x \bullet \sin\omega t = 0 \tag{4.131}$$

For (4.131) to be valid for all x and t,

$$\begin{bmatrix} (\rho A\omega^2 - kGA\alpha^2) & kGA\alpha \\ kGA\alpha & (\rho I\omega^2 - kGA - EI\alpha^2) \end{bmatrix} \begin{bmatrix} C \\ D \end{bmatrix} = \begin{bmatrix} 0 \\ 0 \end{bmatrix} \tag{4.132}$$

To nondimensionalize the terms in the coefficient matrix of (4.132), multiply through the first equation by ℓ^4/EI, the second by ℓ^3/EI and replace the constant C by its nondimensional form C/ℓ:

$$\begin{bmatrix} (\lambda - \epsilon(m\pi)^2) & \epsilon(m\pi) \\ \epsilon(m\pi) & \lambda\left(\frac{r}{\ell}\right)^2 - \epsilon - (m\pi)^2 \end{bmatrix} \begin{bmatrix} \frac{C}{\ell} \\ D \end{bmatrix} = \begin{bmatrix} 0 \\ 0 \end{bmatrix} \tag{4.133}$$

where

$$\lambda = \frac{\omega^2\ell^2\rho A}{EI}, \quad \epsilon = \frac{kGA\ell^2}{EI} = k\frac{G}{E}\left(\frac{\ell}{r}\right)^2 \tag{4.134}$$

are nondimensional frequency and stiffness parameters, respectively, and ℓ/r is the slenderness ratio, where r is the radius of gyration of the cross-section with respect to the neutral axis ($r=\sqrt{I/A}$). For a nontrivial solution, the determinant of the coefficient matrix in (4.133) gives

$$\lambda^2-\lambda\left[\epsilon\left(\frac{\ell}{r}\right)^2+(m\pi)^2\left(\frac{\ell}{r}\right)^2+\epsilon\,(m\pi)^2\right]+\epsilon\,(m\pi)^4\left(\frac{\ell}{r}\right)^2=0 \qquad (4.135)$$

To verify that (4.135) yields classical beam theory results as a limiting case, one may first set $G/E\rightarrow\infty$ to eliminate shear flexibility, whence by (4.134), $\epsilon\rightarrow\infty$, and (4.135) reduces to

$$\lambda\left[\left(\frac{\ell}{r}\right)^2+(m\pi)^2\right]=(m\pi)^4\left(\frac{\ell}{r}\right)^2 \qquad (4.136)$$

It can be seen that, as a beam becomes slender ($\ell/r\rightarrow\infty$), then $\lambda=(m\pi)^4$, which agrees with results of Example 4.1 and Table 4.1.

To include shear deformation effects, but not rotary inertia, one may return to (4.127b) and set its R.H.S. equal to zero. This makes the $\rho I\omega^2$ term zero in (4.132), whence (4.133) yields

$$\lambda=\frac{(m\pi)^4}{1+[(m\pi)/\epsilon^2]}$$

The roots of (4.135) are

$$\lambda_{1,2}=\frac{b}{2}\mp\frac{1}{2}\sqrt{b^2-4c} \qquad (4.137)$$

where

$$b=\epsilon\left(\frac{\ell}{r}\right)^2+(m\pi)^2\left(\frac{\ell}{r}\right)^2+\epsilon\,(m\pi)^2$$

$$c=\epsilon\,(m\pi)^4\left(\frac{\ell}{r}\right)^2 \qquad (4.138)$$

For each value of m used in (4.137), there are two values of λ, and hence two frequencies. Substituting each λ into either of the two equations in (4.133) determines the eigenvector D/C for each corresponding mode shape. The smaller λ, yielding the lower frequency, corresponds to a vibration mode which is predominantly flexural. The larger λ yields a mode which is predominantly thickness–shear. These concepts will be explained more in the numerical example below.

Example 4.10 A beam of rectangular cross-section and length ℓ has both ends simply supported. If the depth of the beam is h, determine the effect of h/ℓ on the frequencies. Show the mode shapes for the case when $h/\ell = 0.4$. Assume $k = 2/3$ and $G/E = 0.4$.

Solution

For a rectangular cross-section, $I = bh^3/12$, where b is the width of the cross-section. Since $I = Ar^2 = bhr^2$, then $r = h/\sqrt{12}$, and $\ell/r = (\ell/h)(h/r) = \sqrt{12}(\ell/h)$. Equation (4.134) yields $\epsilon = 3.2(\ell/h)^2$. Using these in (4.138), the nondimensional frequencies may be calculated by (4.137).

Table 4.4 summarizes the nondimensional fundamental frequency parameters $\sqrt{\lambda} = \omega\ell^2\sqrt{\rho A/EI}$ for a range of h/ℓ ratios. The first values, $\sqrt{\lambda_1}$, correspond to flexural modes. It is seen that for very slender beams ($h/\ell = 0.02$), the bending frequency is only slightly reduced (0.1 percent), but for moderately deep beams ($h/\ell = 0.1$) the frequency reduction is significant (1.9 percent). For very deep beams the decrease in frequency is seen to be large, although the accuracy of the Timoshenko beam theory for $h/\ell = 0.5$ is questionable.

Consider now the second, third, and fourth bending frequencies of the beam having $h/\ell = 0.1$. From Table 4.4 it may be seen that shear deformation and rotary inertia effects reduce these by 6.9, 13.4, and 20.5 percent, respectively, for the beam cross-sections at the node points of the higher modes duplicate simply supported boundary conditions. Thus, even for slender beams, these effects must be taken into consideration for the higher bending frequencies.

Table 4.4 also shows that the thickness–shear frequencies ($\sqrt{\lambda_2}$) are much greater than the fundamental frequencies of bending, even for the deep beams considered. A more meaningful frequency parameter, $\omega_2 h\sqrt{\rho/G}$, for thickness–shear frequencies is also used in Table 4.4. It involves the beam thickness, instead of its length.

Finally, results are also given in Table 4.4 for $k = 0.85$, which is the shear correction factor in Table 4.3 according to Cowper's theory ($v = 0.5E/G - 1 = 0.25$, according to the theory of elasticity). It can be seen that using the higher value of k increases the bending frequencies, but only slightly.

To determine the mode shapes for $h/\ell = 0.4$, substituting $\lambda_1 = 61.56$ and $\lambda_2 = 2384$ into either of the two homogeneous equations of (4.133) yields $D\ell/C =$ 2.162 for the mode which is predominantly flexural, and $D\ell/C = -34.676$ for that which is predominantly thickness–shear. Figure 4.23(a) shows the shape of the beam in its flexural mode, with the amplitude coefficient C set arbitrarily at 0.1ℓ This yields for the slope of the neutral axes, $\partial\omega/\partial\xi = 0.1\pi\ rad = 18.0$ deg at the end ($x = 0$). The bending rotation at this cross-section is then $\Psi = 0.217\ rad = 12.4$ deg, and the rotation of the neutral axis due to shear is $\gamma = 18.0 - 12.4 = 5.6$ deg.

Figure 4.23(b) depicts the thickness-shear mode when C is chosen to be 0.02265ℓ, so that $\psi = -45.0$ deg at the left end. Then $\partial\omega/\partial\xi = 0.0712\ rad = 4.1$ deg, and $\gamma = 49.1$ deg at the end. Thus, the rotation of the neutral axis at the ends (4.1 deg) is small compared with the shear strain. However, if there were no coupling between the two types of modes, the neutral axis would not rotate at all in the thickness–shear mode.

It is interesting and worthwhile to examine the separate contributions of rotary inertia and shear deformation (flexibility) in

k	Frequency	h/ℓ							
		0	0.02	0.05	0.1	0.2	0.3	0.4	0.5
0.67	$\omega_1 \ell^2 \sqrt{\rho A/EI}$	9.870	9.860	9.822	9.685	9.193	8.544	7.846	7.182
	% decrease	0	0.1	0.5	1.9	6.9	13.4	20.5	27.2
	$\omega_2 \ell^2 \sqrt{\rho A/EI}$	∞	15,540	2497.	633.0	166.7	79.74	48.83	34.15
	$\omega_2 h \sqrt{\rho/G}$	2.828	2.837	2.849	2.889	3.043	3.276	3.566	3.897
0.85	$\omega_1 \ell^2 \sqrt{\rho A/EI}$	9.870	9.866	9.828	9.716	9.302	8.726	8.100	7.486
	% decrease	0	0.0	0.4	1.6	5.7	11.6	17.9	24.2
	$\omega_2 \ell^2 \sqrt{\rho A/EI}$	∞	17.500	2810	710.8	185.8	87.93	53.28	36.91
	$\omega_2 h \sqrt{\rho/G}$	–	3.196	3.207	3.244	3.392	3.612	3.981	4.211

TABLE 4.4 Fundamental Frequency Parameters for Simply Supported Rectangular Beams (G/E = 0.4), Considering Shear Deformation and Rotary Inertia

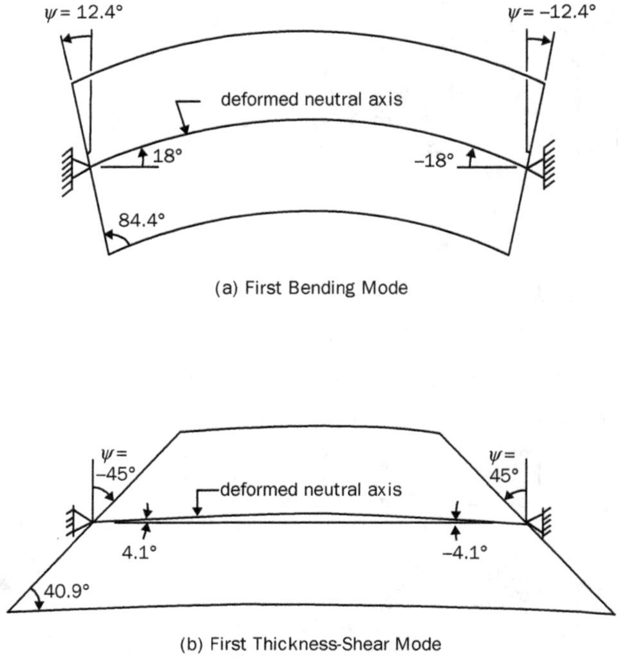

FIGURE 4.23 The shape of the beam in its flexural and thickness-shear modes ($h/\ell = 0.4$).

decreasing beam frequencies. This is done in Table 4.5, where the frequency ratio ω/ω_c is given for a moderately thick ($h/\ell = 0.1$) beam as used in Example 4.10 ($k = 2/3$, $G/E = 0.4$), ω is the frequency including an effect, and ω_c is the "classic" beam frequency (neglecting both shear deformation and rotary inertia); that is $\omega/\omega_c = \omega/(m\pi)^2$. Comparing the first two columns of results it is seen that rotary inertia has much less effect than shear deformation in decreasing frequencies. For $m = 1$ (fundamental mode) rotary inertia decreased the frequency 0.41 percent, whereas shear deformation lowers it 1.51 percent. For the higher modes it is observed that shear deformation is even more important.

In Table 4.5 one also sees that while for $m = 1$ the first thickness–shear frequency is 65 times that of the first bending frequency, for $m = 10$ it is only 2.7 times as high. Moreover, the "bending modes" for the higher m involve considerable shear deformation, as well. This is seen by returning to the first of the two equations in (4.133) and solving for the amplitude ratio.

$$\frac{D}{(C/\ell)} = \frac{\epsilon(m\pi)^2 - \lambda}{\epsilon(m\pi)}$$

	ω/ω_c			
Mode (m)	**Rotary inertia**	**Shear deformation**	**Rotary inertia + shear definition**	
1	0.9959	0.9849	0.9811	63.99
2	.9839	.9435	.9314	16.85
3	.9649	.8847	.8651	8.064
4	.9401	.8183	.7945	4.939
5	.9107	.7514	.7268	3.455
6	.8784	.6884	.6650	2.622
7	.8442	.6310	.6100	2.101
8	.8094	.5799	.5614	1.747
9	.7747	.5347	.5187	1.494
10	.7407	.4948	.4816	1.305

Table 4.5 Separate Effects of Rotary Inertia and Shear Deformation on Simply Supported, Rectangular Beam Frequencies ($h/\ell = 0.1$, $G/E = 0.4$, $k = 2/3$)

The total slope of the neutral axis is $dW/dx = C\alpha = C(m\pi/\ell)$. Therefore, the ratio of the bending slope to the total slope is

$$\frac{\psi}{dW/dx} = \frac{D}{C(m\pi/\ell)} = \frac{\epsilon(m\pi)^2 - \lambda}{\epsilon(m\pi)^2}$$

In Example 4.10, $\epsilon = 320$. For $m = 1$, with $h/\ell = 0.1$, the slope ratio determined by the last equation is 0.938, which is predominantly bending. But for $m = 5$ it is only 0.593, indicating that the shear slope (γ) is almost equal to it. For $m = 10$ the ratio is 0.287, showing dominance of the shear deformation.

In Example 4.10. it was seen that for homogeneous, isotropic beams of moderate depth ($h/\ell = 0.1$), shear deformation and rotary inertia effects cause a small decrease (2 percent) in the fundamental frequency, although frequencies for deeper beams and higher flexural modes are affected more significantly. However, for two types of beams encountered often in modern structural applications, the effects are typically much greater:

1. Sandwich beams composed of stiff outer layers and a soft core (such as foam, or a honeycomb material).
2. Laminated composite beams composed of stiff fibers (e.g., boron, graphite) embedded in a matrix material (e.g., epoxy resin).

In both cases the ratio of the transverse shear stiffness of the beam to its bending stiffness is much less than that of isotropic, homogeneous beams, causing larger shear deformation effects. Rotary inertia effects are also greater for the first type (sandwich beams).

Returning to the isotropic, homogeneous beam and considering other boundary conditions, a more general solution of equation set (4.127) is necessary. We begin by assuming, for free vibrations,

$$w(x,t) = W(x)\sin\omega t$$

$$\psi(x,t) = \Psi(x)\sin\omega t \tag{4.139}$$

Using the nondimensional coordinate $\xi = x/\ell$, and $\partial/\partial x = (1/\ell)/\partial\xi$ etc., (4.127) then become

$$kGA(W'' - \ell\Psi') + \rho A\omega^2\ell^2 W = 0 \tag{4.140a}$$

$$EI\Psi'' + kGA\ell(W' - \ell\Psi) + \rho I\omega^2\ell^2\Psi = 0 \tag{4.140b}$$

Assume solutions to (4.140) as

$$W = Ce^{s\xi}\ , \quad \Psi = De^{s\xi} \tag{4.141}$$

where C, D, and s are constants. Inserting (4.141) into (4.140) results in

$$\begin{bmatrix} (kGAs^2 + \rho A\omega^2\ell^2) & -(kGA\ell)s \\ (kGA\ell)s & (EIs^2 - kGA\ell^2 + \rho I\omega^2\ell^2) \end{bmatrix} \begin{bmatrix} C \\ D \end{bmatrix} = \begin{bmatrix} 0 \\ 0 \end{bmatrix} \tag{4.142}$$

Expanding the determinant of (4.142), we obtain

$$\begin{aligned}(kGAEI)s^4 + (EI\rho A\omega^2\ell^2 + kGA\rho I\omega^2\ell^2)s^2 \\ -(kG\rho A^2\omega^2\ell^4 - \rho^2 AI\omega^4\ell^4) = 0\end{aligned} \tag{4.143}$$

Using the nondimensional parameters in (4.134), then (4.143) may be written as

$$s^4 + \lambda\left[\frac{1}{\epsilon} + \left(\frac{r}{\ell}\right)^2\right]s^2 - \lambda\left[1 - \frac{\lambda}{\epsilon}\left(\frac{r}{\ell}\right)^2\right] = 0 \tag{4.144}$$

This has roots

$$s_{1,2}^2 = -\frac{b}{2} \mp \frac{1}{2}\sqrt{b^2 + 4c} \tag{4.145}$$

where

$$b = \lambda\left[\frac{1}{\epsilon} + \left(\frac{r}{\ell}\right)^2\right], \quad c = \lambda\left[1 - \frac{\lambda}{\epsilon}\left(\frac{r}{\ell}\right)^2\right] \tag{4.146}$$

Clearly $b^2 + 4c = \lambda^2[(1/\epsilon) - (r/\ell)^2]^2 + 4\lambda$ is always positive, therefore the values of s_1^2 and s_2^2 must be real numbers, and cannot be complex. However, s_2^2 can be either positive, or negative, or zero which requires considering three cases.

Using (4.134) and (4.146), the three cases may be summarized as follows:

$$\text{Case I.} \quad s_2^2 > 0,\ c > 0,\ \lambda < k\frac{G}{E}\left(\frac{\ell}{r}\right)^4 \tag{4.147}$$

This case will *typically* occur for the lowest frequencies of the beam. The solutions to (4.140) then are

$$W = C_1 \sin\gamma_1\xi + C_2\cos\gamma_1\xi + C_3\sinh\gamma_2\xi + C_4\cosh\gamma_2\xi \tag{4.148a}$$

$$\ell\Psi = -D_1\cos\gamma_1\xi + D_2\sin\gamma_1\xi + D_3\cosh\gamma_2\xi + D_4\sinh\gamma_2\xi \tag{4.148b}$$

where $\gamma_1{}^2 = -s_1{}^2 = \frac{1}{2}(b + \sqrt{b^2+4c})$

$$\gamma_2{}^2 = s_2{}^2 = \frac{1}{2}(-b + \sqrt{b^2+4c}) \tag{4.149}$$

Equations (4.148) contain eight constants of integrations ($C_1, \ldots, D_4$). However, the system of differential equations (4.140) is only of fourth order. One may obtain the ratios D_i/C_i from either (4.140a) or (4.140b), both of which will yield the same results. Using (4.140a),

$$\frac{D_i}{C_i} = \begin{cases} \dfrac{\rho A\omega^2\ell^2 - kGA\gamma_1^2}{kGA\gamma_1} = \dfrac{\lambda - \epsilon\gamma_1^2}{\epsilon\gamma_1} & (i = 1,2) \\[2ex] \dfrac{\rho A\omega^2\ell^2 - kGA\gamma_2^2}{kGA\gamma_2} = \dfrac{\lambda + \epsilon\gamma_2^2}{\epsilon\gamma_2} & (i = 3,4) \end{cases} \tag{4.150}$$

$$\text{Case II.} \qquad s_2^2 < 0, \quad c < 0, \quad \lambda > k\frac{G}{E}\left(\frac{\ell}{r}\right)^4 \tag{4.151}$$

This case will *always* occur for the highest frequencies of the beam, and especially for the thickness–shear frequencies. For this case, define $\gamma_3^2 = -s_2^2$, so that $\gamma_3^2 > 0$. Then

$$\gamma_3^2 = \frac{b}{2} - \frac{1}{2}\sqrt{b^2+4c} \tag{4.152}$$

The solutions to (4.140) then are:

$$W = C_1\sin\gamma_1\xi + C_2\cos\gamma_1\xi + C_3\sin\gamma_3\xi + C_4\cos\gamma_3\xi \tag{4.153a}$$

$$\ell\Psi = -D_1 \cos\gamma_1\xi + D_2 \sin\gamma_1\xi - D_3 \cos\gamma_3\xi + D_4 \sin\gamma_3\xi \qquad (4.153b)$$

where from (4.140a)

$$\frac{D_i}{C_i} = \begin{cases} \dfrac{\rho A\omega^2\ell^2 - kGA\gamma_1^2}{kGA\gamma_1} = \dfrac{\lambda - \varepsilon\gamma_1^2}{\varepsilon\gamma_1} & (i = 1,2) \\ \dfrac{\rho A\omega^2\ell^2 - kGA\gamma_3^2}{kGA\gamma_3} = \dfrac{\lambda - \varepsilon\gamma_3^2}{\varepsilon\gamma_3} & (i = 3,4) \end{cases} \qquad (4.154)$$

Case III. $\quad s_2^2 = 0\,,\ c = 0,\ \lambda = k\dfrac{G}{E}\left(\dfrac{\ell}{r}\right)^4 = \epsilon\left(\dfrac{\ell}{r}\right)^2$

For this very special case, (4.145) yields the roots $s_{1,2}^2 = -b$, 0 whence the solution is

$$W = C_1 \sin\gamma_1\xi + C_2 \cos\gamma_1\xi + C_3\xi + C_4 \qquad (4.155a)$$

$$\ell\Psi = -D_1 \cos\gamma_1\xi + D_2 \sin\gamma_1\xi + D_3\xi + D_4 \qquad (4.155b)$$

where $\gamma_1 = \sqrt{b}$, and with D_1/C_1 and D_2/C_2 being given by (4.150a). Substituting the last two terms of (4.155a) and (4.155b) into (4.140a), one obtains

$$-\epsilon D_3 + \lambda(C_3\xi + C_4) = 0$$

whence, $C_3 = 0$, and

$$\frac{D_3}{C_4} = \frac{\lambda}{\epsilon} = \left(\frac{\ell}{r}\right)^2 \qquad (4.156)$$

Similarly, (4.140b) yields

$$(D_3\xi + D_4)\left[\lambda\left(\frac{r}{\ell}\right)^2 - \epsilon\right] = 0$$

which is identically satisfied because the bracketed quantity is zero in this case.

To solve a particular problem, two boundary conditions are applied at each end of the beam, as in (4.129), yielding a fourth-order frequency determinant. The roots of this determinant are the nondimensional frequency parameters. Both solution sets (4.148) and (4.153) for W and Ψ must be considered. As indicated above, for a typical situation one may expect (4.148) to apply for the lowest frequencies. However, one must verify for each value of λ whether it is less than or greater than $k(G/E)(\lambda/r)^4$. Corresponding mode shapes are determined in the usual manner; that is, the amplitude ratios C_2/C_1, C_3/C_1, and C_4/C_1 are evaluated from any three of the four equations generated from the boundary conditions.

In the case of symmetry (for example, a beam having identical added masses at each end) the problem is, of course, simplified by considering symmetric and antisymmetric modes. These modes are uncoupled from each other, resulting in two second-order frequency determinants.

Frequency equations and eigenfunctions were derived by Huang [11] for the six most common types of end conditions (C–C, C–SS, C–F, SS–SS, SS–F, F–F). Extensive tables and graphs of frequency correction factors to account for shear deformation and rotary inertia for all six sets of end conditions were also prepared by Huang [12].

The procedure described above for evaluating frequencies and mode shapes is a somewhat lengthy and complicated one, although it is exact. In effect, one must solve *two* eigenvalue problems. The first one determines proper values of the exponents (s) and the coefficient ratios D_i/C_i so that the homogeneous differential equations are satisfied. The second one determines proper values of the frequency parameters (λ) so that the homogeneous boundary conditions are satisfied.

For considering the effects of shear deformation and rotary inertia in more complicated problems, the Rayleigh and Ritz methods are very useful. To be able to use them, we must generalize the potential and kinetic energy functionals of Sec. 4.9. To the strain energy of bending in (4.84) must be added the strain energy due to shear:

$$PE_{\text{shear}} = \frac{1}{2}\int_{\text{vol}} \tau\gamma\, dA\, dx \tag{4.157}$$

Using (4.121) this becomes

$$PE_{\text{shear}} = -\frac{k}{2}\int_0^\ell GA\gamma^2 dx \tag{4.158}$$

The potential energy due to bending results from the bending curvature, $\partial\psi/\partial x$:

$$PE_{\text{bending}} = \frac{1}{2}\int_0^\ell EI\left(\frac{\partial\psi}{\partial x}\right)^2 dx \tag{4.159}$$

Adding this to (4.158) and using (4.122) to eliminate the shear strain in favor of the bending rotation gives the total potential energy as:

$$PE = \frac{1}{2}\int_0^\ell EI\left(\frac{\partial \psi}{\partial x}\right)^2 dx - \frac{k}{2}\int_0^\ell GA\left(\frac{\partial w}{\partial x} - \psi\right)^2 dx \tag{4.160}$$

It is observed that the potential energy is decreased by the presence of the second term, which arises from the shear strain. This has the effect of decreasing the frequencies, as expected.

To the kinetic energy due to translation (4.91) must be added that due to rotation:

$$KE_{\text{rotation}} = \frac{1}{2}\int_0^\ell dJ_y\left(\frac{\partial \psi}{\partial t}\right)^2 dx \tag{4.161}$$

Using (4.118), this becomes

$$KE_{\text{rotation}} = \frac{1}{2}\int_0^\ell \rho I\left(\frac{\partial \psi}{\partial t}\right)^2 dx \tag{4.162}$$

for small rotations. The total kinetic energy is therefore:

$$KE = \frac{1}{2}\int_0^\ell \rho A\left(\frac{\partial w}{\partial t}\right)^2 dx + \frac{1}{2}\int_0^\ell \rho I\left(\frac{\partial \psi}{\partial t}\right)^2 dx \tag{4.163}$$

The effect of the added rotary inertia is to increase the kinetic energy, which decreases the frequencies, as expected.

The Rayleigh method is applied by means of Rayleigh's Quotient (4.96), whereas the Ritz method involves the minimizing equations (4.99). Assuming sinusoidal motion in time the needed functionals become:

$$PE_{max} = \frac{1}{2}\int_0^\ell EI\left(\Psi'\right)^2 dx - \frac{k}{2}\int_0^\ell GA\left(X' - \Psi\right)^2 dx$$

$$KE_{max} = \frac{1}{2}\int_0^\ell \rho AX^2 dx + \frac{1}{2}\int_0^\ell \rho I\Psi^2 dx \tag{4.164}$$

Whereas a reasonable mode shape for Ψ may be assumed, its *amplitude* relative to X is very difficult to estimate, so for practical purposes the Ritz method is better to employ.

The forced vibration response of a Timoshenko beam may be treated by either the eigenfunction superposition or closed-form solution methods described in Sec. 4.8. However, if viscous damping is a factor, it may be desirable to consider damping due to *rotations* of beam cross-sections (boundary shear drag), as well as due to translation [13].

4.13 Curved Beams—Equations of Motion

Curved beams are beams that exhibit some curvature. This section will be limited to curved beams that lie in one plane, and whose radius of curvature is constant. As will be seen, such curvature introduces significant complexity to the beam problem. It raises the order of the differential equations for thin beams from 4 to 6, couples in-plane and transverse motions and complicates boundary conditions.

A curved beam is characterized by its middle surface, which is defined by the polar coordinate α (Fig. 4.24), where

$$\alpha = R\theta \tag{4.165}$$

The constant R identifies the radius of curvature of the middle surface of the beam. The equations derived earlier for straight beams can be generalized to those for curved. The equations presented here are for curved beams subjected to in-plane loading and/or vibrating in the α–z plane.

Middle surface strain (ϵ_0) and curvature change (κ) are

$$\epsilon_0 = \frac{\partial u}{\partial \alpha} + \frac{w}{R}, \qquad \kappa = -\frac{\partial^2 w}{\partial \alpha^2} + \frac{1}{R}\frac{\partial u}{\partial \alpha} \tag{4.166}$$

where u and w are displacements of the beam's middle surface in the α and z directions; respectively.

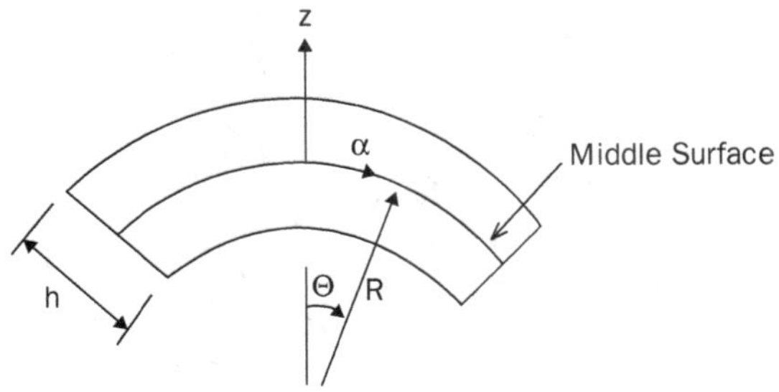

FIGURE 4.24 Parameters used for curved beams.

The strain at an arbitrary point can be found from

$$\epsilon = \epsilon_0 + z \tag{4.167}$$

The force (N) and moment (M) resultants are the integrals of the axial stress over the beam thickness (h):

$$[N,M] = b\int_{-h/2}^{h/2} [1,z]\ \sigma\, dz \tag{4.168}$$

$N = EA\, \epsilon_0$; and $M = (Ebh^3/12)\kappa$

where E is the modulus of elasticity of the material, b is the width of the cross-section (rectangular cross-sections), $A = bh$ is the cross-sectional area. All these parameters are assumed to be constants.

The above equations are valid for cylindrical bending of beams. The equations of motion may be obtained by taking a differential element of a beam having thickness h and midsurface length $d\alpha$ (Fig. 4.25) and summing the external and internal forces in the α and z direction, and the moments causing bending. The resulting equations of motion are

$$\frac{\partial N}{\partial \alpha} + \frac{V}{R} + p_\alpha = \rho A \frac{\partial^2 u}{\partial t^2}$$

$$-\frac{N}{R} + \frac{\partial V}{\partial \alpha} + p_n = \rho A \frac{\partial^2 w}{\partial t^2}$$

$$\frac{\partial M}{\partial \alpha} - V = 0 \tag{4.169}$$

where V is the shear force resultant, and p_α and p_n are external force (or body force) components tangent and normal to the beam midsurface, per unit length, respectively. It is noted that rotary inertia is neglected in the third equation. Solving the third equation in (4.169)

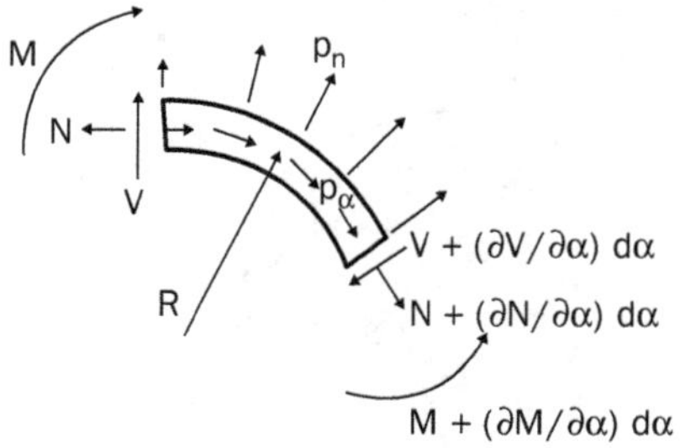

FIGURE 4.25 A differential curved beam element.

for V, and substituting this into the second equation, the equations of motion become

$$\frac{\partial N}{\partial \alpha}+\frac{1}{R}\frac{\partial M}{\partial \alpha}+p_\alpha=\rho A\frac{\partial^2 u}{\partial t^2}$$
$$-\frac{N}{R}+\frac{\partial^2 M}{\partial \alpha^2}+p_n=\rho A\frac{\partial^2 w}{\partial t^2} \tag{4.170}$$

Multiplying the last of (4.170) through by –1 and substituting (4.166) and (4.168) into (4.170), the equations of motion are expressed in terms of the midsurface displacements in matrix form as

$$\begin{bmatrix} L_{11} & L_{12} \\ L_{21} & L_{22} \end{bmatrix}\begin{bmatrix} u \\ w \end{bmatrix}+\begin{bmatrix} -\rho A & 0 \\ 0 & \rho A \end{bmatrix}\frac{\partial^2}{\partial t^2}\begin{bmatrix} u \\ w \end{bmatrix}=\begin{bmatrix} -p_\alpha \\ p_n \end{bmatrix}, \tag{4.171}$$

where

$$L_{11}=EA\frac{\partial^2}{\partial \alpha^2}+\frac{D}{R^2}\frac{\partial^2}{\partial \alpha^2},\quad L_{22}=D\frac{\partial^4}{\partial \alpha^4}+\frac{EA}{R^2},\quad \text{and}$$
$$L_{12}=L_{21}=\frac{EA}{R}\frac{\partial}{\partial \alpha}-\frac{D}{R}\frac{\partial^3}{\partial \alpha^3}$$

where $D = Ebh^3/12$. Note that there exists coupling between in-plane and transverse displacements. Only if the beam is straight ($R = \infty$) that the in-plane and transverse displacements are decoupled.

The strain energy stored in a beam during elastic deformation in terms of the displacements is

$$U=\frac{1}{2}\int_L\left(EA\left(\frac{\partial u}{\partial \alpha}+\frac{w}{R}\right)^2+D\left(-\frac{\partial^2 w}{\partial \alpha^2}+\frac{1}{R}\frac{\partial u}{\partial \alpha}\right)^2\right)d\alpha \tag{4.172}$$

Using the distributed external force components p_α in the tangential (polar) direction, and p_n in the normal direction, the work done by the external forces as the beam displaces is

$$W=\int_L(p_\alpha u_0+p_n w_0)\,d\alpha \tag{4.173}$$

The kinetic energy for the beam is

$$T_k=\frac{1}{2}\rho A\int\left[\left(\frac{\partial u}{\partial t}\right)^2+\left(\frac{\partial w}{\partial t}\right)^2\right]d\alpha \tag{4.174}$$

Boundary conditions may be obtained from a variational formulation. On each boundary, one must specify three conditions:

$$
\begin{aligned}
u &= 0 \qquad \text{or} \qquad N + \frac{M}{R} = 0 \\
w &= 0 \qquad \text{or} \qquad Q = 0 \\
\frac{\partial w}{\partial \alpha} &= 0 \qquad \text{or} \qquad M = 0
\end{aligned} \tag{4.175}
$$

Note that there is an additional term (M/R) in the first boundary condition. This term does not exist for straight or slightly curved beams [14]. Boundaries may also be elastically constrained, with the constraints being represented as translational and rotational springs at the beam edges.

4.14 Curved Beams—Vibration Analysis

The simple support boundary conditions may take two forms for curved beams at a boundary, namely

$$
\begin{aligned}
&S1: \quad w_0 = N = M = 0 \quad \text{at} \qquad a = -\frac{\ell}{2}, \frac{\ell}{2} \\
&S2: \quad w_0 = u_0 = M = 0 \quad \text{at} \qquad a = -\frac{\ell}{2}, \frac{\ell}{2}
\end{aligned} \tag{4.176a}
$$

Similarly, the free boundary conditions may take the following forms:

$$
\begin{aligned}
&F1: \quad Q = N = M = 0 \qquad \text{(Completely Free)} \\
&F2: \quad Q = u_0 = M = 0
\end{aligned} \tag{4.176b}
$$

and the clamped boundary conditions can take the following forms:

$$
\begin{aligned}
&C1: \quad w_0 = N = \frac{\partial w}{\partial \alpha} = 0 \\
&C2: \quad w_0 = u_0 = \frac{\partial w}{\partial \alpha} = 0 \qquad \text{(Completely Clamped)}
\end{aligned} \tag{4.176c}
$$

Simply supported curved beams (with S1 boundary condition) will be studied in this section. For such beams, straightforward exact solution can be found. The above S1 boundary conditions are exactly satisfied at both ends ($\alpha = -\ell/2$ and $\alpha = \ell/2$) by choosing:

$$[u,w] = \sum_{m=1}^{M} [A_m \sin(\alpha_m \alpha),\ C_m \cos(\alpha_m \alpha)] \sin(\omega t), \tag{4.177}$$

where $\alpha_m = m\pi/\ell$, m is an odd number and A_m and C_m are arbitrary constants.

The external forces (important in forced vibration analysis) may be expanded in a Fourier series in α:

$$[p_\alpha, p_z] = \sum_{m=1}^{m} [p_{\alpha m} \sin(\alpha_m \alpha), p_{zm} \cos(\alpha_m \alpha)] \sin(\omega t), \tag{4.178}$$

where

$$p_{\alpha m} = \frac{2}{\ell} \int p_\alpha \sin(\alpha_m \alpha) d\alpha \quad \text{and} \quad p_{zm} = \frac{2}{\ell} \int p_z \cos(\alpha_m \alpha) d\alpha$$

Substituting (4.177) and (4.178) into the equations of motion written in terms of displacement (4.171) yields

$$\begin{bmatrix} C_{11} & C_{12} \\ C_{21} & C_{22} \end{bmatrix} \begin{bmatrix} A_m \\ C_m \end{bmatrix} + \omega^2 \begin{bmatrix} \rho A & 0 \\ 0 & -\rho A \end{bmatrix} \begin{bmatrix} A_m \\ C_m \end{bmatrix} + \begin{bmatrix} p_{\alpha m} \\ -p_{zm} \end{bmatrix} = \begin{bmatrix} 0 \\ 0 \end{bmatrix} \tag{4.179}$$

where:

$$C_{11} = -\alpha_m^{\ 2} \left[EA + D/R^2 \right]$$

$$C_{22} = D\alpha_m^{\ 4} + \left(EA/R^2 \right)$$

$$C_{21} = -C_{12} = \alpha_m (EA/R) + \alpha_m^{\ 3} (D/R)$$

These equations are valid for problems of forced vibrations. The static problem results when the frequency is set to zero. The free vibration problem arises by setting the pressure terms equal to zero.

Table 4.6 shows nondimensional frequency parameters for a very slender ($\ell/h = 100$) curved beam with S1 boundary conditions. It is observed that since the in-plane displacement is allowed to be free with this boundary condition, initial curvature has limited effect on these beams when they exhibit shallow curvature. Interestingly, increasing curvature does decrease the bending mode frequencies

R/ℓ	Bending modes $m = 1$	2	3	4	Longitudinal $m = 1$
	$\omega \ell^2 \sqrt{\rho A/EI}$				$\omega \ell \sqrt{\rho/E}$
–	9.8696	39.478	88.826	157.91	3.1416
10	9.8549	39.475	88.876	158.11	3.1432
5	9.8102	39.431	88.830	158.06	3.1480
3.33	9.7364	39.356	88.757	157.99	3.1559
2	9.4993	39.116	88.516	157.75	3.1811
1.25	8.9473	38.538	87.935	157.16	3.2419
1.0	8.4516	38.000	87.335	156.42	3.2970

TABLE 4.6 Exact Frequency Parameters $\omega \ell^2 \sqrt{\rho A/EI}$ for Curved Thin Beams with S1 Boundary Conditions, $\ell/h = 100$.

for these beams with S1 boundaries. For $m = 1$, longitudinal mode frequencies are also listed in terms of the frequency parameter used in Chap. 3 for comparison purposes.

Exact solutions can be found for other boundary condition, but will not be sought here. Instead, approximate solutions using the Ritz method will be obtained.

The deformation is assumed to be sinusoidal with time. Displacements are thus assumed as

$$u(\alpha,t) = U(\alpha)\sin \omega t,$$
$$w(\alpha,t) = W(\alpha)\sin \omega t \tag{4.180}$$

Various functions can be used for U and W. Among the most commonly used functions are trigonometric functions, beam functions and algebraic polynomials. Beam functions are actually the free vibration mode shapes obtained in the analysis of straight beams, seen earlier in this chapter.

Algebraic polynomial trial functions are used in the analysis, as were used earlier in this chapter for straight beams, because they form a mathematically complete set of functions, which guarantees convergence to the exact solution as the number of terms taken increases. They are also relatively simple to use in the algebraic manipulation and computer programming subsequently required and can be differentiated and integrated exactly in the energy functionals needed. Using algebraic polynomials, one can solve for all possible combinations of boundary conditions for these beams.

Thus the displacement functions U and W are written in terms of the nondimensional coordinate ξ as

$$U(\xi)=(1-\xi)^{j_o}\sum_{i=i_0}^{I}C_i\xi^i$$

$$W(\xi)=(1-\xi)^{m_o}\sum_{\ell=\ell_o}^{L}D_\ell\xi^\ell \tag{4.181}$$

where $\xi=\alpha/\ell$, and C_i and D_i are arbitrary coefficients to be determined subsequently.

The Ritz method requires the satisfaction of the geometric (or "forced") boundary conditions only. Thus by suitable selection of i_0 and ℓ_0 one can solve for any boundary conditions at $\xi = 0$, and by suitable selection of j_0 and m_0, one can solve for any boundary conditions at the other end (i.e., $\xi = 1$). For example, if one chooses the cantilever boundary conditions, where the edge at $\xi = 0$ is clamped and that at $\xi = 1$ is free, then one should satisfy $u = w = dw/d\alpha = 0$ at $\xi = 0$. This is done by choosing i_0 as 1 to satisfy $u_0 = 0$ and $\ell_0 = 2$ to satisfy $w = dw/d\alpha = 0$. One needs to satisfy no geometric boundary conditions at $\xi = 1$, which results in the selection of j_0 and m_0 as zeros. Table 4.7 shows a convergence study done with the Ritz method with the first column describing the degrees of freedom (d.o.f.) used. A reasonably close agreement for the first three frequencies is observed with the exact solution (Table 4.6) when a 2×8 solution is obtained.

Table 4.8 shows results obtained for the S2 boundary conditions. Note here that the frequencies are much higher than those for the S1 case, due to the axial end constraints and a slight increase in curvature results in significant change in the frequency parameters. The table also shows results for the completely clamped case (i.e. C2 boundary condition), where a similar observation is made.

d.o.f.	$m = 1$	2	3	4
2×6	9.500	40.98	93.00	215.8
2×7	9.500	39.12	92.91	167.8
2×8	9.500	39.12	88.53	164.8
2×9	9.500	39.12	88.52	157.9
Exact	9.499	39.12	88.52	157.8

TABLE 4.7 Convergence of the Frequency Parameters for Simply Supported S1 Curved Thin Beam, $\ell/R = 0.5$, $\ell/h = 100$

	S2 boundary conditions				C2 boundary conditions			
R/ℓ	$m = 1$	2	3	4	$m = 1$	2	3	4
–	9.870	39.48	88.83	157.9	22.37	61.67	120.9	199.9
100	10.35	39.48	88.83	157.9	22.56	61.67	120.9	199.9
20	18.43	39.47	88.98	157.9	26.59	61.67	121.1	199.9
10	32.47	39.45	89.50	157.9	36.31	61.68	121.6	199.8
5	39.38	60.15	93.10	157.8	60.16	61.57	124.2	199.7
2	38.86	82.28	157.3	165.3	61.03	103.0	168.4	199.1
1	37.09	82.18	155.5	237.1	59.16	107.9	197.0	268.5

TABLE 4.8 Curvature Effects on the Frequency Parameters for Curved Beam, $\ell/h = 100$

Additional complications for curved beams that can be considered are shear deformation and rotary inertia. In addition, beams made of composite materials can be treated [14].

References

1. D. Young and R. P. Felgar, Jr., "Tables of characteristic functions representing normal modes of vibration of a beam," University of Texas Publication No. 4913, 1949, 31 pp.
2. J. C. MacBain and J. Genin, "Natural frequencies of a beam considering support characteristics," *J. Sound Vib.* 27 (1973): 197–206.
3. A. W. Leissa and M. Sonalla, "Nonperiodic vibration of a cantilever beam subjected to various initial conditions," *J. Sound Vib.* 150 (1991): 83–99.
4. A. W. Leissa, "Closed form exact solutions for the steady state vibrations of continuous systems subjected to distributed exciting forces," *J. Sound Vib.* 134 (1989): 435–53.
5. C. R. Wylie, Jr., *Advanced Engineering Mathematics*, McGraw-Hill Book Co., 1951, 640 pp.
6. S. P. Timoshenko and J.M. Gere, *Theory of Elastic Stability*, McGraw-Hill Book Co., 1961, 541 pp.
7. Lord Rayleigh, *Theory of Sound*, vol. I, 1st ed., MacMillan Co., 1877, 480 pp., reprinted by Dover Publications, 1945.
8. S. P. Timoshenko, "On the correction for shear of the differential equation for transverse vibrations of prismatic bars," *Phil. Mag.*, Ser. 6, 41 (1921): 744–46.
9. S. P. Timoshenko, "On the transverse vibrations of bars of uniform crosssections," *Phil. Mag.*, Ser. 6, 43 (1922): 125–31.
10. G. R. Cowper, "The shear coefficient in Timoshenko's beam theory," *J. Appl. Mech.* 1966: 335–40.
11. T. C. Huang, "The effect of rotary inertia and of shear deformation on the frequency and normal mode equations of uniform beams with simple end conditions," *J. Appl. Mech.*, 1961: 579–84.
12. T. C. Huang, "Eigenvalues and modifying quotients of vibration of beams," Report No. 25, Engr. Exper. Station, Univ. of Wisconsin, 1964, 65 pp.
13. A. W. Leissa and M. O. Hwee, "Forced vibrations of Timoshenko beams with viscous damping," *Developments in Mechanics* (Proceedings of the Ninth Midwestern Mechanics Conference, Madison, Wisconsin), 1965: 71–81.
14. M. S. Qatu, *Vibrations of Laminated Shells and Plates*, Elsevier, 2004, 409 pp.

Problems

1 A. Derive the generalization of (4.8) which includes transverse gravitational effects.

B. Solve the derived equation, apply the boundary conditions for a beam having both ends simply supported, and determine the free vibration mode shapes with respect to the *static equilibrium position.*

C. Generalize the conclusion reached in Part B to a beam of variable cross-section having arbitrary boundary conditions, and carefully explain how you reach this conclusion.

2 A. Derive the characteristic equation for Part A of Example 4.3 more easily by choosing the coordinate origin at the free end of the beam.

B. Plot the mode shape for the beam having the smaller root ($K = 0.2179$) and the same frequency ($\beta^2 = 1.7580$). Compare this mode shape with those of the F–F beam and with that plotted for the larger root ($K = 4.8719$) in Example 4.3.

C. Prove whether or not the higher frequencies and mode shapes for all finite K approach those of the free–free beam as the mode number is increased.

3 A beam has both ends simply supported, and has rotational springs of stiffness K also at each end (Fig. 4.26).

A. Derive the characteristic equation(s) for the free vibration frequencies (Hint: Consider symmetry and antisymmetry of the modes to reduce the work.)

B. Evaluate the first two nondimensional frequencies for $K\ell/EI = 0.1, 1, 10,$ and 100. Make sketches showing how they change as $K\ell/EI$ varies from 0 to ∞ (consider a logarithmic scale for $K\ell/EI$).

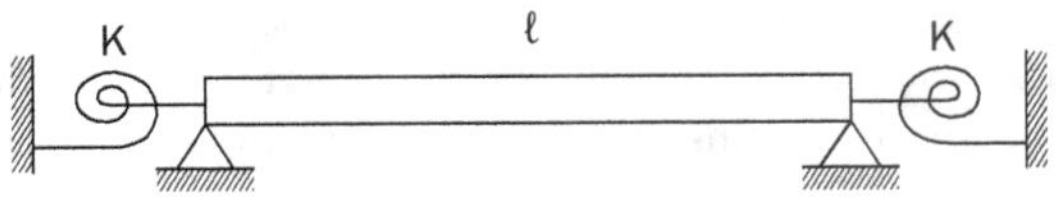

FIGURE 4.26 Problem 3.

4 A cantilever beam of length ℓ has a mass M and a translational spring of stiffness k attached to its free end (Fig. 4.27). Prove whether or not the free vibration eigenfunctions are orthogonal. Use any method you wish.

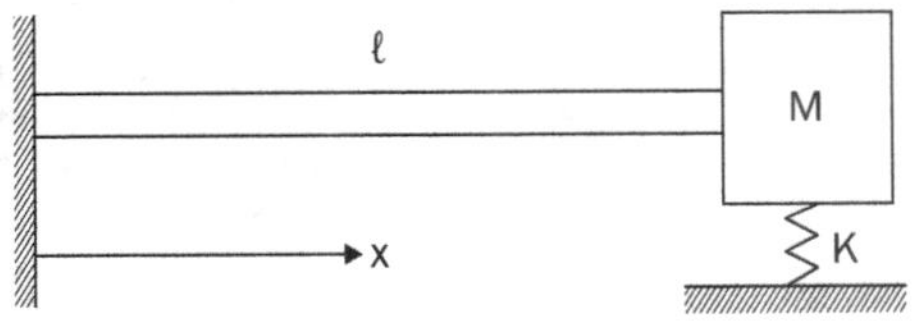

FIGURE 4.27 Problem 4.

5 A beam has both ends clamped. A uniformly distributed static loading of q_0 is applied. Suddenly the load is removed. Determine the subsequent $w(x,t)$.

6 In Sec. 2.8 it was proved that, for the underdamped free vibrations of a string immersed in a viscous fluid, the amplitudes of all modes decayed at the same rate. Prove whether or not this is true for beams vibrating freely in a viscous fluid (for arbitrary boundary conditions). If it is not generally true, are there any boundary conditions for which it is true?

7 A. Verify (4.64).

B. Expand the determinant of (4.64) to arrive at a frequency equation. Use identities to simplify it. Show that for $r = 1$ the frequency equation factors into (4.32a) and (4.35a).

C. Let $r = 1.5$. Find the first two frequencies and plot the first two mode shapes.

8 Determine the first four frequencies of a continuous beam of length 3ℓ supported along three equal spans of length ℓ, as shown (Fig. 4.28). Plot also the corresponding mode shapes. (Hint: Consider the symmetry present.)

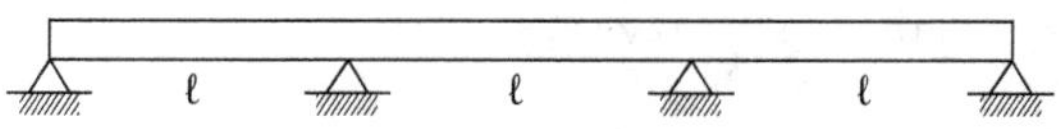

FIGURE 4.28 Problem 8.

9 A C–C beam has a translational spring of stiffness k attached at an intermediate point (Fig. 4.29).

A. Set up a *fourth order*, determinant for finding the frequencies by using a separate coordinate origin at each end, and either by using displacement functions which satisfy the clamped B.C. exactly, or by reducing eight equations to four by adding or subtracting equations.

B. For the special case when $\ell_1 = \ell_2 = \ell/2$, expand the determinant to obtain frequency equations.

C. For the special case of Part B, make a plot of the first two nondimensional frequencies $\omega\ell^2\sqrt{\rho A/EI}$ versus $k\ell^3/EI$ to show how they change with increasing spring stiffness.

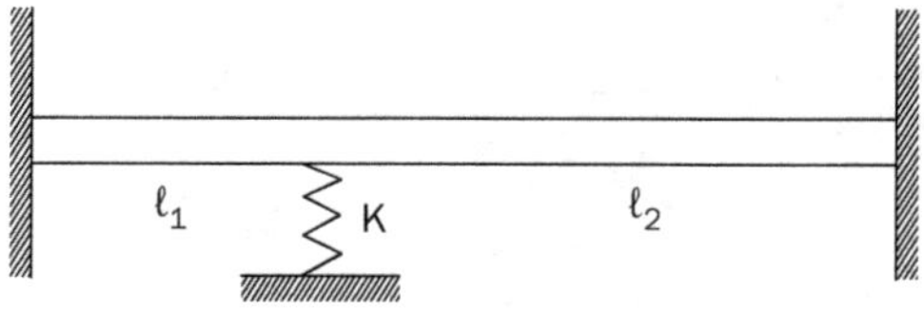

FIGURE 4.29 Problem 9.

10 The frame shown (Fig. 4.30) may be regarded as two beams of lengths ℓ_1 and ℓ_2, welded together at a right angle. The longitudinal stiffness of each beam may be considered as infinite in comparison with the bending stiffnesses. Both ends of the frame are hinged.

A. Choosing a coordinate origin at the hinged end of each beam, set up the characteristic determinant for determining the frequencies of the frame. The reduced determinant will be of fourth order. Let the ρ, E, I, and A of each beam be equal, and express the determinant in terms of $\beta_1 = \omega \ell_1^2 \sqrt{\rho A / EI}$ and $\ell^* = \ell_2/\ell_1$.

B. Determine β_1 when $\ell^* = 2$, and compare with similar known results for single straight beams of length ℓ_1, ℓ_2, and $\ell_1 + \ell_2$.

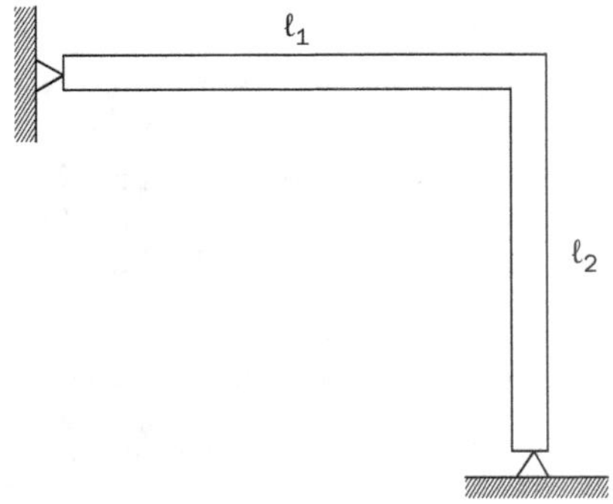

FIGURE 4.30 Problem 10.

11 A. Let $\ell_1 = \ell_2$ in Problem 10. By taking into consideration the 45 deg symmetry axis passing through the corner of the frame which exists in this case, derive the two second-order characteristic determinants which yield all frequencies. Expand them to get frequency equations. Justify carefully and logically all boundary conditions used.

B. Determine the lowest two frequencies. Sketch the corresponding mode shapes (no calculations needed).

C. Suppose the longitudinal stiffnesses of the beams is *finite*, perhaps even the same order of magnitude as the transverse stiffnesses. Write down the system of differential equations and boundary conditions that would be required to solve this problem, but do not attempt to solve them.

12 A clamped–clamped beam has a concentrated force $Q_0 \sin \Omega t$ acting at its middle (Fig. 4.31). Assume no damping.

A. Use the eigenfunction superposition method to determine $w(x,t)$. Represent the concentrated force in one of two ways:

(1) A direct delta function

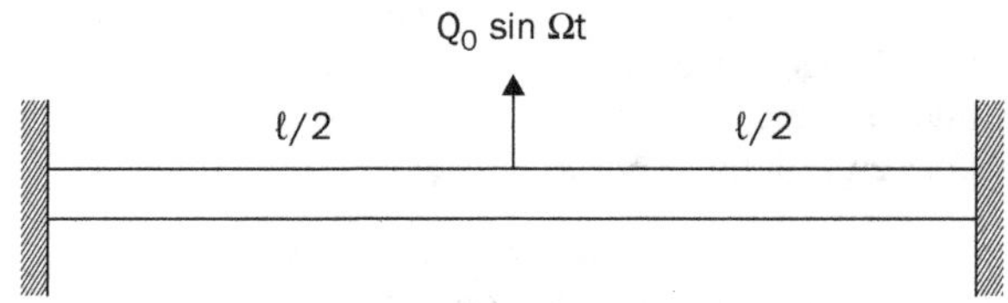

FIGURE 4.31 Problem 12.

(2) A uniform distributed pressure $p_0 = Q_0/b$ along a length b in the middle of the beam. Then take the limit as $b/\ell \to 0$.

Hint: Take advantage of the symmetry present.

B. Evaluate $w(x,t)$ at the middle of the beam for $\Omega/\omega_1 = 3$. Compare this amplitude with the static deflection.

13 Solve problem 12 by the closed-form method. Plot the vibratory displacement shapes $X(x)/\delta$ of the beam for $\Omega/\omega_1 = 1.2$, 3, and 5.2, where δ is the static displacement at the center and ω_1 is the lowest free vibration frequency. Explain *why* the shapes appear as they do.

14 Use the Rayleigh method to find approximate values of the fundamental frequency of a C–SS beam, and compare with the exact value.

A. Use a trial function which satisfies only the geometric B.C.

B. Use one which satisfies all the B.C.

C. Determine the "inertia loading" caused by the two functions.

D. Use a trial function which will cause a reasonable inertia loading.

15 Use the Rayleigh method to obtain an estimate of the fundamental frequency of a cantilever beam of length ℓ and linear taper (Fig. 4.32). The cross-section of the beam is circular, having a diameter which varies as $d = d_0(x/\ell)$ in terms of the coordinate origin shown.

Set up Rayleigh's Quotient by two approaches, one with an assumed mode shape $X(x)$ and the other with $Y(y)$. Compare the expected work required by

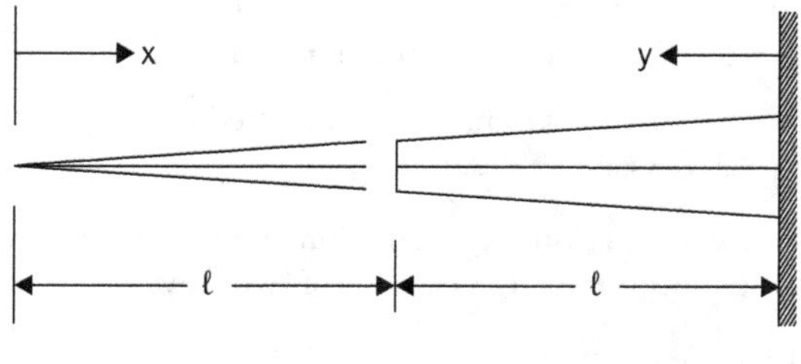

FIGURE 4.32 Problem 15.

the two approaches. Plot the two functions X and Y, as well as their "inertia loadings." Use the one you consider best to get the numerical value for the frequency.

16 Use the Ritz method to determine the first two frequencies of a C–C beam with reasonable accuracy.

17 A beam has both ends clamped and is subjected to an axial force T which is uniform along its length.

A. Make a sketch of the beam and its boundary constraints or explain otherwise how this physical situation could arise.

B. Derive the frequency equation, in nondimensional form, which would yield the fundamental frequency.

C. Make a plot of ω/ω_0 versus T/P_{cr} where ω_0 is the fundamental frequency of the beam when $T = 0$, and P_{cr} is the critical value of buckling load. Plot this curve carefully as $T/P_{cr} \to -1$. Compare this curve with that for the beam having both ends simply supported (both curves intersect the ordinate at +1 and the abscissa at −1).

D. On a single graph make a plot of the two mode shapes X/X_{max} in the special cases when $T/P_{cr} = 0$ and −1, and compare. How do they compare in the case of the SS–SS beam?

18 A beam of length ℓ is welded on one end to a rigid circular ring having an inside radius, R, and its other end is free (Fig. 4.33). The ring rotates with constant angular speed Ω about its center. Considering free vibrations of the beam in the vertical direction (i.e., in the direction, z, of the angular velocity vector, $\bar{\Omega}$):

A. Determine $\omega\ell^2\sqrt{\rho A/EI}$ for the fundamental frequency of the beam as a function of ℓ/R and Ω/ω_0, where ω_0 is the beam frequency when $\Omega = 0$.

B. Using these results, plot $\omega\ell^2\sqrt{\rho A/EI}$ versus Ω/ω_0 for $\ell/R = 0.5, 1, 1.5$.

C. There are at least *two* reasons why vibrations in the sideways (i.e. circumferential) direction would be more difficult to analyze. Explain why.

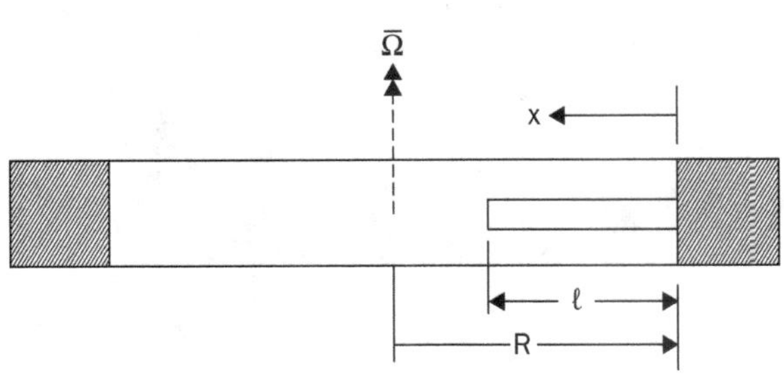

FIGURE 4.33 Problem 18.

19 Use the following method of finding a general solution to (4.127):

A. Uncouple (4.127a) and (4.127b) to find a single differential equation in Ψ, to complement (4.128).

B. Derive solutions for w and Ψ from the uncoupled equations. Convert these solutions into the same forms as (4.148) and (4.153).

C. Substitute the solutions obtained into (4.127b) to determine relationships between the constants of integration, so that the final solution only has four such constants.

20 A beam of circular cross-section has both ends free. Its diameter is d and its length is ℓ. A free vibration analysis is desired which includes the effects of shear deformation and rotary inertia.

A. Derive the frequency equation for symmetric modes (only) in terms of $\omega\ell^2\sqrt{\rho A/EI}, \ell/d$ and G/E.

B. Show that $\omega = 0$ is one solution to the frequency equation, and prove that this corresponds to a rigid body mode.

C. Assume that the beam is made of titanium. Make a plot of $\omega\ell^2\sqrt{\rho A/EI}$ versus ℓ/R for $3 < \ell/R < 20$ for the first two, non-zero, symmetric mode frequencies.

D. Evaluate the mode shape $W(x)$ and $\Psi(x)$ of the second, nonzero, symmetric mode frequency of the titanium beam when $\ell/d = 5$. Use this to draw the beam in its deformed shape in this mode.

21 Use the Ritz method to obtain an estimate of the effects of shear flexibility and rotary inertia on the fundamental frequency of a cantilever beam of length ℓ.

A. Choose suitable function $X(x) = C_1\Phi_1(x)$ and $\Psi(x) = C_2\Phi_2(x)$ where C_1 and C_2 are constants. Obtain the frequency determinant from which $\omega\ell^2\sqrt{\rho A/EI}$ could be obtained. The other parameters are k, G/E, and ℓ/R.

B. Let $k = 5/6$ (rectangle) and $G/E = 0.4$ (e.g., steel). Evaluate $\omega_1\ell^2\sqrt{\rho A/EI}$ (ω_1 is the fundamental frequency) for $h/\ell = 0, 0.1, 0.4$, and compare the percent decreases with those shown in Table 4.4.

CHAPTER 5
Membrane Vibrations

A membrane is a structural element which is relatively thin in one direction compared with the other two, and is flat. It is stretched in its plane, thereby withstanding tensile stresses. However, it is perfectly flexible with respect to bending. It may be regarded as a two-dimensional generalization of the string, which was studied in Chap. 2. Indeed, the equations and methods used here will typically have strong similarities to ones found in Chap. 2. The major difference is that while the transverse vibration will be described by the displacement, w, it will be a function of two space variables (such as x and y) and time—i.e., $w = w(x,y,t)$—whereas, string vibrations involve only one space variable.

Membrane have many applications in various fields. In music and acoustics, membranes constitute major components in many musical instruments (e.g., drums). In addition, membrane surfaces constitute components of microphones, speakers, and other devices.

In the physical sciences, membranes may be used to study two dimensional wave mechanics and propagation. The fundamental equations of wave propagation in two dimensions are the same as the membrane vibration equations.

In bioengineering, many human tissues are considered as membranes. The vibration characteristics of an eardrum are important in understanding hearing. The design of hearing aid devices involves incorporating membranes in them.

The transverse vibrations of membrane are studied in this chapter. The equations of motion will be first derived. The free vibrations of rectangular, circular, annular, and sectorial membranes will then be examined. Some interesting mode shapes of free vibration will be discovered. Vibration of membranes under various initial conditions will be studied. Forced vibration will be treated using both modal expansion and closed-form solutions. The approximate methods of Rayleigh and Ritz will be introduced and used to solve problems which do not allow exact solutions, or when such solutions are not easy to obtain.

5.1 Equation of Motion for Transverse Vibrations

The membrane is assumed to be stretched with sufficiently large stresses so that, like the string, if the subsequent transverse vibrational displacement is kept small, the stresses will remain essentially constant during vibration. In practice, this usually consists of stretching a membrane uniformly over a support frame as, for example, in the case of a drum. However, the tensile stress need not be uniform, and in reality it never is. Moreover, inplane shear stresses, in addition to tensile stresses, may be applied at the support frame (the membrane boundary). Such stresses are limited only by the requirement that at no point, and in no direction, within the membrane region will there be compressive stress. If a compressive stress component were to exist, the membrane would wrinkle because of its complete lack of bending stiffness.

Figure 5.1 depicts a membrane which is stretched over a boundary of arbitrary curvilinear shape. The stress applied externally is shown as a normal stress (σ_n) which may vary along the boundary. In addition a variable shear stress, which is not shown, may also act in the plane (xy) of the membrane and tangent to its boundary. These stresses are assumed not to vary with time. The membrane is shown in its static equilibrium position.

Consider now the transverse motion of an infinitesimal element of the membrane. It has dimensions $dx \times dy$ in the plane of the membrane, as shown in Fig. 5.1, and a thickness h. Figure 5.2 is a three-dimensional sketch of its middle surface in a typical displaced position. Shown on the element are the membrane *stress resultants* T_x, T_{xy}, etc., acting on each side of the element. They each have the dimensions of force per unit tangent length, and are related to the

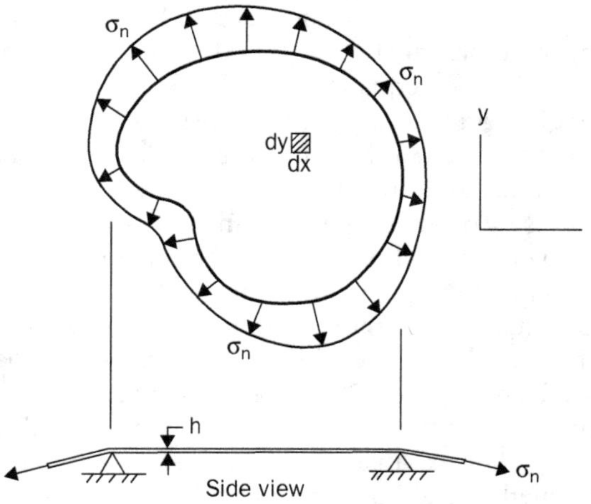

FIGURE 5.1 Membrane of arbitrary boundary shape subjected to nonuniform tensile stress.

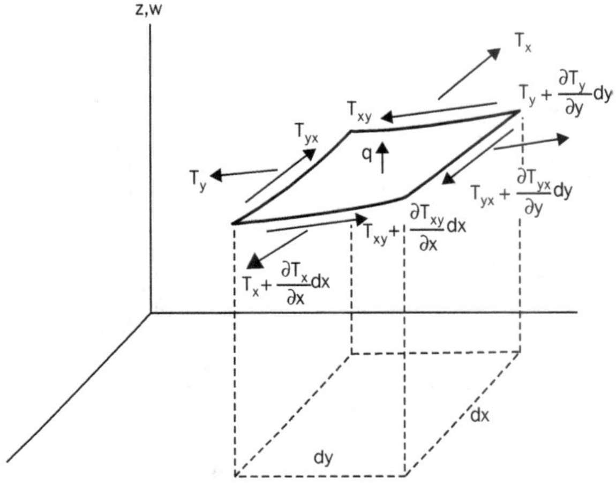

FIGURE 5.2 Transversely displaced membrane element with stress resultants.

stresses by $T_x = \sigma_x h$, $T_{xy} = \tau_{xy} h$, etc. T_x and T_y are *normal* stress resultants and T_{xy} and T_{yx} are *shear* stress resultants. In Fig. 5.2, they are drawn in their positive directions in accordance with traditional notation used in the theory of elasticity (cf. [1]). Also shown is a pressure (q) having units of force/area, distributed over the surface of the membrane.

Assuming that the membrane slopes at all points and in all directions are small during the vibratory motion, and summing forces in the transverse (z) direction:

$$
\begin{aligned}
&- T_x \frac{\partial w}{\partial x} dy + \left(T_x + \frac{\partial T_x}{\partial x} dx \right)\left(\frac{\partial w}{\partial x} + \frac{\partial}{\partial x}\frac{\partial w}{\partial x} dx \right) dy \\
&- T_y \frac{\partial w}{\partial y} dx + \left(T_y + \frac{\partial T_y}{\partial y} dy \right)\left(\frac{\partial w}{\partial y} + \frac{\partial}{\partial y}\frac{\partial w}{\partial y} dy \right) dx \\
&- T_{xy} \frac{\partial w}{\partial y} dy + \left(T_{xy} + \frac{\partial T_{xy}}{\partial x} dx \right)\left(\frac{\partial w}{\partial y} + \frac{\partial}{\partial x}\frac{\partial w}{\partial y} dx \right) dy \\
&- T_{yx} \frac{\partial w}{\partial x} dx + \left(T_{yx} + \frac{\partial T_{yx}}{\partial y} dy \right)\left(\frac{\partial w}{\partial x} + \frac{\partial}{\partial y}\frac{\partial w}{\partial x} dy \right) dx \\
&+ q\,dx\,dy = (\rho h\, dx\,dy)\frac{\partial^2 w}{\partial t^2}
\end{aligned}
\tag{5.1}
$$

where now ρ is mass density per unit volume of the material, and h is the membrane thickness. In the terms of (5.1), the sine of the angle made with the xy-plane was replaced by its tangent, as was done for

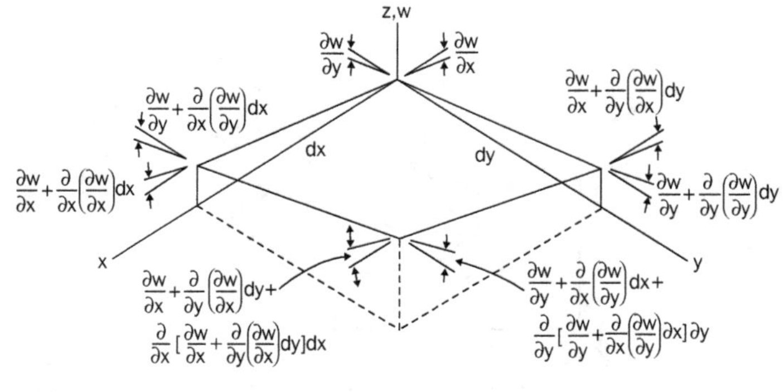

FIGURE 5.3 Positive slopes and slope changes for a displaced element.

the string in Chap. 2. It can be seen in (5.1) that not only are the stress resultants changing in magnitude, in general, as one moves across the element from one edge to an opposite edge, but they are changing in direction. These direction changes must be incorporated into the z-direction components of the stress resultants as in (5.1). A sketch showing all the positive slopes $(\partial w/\partial x)$ and $(\partial w/\partial y)$ and their positive rates of change is shown in Fig. 5.3.

A more careful derivation of (5.1) would, for example, in the first term, replace T_x by its *average* value $\{T_x + [T_x + (\partial T_x/\partial y)dy]\}/2$ along the edge, and $\partial w/\partial x$ by its average value $\{\partial w/\partial x + [\partial w/\partial x + (\partial(\partial w/\partial x)/\partial y) dy]\}/2$. However, all the additional terms thus introduced are found to be higher order terms, yielding (5.1).

Taking the first line of (5.1), which arises from T_x, and expanding the second term, one obtains

$$-T_x\frac{\partial w}{\partial x}dy+\left(T_x\frac{\partial w}{\partial x}+T_x\frac{\partial^2 w}{\partial x^2}dx+\frac{\partial T_x}{\partial x}\frac{\partial w}{\partial x}dx+\frac{\partial T_x}{\partial x}dx\cdot\frac{\partial^2 w}{\partial x^2}dx\right)dy \quad (5.2)$$

The first two terms are first-order differential quantities, but cancel each other out. Two of the remaining terms contain $dxdy$, and are therefore second-order differentials, the same order as the pressure and inertia terms in (5.1). The last term in (5.2) contains $(dx)^2dy$, which is a third-order differential. This term is therefore discarded, for it contributes nothing to (5.1). Using the same arguments on lines 2, 3, and 4 of (5.1), and dividing through by the area $dxdy$, it becomes

$$\left(T_x\frac{\partial^2 w}{\partial x^2}+\frac{\partial T_x}{\partial x}\frac{\partial w}{\partial x}\right)+\left(T_y\frac{\partial^2 w}{\partial y^2}+\frac{\partial T_y}{\partial y}\frac{\partial w}{\partial y}\right)$$
$$+\left(T_{xy}\frac{\partial^2 w}{\partial x\partial y}+\frac{\partial T_{xy}}{\partial x}\frac{\partial w}{\partial y}\right)+\left(T_{yx}\frac{\partial^2 w}{\partial x\partial y}+\frac{\partial T_{yx}}{\partial y}\frac{\partial w}{\partial x}\right)+q=\rho h\frac{\partial^2 w}{\partial t^2} \quad (5.3)$$

This may also be written as

$$\frac{\partial}{\partial x}\left(T_x \frac{\partial w}{\partial x}\right) + \frac{\partial}{\partial y}\left(T_y \frac{\partial w}{\partial y}\right) + \frac{\partial}{\partial x}\left(T_{xy} \frac{\partial w}{\partial y}\right)$$
$$+ \frac{\partial}{\partial y}\left(T_{yx} \frac{\partial w}{\partial x}\right) + q = \rho h \frac{\partial^2 w}{\partial t^2} \tag{5.4}$$

In this form, it is seen that the incremental changes of the z-components of the stress resultants as one moves across the element are what contribute to the equation of motion.

Summing moments about an axis parallel to z, and through the center of mass of the element shows that $T_{xy} = T_{yx}$ (the rotational inertia of the element contributes only a higher order differential term). Summing forces in the x and y directions, and assuming that there are no significant accelerations or body forces (e.g., gravity, centrifugal) in the x and y directions, one obtains the classical equilibrium equations of plane elasticity.

$$\frac{\partial T_x}{\partial x} + \frac{\partial T_{xy}}{\partial y} = 0 \tag{5.5a}$$

$$\frac{\partial T_{xy}}{\partial x} + \frac{\partial T_y}{\partial y} = 0 \tag{5.5b}$$

Multiplying (5.5a) by $\partial w/\partial x$, and (5.5b) by $\partial w/\partial y$, and then subtracting both from (5.3), the latter becomes

$$T_x \frac{\partial^2 w}{\partial x^2} + 2T_{xy} \frac{\partial^2 w}{\partial x \partial y} + T_y \frac{\partial^2 w}{\partial y^2} + q = \rho h \frac{\partial^2 w}{\partial t^2} \tag{5.6}$$

It is important to note that in this form, T_x, T_y, and T_{xy} may all be functions of x and y. T_x, T_y, and T_{xy} are determined by solving first a plane elasticity problem. Moreover, ρ and h may each be functions of x and y (material nonhomogeneity and variable thickness).

The most widely used form of (5.6) is the special case wherein the inplane shear stress is zero, and the remaining tensile stress is constant and the same in all directions ($T_x = T_y = T$):

$$T\nabla^2 w + q = \rho h \frac{\partial^2 w}{\partial t^2} \tag{5.7}$$

where ∇^2 is the Laplacian operator

$$\nabla^2 = \frac{\partial^2}{\partial x^2} + \frac{\partial^2}{\partial y^2} \tag{5.8}$$

in rectangular coordinates. The stress state $T_x = T_y = T =$ constant satisfies all the required equations of plane elasticity, and therefore is acceptable. In many practical applications involving membranes this is the situation, or very nearly so. Moreover, ρ and h are nearly always assumed to be constant in (5.7). For free, undamped vibrations, (5.7) becomes

$$\boxed{T\nabla^2 w = \rho h \frac{\partial^2 w}{\partial t^2}} \tag{5.9}$$

which is the classical, two-dimensional wave equation.

5.2 Free Vibrations of Rectangular Membranes

A rectangular membrane with planform dimensions $a \times b$ is shown in Fig. 5.4. Assume that uniform tension is applied to it in all directions, so that free vibrations are governed by the equation of motion (5.9). To determine the natural frequencies and mode shapes, we will proceed in the usual manner, that is, a solution to (5.9) will first be found, and then the boundary conditions will be applied.

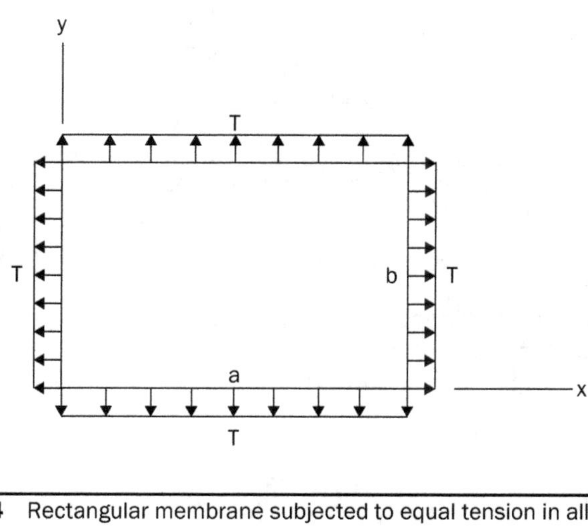

FIGURE 5.4 Rectangular membrane subjected to equal tension in all directions.

Using the method of separation of variables, a solution to (5.9) is assumed in the form

$$w(x,y,t) = X(x) \bullet Y(y) \bullet \Phi(t) \tag{5.10}$$

Substituting this into (5.9), and dividing by $XY\Phi$ results in

$$\frac{X''}{X} + \frac{Y''}{Y} = \left(\frac{\rho h}{T}\right)\frac{\Phi''}{\Phi} \tag{5.11}$$

Each of the three terms in (5.11) is a function of a different variable (x, y, or t), therefore, the only way in which (5.11) may be valid is if each term is equal to a constant. Let these constants be $-\alpha^2$, $-\beta^2$, and $-\gamma^2$. Then, (5.11) yields

$$X'' + \alpha^2 X = 0 \tag{5.12a}$$

$$Y'' + \beta^2 X = 0 \tag{5.12b}$$

$$\Phi'' + \left(\frac{T}{\rho h}\right)\gamma^2 \Phi = 0 \tag{5.12c}$$

and

$$\alpha^2 + \beta^2 = \gamma^2 \tag{5.13}$$

In anticipation of the solution form of (5.12c), replace $(T/\rho h)\gamma^2$ by the constant ω^2, which will be, of course, the circular frequency. Solutions to (5.12) are then

$$X = A \sin \alpha x + B \cos \alpha x \tag{5.14a}$$

$$Y = C \sin \beta y + D \cos \beta y \tag{5.14b}$$

$$\Phi = E \sin \omega t + F \cos \omega t \tag{5. 14c}$$

where

$$\alpha^2 + \beta^2 = \left(\frac{\rho h}{T}\right)\omega^2 \tag{5.15}$$

Boundary conditions where all edges are fixed will be considered. The boundary conditions are therefore (see Fig. 5.4)

$$w(0,y,t) = w(a,y,t) = w(x,0,t) = w(x,b,t) = 0 \tag{5.16}$$

whence

$$X(0) = X(a) = Y(0) = Y(b) = 0 \tag{5.17}$$

Substituting (5.14a) and (5.14b) into (5.17) gives $B = D = 0$ and

$$\alpha = \frac{m\pi}{a} \qquad (m = 1, 2, \ldots) \tag{5.18a}$$

$$\beta = \frac{n\pi}{b} \qquad (n = 1, 2, \ldots) \tag{5.18b}$$

The frequency is thus determined from (5.15) to be

$$\omega = \pi\sqrt{\frac{T}{\rho h}\left[\left(\frac{m}{a}\right)^2 + \left(\frac{n}{b}\right)^2\right]} \tag{5.19}$$

Or, in nondimensional form,

$$\lambda = \omega a\sqrt{\frac{\rho h}{T}} = \pi\sqrt{m^2 + \left(\frac{a}{b}\right)^2 n^2} \qquad (m, n = 1, 2, 3, \ldots, \infty) \tag{5.20}$$

It is seen that the nondimensional frequency parameter $\omega a\sqrt{\rho h/T}$ depends on the aspect ratio (a/b) of the membrane and that, for any a/b, there is a doubly infinite set of frequencies depending on the choices of m and n.

Combining the remaining constants A, C, E, and F in (5.14), it may now be said that the membrane is capable of vibrating freely with a natural frequency ω_{mn} and in a corresponding mode shape $W_{mn}(x,y)$ where

$$W_{mn}(x, y) = \sin\frac{m\pi x}{a} \bullet \cos\frac{n\pi y}{b} \tag{5.21}$$

The complete motion of the membrane in this mode shape is described by

$$w_{mn}(x, y, t) = W_{mn}(x, y)(A_{mn} \sin\omega_{mn}t + B_{mn}\cos\omega_{mn}t) \tag{5.22}$$

where A_{mn} and B_{mn} are determined from the initial conditions.

Looking at (5.20), it is clear that the fundamental frequency of a rectangular membrane, regardless of a/b, will be for a mode shape having one half-sine wave in each direction ($m = n = 1$). The second frequency will then be for $m, n = 1, 2$ or $2, 1$, depending on a/b. Beyond that, the sequence of modes becomes unclear, and one must evaluate (5.20) for various m and n to determine it. For example, the first ten nondimensional frequency parameters λ for a membrane with $a/b = 1.5$ are listed in ascending sequence in Table 5.1. The ratio ω/ω_1 is also given, where ω_1 is the fundamental (lowest) value. This shows that the first three frequencies are rather widely separated, but some others are close together (e.g., ω_{32} and ω_{41}).

It is also interesting to compare the "density" of the frequency spacing for membranes with those of other structural elements. For $a/b = 1.5$, 12 frequencies are found within the range of frequency ratios $1 < \omega/\omega_1 \leq 3$ (those in Table 5.1, plus $\omega_{51}$and ω_{33}). For a string [see (2.23)], there are only three. For a beam having any of the classical boundary conditions (see Table 4.1), there are only two at most (for C–C and F–F beams). Thus, one can imagine that the *experimental* determination of resonant frequencies, especially when ten or more are needed, is more difficult for a membrane than for a string or beam. In particular, one is more likely to miss one of two close frequencies.

Figure 5.5 shows the first nine "nodal patterns" for the rectangular membrane with $a/b = 1.5$. These consist of planform views of the membrane showing the node lines (lines which remain stationary while all other parts of the membrane vibrate) for each of the mode shapes. Since, in this case, the half-waves between node lines are all

Mode number	m	n	$\lambda = \omega a\sqrt{\rho h/T}$	ω/ω_1
1	1	1	5.664	1
2	2	1	7.854	1.387
3	1	2	9.935	1.754
4	3	1	10.538	1.861
5	2	2	11.327	1.999
6	3	2	13.329	2.353
7	4	1	13.421	2.370
8	1	3	14.482	2.557
9	2	3	15.471	2.731
10	4	2	15.708	2.773

TABLE 5.1 The First Ten Frequencies of a Membrane with Aspect Ratio (a/b) of 1.5

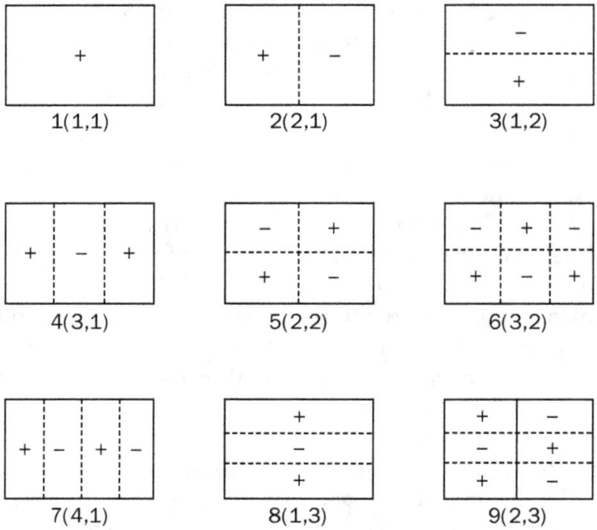

FIGURE 5.5 First nine nodal patterns for $a/b = 1.5$.

sine waves, the nodal patterns readily describe the mode shapes. Plus and minus signs are used in Fig. 5.5 to emphasize that one region is moving towards the reader (+) while another is simultaneously moving away (–); these are not necessary and are usually omitted from nodal pattern drawings. The node lines in Fig. 5.5 are all straight, but this is not the usual case, especially for nonrectangular membranes.

It is possible to have two mode shapes having the same frequency for a particular membrane. This is obvious enough for a square ($a/b = 1$) configuration, where $\omega_{12} = \omega_{21}$, $\omega_{13} = \omega_{31}$, $\omega_{23} = \omega_{32}$, etc. However, it may be shown from (5.20), for example, that $\omega_{12} = \omega_{31}$ if $a/b = \sqrt{8/3} = 1.633$. In such cases the frequencies are said to be "degenerate." A plot of λ^2 versus $(a/b)^2$ by means of (5.20), for all values of m and n, would show an infinite number of curve crossings, each of which corresponds to a degeneracy.

If two modes have the same frequency, then it is possible to start the motion so that both modes are present during the free vibration. Consider the square membrane and its degenerate frequencies $\omega_{12} = \omega_{21}$. Superposition of the two mode shapes may be taken as

$$W(x,y) = C_{12} \sin\frac{\pi x}{a} \cdot \sin\frac{2\pi y}{a} + C_{21} \sin\frac{2\pi x}{a} \cdot \sin\frac{\pi y}{a} \tag{5.23}$$

If $C_{21} = -C_{12}$ (*subtracting*, the two modes), then (5.23) indicates that along the line $y = x$, which is one diagonal of the square, $W(x, y) = 0$. Similarly, if $C_{21} = +C_{12}$ (*adding* the two modes), then, the other diagonal, line $y = a - x$ has no displacement. Thus, superposition of the modes yields two other mode shapes which have node lines along one diagonal or the other, as shown in Fig. 5.6(a).

Figure 5.6(b) shows the combined nodal patterns arising from the subtraction ($C_{31} = -C_{13}$) and addition ($C_{31} = +C_{13}$) of the (3, 1) and (1, 3) modes. The first causes both diagonal nodal lines to appear in the combined mode. The second yields a nodal line which is similar to (but slightly different from) a circle.

The superpositions shown in Fig. 5.6 involve *equal* amounts of the two modes being superimposed. Figure 5.7 shows some interesting nodal patterns which arise when one superimposes two degenerate modes *unequally*. As one changes the ratio C_{mn}/C_{nm} of the two amplitudes, a continuous change in the node line of the combined

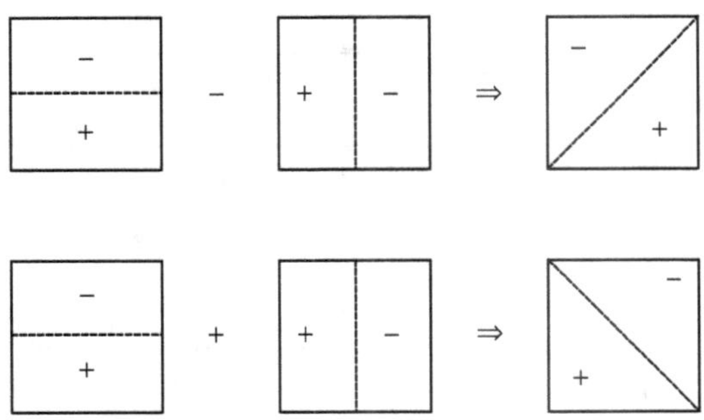

(a) (1,2) ∓ (2,1) modes

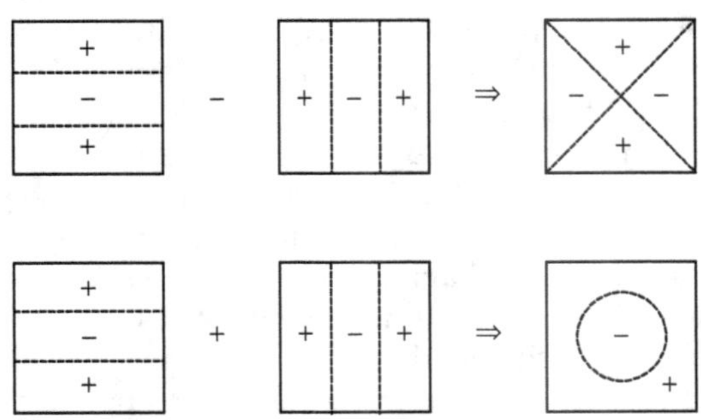

(b) (1,3) ∓(3,1) modes

FIGURE 5.6 Nodal patterns for superimposed degenerate modes of square membranes.

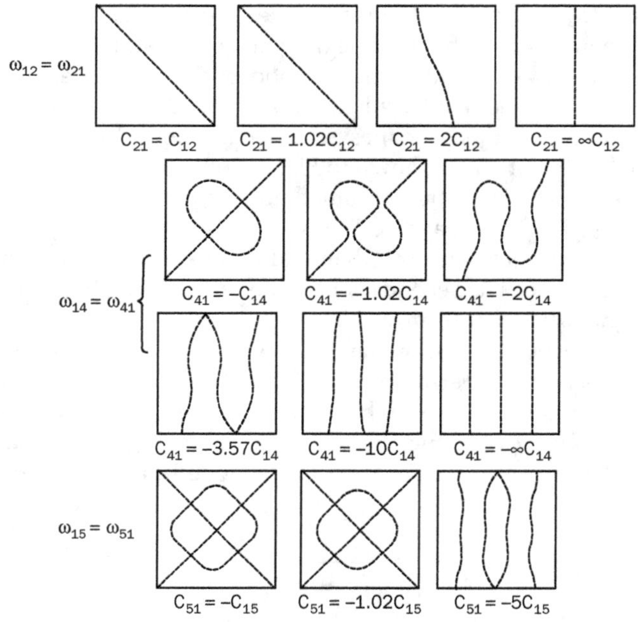

FIGURE 5.7 Some nodal patterns which arise when one superimposes two degenerate modes unequally.

mode is seen to take place, and some node lines quite complicated in shape are seen to arise.

Finally, before leaving the topic of degenerate modes and their combined nodal patterns, it should be pointed out that because the node lines satisfy $W(x, y) = 0$, they may be regarded as boundaries of other membrane shapes vibrating with the same frequency. Thus, for example, the superimposed (1,2)∓(2,1) modes in Fig. 5.6(a) are also those of the isosceles right triangle. The fundamental frequency of the isosceles right triangle membrane is therefore $\omega a\sqrt{\rho h/T} = 7.025$, where "$a$" now is the length of the short sides of the triangle. The second combined mode in Fig. 5.6(b) solves not only the problem of the membrane having the "nearly circular" boundary shape, but also that of a square membrane having a "nearly circular" inner boundary, as well (where, of course, the inner boundary has one particular size and location), with both boundaries being fixed ($w = 0$).

The problems discussed above were for rectangular membranes having all edges fixed. More general boundary conditions, such as an edge having an attached mass and/or a distributed spring, are possible, but are difficult to achieve physically. Such boundary conditions may be dealt with by methods similar to those for strings (see Sec. 2.6).

5.3 Circular Membranes

The equation of motion (5.9) for the free vibrations of membranes subjected to uniform tension (T) was solved in rectangular coordinates in Sec. 5.2, and was applied to membranes having rectangular boundaries. An exact solution to it is also possible in polar coordinates, which may be conveniently applied to membranes having circular boundaries.

Let us begin by assuming for free, undamped vibrations

$$w(r,\theta,t) = W(r,\theta)\sin(\omega t + \phi) \tag{5.24}$$

where, as before, ω is a natural frequency and ϕ is a phase angle (which depends on the initial conditions). Substituting (5.24) into (5.9) results in

$$\nabla^2 W + k^2 W = 0 \tag{5.25}$$

where k^2 is defined by

$$k^2 = \frac{\rho h \omega^2}{T} \tag{5.26}$$

Figure 5.8 shows a circular membrane of radius "a" and an associated polar coordinate system (r, θ) having its origin at the membrane center. The Laplacian operator ∇^2 in polar coordinates is

$$\nabla^2 = \frac{\partial^2}{\partial r^2} + \frac{1}{r}\frac{\partial}{\partial r} + \frac{1}{r^2}\frac{\partial^2}{\partial \theta^2} \tag{5.27}$$

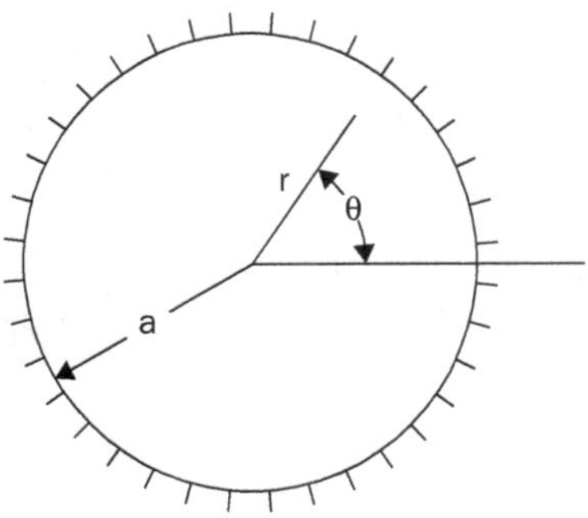

FIGURE 5.8 Circular membrane and polar coordinates.

A solution to (5.25) is therefore assumed as

$$W(r,\theta) = W_n(r)\cos n\theta \qquad (n = 1, 2, \ldots) \tag{5.28}$$

Not only does this "fit" (5.25), but the integer values of n also give the desired *periodicity* needed by the solution in the case of the circular membrane, that is, $W(r, \theta) = W(r, \theta + 2\pi)$. Using (5.27) and (5.28) in (5.25) yields

$$\frac{d^2W_n}{dr^2} + \frac{1}{r}\frac{dW_n}{dr} + \left(k^2 - \frac{n^2}{r^2}\right)W_n = 0 \tag{5.29}$$

It should be mentioned at this point that another solution of (5.25) is possible which uses sin $n\theta$ in place of cos $n\theta$ in (5.28). When substituted into (5.25), it also results in (5.29), except that it only gives nontrivial solutions for $n > 0$ and misses the axisymmetric ($n = 0$) modes. It is not needed in the determination of the complete set of free vibration frequencies and mode shapes of circular membranes having *uniform* boundary conditions (i.e., boundary conditions not changing with θ), to which the present section is limited. For these problems, the cos $n\theta$ functions are sufficient.

Equation (5.29) is the well-known Bessel's equation which we encountered previously in the longitudinal vibrations of tapered bars (Sec. 3.5). Its solution is

$$W_n(r) = A_n J_n(kr) + B_n Y_n(kr) \tag{5.30}$$

where J_n and Y_n are Bessel functions of the first and second kinds, respectively, of "order" n, and having the argument kr. The constants of integration are A_n and B_n.

From the properties of Bessel functions (see Appendix B), we know that the $Y_n(kr)$ are infinite at $r = 0$ for all n. Therefore, to enforce the *regularity condition* that the displacement is finite at $r = 0$, it is necessary to set $B_n = 0$. At the boundary, $w(a, \theta, t) = 0$, which implies that $W_n(a) = 0$ or, from (5.30),

$$J_n(\lambda) = 0 \tag{5.31}$$

where $\lambda = ka$.

Equation (5.31) has an infinite set of roots, ka, which are eigenvalues, for each value of n. From (5.26), it is seen that these are the nondimensional frequency parameters:

$$\lambda = \omega a\sqrt{\frac{\rho h}{T}} \tag{5.32}$$

The first five roots (nondimensional frequencies) of (5.31) for $n = 0, \ldots, 5$ are listed in Table 5.2.

Combining (5.28) and (5.30), the eigenfunctions describing the mode shapes are

$$W(r,\theta) = J_n(kr)\cos n\theta = J_n\left(\lambda\frac{r}{a}\right)\cos n\theta \qquad (n = 1, 2, \ldots) \qquad (5.33)$$

For $n = 0$, (5.31) has the roots $\lambda = 2.4048, 5.5201, 8.6537, 11.7915, \ldots$ as seen in the plot of $J_0(z)$ in Appendix B. These correspond to the axisymmetric modes (i.e., no variation of w with θ), and have 0, 1, 2, 3 interior circles as node lines. The mode shapes are exactly shown by the plot of $J_0(z)$ in Appendix B, where zero values of $J_0(z)$ correspond to the locations of the membrane boundary and nodal circles. Thus, the first mode shape, along every radial line, resembles a half-cosine wave; however, it has an inflection point (point of curvature change) well before the boundary ($z = 2.4048$). The second mode shape has a nodal circle located at $r/a = 2.4048/5.5201 = 0.4356$. During its vibratory motion in this mode shape, the membrane displaces 2.48 times as much at its center than its maximum value in its outer portion. The third mode shape has nodal circles located at $r/a = 2.4048/8.6537 = 0.2779$ and at $r/a = 5.5201/8.6537 = 0.6379$, with the amplitudes of the three half-waves of the mode shape decreasing from the center outwards. Similarly, the fourth mode shape has nodal circles at $r/a = 0.2039, 0.4681$, and 0.7339.

For $n > 0$, the mode shapes have n circumferential waves in them, with n diametral node lines. The first five roots of (5.31), corresponding to the first five nondimensional frequency parameters, are listed in Table 5.2 for $n = 0, \ldots, 5$ ([2], p. 409). Higher roots of (5.31) become separated by π (=3.1416) for every n.

Mode shapes along any radial line for $n = 1$ and $n = 2$ may be seen in the plots of $J_1(z)$ and $J_2(z)$ in Appendix B. Nodal patterns for the first nine frequencies of circular membranes are shown in Fig. 5.9.

Number of root*	*n* (nodal diameters)					
	0	1	2	3	4	5
1	2.405	3.832	5.136	6.380	7.588	8.771
2	5.520	7.016	8.417	9.761	11.065	12.339
3	8.654	10.173	11.620	13.015	14.373	15.700
4	11.792	13.324	14.796	16.223	17.616	18.980
5	14.931	16.471	17.960	19.409	20.827	22.218

*Number of nodal circles plus one.

Table 5.2 Roots of $J_n(\lambda) = 0$ (Nondimensional Frequencies $\omega a\sqrt{\rho h/T}$ for Circular Membranes)

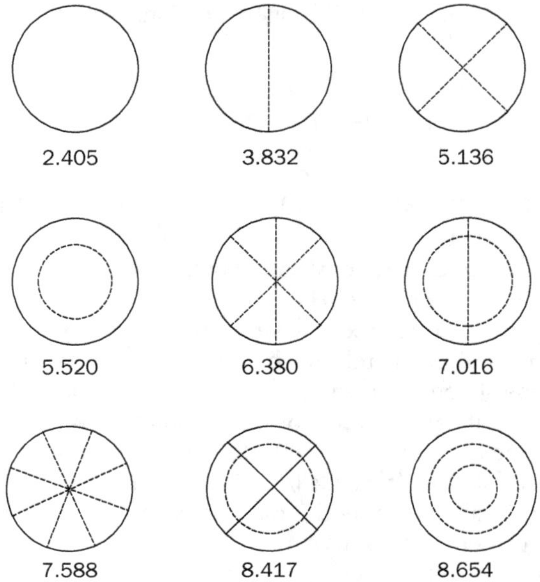

FIGURE 5.9 Nodal patterns and nondimensional frequencies $\omega a\sqrt{\rho h/T}$ for the first nine modes of circular membranes.

5.4 Annular and Sectorial Membranes

An annular membrane is bounded by two concentric circles. A membrane having both its outer boundary ($r = a$) and inner boundary ($r = b$) fixed is shown in Fig. 5.10.

One could use the results obtained in Sec. 5.3 to find the frequencies for annular membranes with certain b/a ratios. For example, consider the second axisymmetric ($n = 0$) mode of a complete circular membrane. This mode has one interior nodal circle, as depicted in Fig. 5.9, and a frequency $\lambda = \omega a\sqrt{\rho h/T} = 5.520$. Furthermore, as discussed in Sec. 5.3, the nodal circle is located at $r/a = 0.436$. Because the boundary condition of a fixed edge ($w = 0$) is duplicated along any node line of a membrane, $\lambda = 5.520$ is also the first axisymmetric frequency parameter of an annular membrane having $b/a = 0.436$. From the third axisymmetric mode of the complete membrane (see Fig. 5.9) one determines two additional results for annular membranes:

1. The fundamental frequency for $b/a = 5.520/8.654 = 0.638$.
2. The second axisymmetric mode frequency for $b/a = 2.405/8.654 = 0.278$.

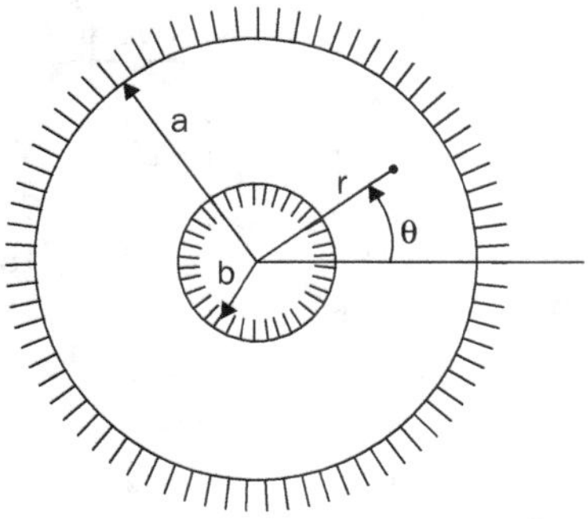

FIGURE 5.10 Annular membrane.

Both modes have the frequency parameter $\lambda = 8.654$. Similarly, from Fig. 5.9 one finds the lowest $n = 1$ frequency for an annular membrane having $b/a = 3.832/7.016 = 0.546$ to be $\lambda = 7.016$.

The procedure described above yields certain individual frequencies for annular membranes having limited values of b/a. A straightforward procedure exists for finding the complete frequency spectrum for arbitrary b/a. Returning to the exact solution of the equation of motion in polar coordinates, expressed by (5.24), (5.28), and (5.30), the $Y_n(kr)$ term in (5.30) may now be retained, for we no longer have concern about the behavior of $W_n(r)$ at $r = 0$ or, for that matter, anywhere outside of the interval $b \le r \le a$. Applying the boundary conditions of zero displacement at both $r = a$ and $r = b$ yields

$$\begin{bmatrix} J_n(ka) & Y_n(ka) \\ J_n(kb) & Y_n(kb) \end{bmatrix} \begin{bmatrix} A_n \\ B_n \end{bmatrix} = \begin{bmatrix} 0 \\ 0 \end{bmatrix} \tag{5.34}$$

As usual, for a nontrivial solution, the determinant of the coefficient matrix of (5.34) is set equal to zero, resulting in the frequency equation

$$J_n(\lambda)\, Y_n\left(\frac{b}{a}\lambda\right) - J_n\left(\frac{b}{a}\lambda\right)\, Y_n(\lambda) = 0 \tag{5.35}$$

where again, $\lambda = ka$. For any given value of b/a, all frequencies of the annular membrane are obtainable from (5.35). The first three frequencies (corresponding to 0, 1, and 2 interior nodal circles) are listed in Table 5.3 (adapted from Ref. [2], p. 415) for the $n = 0$ (axisymmetric) and $n = 1$ modes of annular membranes having various b/a ratios, especially for $b/a < 0.5$. It is interesting to note that, as b/a approaches zero, which corresponds to a point support, the

		Number of root		
n	b/a	1	2	3
0	0.80	15.698	31.411	47.121
	0.60	7.828	15.695	23.553
	0.40	5.183	10.443	15.688
	0.20	3.816	7.786	11.732
	0.10	3.314	6.858	10.377
	0.02	2.884	6.136	9.376
	0.00	2.405	5.520	8.654
1	0.80	15.738	31.431	47.134
	0.60	7.930	15.747	23.588
	0.40	5.391	10.558	15.766
	0.20	4.236	8.055	11.927
	0.10	3.941	7.331	10.748
	0.02	3.836	7.031	10.205
	0.00	3.832	7.016	10.173

TABLE 5.3 Roots of Equation (5.35) (Nondimensional Frequencies $\omega a\sqrt{\rho h/T}$ for Annular Membranes).

frequency for n = 0 becomes the same as for the unsupported membrane. This is because membranes are incapable of transmitting transversely applied concentrated forces without violating the linearizing assumptions made in deriving the theory. This is true for any membrane and for any shape. For further discussion of this, see Ref. [3].

A sectorial membrane has the shape of a sector of a circle (Fig. 5.11), with a sector angle, α. Quite a few frequencies for such membranes are obtainable from the data for complete circular membranes previously given in Table 5.2, because radial node lines duplicate the boundary conditions of zero displacement on the sides of the corresponding sector. Thus, for example, studying the nodal patterns of Fig. 5.9, and using additional data from Table 5.2, one sees that the frequencies of the symmetric modes (symmetric about $\theta = \alpha/2$) of a semicircular membrane are, in order, $\lambda = \omega a\sqrt{\rho h/T} = 3.832$, 6.380, 7.016, 8.771, 9.761, ... Similarly, antisymmetric mode frequencies are λ = 5.136, 7.588, 8.417, ... In the same manner the fundamental frequencies for α = 90°, 60°, 45°, and 36° are observed to be λ = 5.136, 6.380, 7.588, and 8.7771, respectively.

To obtain solutions for *arbitrary* sector angles, it is necessary to return to the equation of motion (5.25) and assume a displacement form.

$$W(r,\theta) = W_\nu(r)\sin \nu\theta \tag{5.36}$$

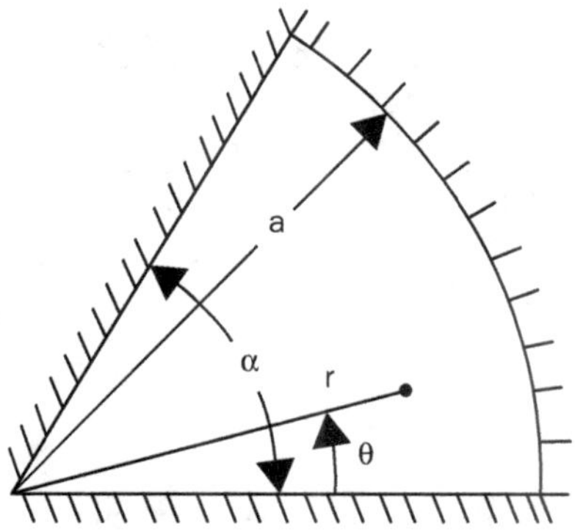

Figure 5.11 Sectorial membrane.

instead of (5.28) where ν, usually, is not an integer. Substituting (5.36) into (5.25) yields (5.29) again, but with n replaced by ν. The solution to this ordinary differential equation is

$$W_\nu(r) = A_\nu J_\nu(kr) + B_\nu Y_\nu(kr) \tag{5.37}$$

Thus, the Bessel functions are now of non-integer order.

Consider first the boundary conditions along the radial edges. The condition $W(r, 0) = 0$ is clearly satisfied by (5.36), regardless of ν. Enforcing the boundary condition along $\theta = \alpha$ yields $\sin \nu\alpha = 0$, whence

$$\nu = \frac{n\pi}{\alpha} \qquad (n = 1, 2, \ldots) \tag{5.38}$$

Applying next the regularity condition at $r = 0$ requires that $B_\nu = 0$. Finally, applying the condition $W(a,\theta) = 0$ yields the frequency equation

$$J_\nu(\lambda) = 0 \tag{5.39}$$

which is the same as (5.31), except that ν can be any real number, integer, or non-integer. For every sector angle, α, there are an infinite number of ν, given by (5.38). Symmetric modes result from $n = 1, 3\ldots$ antisymmetric modes from $n = 2, 4\ldots$ For every ν, there are an infinite number of nondimensional frequencies, λ.

Frequencies arising from the half-integer orders of v are given in Table 5.4. These are taken from Ref. [4], where zeros of (5.39) may be found for $1/2 < \nu < 39/2$. As discussed above, results for the intermediate values of $\alpha = 180°, 90°, 60°, 45°$, and $36°$, corresponding to integer ν, may be taken from Table 5.2. The case of $\nu = 1/2$, corresponding to $\alpha = 360°$ in Table 5.4, is an especially interesting

Number of root*	Sector angle, α (value of v)					
	360° (1/2)	120° (3/2)	72° (5/2)	51.43° (7/2)	40° (9/2)	32.73° (11/2)
1	3.142	4.493	5.763	6.988	8.183	9.356
2	6.283	7.725	9.095	10.418	11.705	12.967
3	9.425	10.904	12.323	13.698	15.040	16.355

*Number of nodal circles plus one.

TABLE 5.4 Roots of $J_v(\lambda) = 0$ for Half-Integer Values of v (Nondimensional Frequencies $\omega a\sqrt{\rho h/T}$ for Sectorial Membranes).

one. In this case the boundaries are at $\theta = 0°$ and $\theta = 360°$, which are the same radial line, and the problem of a complete circular membrane supported along a single radial line is thereby represented. It is also interesting that in this case the roots are separated exactly by π, which indicates also that the nodal circles are uniformly spaced.

Finally, the solutions in polar coordinates are also applicable to *annular sectorial* membranes. Such configurations are bounded by two circle arcs ($r = b$, $r = a$) and two radial lines ($\theta = 0$, $\theta = \alpha$). Considerable data for such shapes are available in Tables 5.2 and 5.3. For example, the fundamental frequency for a membrane with $\alpha = 60°$ (see $n = 3$ in Table 5.2) and $b/a = 9.761/13.015 = 0.750$ is $\lambda = 13.015$. Frequencies for arbitrary α and b/a may be found from (5.35), replacing n by v, where v depends on α as given in (5.38).

5.5 Initial Conditions

The free vibration of a membrane begins with a set of initial conditions specifying the displacement and velocity at $t = 0$. The procedure for determining the subsequent motion is, in principle, the same as that followed for strings (Sec. 2.3) and beams (Sec. 4.6). That is, one expresses the displacement as a sum of the vibratory responses of the individual modes and determines the amplitudes of the modes, taking advantage of the orthogonality of the eigenfunctions. The only significant new feature for the membrane problem is that it is two-dimensional instead of one-dimensional.

Taking first the *rectangular* membrane, let the initial displacement and velocities be given functions $f(x,y)$ and $g(x,y)$, respectively, so that

$$w(x,y,0) = f(x,y) \tag{5.40a}$$

$$\frac{\partial w}{\partial t}(x,y,0) = g(x,y) \tag{5.40b}$$

From (5.21) and (5.22), the general motion of a freely vibrating membrane is

$$w(x,y,t) = \sum_{m=1}^{\infty}\sum_{n=1}^{\infty} \sin\frac{m\pi x}{a}\sin\frac{n\pi y}{b}(A_{mn}\sin\omega_{mn}t + B_{mn}\cos\omega_{mn}t) \tag{5.41}$$

Substituting (5.41) into the left-hand-sides of (5.40) gives

$$\sum_{m=1}^{\infty}\sum_{n=1}^{\infty} B_{mn}\sin\frac{m\pi x}{a}\bullet\sin\frac{n\pi y}{b} = f(x,y) \tag{5.42a}$$

$$\sum_{m=1}^{\infty}\sum_{n=1}^{\infty} A_{mn}\omega_{mn}\sin\frac{m\pi x}{a}\bullet\sin\frac{n\pi y}{b} = g(x,y) \tag{5.42b}$$

Multiplying both sides of (5.42) by $\sin(i\pi x/a)\cdot\sin(j\pi y/b)$, where i and j also are integers, and integrating over the area of the membrane, it is seen that

$$\int_0^a\int_0^b \sin\frac{m\pi x}{a}\bullet\sin\frac{n\pi y}{b}\bullet\sin\frac{i\pi x}{a}\bullet\sin\frac{j\pi y}{b}dx\,dy$$
$$= \begin{cases} 0, & \text{if } i \neq m \text{ or } j \neq n \\ \dfrac{ab}{4}, & \text{if } i = m \text{ or } j = n \end{cases} \tag{5.43}$$

Thus, the eigenfunctions of free vibration for the rectangular membrane are orthogonal over the membrane area. Using (5.43), then, result in the formulas for calculating the A_{mn} and B_{mn}. In this case they are coefficients of double Fourier series:

$$B_{mn} = \frac{4}{ab}\int_0^a\int_0^b f(x,y)\sin\frac{m\pi x}{a}\sin\frac{n\pi y}{b}dxdy \tag{5.44a}$$

$$A_{mn} = \frac{4}{\omega_{mn}ab}\int_0^a\int_0^b g(x,y)\sin\frac{m\pi x}{a}\sin\frac{n\pi y}{b}dxdy \tag{5.44b}$$

The motion of a rectangular membrane given arbitrary initial displacement and velocity (consistent, of course, with the boundary constraints) is thus given by (5.41), where the A_{mn} and B_{mn} are determined from (5.44).

For a *circular* membrane the initial conditions may be stated as

$$w(r,\theta,0) = f(r,\theta) \tag{5.45a}$$

$$\frac{\partial w}{\partial t}(r,\theta,0) = g(r,\theta) \tag{5.45b}$$

In general, $f(r, \theta)$ and $g(r, \theta)$ are functions which are neither even nor odd with respect to θ. Therefore, in general, it is necessary to use both the $\cos n\theta$ and the $\sin n\theta$ solution forms for the eigenfunctions. That is, the general motion is expressed as

$$\begin{aligned} w(r,\theta,t) &= \sum_{m=1}^{\infty}\sum_{n=0}^{\infty} J_n\left(\lambda_{mn}\frac{r}{a}\right)\cos n\theta(A_{mn}\sin\omega_{mn}t + B_{mn}\cos\omega_{mn}t) \\ &+ \sum_{m=1}^{\infty}\sum_{n=1}^{\infty} J_n\left(\lambda_{mn}\frac{r}{a}\right)\sin n\theta(C_{mn}\sin\omega_{mn}t + D_{mn}\cos\omega_{mn}t) \end{aligned} \tag{5.46}$$

where the $A_{mn}, \ldots, D_{mn}$ are constants to be determined, λ_{mn} is the mth eigenvalue of (5.31), and ω_{mn} is the corresponding natural frequency. Considerable simplification of the general form (5.46) can take place in special cases. For example, if the membrane is released from rest from an initial displaced shape which is axisymmetric, then (5.46) simplifies to

$$w(r,\theta,t) = \sum_{m=1}^{\infty} B_{mo}J_o\left(\lambda_{mn}\frac{r}{a}\right)\cos\omega_{mo}t \tag{5.47}$$

The eigenfunctions $J_n(\lambda_{mn}r/a)\cos n\theta$ and $J_n(\lambda_{mn}r/a)\sin n\theta$ are orthogonal over the area of the membrane, and thus the $A_{mn}, \ldots, D_{mn}$ of (5.46) may be calculated in a manner similar to that used earlier for the rectangular membrane. However, before carrying out the procedure, it would be desirable to say something more about the orthogonality.

Equation (5.29) is a form of the more general differential equation

$$\frac{d}{dz}\left(\phi_1\frac{d\psi}{dz}\right) + (\phi_2 + \mu\phi_3)\psi = 0 \tag{5.48}$$

where ψ, ϕ_1, ϕ_2, and ϕ_3 are arbitrary continuous functions of z (the first derivative of ψ must also be continuous), and μ is a constant. Let boundary conditions be applied at the ends of the interval $a \le z \le b$. In their most general forms, the boundary conditions may be written as

$$c_1\psi(a) + c_2\frac{d\psi}{dz}(a) = 0 \tag{5.49a}$$

$$c_3\psi(b) + c_4\frac{d\psi}{dz}(b) = 0 \tag{5.49b}$$

where c_1, … , c_4 are constants. Equation (5.48) and the boundary conditions (5.49) comprise the famous Sturm–Liouville problem of applied mathematics. Solutions to it are an infinite set of eigenvalues, μ_m, and a corresponding set of eigenfunctions, $\psi_m(z)$. It can be proven (cf. [5], p. 231) that the ψ_m are orthogonal with respect to the "weight function" $\phi_3(z)$ over the interval $a \le z \le b$. That is,

$$\int_a^b \psi_m \psi_i \phi_3 dz = 0, \quad \text{if } m \neq i \tag{5.50}$$

Returning to the circular membrane vibration problem, it is clear that if (5.29) is multiplied through by r, it is a special case of (5.48) with $\phi_1 = r$, $\phi_2 = -n^2/r$, $\phi_3 = r, z = r$, and $\mu = k^2$. It can also be shown [5] that if the regularity condition at $r = 0$ replaces (5.49b), then the orthogonality stated by (5.50) is valid over the interval $0 \le z \le a$. Thus, for the vibrating circular membrane

$$\int_0^a \int_0^{2\pi} J_n\left(\lambda_{mn}\frac{r}{a}\right) J_j\left(\lambda_{ij}\frac{r}{a}\right) \cos n\theta \cos j\theta \, r dr\, d\theta = 0, \quad \text{if } i \neq m \text{ or } j \neq n \tag{5.51}$$

and similarly for integrands involving $\sin n\theta$. Thus, substituting (5.46) into (5.45), multiplying through by $J_j(\lambda_{ij} r/a)\cos j\theta$, integrating over the area (where the elemental area is $rdr\, d\theta$), and using (5.51) and

$$\int_0^{2\pi} \cos^2 n\theta \, d\theta = \pi, \qquad (n = 1, 2, \ldots) \tag{5.52}$$

one obtains formulas for the coefficients B_{mn} and A_{mn} as:

$$B_{mn} = \frac{\int_0^a \int_0^{2\pi} f(r,\theta) J_n(\lambda_{mn} r/a) \cos n\theta \, r dr\, d\theta}{\pi \int_0^a [J_n(\lambda_{mn} r/a)]^2 \, r dr} \tag{5.53a}$$

$$A_{mn} = \frac{\int_0^a \int_0^{2\pi} g(r,\theta) J_n(\lambda_{mn} r/a) \cos n\theta \, r dr\, d\theta}{\pi \omega \int_0^a [J_n(\lambda_{mn} r/a)]^2 \, r dr} \tag{5.53b}$$

(except for $n = 0$, where π in the denominator is replaced by 2π), and similar formulas for C_{mn} and D_{mn} (replacing $\cos n\theta$ by $\sin n\theta$). Evaluation of the integrals in (5.53) is formidable, typically requiring numerical integration.

Because of the orthogonality of the Bessel functions over the interval $b \le r \le a$, as stated in (5.50), it is clear that (5.53) may also be applied to annular membranes simply by changing lower limits of the integrations on r, replacing 0 by b. Similar expressions may be used for sectorial membranes by changing limits of the integration on θ, and using the more general v in place of n.

5.6 Forced Vibrations

In the previous chapters, two methods were presented for the exact solution of steady-state, forced vibration problems: eigenfunction superposition and closed-form solution. Both methods are applicable to membranes. They will be summarized below for rectangular and circular membranes. Only viscous damping will be considered. The governing equation of motion is (5.7), generalized to

$$T\nabla^2 w + q = \rho h \frac{\partial^2 w}{\partial t^2} + c\frac{\partial w}{\partial t} \tag{5.54}$$

where c is a viscous damping coefficient.

For a *rectangular* membrane, using the eigenfunction superposition method, the exciting pressure is first expressed in terms of modal components:

$$q(x,y,t) = \sum_{m=1}^{\infty}\sum_{n=1}^{\infty} q_{mn}(t)\sin\frac{m\pi x}{a} \bullet \sin\frac{n\pi y}{b} \tag{5.55}$$

The displacement response is also characterized by the response of the modes:

$$w(x,y,t) = \sum_{m=1}^{\infty}\sum_{n=1}^{\infty} \Phi_{mn}(t)\sin\frac{m\pi x}{a} \bullet \sin\frac{n\pi y}{b} \tag{5.56}$$

Taking advantage of the orthogonality of the eigenfunctions, as in Sec. 5.5, the coefficients q_{mn} are obtained

$$q_{mn}(t) = \frac{4}{ab}\int_0^a\int_0^b q(x,y,t)\sin\frac{m\pi x}{a}\sin\frac{n\pi y}{b}\,dx\,dy \tag{5.57}$$

Substituting (5.55) and (5.56) into (5.54), and equating coefficients of like terms in x and y, one obtains the doubly infinite set of equations of motion for the individual modes:

$$\rho h\Phi''_{mn} + c\Phi'_{mn} + T(\alpha_m^2 + \beta_n^2)\Phi_{mn} = q_{mn}(t) \qquad (m,n = 1, 2, 3, \ldots) \tag{5.58}$$

where $\alpha m = m\pi/a$, $\beta_n = n\pi/b$. Equation (5.58) is the same form as (2.82) (for strings), and its solution may be taken directly from the results of Sec. 2.9.

To find a closed form solution to the rectangular membrane problem, let us first reduce the complexity of the problem by assuming that there is no damping, and that the exciting pressure is sinusoidal in time. Then, (5.54) becomes

$$T\nabla^2 w + Q(x,y)\sin\Omega t = \rho h \frac{\partial^2 w}{\partial t^2} \tag{5.59}$$

For more general, periodic forcing functions, the exciting pressure may be expressed by a Fourier series in time (see Sec. 2.9), for which (5.59) is representative of a single term of the series. Assuming a solution in the form

$$w(x,y,t) = W(x,y)\sin\Omega t \tag{5.60}$$

(5.59) reduces to

$$T\nabla^2 w + \rho h\Omega^2 W + Q = 0 \tag{5.61}$$

Assume further that $Q(x,y)$ is expressible as a Fourier sine series in x, that is,

$$Q(x,y) = \sum_{m=1}^{\infty} Q_m(y)\sin\alpha_m x \tag{5.62}$$

where $$Q_m(y) = \frac{2}{a}\int_0^a Q(x,y)\sin\,\alpha_m x dx \tag{5.63}$$

and $\alpha m = m\pi/a$. Similarly, take W in the form

$$W(x,y) = \sum_{m=1}^{\infty} W_m(y)\sin\alpha_m x \tag{5.64}$$

which satisfies the boundary conditions at $x = 0, a$ exactly. Substituting (5.62) and (5.64) into (5.61), and equating coefficients of like terms in x,

$$TW_m'' + (\rho h\Omega^2 - T\alpha_m^2)W_m = -Q_m(y) \qquad (m = 1, 2, \ldots) \tag{5.65}$$

This has three forms of solution, depending on whether $\rho h\Omega^2 - T\alpha_m^2$ is positive, negative, or zero, that is, depending on the magnitude of the forcing frequency and on m. Applying the boundary conditions at $y = 0, b$ determines the two constants of integration arising from (5.65). Further details are available in Ref. [6].

If damping is to be considered in the closed form analysis, then (5.60) must be generalized to recognize the out-of-phase motion:

$$w(x,y,t) = W_1(x,y)\sin\Omega t - W_2(x,y)\cos\Omega t \tag{5.66}$$

Substituting this into (5.54) with sinusoidal excitation results in

$$T\nabla^2 W_1 + \rho h\Omega^2 W_1 - c\Omega W_2 = -Q(x,y) \tag{5.67a}$$

$$T\nabla^2 W_2 + \rho h\Omega^2 W_2 + c\Omega W_1 = 0 \tag{5.67b}$$

Solving (5.67b) for W_1 and substituting it into (5.67a) yields a fourth order partial differential equation in W_2. Assuming further a $\sin a_m x$ variation for W_1, W_2, and Q, as in (5.62) and (5.64), yields a fourth-order ordinary differential equation similar in form to (2.111). Solution of the problem is completed in the manner followed in Sec. 2.10 subsequent to (2.111).

For a *circular* membrane similar procedures are followed, except that solutions are in polar coordinates and that Bessel functions are required. Thus, with the eigenfunction superposition method one expresses an arbitrary forcing pressure as

$$q(r,\theta,t) = \sum_{m=1}^{\infty}\sum_{n=0}^{\infty} q_{mn}^{(1)}(t) J_n\left(\lambda_{mn}\frac{r}{a}\right)\cos n\theta + \sum_{m=1}^{\infty}\sum_{n=1}^{\infty} q_{mn}^{(2)}(t) J_n\left(\lambda_{mn}\frac{r}{a}\right)\sin n\theta \tag{5.68}$$

and the displacement response in the same form, with unknown functions of time $\Phi_{mn}^{(1)}$ and $\Phi_{mn}^{(2)}$. The coefficients $q_{mn}^{(1)}$ and $q_{mn}^{(2)}$ are found from quotients involving the integrals of Bessel functions, as in (5.53a). This yields uncoupled ordinary differential equations in $\Phi_{mn}^{(1)}$ and $\Phi_{mn}^{(2)}$ in the form of (5.58).

The closed form solution procedure for a circular membrane will be demonstrated by an example below.

Example 5.1 Determine the displacement response of a circular membrane subjected to a uniform pressure which is sinusoidal in time:

$$q = q_0 \sin\Omega t$$

Neglect damping.

Solution

Uniform pressure can excite only the axisymmetric ($n = 0$) motions. Therefore, assume a solution as

$$w(r,t) = W(r)\sin\Omega t$$

Substituting this and the exciting pressure into (5.54), with $c = 0$, gives

$$T\left(\frac{d^2W}{dr^2} + \frac{1}{r}\frac{dW}{dr}\right) + \rho h\Omega^2 W = -q_o$$

or

$$\frac{d^2W}{dr^2} + \frac{1}{r}\frac{dW}{dr} + k^2 W = -\frac{q_o}{T}$$

where $k^2 = \rho h\Omega^2/T$. This is seen to be Bessel's equation (5.29) for $n = 0$, with an added R.H.S. for which a particular solution must be found. Therefore, the complete solution is

$$W(r) = A_0 J_0(kr) + B_0 Y_0(kr) - \frac{q_0}{k^2 T}$$

Considering the regularity condition at $r = 0$, the constant B_0 must be taken as zero. The boundary condition $W(a) = 0$ gives

$$A_0 = \frac{q_0}{k^2 T J_0(ka)}$$

A closed form solution to the problem is therefore

$$w(r,t) = \left[A_0 J_0(kr) - \frac{q_0}{k^2 T}\right]\sin\Omega t$$

with A_0 depending on ka, as shown above. Furthermore, since

$$ka = \Omega a\sqrt{\frac{\rho h}{T}} = \left(\frac{\Omega}{\omega_{mo}}\right)\lambda_{mo}$$

where ω_{mo} is the mth axisymmetric natural frequency, and λ_{mo} is the corresponding eigenvalue (2.4048, 5.5201, etc.), then as the exciting frequency (Ω) approaches any natural frequency, ka approaches λ_{mo}, and $J_0(\lambda_{mo})$ approaches zero, as may be seen from (5.31). Thus, resonances occur at all the axisymmetric natural frequencies.

Comparing the eigenfunction superposition (ES) and closed form (CF) methods for accomplishing the exact solutions of forced vibration problems for membranes, the following summary statements may be made:

1. For forcing pressures having general spatial distribution, the membrane problem, which is two-dimensional, requires summing a double series of displacement functions with the ES method, whereas the CF approach requires only a single series.
2. When damping is neglected, the ES procedure is easy to apply for the rectangle, but complicated for the circular case. The CF method is easy for both problems.

3. When damping is included, the ES approach is only slightly more complicated, whereas the CF method becomes difficult.

Thus, both methods are useful, and each has classes of problems for which it is better suited than the other.

5.7 Energy Functionals; Rayleigh and Ritz Methods

As seen in the preceding chapters, the Rayleigh and Ritz methods are effective procedures for determining free vibration frequencies and mode shapes of structural elements having complications due to variations in axial tension, density, cross-sectional area, or elastic modulii. In such cases the differential equations of motion have variable coefficients and, in general, have no known exact solutions.

Considering the equation of motion (5.6) for a membrane, it was observed that T_x, T_y, T_{xy}, ρ, and h may each be functions of x and y. For such problems the Rayleigh and Ritz techniques may also be well suited. Moreover, for the most simple problems governed by (5.9), boundaries other than rectangular or circular may be encountered, which can also be dealt with by Rayleigh and Ritz methods. With these approaches the energy functionals are required.

The potential energy of a stretched membrane in a position displaced from its flat equilibrium position may be determined in a manner similar to that used for the string (see Sec. 2.11). That is, considering the displaced infinitesimal element shown in Fig. 5.2, the change in the potential energy of the element due to T_x is

$$d(PE)_x = \frac{1}{2} T_x dy \left(\frac{\partial w}{\partial x} \right)^2 dx \tag{5.69}$$

which resembles that of the string (2.124), except that T_x is force/length, which must be multiplied by the length (dy) along which it acts. Similarly, due to T_y,

$$d(PE)_y = \frac{1}{2} T_y dx \left(\frac{\partial w}{\partial y} \right)^2 dy \tag{5.70}$$

In addition, local changes in slope cause shear strains. If the membrane is also subjected to an initial shear stress $\tau_{xy} = T_{xy}/h$, which is sufficiently large to remain unchanged during the vibratory displacement, then the potential energy stored in the deformed element is $\tau_{xy}\gamma_{xy}dV$, where γ_{xy} is the shear strain due to the vibratory displacement, and $dV = hdxdy$ is the volume of the element.

Figure 5.12 shows an infinitesimal element which was rectangular in shape before being displaced, with lengths dx and dy. In

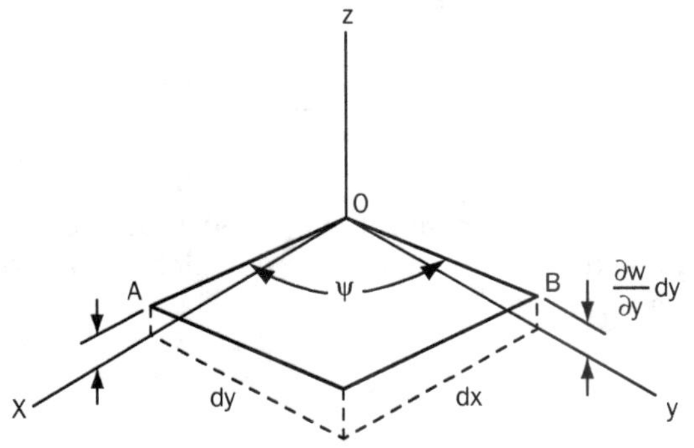

Figure 5.12 Shear strain in a membrane element.

general, the element undergoes a transverse displacement, w (not shown), slope changes $\partial w/\partial x$ and $\partial w/\partial y$, and curvature changes (which yield higher order terms in the present analysis). Due to the slopes, the angle AOB = ψ originally 90°, is changed, with resulting shear strain, γ_{xy}.

Using unit vectors i, j, k in the x, y, z directions, respectively, and considering small slopes, the vectors $\overline{OA}$ and $\overline{OB}$ taken along the edges of the deformed element are

$$\overline{OA} = dx\,\bar{i} + \frac{\partial w}{\partial x}\,dx\,\bar{k}$$

$$\overline{OB} = dx\,\bar{j} + \frac{\partial w}{\partial y}\,dy\,\bar{k} \tag{5.71}$$

The cosine of the angle between the two vectors is obtained from their dot product:

$$\overline{OA} \bullet \overline{OB} = |OA||OB|\cos\psi = \frac{\partial w}{\partial x}\frac{\partial w}{\partial y}dxdy \tag{5.72}$$

where $|OA|$ and $|OB|$ are their magnitudes:

$$|OA| = dx\sqrt{1+\left(\frac{\partial w}{\partial x}\right)^2}$$

$$|OB| = dy\sqrt{1+\left(\frac{\partial w}{\partial y}\right)^2} \tag{5.73}$$

For small slopes, (5.72) yields

$$\cos\psi = \frac{\partial w}{\partial x}\frac{\partial w}{\partial y} \tag{5.74}$$

The shear strain in the element is tan (90 – ψ) which, for small values of 90 – ψ, can be replaced by sin (90 – ψ), which equals cos ψ. Thus, (5.74) becomes

$$\gamma_{xy} = \frac{\partial w}{\partial x}\frac{\partial w}{\partial y} \tag{5.75}$$

The potential energy in the element due to shear strain is therefore

$$\begin{aligned} d(PE)_{xy} &= \tau_{xy}\,\gamma_{xy}\,dV \\ &= \frac{T_{xy}}{h}\frac{\partial w}{\partial x}\frac{\partial w}{\partial y}(h\,dx\,dy) \\ &= T_{xy}\frac{\partial w}{\partial x}\frac{\partial w}{\partial y}\,dx\,dy \end{aligned} \tag{5.76}$$

and the total potential energy in the displaced membrane is

$$PE = \frac{1}{2}\iint_A \left[T_x\left(\frac{\partial w}{\partial x}\right)^2 + T_y\left(\frac{\partial w}{\partial y}\right)^2 + 2T_{xy}\frac{\partial w}{\partial x}\frac{\partial w}{\partial y}\right] dxdy \tag{5.77}$$

where the integration is taken over the area (A) of the membrane.

The kinetic energy of an element undergoing vibratory displacement is

$$d(KE) = \frac{\rho h}{2}\left(\frac{\partial w}{\partial t}\right)^2 dxdy \tag{5.78}$$

where again, ρ is mass per unit volume. The kinetic energy of the complete vibrating membrane is therefore

$$KE = \frac{1}{2}\iint_A \rho h\left(\frac{\partial w}{\partial t}\right)^2 dxdy \tag{5.79}$$

where, in general, ρ and h may vary with x and y.

The procedures followed in using the Rayleigh or Ritz methods for a vibrating membrane are the same as those for a string (see

Secs. 2.12 and 2.13). That is, one obtains maximum values of *PE* and *KE* during a vibratory cycle. Taking

$$w(x,y,t) = W(x,y)\sin(\omega t + \phi) \tag{5.80}$$

$$\boxed{PE_{\max} = \frac{1}{2}\iint_A \left[T_x\left(\frac{\partial W}{\partial x}\right)^2 + T_y\left(\frac{\partial W}{\partial y}\right)^2 + 2T_{xy}\frac{\partial W}{\partial x}\frac{\partial W}{\partial y}\right] dxdy} \tag{5.81}$$

$$\boxed{KE_{\max} = \frac{\omega^2}{2}\iint_A \rho h W^2 dxdy} \tag{5.82}$$

The Rayleigh method requires assuming a reasonable representation for the mode shape desired, $W(x,y)$, and substituting it into Rayleigh's Quotient, (2.133), where $KE_{max} = \omega^2 KE^*_{max}$. The Ritz method involves assuming a set of trial (or admissible) functions to represent the mode shape,

$$W(x,y) = \sum_{i=1}^{I} C_i \phi_i(x,y) \tag{5.83}$$

and employing the frequency minimizing equations (2.136).

The Rayleigh and Ritz methods may be applied to membranes of arbitrary boundary shape by assuming admissible functions as

$$W(x,y) = G(x,y)\sum_{i=0}^{I}\sum_{j=0}^{J} C_{ij}x^i y^j \tag{5.84}$$

where $G(x,y)$ is the equation of the boundary written in *implicit* form. For example, consider the triangular boundary determined by any three intersecting straight lines $y = a_k x + b_k$ $(k = 1, 2, 3)$ where a_k and b_k are arbitrary. Then,

$$G(x,y) = (y - a_1x - b_1)(y - a_2x - b_2)(y - a_3x - b_3) \tag{5.85}$$

For an elliptical boundary having semi-major and semi-minor axes a and b, respectively.

$$G(x,y) = \frac{x^2}{a^2} + \frac{y^2}{b^2} - 1 \tag{5.86}$$

For a membrane fixed along an outer triangular boundary and along an inner boundary, which is elliptical (or conversely), $G(x,y)$ is a product of the two functions given in (5.85) and (5.86). Using the Rayleigh method, a single term of (5.84) could be taken (e.g., $i = j = 0$ for the fundamental mode). Typically, frequencies obtained by this simple approach will not be very accurate. However, if the Ritz method is used, and sufficiently large I and J are taken in (5.84), accurate frequencies are guaranteed, for the set of admissible functions used in (5.84) is complete.

Example 5.2 A membrane has boundaries consisting of the two straight lines $x = a$ and $y = 0$, and the curve $y/a = (x/a)^n$, as depicted in Fig. 5.13. It is subjected to a uniform tensile stress resultant T normal to all boundaries.

A. By means of the Rayleigh method, derive a formula which one could use to obtain an approximate value of the fundamental frequency for a membrane having arbitrary values of a and n.
B. Use the formula of Part A to calculate $\omega a\sqrt{\rho h/T}$ for $n = 0.1, 0.5, 1, 2,$ and 10. Plot the membrane boundaries for each case. Do the frequencies seem reasonable for the various shapes? Make comparisons with known frequencies of other shapes, where possible, to verify that the results are reasonable.

Solution

Part A. Using nondimensional coordinates $\xi = x/a$, $\eta = y/a$, we choose

$$W(\xi,\eta) = \eta(\xi - 1)(\eta - \xi^n)$$

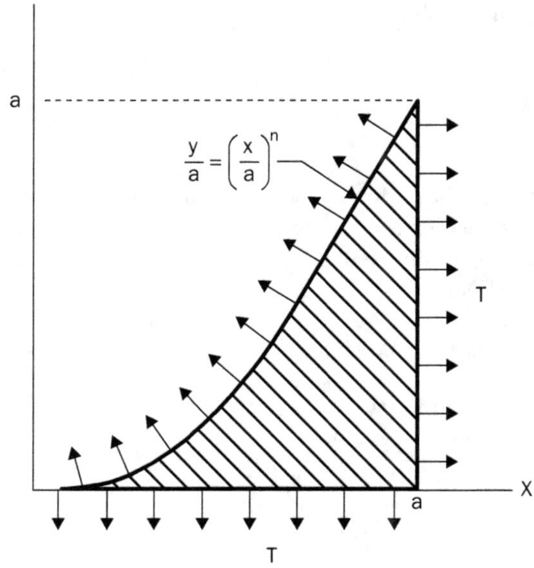

FIGURE 5.13 A membrane with two sides straight and one curved.

which satisfies $W = 0$ on all boundaries. The energy functionals are evaluated by (5.81) and (5.82) as

$$PE_{max} = \frac{T}{2}\left[\frac{2n}{15(5n-1)(5n+1)} + \frac{2}{3(3n+1)(3n+2)(3n+3)}\right]$$

$$KE^*_{max} = \frac{\rho h a^2}{2}\frac{1}{15(5n+1)(5n+2)(5n+3)}$$

The frequency squared is therefore obtained from (2.133) as

$$\frac{\omega^2 a^2 \rho h}{T} = \left[\frac{2n}{(5n-1)(5n+1)} + \frac{10}{(3n+1)(3n+2)(3n+3)}\right](5n+1)(5n+2)(5n+3)$$

Part B. Plots of the membrane shapes for $n = 0.1, 0.5, 1, 2$, and 10 are made in Fig. 5.14. For $n = 1$, the shape is a right triangle. For $n = 10$, the shape consists of two narrow, perpendicular lobes (shown shaded). For $n = 0.5$ and $n = 2$, the curved boundaries are parabolas. For $n = 0.1$, a nearly square shape results. The circle arc $(x/a)^2 + (y/a)^2 = 1$ is also shown for comparison.

Nondimensional frequencies calculated from the formula derived above are shown in Table 5.5 for $n = 0$, 0.2, and ∞, as well as the values in the problem statement. Exact results for $n = 0$ (square), $n = 1$ (right triangle), and a quarter circle sector (see dashed curve of Fig. 5.14) are also provided for comparison, taken from data presented earlier in this chapter. In the case of the right triangular shape, the Rayleigh solution is reasonably accurate (6.6 percent error). The frequency for the parabolic half-segment ($n = 0.5$) is also reasonable,

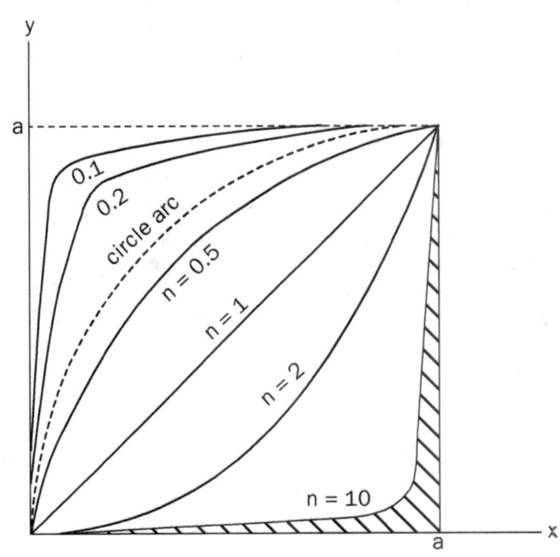

FIGURE 5.14 Membrane shapes having various curves $y/a = (x/a)^n$ for one boundary.

n	Shape description	$\omega a\sqrt{\rho h/T}$	
		Rayleigh	**Exact**
$\to 0$	Square	3.16	4.44
0.1	Nearly square	3.65	-
0.2	-	∞	-
-	Quarter circle sector	-	5.14
0.5	Parabolic half-segment	6.20	-
1	Right triangle	7.48	7.02
2	Parabolic spandrel	10.17	-
10	Two narrow lobes	34.17	-
$\to\infty$	No area	∞	∞

Table 5.5 Frequencies of the Membranes of the Shape Shown in Fig. 5.13

falling between those of the quarter circle sector and the right triangle, as Fig. 5.14 suggests it should. The result for $n = 0.2$ was added to Table 5.5 to show the singularity in the frequency formula in that case. For $n = 0$ the Rayleigh frequency should be an upper bound on the exact frequency, but is not, which indicates serious trouble. The trouble occurs because the first term in the result for $PE_{\max}$ shown above becomes negative for $n < 0.2$. The first term is the integral of $(\partial W/\partial x)^2$ in the potential energy functional, and it should not be negative.

Thus, the frequency formula derived by the Rayleigh method appears to yield reasonable results for $n > 0.5$, but should definitely not be used for n in the vicinity of 0.2 or less than 0.2. This example shows clearly the importance of looking at the *reasonableness* of results obtained from any theoretical approach.

Example 5.3 A square membrane (dimensions $a \times a$) is subjected to uniform tension T_x and T_y in the x and y directions, as shown by Fig. 5.15. In addition shear stress resultants T_{xy} are applied to the four edges as shown. Simple plane elasticity tells us T_x, T_y, and T_{xy} are constant throughout the membrane. The membrane is fixed along all four edges.

In the case of hydrostatic tension ($T_y = T_x$), make a study of what happens to the free vibration frequencies and mode shapes as T_{xy} is added and increased.

Solution

No exact solution for this problem is known if T_{xy} is present. The Ritz method is used, with the assumed displacement functions.

$$W(x,y) = \sum_{m=1}^{M}\sum_{n=1}^{N} C_{mn} \sin\frac{m\pi x}{a} \sin\frac{n\pi y}{b}$$

In addition to satisfying the zero displacement boundary conditions exactly, they form a mathematically complete set. This guarantees that, in the limit, as M and N become large, the exact solution is approached.

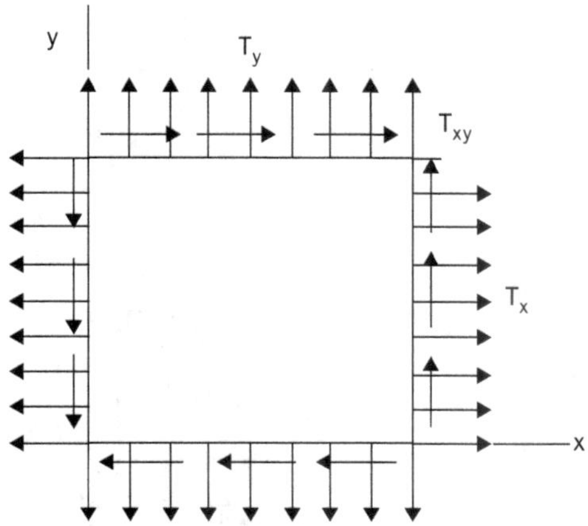

Figure 5.15 Uniform shear (T_{xy}) superimposed upon hydrostatic tension ($T_x = T_y$).

Using the Ritz minimizing equations (2.136) or (4.99), with PE_{max} and KE_{max} for the membrane given by (5.81) and (5.82), respectively, one obtains a set of $M \times N$ simultaneous, homogeneous equations in terms of the coefficient C_{mn}. For a nontrivial solution, the determinant of the coefficient matrix is set equal to zero. The roots of the determinants are $M \times N$ nondimensional frequencies.

Table 5.6 lists the first six nondimensional frequencies $\omega a\sqrt{\rho/T_x}$ as T_{xy}/T_x increases from 0 to 1. They were obtained using $M = N = 8$ in $W(x,y)$, yielding a determinant of order 64. Comparisons with very precise solutions arising from using determinants of order $M \times N = 16 \times 16 = 25$ 6 show that all frequencies in Table 5.6 are within 0.2 percent of their converged exact values. As T_{xy}/T_x increases, increased numbers of terms in the series are required for accurate frequencies. For $T_{xy} = 0$, only one term is needed for the exact solution. For example, the first nondimensional frequency is then $\sqrt{2}\pi = 4.443$, as given by (5.20), and listed in Table 5.6. The table shows that the frequencies eventually all decrease as T_{xy} increases. For $T_{xy}/T_x > 1$, the membrane would be in compression along one plane, it would buckle, and any vibration results would be meaningless. In Table 5.6 the largest terms (mn) at the double sine series used in the solution are shown in parentheses.

Figure 5.16 shows the two-dimensional contour plots (lines of constant displacement) for the various frequencies. Points of maximum and minimum displacements are identified by plus and minus signs. The first four mode shapes are shown for shear load ratios, $T_{xy}/T_x = 0$, 0.2, 0.5, 0.85, and 1. Looking at the membrane shape and the edge loading (Fig. 5.15), one sees that both diagonals of the square are diagonals of the loading symmetry. Therefore, mode shapes are either symmetric or antisymmetric with respect to both diagonals. This is readily seen in the contour plots of Fig. 5.16.

As the shear loading is increased, Fig 5.16 shows that the contour lines become increasingly skewed. In some cases node lines ($w = 0$) separate, as can be seen for the fourth mode as T_{xy}/T_x changes from 0 to 0.2. For $T_{xy}/T_x = 0.01$, the node lines of the fourth mode would *seem* to be the same as for $T_{xy}/T_x = 0$. But they would not quite cross, instead having kinks very close to the membrane midpoint.

For more results for this problem, and also for membranes loaded with linearly varying edge tensions, see Ref. [7].

T_{xy}/T_x	Mode number					
	1	**2**	**3**	**4**	**5**	**6**
0.00	4.443 (11)	7.025 (12)	7.025 (21)	8.886 (22)	9.935 (13)	9.935 (31)
0.10	4.438 (11)	6.916 (12)	7.118 (21)	8.837 (22)	9.922 (13)	9.962 (31)
0.20	4.425 (11)	6.790 (12)	7.197 (21)	8.701 (22)	9.886 (13)	10.030 (31)
0.50	4.322 (11)	6.285 (12)	7.337 (21)	7.934 (22)	9.603 (13)	9.606 (23)
0.75	4.126 (11)	5.620 (12)	6.864 (22)	7.306 (21)	8.099 (23)	9.062 (31)
0.85	3.989 (11)	5.211 (12)	6.221 (22)	7.212 (34)	7.221 (21)	8.149 (44)
0.98	3.591 (11)	4.174 (23)	4.647 (11)	5.110 (45)	5.584 (55)	6.114 (56)
1.00	3.376 (11)	3.677 (34)	4.015 (11)	4.411 (21)	4.889 (11)	5.488 (12)

Table 5.6 Nondimensional Frequencies $\omega a\sqrt{\rho/T_x}$ for Biaxial Membrane Tension with Added Shear

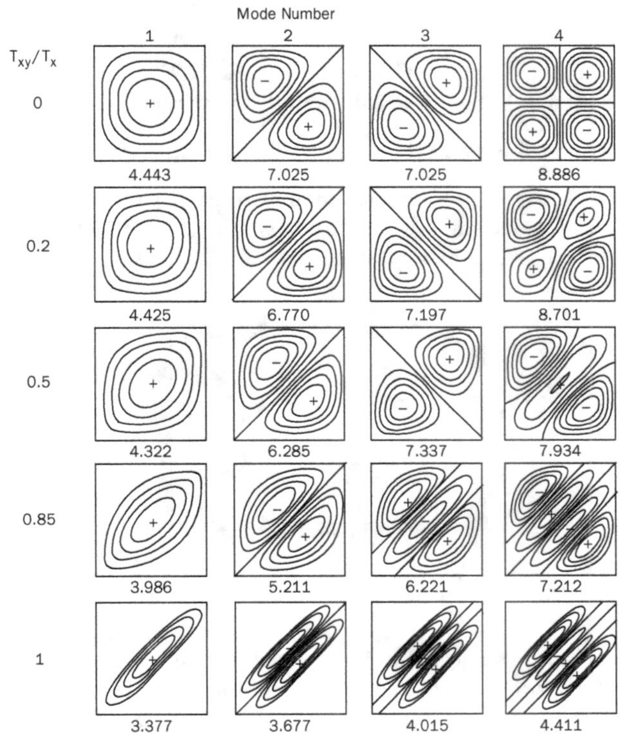

FIGURE 5.16 Contour plots for Example 5.3.

References

1. S. P. Timoshenko and J. N. Goodier, *Theory of Elasticity*, 3rd ed., McGraw-Hill Book Co., 1970.
2. F. W. J. Olver, "Bessel functions of integer order," Chap. 9 in *Handbook of Mathematical Functions*, ed. M. Abramowitz and I. A. Stegun, U.S. Government Printing Office, 1964.
3. A. W. Leissa. "Singularity considerations in membrane, plate and shell behaviors," *Int. J. Solids Struct.*, 38 (2001): 3341–53.
4. H. A. Antosiewicz, "Bessel functions of fractional order," Chap. 10 in *Handbook of Mathematical Functions*, ed. M. Abramowitz and I. A. Stegun, U.S. Government Printing Office, 1964.
5. C. R. Wylie, Jr., *Advanced Engineering Mathematics*, 1st ed. McGraw-Hill Book Co., 1951.
6. A. W. Leissa, "Closed form exact solutions for the steady state vibrations of continuous systems subjected to distributed exciting forces," *J. Sound Vib.*, 134 (1989): 435–53.
7. A. W. Leissa and A. Ghamat-Rezaei, "Vibrations of rectangular membranes subjected to shear and nonuniform tensile stresses," *J. Acoust. Soc. Amer.*, 88 (1990): 231–8.

Problems

1 A rectangular membrane with aspect ratio $(a/b) = 1.5$ is subjected to uniform tension (T), and is fixed all around at its boundaries (Fig. 5.17).

A. Determine the lowest three *degenerate* frequencies $\omega a\sqrt{\rho h/T}$.

B. For each of the three frequencies found in Part A, plot the nodal patterns $(w = 0)$ with mode amplitude ratios of 0.1, 0.5, 1, 2, and 10. (In Fig. 5.7, for example, one sees ratio of 1, 1.02, 2, and ∞ for C_{21}/C_{12}.)

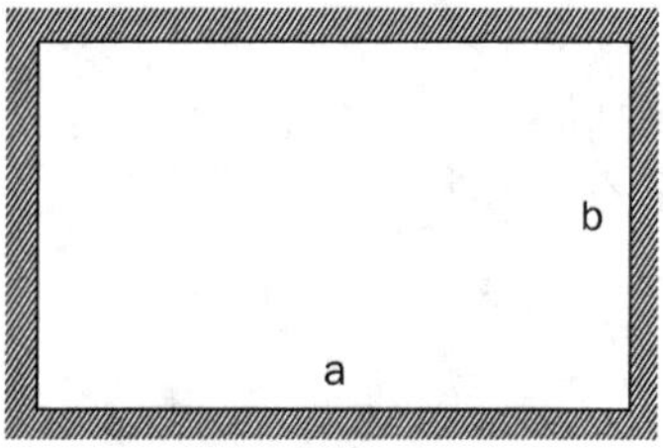

Figure 5.17 Problem 1.

2 A membrane is subjected to inplane stresses which may be expressed in terms of polar coordinate components. A differential element with stress resultant components T_r, T_θ, $T_{r\theta}$, and $T_{\theta r}$ is shown below (Fig. 5.18). In general, the magnitudes of the components vary with r and θ, so that differential changes in them occur across the element (shown by plus signs). The equation of motion (5.6) was derived early in this chapter in rectangular coordinates, using rectangular stress resultants T_x, T_y, and T_{xy}. Do the same in polar coordinates using the polar stress resultants.

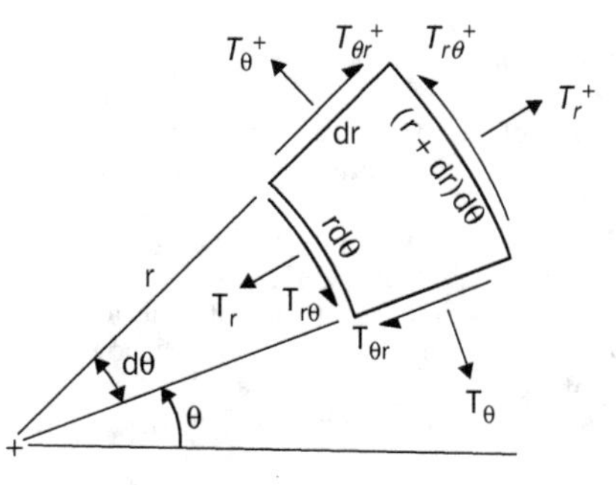

Figure 5.18 Problem 2.

3 A rectangular membrane with aspect ratio $a/b = 2$, has uniform tension (T) stretching it along each of its four edges. The membrane is subjected to a uniformly distributed transverse pressure acting sinusoidally with time

$$q = q_0 \sin \Omega t$$

where q_0 and Ω are constants. There is no damping present.

A. Determine the transverse displacement (w) as the membrane as a function of its coordinates (x, y) and time (t).

B. Determine the displacement $w_0(x, y)$ of the membrane when it is subjected to a *static* pressure w_0.

C. At a point $x/a = y/b = 1/4$ on the membrane, make a plot of W/δ versus Ω/ω_1, where W and δ are dynamic and static displacements of the point, and ω_1 is the first natural frequency.

D. Without making any further calculations, *discuss* how and why the plot of Part C would be different if one looked instead at the point $x/a = y/b = 1/2$.

4 Equations (5.81) and (5.82) are the energy functionals in rectangular coordinates, which are useful for solving membrane vibration problems by Rayleigh and Ritz methods. Derive the corresponding functionals PE_{max} and KE_{max} in polar coordinates (r, θ) in the case where $T_x = T_y = T$ and $T_{xy} = 0$. (Hint: One could use chain rule differentiation: $\partial w/\partial x = [(\partial w/\partial r)\ (\partial r/\partial x)] + [(\partial w/\partial \theta)\ (\partial \theta/\partial x)]$, etc.)

5 A. A symmetrical trapezoidal membrane shown (Fig. 5.19) is subjected to uniform tension in all directions. Use the Ritz method twice, each time with a two-term admissible function as in (5.84), to obtain approximate formulas for the fundamental frequencies, $\omega c\sqrt{\rho h/T}$. (Hint: choose the coordinate origin at the intersection A of the extended lines of the sides to simplify the algebra.)

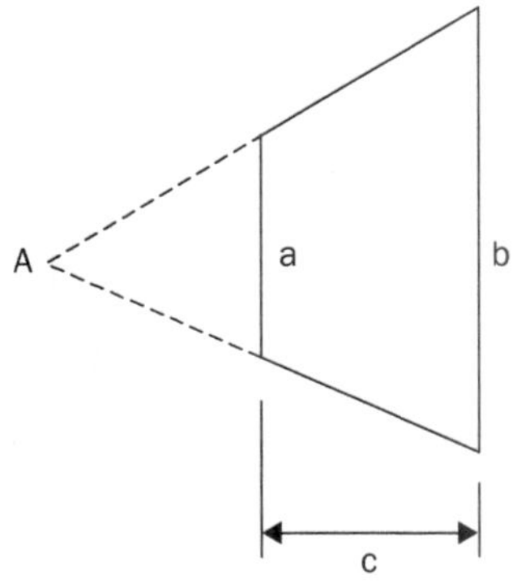

FIGURE 5.19 Problem 5.

B. Plot $\omega c\sqrt{\rho h/T}$ versus c/a for both solutions on a single graph, with the *area of the membrane* remaining constant as c/a changes. Do the two curves cross each other? If so, explain why.

6 It is desired to create reasonably accurate *fundamental* frequencies (only) for a class of membranes having two-fold symmetry in their shapes, with their boundaries defined by $\xi^n + \zeta^n=1$, where $\xi = x/a$ and $\zeta = y/b$. A representative shape for $n = 2$ (i.e., an ellipse) is shown in the sketch (Fig. 5.20). Assume that the classical membrane theory applies, that is, boundaries are fixed, with a uniform tension (T) applied, uniform thickness and mass density, small slopes, etc.

A. Plot the membrane shapes corresponding to all combinations of $n = 1$, 2, 4 with $a/b = 1, 2, 5$ (that is 9 shapes).

B. Use the Rayleigh method with a single admissible function to obtain the crude approximations for the fundamental frequency parameters $\omega b\sqrt{\rho h/T}$ for all combinations of the boundary shapes shown in Part A. (First do a *general* analysis for *arbitrary* n and a/b; then, substitute in the various n and a/b.)

C. Improve on the frequencies found in Part B by carrying out a Ritz analysis with *three* admissible functions.

D. Discuss the results of Part B and C. Did you find any interesting things? How close to the exact frequencies do you think the various frequencies are in Part C? How do they compare with *known* exact frequencies?

Hints:

(1) In Part C, differentiating with respect to the coefficients (C_{ij}) before carrying out the area integrations should reduce the algebraic labor involved.

(2) Use a symbolic logic program (e.g., MAPLE®, MACSYMA®) to reduce your effort and eliminate mistakes.

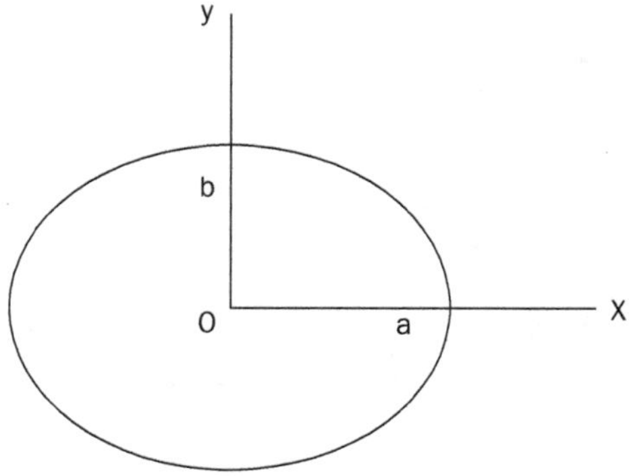

FIGURE 5.20 Problem 6.

CHAPTER 6

Plate Vibrations

From a geometric viewpoint, a plate is very much like a membrane. That is, like a membrane, a plate is a structural element that is relatively thin in one direction compared with the other two, and is flat. However, the plate has bending stiffness, whereas the membrane does not. Typically, the flexural (bending) stiffness arises because a plate is considerably thicker than a membrane relative to its other dimensions.

As seen in Chap. 5, the sole stiffness in transverse vibrations of membranes arises from the in-plane initial stresses. The basic stiffness of a plate is due to its resistance to bending, although additional stiffness may be due to in-plane initial stresses (see Sec. 6.8). Whereas the membrane is the two-dimensional (2D) generalization of the string (Chap. 2), the plate is the 2D generalization of the beam (Chap. 4).

Although a plate is typically thicker than a membrane, the ratio of its thickness to its average lateral dimensions is usually taken not to exceed 1/20 to represent its fundamental vibration mode reasonably accurately with classical thin plate theory. For thicker plates, or higher frequency modes, one needs to consider shear deformation (i.e., the shear flexibility, in addition to the bending flexibility) and rotary inertia effects (see Sec. 6.8).

As in the case of the membrane, our study will be limited to the *transverse* vibration modes, because the in-plane vibration modes have significantly higher frequencies (at least one order of magnitude) than those of transverse vibration. The in-plane vibration analysis of a plate is the same as that for a membrane, and the frequencies and the mode shapes do not depend on the initial in-plane stresses.

Plates are important structural elements. They may exist in many applications. In civil engineering, flat panels exist in various steel or concrete structures (e.g., floor slabs). They may be of various shapes (rectangular, circular, rhombic, triangular, trapezoidal, and others). Plates also occur in aerospace (e.g., aircraft, missile) and naval (e.g., ship, submarine) structures. In mechanical engineering, plates can be seen as rotor disks in brake systems, parts of various clutch and other components. They can also exist as flat panels in machine housings. Plates are also used as housings of many electrical and electronic equipment. Compact discs (CDs) and similar devices (DVDs) are actually circular plates.

Although shells (to be studied in the next chapter) have, indeed, wider application in almost every engineering field, plates also present a reasonable introduction to the analysis of those complex structures. Plates give us the chance to study the influence of bending in two dimensions without having tangential stresses caused by transverse vibratory motion.

The equations of motion will first be derived in this chapter. Subsequently free vibration of rectangular, circular, annular, and sectorial plates will be studied using exact solutions. In addition, energy terms will be developed for use in Raleigh and Ritz analyses. Treatment of elliptical, triangular, and trapezoidal plates is then also made. Many other possible complexities in the analyses of plates will be introduced. Nonlinear vibrations of plates and vibrations of orthotropic and laminated composite plates will also be described as examples of these numerous complexities.

6.1 Equation of Motion for Transverse Vibrations

Figure 6.1 shows a plate of thickness h and arbitrary shape in its static equilibrium position. As a rather general representation it has one portion of its boundary clamped, another portion simply supported (i.e., hinged or knife-edge support), and a third portion completely free. In the top (plan) view, the simply supported edge is indicated by a dashed line parallel to and inside of the boundary. This is a conventional way of showing such an edge constraint. Rectangular coordinates are shown, where x and y are the in-plane coordinates and z is transverse. The origin of the coordinate system is at the mid-plane of the plate, so that the top and bottom surfaces of the plate are at $z = \pm h/2$ when it is in equilibrium.

In Fig. 6.1 a typical rectangular plate element is shown with in-plane dimensions $dx \times dy$. Its transverse dimension is h, the plate thickness. Figure 6.2 is a three-dimensional sketch of the element in a typical, displaced position and deformed shape. Shown on the element, by vectors with single arrows, are transverse shearing forces (per unit length along an edge) Q_x and Q_y, and their incremental changes, acting along its edges, and a distributed pressure (per unit surface area) q which is applied to a lateral surface. Also shown, by vectors with double arrows, are bending moment resultants (M_x and M_y) and twisting moment resultants (M_{xy} and M_{yx}), and their incremental changes, which have dimensions of moment per unit length along an edge. As in elementary beam theory, the shearing forces are the integrals over the plate thickness of transverse shear stresses (τ_{xz} and τ_{yz}), which may be assumed to vary parabolically across the plate thickness from zero at one face, to a maximum at the deformed mid-surface, to zero at the other face. The bending moments

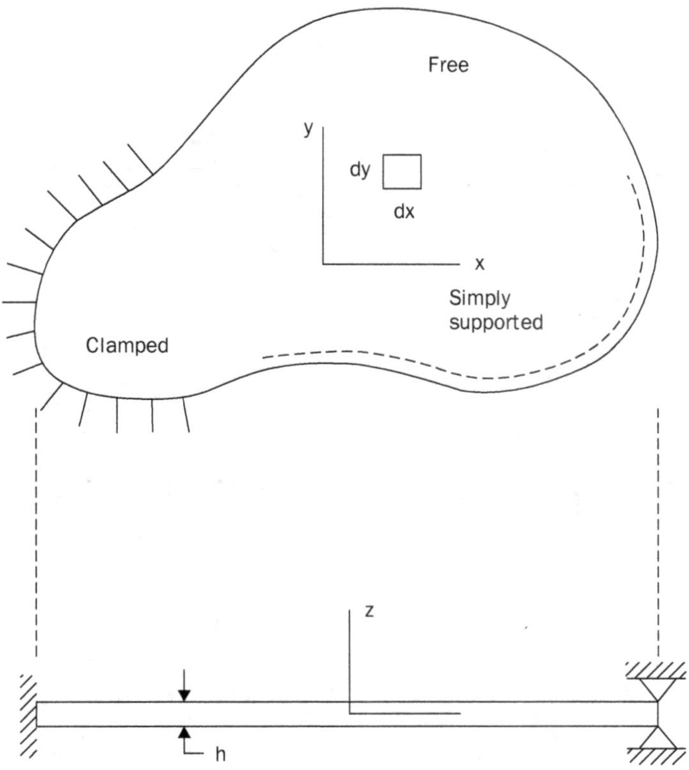

FIGURE 6.1 Plate of arbitrary shape and edge conditions.

are caused by bending stresses (σ_x and σ_y) varying linearly from tension at one face, to zero at the mid-surface, to compression at the other face, as seen in Fig. 6.2. Similarly, the twisting moments are due to shear stresses (τ_{xy} and τ_{yx}), also varying linearly as shown. From moment equilibrium about the z-axis for a typical 3D element ($dx \times dy \times dz$), $\tau_{yx} = \tau_{xy}$. Therefore $M_{yx} = M_{xy}$ everywhere.

The deformed middle surface of the plate is characterized in the same manner as that of a membrane surface, shown previously in Fig. 5.3. Its displacement in the z-direction is w. Assuming small slopes ($\partial w/\partial x$ and $\partial w/\partial y$) and utilizing Figs. 5.3 and 6.2, summing forces in the z-direction yields

$$-Q_x dy + \left(Q_x + \frac{\partial Q_x}{\partial x}dx\right)dy - Q_y dx + \left(Q_y + \frac{\partial Q_y}{\partial y}dy\right)dx$$
$$+ q\,dx\,dy = \rho h\,dx\,dy\frac{\partial^2 w}{\partial t^2} \tag{6.1}$$

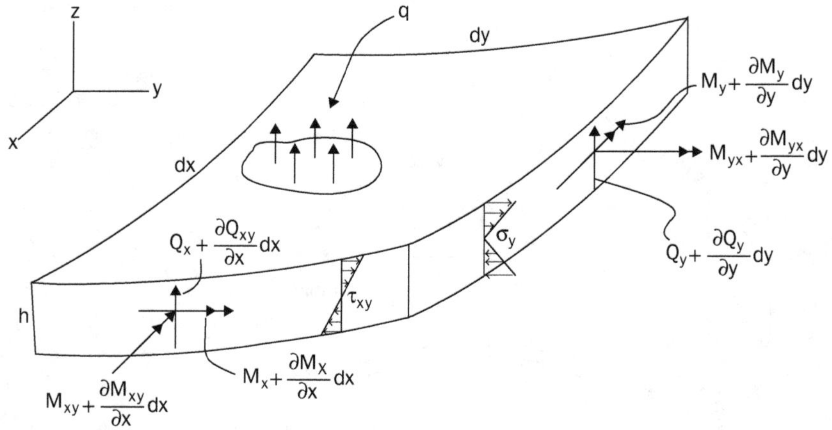

FIGURE 6.2 Displaced and deformed plate element, with forces and moments acting on it.

where ρ is mass density per unit volume. Cancelling terms in (6.1) and dividing through the equation by the area $dxdy$ yields

$$\frac{\partial Q_x}{\partial x} + \frac{\partial Q_y}{\partial y} + q = \rho h \frac{\partial^2 w}{\partial t^2} \tag{6.2}$$

Summing moments about an axis parallel to y through the center of the element,

$$\begin{aligned} M_x dy - \left(M_x + \frac{\partial M_x}{\partial x} dx \right) dy + M_{yx} dx - \left(M_{yx} + \frac{\partial M_{yx}}{\partial y} dy \right) dx \\ + Q_x dy \bullet \frac{dx}{2} + \left(Q_x + \frac{\partial Q_x}{\partial x} dx \right) dy \bullet \frac{dx}{2} = \rho \left(\frac{h^3}{12} dxdy \right) \frac{\partial^2}{\partial t^2} \left(\frac{\partial w}{\partial x} \right) \end{aligned} \tag{6.3}$$

where the term on the R.H.S. is the rotary inertia of the plate element. For classical, thin plate theory, this is typically small and is neglected. Doing so, then setting $M_{yx} = M_{xy}$, dropping the higher order term involving $\partial Q_x/\partial x$, and simplifying (6.3) results in

$$Q_x - \frac{\partial M_x}{\partial x} - \frac{\partial M_{xy}}{\partial y} = 0 \tag{6.4}$$

Similarly, summing moments about an axis parallel to x gives

$$Q_y - \frac{\partial M_{xy}}{\partial x} - \frac{\partial M_y}{\partial y} = 0 \tag{6.5}$$

Let us now consider the *kinematics* of the plate deformation. The assumption of elementary (Euler–Bernoulli) beam theory that "plane cross-sections remain plane during deformation" is generalized to a thin plate to apply as follows:

> Normals to the mid-surface of the undeformed plate remain straight and normal to the mid-surface, and unstretched in length, during deformation.

This simple deformation characterization is attributed to G. Kirchhoff, and is called the "Kirchhoff hypothesis." One consequence of this assumption for a plate which is homogeneous through its thickness is that its mid-surface does not develop in-plane stresses σ_x, σ_y, and τ_{xy} during deformation. Components of the displacement in the x and y directions (u and v, respectively) are then related to the rotation components of the normal by

$$u = -z\frac{\partial w}{\partial x}, \quad v = -z\frac{\partial w}{\partial y} \tag{6.6}$$

That is, an in-plane component of displacement at a point is proportional to its distance away from the mid-surface, and the amount a normal from it to the mid-surface rotates ($\partial w/\partial x$ or $\partial w/\partial y$).

The in-plane strains caused by the u and v displacement components are

$$\epsilon_x = \frac{\partial u}{\partial x}, \quad \epsilon_y = \frac{\partial v}{\partial y}, \quad \gamma_{xy} = \frac{\partial v}{\partial x} + \frac{\partial u}{\partial y} \tag{6.7}$$

where ϵ_x and ϵ_y are normal strains, and γ_{xy} is the *engineering* shear strain (different from the tensorial shear strain by a factor of two). Substituting (6.6) into (6.7) gives

$$\epsilon_x = -z\frac{\partial^2 w}{\partial x^2}, \quad \epsilon_y = -z\frac{\partial^2 w}{\partial y^2}, \quad \gamma_{xy} = -2z\frac{\partial^2 w}{\partial x \partial y} \tag{6.8}$$

Equations (6.8) show that the normal strains are proportional to the linearized (small slope) *curvatures* $\partial^2 w/\partial x^2$ and $\partial^2 w/\partial y^2$ of the plate mid-surface, and that the shear strain is proportional to the *twist* $\partial^2 w/\partial x\partial y$.

For an isotropic plate material the strains are related to the stresses by

$$\epsilon_x = \frac{1}{E}(\sigma_x - \nu\sigma_y), \quad \epsilon_y = \frac{1}{E}(\sigma_y - \nu\sigma_x), \quad \gamma_{xy} = \frac{\tau_{xy}}{G} \tag{6.9}$$

where E is modulus of elasticity (Young's modulus), ν is Poisson's ratio, and G is the shear modulus, related to E and ν by $G = E/2(1 + \nu)$. Inverting (6.9) gives

$$\sigma_x = \frac{E}{1-\nu^2}(\epsilon_x + \nu\epsilon_y), \quad \sigma_y = \frac{E}{1-\nu^2}(\epsilon_y + \nu\epsilon_x), \quad \tau_{xy} = \frac{E}{2(1+\nu)}\gamma_{xy} \tag{6.10}$$

The bending moments are obtained by integrating the moments of the in-plane stresses over the plate thickness. That is,

$$M_x = \int_{-h/2}^{h/2} \sigma_x z dz, \quad M_y = \int_{-h/2}^{h/2} \sigma_y z dz, \quad M_{xy} = \int_{-h/2}^{h/2} \tau_{xy} z dz \tag{6.11}$$

Substituting (6.8) and (6.10) into (6.11) and carrying out the integrations yields

$$M_x = -D(\kappa_x + \nu\kappa_y), \quad M_y = -D(\kappa_y + \nu\kappa_x), \quad M_{xy} = -D(1-\nu)\kappa_{xy} \tag{6.12}$$

where

$$\kappa_x = \frac{\partial^2 w}{\partial x^2}, \quad \kappa_y = \frac{\partial^2 w}{\partial y^2}, \quad \kappa_{xy} = \frac{\partial^2 w}{\partial x \partial y} \tag{6.13}$$

are the linearized (small slope) curvatures and twist of the plate mid-surface, used once before in (6.8), and

$$D = \frac{Eh^3}{12(1-\nu^2)} \tag{6.14}$$

is the so-called "flexural rigidity" of the plate. If one were to use the moment of inertia of beam theory (per unit width), it would be $Eh^3/12$. Because for an isotropic material, ν must fall between the limits $0 \le \nu \le 0.5$ (from energy and incompressibility considerations), the factor $1-\nu^2$ in (6.14) is typically close to unity.

At this point, all the needed parts of the classical plate theory are in hand, and they may be combined to obtain the desired form of the equation of motion. Substituting (6.4) and (6.5) into (6.2) gives

$$\frac{\partial^2 M_x}{\partial x^2}+\frac{\partial^2 M_y}{\partial y^2}+2\frac{\partial^2 M_{xy}}{\partial x \partial y}+q=\rho h\frac{\partial^2 w}{\partial t^2} \tag{6.15}$$

Then, substituting (6.12) and (6.13) into (6.15), and assuming that D is *constant* (limiting us to homogeneous plates of constant thickness), yields

$$D\left(\frac{\partial^4 w}{\partial x^4}+2\frac{\partial^4 w}{\partial x^2 \partial y^2}+\frac{\partial^4 w}{\partial y^4}\right)+\rho h\frac{\partial^2 w}{\partial t^2}=q \tag{6.16}$$

This may be abbreviated as

$$D\nabla^4 w+\rho h\frac{\partial^2 w}{\partial t^2}=q \tag{6.17}$$

where ∇^4 (del fourth) is the biharmonic differential operator, defined as

$$\nabla^4=\nabla^2(\nabla^2),\ \nabla^2=\frac{\partial^2}{\partial x^2}+\frac{\partial^2}{\partial y^2} \tag{6.18}$$

and ∇^2 (del squared) is the Laplacian operator, used previously for membranes in Chap. 5.

Before solving any plate vibration problems, let us also state the possible types of classical boundary conditions that may exist for the thin plate theory developed above. For boundaries which are parallel to the x-direction, the possible boundary conditions are

$$\begin{aligned}&\text{either}\quad w=0\quad\text{or}\quad V_y=0\\ \text{and,}\ &\text{either}\quad \frac{\partial w}{\partial y}=0\quad\text{or}\quad M_y=0\end{aligned} \tag{6.19}$$

Similarly, along edges parallel to the y-direction they are

$$\begin{aligned}&\text{either}\quad w=0\quad\text{or}\quad V_x=0\\ \text{and,}\ &\text{either}\quad \frac{\partial w}{\partial x}=0\quad\text{or}\quad M_x=0\end{aligned} \tag{6.20}$$

Let us consider the possible combinations of (6.19) along an edge x = constant. The ones which represent physical reality reasonably well are

$$\text{clamped edge: } w = \frac{\partial w}{\partial y} = 0 \tag{6.21a}$$

$$\text{simply supported edge: } w = M_y = 0 \tag{6.21b}$$

$$\text{free edge: } M_y = V_y = 0 \tag{6.21c}$$

The fourth possible combination of (6.19) would be $\partial w/\partial y = V_y = 0$, but this cannot be approached in physical problems for the same reasons described for beams [see (4.23) and its subsequent discussion].

The quantities V_y and V_x in (6.19), (6.20), and (6.21) are the so-called Kelvin-Kirchhoff edge reactions. In mathematical terms, they are expressed as

$$V_x = Q_x + \frac{\partial M_{xy}}{\partial y}$$

$$V_y = Q_y + \frac{\partial M_{xy}}{\partial x} \tag{6.22}$$

Substituting (6.4), (6.5), (6.12), and (6.13) into (6.22), the latter become

$$V_x = -D\left[\frac{\partial^3 w}{\partial x^3} + (2-\nu)\frac{\partial^3 w}{\partial x \partial y^2}\right] \tag{6.23a}$$

$$V_y = -D\left[\frac{\partial^3 w}{\partial y^3} + (2-\nu)\frac{\partial^3 w}{\partial x^2 \partial y}\right] \tag{6.23b}$$

Thus, they are a combination of the transverse shear force and the rate change of the twisting moment in the direction tangent to the edge. That $V_x = 0$ or $V_y = 0$ are the correct boundary conditions along free edges was explained by physical reasoning by Lord Kelvin. These were also shown by Kirchhoff to be the mathematically correct boundary conditions when derived by variational calculus. (For good explanations of both rationales, see the famous book by Timoshenko and Woinowsky-Krieger [1, pp. 84, 91].) That the internal shear forces Q_x and Q_y suddenly are replaced by V_x and V_y as one moves from within the plate to its edge is one of the anomalies of the classical thin plate theory.

Finally, it should be mentioned that the classical, thin plate theory gives a concentrated, transverse force $R = 2M_{xy}$ at each rectangular corner of a plate ([1], p. 85). Thus, at the intersection of two *free* edges, the supplemental condition $M_{xy} = 0$ must be satisfied.

6.2 Free Vibrations of Rectangular Plates; Exact Solutions

Assuming complete support per edge for rectangular plates, there exist 21 distinct cases which involve all possible combinations of classical boundary conditions (i.e., clamped, simply supported, or free) which are continuous along each edge. For the six cases having two opposite edges simply supported, exact solutions exist which are extensions of the approach used by Voigt [2] more than a century ago. For the remaining 15 cases, approximate methods must be used. Some approximate solutions are discussed later in this chapter (Sec. 6.6), after the Rayleigh and Ritz methods are taken up.

Consider a rectangular plate of lateral dimensions $a \times b$, as shown in Fig. 6.3, having its edges $x = 0$ and $x = a$ simply supported, whereas the other two edges ($y = 0$ and $y = b$) are each either clamped, simply supported, or free. For free, undamped vibrations, the governing equation of motion (6.17) reduces to

$$D\nabla^4 w + \rho h \frac{\partial^2 w}{\partial t^2} = 0 \tag{6.24}$$

Assuming the expected form of sinusoidal time response,

$$w(x,y,t) = W(x,y)\sin(\omega t + \phi) \tag{6.25}$$

(6.24) becomes

$$(\nabla^4 - k^4)W = 0 \tag{6.26}$$

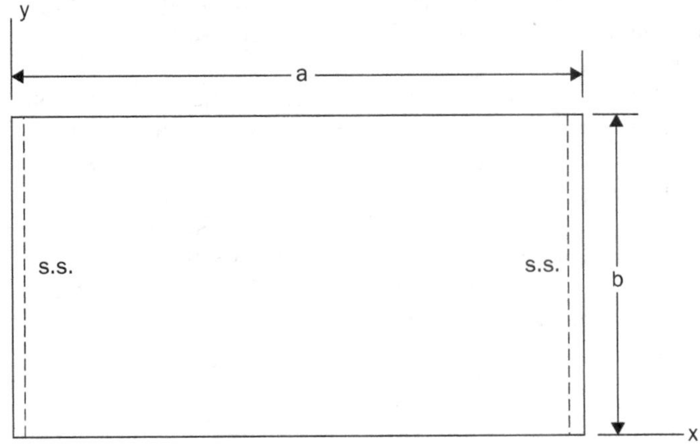

FIGURE 6.3 Rectangular plate with opposite edges simply supported.

where $k^4 \equiv \rho h\omega^2/D$. The differential operator $(\nabla^4 - k^4)$ may be factored into $(\nabla^2 n + k^2)(\nabla^2 - k^2)$, whence the solution to (6.26) is the sum of the solutions arising from its two parts,

$$(\nabla^2 + k^2)W = 0 \tag{6.27a}$$

$$(\nabla^2 - k^2)W = 0 \tag{6.27b}$$

The simply supported boundary conditions $w = M_x = 0$ along $x = 0$ and $x = a$ are satisfied exactly if one chooses

$$W(x, y) = Y(y)\sin\alpha x \tag{6.28}$$

where $\alpha = m\pi/a$ and $m = 1, 2, 3, \ldots$. Substituting (6.28) into (6.27) results in

$$Y'' + (k^2 - \alpha^2)Y = 0 \tag{6.29a}$$

$$Y'' - (k^2 + \alpha^2)Y = 0 \tag{6.29b}$$

If $k^2 > \alpha^2$, then the solution of (6.29a) involves trigonometric functions; that of (6.29b) involves hyperbolic functions. Thus, a solution to (6.26) is

$$\begin{aligned} W(x, y) = (&C_1 \sin\sqrt{k^2 - \alpha^2 y} + C_2 \cos\sqrt{k^2 - \alpha^2 y} \\ &+ C_3 \sinh\sqrt{k^2 + \alpha^2 y} + C_4 \cosh\sqrt{k^2 + \alpha^2 y})\sin\alpha x \end{aligned} \tag{6.30}$$

It may be noted that a solution to (6.24) *cannot* be achieved by the classical separation of variables approach used in the previous chapter [cf. (5.10)].

The free vibration problem is solved by substituting (6.30) into the appropriate two boundary conditions at $y = 0$ and at $y = b$. This yields four simultaneous, linear, algebraic equations in terms of the four unknowns C_1, C_2, C_3, and C_4 which are homogeneous (i.e., zero right-hand-sides). For a nontrivial solution one sets the fourth-order determinant of the coefficient matrix equal to zero, and solves for the eigenvalues, which may be taken as ka. These are related to the nondimensional frequencies by

$$\lambda^2 = (ka)^2 = \omega a^2\sqrt{\frac{\rho h}{D}}$$

The mode shape corresponding to a particular free vibration frequency is obtained in the usual manner by substituting the value of ka back into the four homogeneous equations, and solving any three of them for the amplitude ratios C_2/C_1, C_3/C_1, and C_4/C_1.

Example 6.1 Determine the frequencies and mode shapes of free vibration for a rectangular plate having all edges simply supported.

Solution

From (6.21b), the boundary conditions at $y = 0$ and b may be stated as

$$w(x,0,t) = 0 \rightarrow Y(0) = 0$$

$$M_y(x,0,t) = 0 \rightarrow \frac{\partial^2 w}{\partial y^2}(x,0,t) + \nu\frac{\partial^2 w}{\partial x^2}(x,0,t) = 0 \rightarrow Y''(0) = 0$$

$$w(x,b,t) = 0 \rightarrow Y(b) = 0$$

$$M_y(x,b,t) = 0 \rightarrow \frac{\partial^2 w}{\partial y^2}(x,b,t) + \nu\frac{\partial^2 w}{\partial x^2}(x,b,t) = 0 \rightarrow Y''(b) = 0$$

It may be noted that if $w = 0$ along the edges $y = 0$ and b, this guarantees that all derivatives of w in the x-direction ($\partial w/\partial x$, $\partial^2 w/\partial x^2$, etc.) along those edges are zero. Substituting (6.30) into the first two boundary conditions results in $C_2 = C_4 = 0$. The last two become

$$\begin{bmatrix} \sin\phi_1 & \sinh\phi_2 \\ -\phi_1^2\sin\phi_1 & \phi_2^2\sinh\phi_2 \end{bmatrix}\begin{bmatrix} C_1 \\ C_3 \end{bmatrix} = \begin{bmatrix} 0 \\ 0 \end{bmatrix}$$

where $\phi_1 \equiv (b/a)\sqrt{\lambda^2 - m^2\pi^2}$ and $\phi_2 \equiv (b/a)\sqrt{\lambda^2 + m^2\pi^2}$. Setting the determinant of the coefficient matrix equal to zero, and then expanding it, yields

$$(\phi_1^2 + \phi_2^2)\sin\phi_1 \bullet \sinh\phi_2 = 0$$

The nontrivial roots of this equation are

$$\sin\phi_1 = 0 \rightarrow \phi_1 = n\pi \quad (n = 1, 2, \ldots, \infty)$$

whence $\omega a^2\sqrt{\dfrac{\rho h}{D}} = \pi^2\left[m^2 + \left(\dfrac{a}{b}\right)^2 n^2\right] \quad (m, n = 1, 2, 3, \ldots, \infty)$

Substituting $\phi_1 = n\pi$ into either of the two homogeneous equations in C_1 and C_3 above gives $C_3 = 0$. Thus, the eigenfunctions describing the mode shape are

$$W_{mn}(x,y) = \sin\frac{m\pi x}{a}\sin\frac{n\pi y}{b}$$

The mode shapes for these boundary conditions are seen to be exactly the same as those for a rectangular membrane [see (5.21)].

Example 6.2 Obtain the frequency equation for an SS–C–SS–F rectangular plate (i.e., two opposite sides simply supported, another one clamped, and one free).

Solution

The boundary conditions (6.21a) and (6.21c), together with equations (6.12) and (6.23), yield

$$Y(0)=0$$
$$Y'(0)=0$$
$$Y''(b)-\nu\alpha^2 Y(b)=0$$
$$Y'''(b)-(2-\nu)\alpha^2 Y'(b)=0$$

Substituting (6.30) into these four boundary conditions yields a fourth-order frequency determinant. Expanding the determinant and using some trigonometric identities results in

$$\phi_1\phi_2\left[\lambda^4-m^4\pi^4(1-v)^2\right]+\phi_1\phi_2\left[\lambda^4+m^4\pi^4(1-v)^2\right]\cos\phi_1\cosh\phi_2$$
$$+m^2\pi^2\left(\frac{b}{a}\right)^2\left[\lambda^4(1-2v)-m^4\pi^4(1-v)^2\right]\sin\phi_1\sinh\phi_2=0$$

where ϕ_1 and ϕ_2 are defined in Example 6.1. The nondimensional frequency parameter $\lambda^2=\omega a^2\sqrt{\rho h/D}$ is contained above not only in λ^4, but also in ϕ_1 and ϕ_2.

It is interesting to note from the two examples above that the nondimensional frequency parameter $\lambda^2=\omega a^2\sqrt{\rho h/D}$ does not directly depend on the Poisson's ratio (ν) of the plate material unless at least one of the plate edges is free. This observation may be extrapolated to any rectangular plate. However, while λ^2 does not depend on ν for, say, a plate with all edges simply supported, the frequency does, for $D=h^3/12(1-\nu^2)$ contains ν.

The displacement function W given by (6.30) is based on the assumption that $k^2>\alpha^2$. If $k^2<\alpha^2$, then, (6.29a) may be rewritten as

$$Y''-(\alpha^2-k^2)Y=0 \tag{6.31}$$

Because the parenthetical quantity is now positive, the solution to (6.31) is

$$Y=A\sinh\sqrt{\alpha^2-k^2}\,y+B\cosh\sqrt{\alpha^2-k^2}\,y \tag{6.32}$$

Thus, in this case, the first two terms of (6.30) are replaced by those of (6.32). In solving a free vibration problem one should ordinarily employ both solution cases to obtain sets of eigenvalues for each case. Then one must verify that ka is appropriately either greater than or less than $m\pi$ for an eigenvalue to be valid. In Ref. [3], it is proved that proper eigenvalues for the case with $k^2>\alpha^2$ ($ka>m\pi$) will exist, for all six rectangular plate problems having two opposite sides simply supported, but that the case $k^2<\alpha^2$ is only valid for the three problems having one or more free sides (SS–C–SS–F, SS–SS–SS–F, SS–F–SS–F). For the SS–C–SS–F plate of Example 6.2, when $\nu = 0.3$, it was found that the case $k^2 < \alpha^2$ applies for $mb/a > 7.353$.

Three of the six problems for plates having two opposite edges simply supported have two-fold symmetry. These are the SS–C–SS–C, SS–SS–SS–SS, and SS–F–SS–F plates. For such situations the free vibration mode shapes will be either symmetric or antisymmetric with respect to (w.r.t.) not only the symmetry axis parallel to the y-coordinate (i.e., the line $x = a/2$ in Fig. 6.3), but also w.r.t. the other symmetry axis ($y = b/2$ in Fig. 6.3). As in all eigenvalue problems having symmetry, one may simplify the solution by taking advantage of the symmetry. Locating the x-coordinate along the symmetry axis, so that the two plate edges are at $y = \pm b/2$, one may use the *even* part of the displacement function (6.30) to determine the symmetric mode frequencies and mode shapes, and the *odd* part of (6.30) for the antisymmetric ones. This is demonstrated in Example 6.3 for the SS–C–SS–C plate.

Example 6.3 For the SS–C–SS–C plate, determine the frequency equations which arise when symmetry of the modes in the y-direction is considered, and compare them with the equation arising from the coordinate axis of Fig. 6.3.

Solution

For the symmetric modes, locating the x-coordinate along the symmetry axis allows one to use (for $k^2 > \alpha^2$) the *even* part of (6.30)

$$Y(y) = C_2 \cos\sqrt{k^2 - \alpha^2}\,y + C_4 \cosh\sqrt{k^2 + \alpha^2}\,y$$

Applying the boundary conditions $Y(+b/2) = Y'(+b/2) = 0$ yields a second-order frequency determinant. Expanding this gives

$$\phi_1 \tan\frac{\phi_1}{2} + \phi_2 \tanh\frac{\phi_2}{2} = 0$$

where ϕ_1 and ϕ_2 are as in Example 6.1. It is noted that the boundary conditions at $y = -b/2$ are also satisfied by the even displacement function. Similarly, using the odd part of the displacement function (6.30) with the two boundary conditions yields

$$\phi_2 \tan\frac{\phi_1}{2} - \phi_1 \tanh\frac{\phi_2}{2} = 0$$

for the antisymmetric modes. If one employs the coordinate system shown in Fig. 6.3 with boundary conditions $Y(0) = Y'(0) = Y(b) = Y'(b) = 0$ and expands the resulting fourth-order frequency determinant, the following eigenvalue equation is generated:

$$\phi_1\phi_2(\cos\phi_1 \cosh\phi_2 - 1) - m^2\pi^2\left(\frac{b}{a}\right)^2 \sin\phi_1 \sinh\phi_2 = 0$$

This is clearly more complicated than either of the two eigenvalue equations for the symmetric or antisymmetric modes. Yet, this more complicated equation contains exactly the same eigenvalues as the two more simple equations. This can be shown by substituting trigonometric and hyperbolic function identities for angles $2(\phi_1/2)$ and $2(\phi_2/2)$ above. The resulting equation will be the product of the two more simple equations shown earlier.

There is no need to consider the case $k^2 < \alpha^2$. As discussed earlier, no viable eigenvalues arise for this case with these boundary conditions.

Table 6.1 lists the first three nondimensional frequencies of square plates for all six cases of plates with two opposite edges simply supported, for $v = 0.3$. These data are taken from Ref. [3], where one can find the first nine frequencies for all six cases with $a/b = 0.4$, 2/3, 1, 1.5, and 2.5. Frequencies in Table 6.1 are listed in order, from the stiffest situation (SS–C–SS–C) to the most flexible one (SS–F–SS–F). Mode shapes are identified in the table by 11, 21, 12, and 13, which denote the numbers of half-waves in each direction. For example, the second mode for the SS–C–SS–F plate (denoted as the 12-mode) has one half-wave in the x-direction and two in the y-direction. For all six cases, of course, the half-waves in the x-direction are simple sine functions, so the node lines lying in the y-direction are all straight. However, as has been shown above, the half-waves in the y-directions are combinations of trigonometric and hyperbolic functions. Thus, the node lines lying in the x-directions are not straight (except for the SS–SS–SS–SS case, which entails only sine functions in the y-direction). Along the node lines $x = a/m$ one finds not only that $w = 0$, but also that the curvature $\partial^2 w/\partial x^2 = 0$. Therefore, these straight node lines duplicate simple support boundary conditions. Then, for example, the frequency of 54.7431 for the second mode of the SS–C–SS–C square plate in Table 6.1 is also the fundamental frequency of an SS–C–SS–C rectangular plate with dimensions $a/2 \times b$.

One finds in Table 6.1 that the first three modes of the SS–F–SS–F square plate all have only one half-wave in the x-direction. The

	Mode		
Edge conditions	**1**	**2**	**3**
SS–C–SS–C	28.9509 (11)	54.7431 (21)	69.3270 (12)
SS–C–SS–SS	23.6463 (11)	51.6743 (21)	58.6464 (12)
SS–SS–SS–SS	19.7392 (11)	49.3480 (21)	49.3480 (12)
SS–C–SS–F	12.6874 (11)	33.0651 (12)	41.7019 (21)
SS–SS–SS–F	11.6845 (11)	27.7563 (12)	41.1967 (21)
SS–F–SS–F	9.6314 (11)	16.1348 (12)	36.7256 (13)

TABLE 6.1 Nondimensional Frequencies $\omega a^2\sqrt{\rho h/D}$ for Square Plates ($a/b = 1$) with Two Opposite Sides Simply Supported ($\nu = 0.3$)

21 mode has the fourth frequency, with $\omega a^2\sqrt{\rho h/D} = 38.9450$. It is also interesting to compare the first three frequencies of the SS–F–SS–F plate with those obtained from beam theory, taking the beam as having a rectangular cross-section of width b and depth h, and having its ends simply supported. From the data for the SS–SS beam in Table 4.1, one obtains $\omega a^2\sqrt{\rho/Eh^2} = 2.849$ and 11.396 for the first two frequencies. From $\omega a^2\sqrt{\rho h/D} = 9.6314$ and 38.9450 for the SS–F–SS–F plate one obtains $\omega a^2\sqrt{\rho/Eh^2} = 2.915$ and 11.785 for $\nu = 0.3$. The differences are due to the fact that the plate undergoes *anticlastic* bending (curvatures of opposite signs in the x and y directions), whereas the beam does not consider anticlastic curvature. In the plate, the vibratory curvatures cause transverse bending moment M_y. This adds to the stiffness, and results in the plate frequencies being slightly greater than those from beam theory. For $\nu = 0$ the transverse curvature $\partial^2 w/\partial y^2$ and bending moment M_y both vanish for the plate, and the frequencies for the 11, 21, 31,... modes are identical to those of the SS–SS beam.

Degenerate free vibration modes (two different mode shapes having the same frequency) exist for plates as they do for membranes (see discussion in the last part of Sec. 5.2).

6.3 Circular Plates

Circular plates are most readily analyzed in polar coordinates (cf. Fig. 5.8). Although the plate equations were derived in rectangular coordinates in Sec. 6.1, they may be transformed into polar coordinates by using the relationships

$$x = r\cos\theta, \quad y = r\sin\theta, \quad x^2+y^2 = r^2 \tag{6.33}$$

and chain rule differentiations such as

$$\frac{\partial w}{\partial x} = \frac{\partial w}{\partial r}\frac{\partial r}{\partial x} + \frac{\partial w}{\partial \theta}\frac{\partial \theta}{\partial x} = \cos\theta\frac{\partial w}{\partial r} - \frac{1}{r}\sin\theta\frac{\partial w}{\partial \theta} \tag{6.34}$$

Carrying out such operations, one finds that the Laplacian operator ∇^2 in (6.18) is transformed to

$$\nabla^2 = \frac{\partial^2}{\partial r^2} + \frac{1}{r}\frac{\partial}{\partial r} + \frac{1}{r^2}\frac{\partial^2}{\partial \theta^2} \tag{6.35}$$

as was given previously in (5.27). The slopes of the mid-surface displacement in radial and circumferential directions are $\partial w/\partial r$ and

$(1/r)\,\partial w/\partial\theta$, respectively, whereas the curvatures (κ_r and κ_θ) and twist ($\kappa_{r\theta}$) are

$$\kappa_r = \frac{\partial^2 w}{\partial r^2}, \quad \kappa_\theta = \frac{1}{r}\frac{\partial w}{\partial r} + \frac{1}{r^2}\frac{\partial^2 w}{\partial \theta^2}, \quad \kappa_{r\theta} = \frac{\partial}{\partial r}\left(\frac{1}{r}\frac{\partial w}{\partial \theta}\right) \tag{6.36}$$

If one takes a triangular plate element having faces normal to the x, y, and r directions (Fig. 6.4), sums moments (vectorially) in the direction of M_r, substitutes (6.12) and (6.13), and employs chain rule differentiation as in (6.34) there results expressions for M_r and $M_{r\theta}$. Choosing another element with its face normal to the θ direction gives M_θ (and again $M_{r\theta}$). These expressions are

$$M_r = -D(\kappa_r + \nu\kappa_\theta), \; M_\theta = -D(\kappa_\theta + \nu\kappa_r), \; M_{r\theta} = -D(1-\upsilon)\kappa_{r\theta} \tag{6.37}$$

Summing transverse shearing forces on the same two elements, and carrying out the transformations, yields

$$Q_r = -D\frac{\partial}{\partial r}(\nabla^2 w), \quad Q_\theta = -D\frac{1}{r}\frac{\partial}{\partial \theta}(\nabla^2 w) \tag{6.38}$$

and the Kelvin–Kirchhoff edge reactions are

$$V_r = Q_r + \frac{1}{r}\frac{\partial M_{r\theta}}{\partial \theta}, \quad V_\theta = Q_\theta + \frac{\partial M_{r\theta}}{\partial r} \tag{6.39}$$

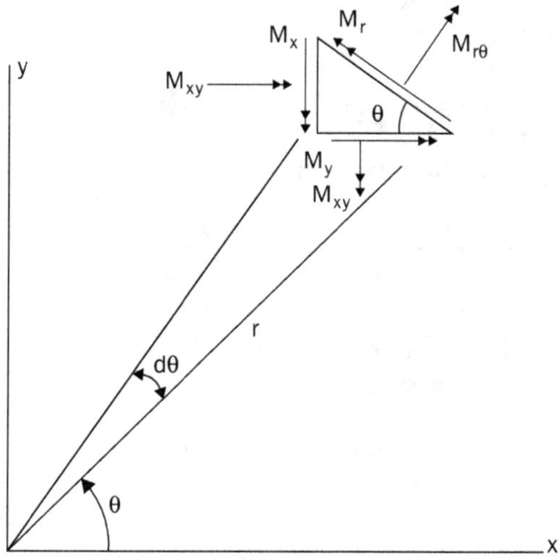

FIGURE 6.4 Triangular plate element with bending and twisting moments in rectangular and polar coordinates (positive moments as shown).

More background information on these relationships in polar coordinates is found in Refs. [1] and [4].

Consider now a solid circular plate of radius "a." Let the origin of the polar coordinate system be at the center of the plate (Fig. 5.8 applies to the circular plate, as well as to the membrane). Assuming the same sinusoidal response in time as in (6.25) for free vibrations, (6.26) and (6.27) are also valid in polar coordinates, except that now $W = W(r,\theta)$ and ∇^2 is given by (6.35). Further, as for the circular membrane (see Sec. 5.1) it is assumed that

$$W(r,\theta) = W_n(r)\cos n\theta \tag{6.40}$$

Substituting this into (6.27a) and (6.27b) yields

$$\frac{d^2W_n}{dr^2} + \frac{1}{r}\frac{dW_n}{dr} - \left(\frac{n^2}{r^2} - k^2\right)W_n = 0 \tag{6.41a}$$

$$\frac{d^2W_n}{dr^2} + \frac{1}{r}\frac{dW_n}{dr} - \left(\frac{n^2}{r^2} + k^2\right)W_n = 0 \tag{6.41b}$$

The solution to (6.41a) was already seen in membrane vibrations by (5.30), involving ordinary Bessel functions of the first and second kinds, of order n, and having *real* arguments kr. The solution to (6.27b) can be the same, except that the arguments of these Bessel functions are *imaginary* (ikr). Alternatively, this part of the solution may be expressed in terms of "modified" Bessel functions $I_n(kr)$ and $K_n(kr)$, where kr is real. These functions are described in Appendix B.

Thus the complete solution to the fourth-order differential equation (6.26) in polar coordinates is

$$W(r,\theta) = \left[A_n J_n(kr) + B_n Y_n(kr) + C_n I_n(kr) + D_n K_n(kr)\right]\cos n\theta$$
$$(n = 0, 1, 2, \ldots) \tag{6.42}$$

where A_n, B_n, C_n, and D_n are constants of integration.

For the solid circular plate with the coordinate origin within the plate, B_n must be zero because $Y_n(0) = -\infty$, as it was seen for the membrane. Similarly, D_n must be zero because $K_n(0) = +\infty$, as seen in the plots shown in Appendix B. Therefore (6.42) reduces to

$$W(r,\theta) = \left[A_n J_n(kr) + C_n I_n(kr)\right]\cos n\theta \tag{6.43}$$

Suppose now the circular boundary is *clamped*. Then the boundary conditions are

$$w(a,\theta,t)=0, \quad \frac{\partial w}{\partial r}(a,\theta,t)=0 \tag{6.44}$$

whence $W_n(a) = 0, \quad W_n'(a) = 0$ (6.45)

Substituting (6.43) into (6.45) gives

$$\begin{bmatrix} J_n(\lambda) & I_n(\lambda) \\ J_n'(\lambda) & I_n'(\lambda) \end{bmatrix} \begin{bmatrix} A_n \\ C_n \end{bmatrix} = \begin{bmatrix} 0 \\ 0 \end{bmatrix} \tag{6.46}$$

where $\lambda^2 = (ka)^2 = \omega a^2 \sqrt{\rho h / D}$, and where the primes in (6.46) indicate differentiation w.r.t. the argument of the function, in this case kr. Using the relationships

$$\begin{aligned} \lambda J_n'(\lambda) &= nJ_n(\lambda) - \lambda J_{n+1}(\lambda) \\ \lambda I_n'(\lambda) &= nI_n(\lambda) + \lambda I_{n+1}(\lambda) \end{aligned} \tag{6.47}$$

from Appendix B, and expanding the frequency determinant from (6.46) gives

$$J_n(\lambda)I_{n+1}(\lambda) + I_n(\lambda)J_{n+1}(\lambda) = 0 \tag{6.48}$$

The roots (λ) of (6.48) are available from several sources (cf. [4]–[6]). Values of $\lambda^2 = \omega a^2 \sqrt{\rho h / D}$ for modes having n nodal diameters and s interior nodal circles are presented in Table 6.2 for $0 \le n \le 5$ and $0 \le s \le 3$. Additional frequencies (up to n = 14 and s = 9) are available in Ref. [4].

	s			
n	**0**	**1**	**2**	**3**
0	10.22	39.77	89.10	158.18
1	21.26	60.82	120.1	199.1
2	34.88	84.58	153.8	242.7
3	51.04	111.0	190.3	289.2
4	69.67	140.1	229.5	338.4
5	90.74	171.8	271.4	390.4

TABLE 6.2 Nondimensional Frequencies $\omega a^2 \sqrt{\rho h / D}$ for a Clamped Circular Plate

The amplitude ratio C_n/A_n for a vibration mode is determined from either of the two equations in (6.46). Substituting back into (6.43) yields eigenfunctions describing the mode shapes as

$$W_{mn}(r,\theta)=\left[J_n\left(\lambda_{mn}\frac{r}{a}\right)-\frac{J_n(\lambda_{mn})}{I_n(\lambda_{mn})}I_n\left(\lambda_{mn}\frac{r}{a}\right)\right]\cos n\theta$$

$$(m = 1, 2, \ldots, \infty; \quad n = 1, \ldots, \infty) \tag{6.49}$$

where λ_{mn} denotes the mth eigenvalue from (6.48) for a given n. Locations of interior node circles (lines for which $w = 0$ during free vibration) for mode shapes are presented in Table 6.3, taken from Refs. [4] and [7], corresponding to the frequencies of Table 6.2. Nodal circles for plates do not duplicate simply supported conditions for, while $w = 0$ and $\partial^2 w/\partial\theta^2 = 0$ along a nodal circle, $\partial w/\partial r$ and $\partial^2 w/\partial r^2$ are not zero there and, hence, $M_r \neq 0$ there [see (6.36) and (6.37)].

Leissa and Narita [8] presented extensive frequency data for *simply supported* circular plates (frequencies for $0 \le n \le 10$; $0 \le s \le 10$ and $\nu = 0, 0.1 \ldots, 0.5$). In that case, the boundary conditions are

$$W(a, \theta, t) = 0, \quad M_r(a, \theta, t) = 0 \tag{6.50}$$

Substituting (6.43) into these, and using (6.47), yields the frequency equation

$$\frac{J_{n+1}(\lambda)}{J_n(\lambda)}+\frac{I_{n+1}(\lambda)}{I_n(\lambda)}=\frac{2\lambda}{1-v} \tag{6.51}$$

where $\lambda = ka$, as before. Some of the data available [8] is summarized in Table 6.4, where the effect of Poisson's ratio is also seen. In this table the nondimensional frequency parameter $\omega a^2\sqrt{\rho/Eh^2} = \lambda^2\sqrt{12/(1-\nu^2)}$ is given, which does not contain ν, so that the direct effect of ν on ω may be seen. The data shows that the effect of ν on ω

	s		
n	**1**	**2**	**3**
0	0.373	0.583, 0.255	0.688, 0.439, 0.191
1	.490	.640, .350	.721, .497, .272
2	.559	.679, .414	.746, .540, .330
3	.606	.708, .462	.765, .574, .375
4	.641	.730, .501	.781, .601, .412
5	.669	.749, .532	.787, .618, .439

TABLE 6.3 Radii of Interior Nodal Circles (r/a) for a Clamped Circular Plate

n	v	s			
		0	1	2	3
0	0	1.283	8.476	21.31	39.84
	0.3	1.493	8.994	22.44	41.86
	0.5	1.738	9.982	24.79	46.18
1	0	3.897	13.90	29.57	50.95
	0.3	4.206	14.67	31.10	53.50
	0.5	4.714	16.23	34.33	59.00
2	0	7.287	20.14	38.67	62.90
	0.3	7.751	21.22	40.64	66.03
	0.5	8.615	23.44	44.84	72.80
3	0	11.43	27.20	48.60	75.68
	0.3	12.09	28.61	51.04	79.43
	0.5	13.39	31.59	56.30	87.56

TABLE 6.4 Nondimensional Frequencies $\omega a^2\sqrt{\rho h/D}$ for Simply Supported Circular Plates with v = 0, 0.3, 0.5

can be considerable for all modes. The greatest effect is on the fundamental frequency ($n = 0$, $s = 0$). In this case, if v changes from 0 to 0.5, the frequency increases by 35 percent. The increase of ω with v is primarily due to the increased circumferential stiffening with increased v. Mode shapes are again given by (6.49).

Extensive data for *completely free* circular plates was published by Itao and Crandall [9], who gave the lowest 701 frequencies for $v = 0.33$.

6.4 Annular and Sectorial Plates

Figure 5.10 describes an *annular* plate, as well as an annular membrane. However, each edge of the plate may have any of the three classical boundary conditions (clamped, simply supported, or free). This gives rise to nine distinct cases having all possible combinations of these three boundary conditions.

Exact solutions for the frequencies and mode shapes of annular plates may be found by taking the general solution (6.42) to the equation of motion in polar coordinates, applying two boundary conditions at each edge, and thereby generating four homogeneous, linear algebraic equations in A_n, B_n, C_n, and D_n. The eigenvalues ka are obtained from the fourth-order determinant of the homogeneous equation. Amplitude ratios B_n/A_n, C_n/A_n, and D_n/A_n are obtained by back-substitution of the eigenvalues in the usual manner.

As an example, consider the annular plate having both edges clamped. Applying the boundary conditions $w = \partial w/\partial r = 0$ at $r = a$ and $r = b$ and finding the roots of the frequency determinant yields nondimensional frequencies $(ka)^2 = \omega a^2\sqrt{\rho h/D}$ as given in Table 6.5.

s	n	b/a 0.1	0.3	0.5	0.7	0.9
0	0	27.3	45.2	89.2	248	2237
	1	28.4	46.6	90.2	249	2238
	2	36.7	51.0	93.3	251	-
	3	51.2	60.0	99.0	256	2243
1	0	75.3	125	246	686	6167
	1	78.6	127	248	686	6167
	2	90.5	134	253	689	-
	3	112	145	259	694	6174

TABLE 6.5 Nondimensional Frequencies $\omega a^2\sqrt{\rho h/D}$ for Annular Plates Having Both Edges Clamped

These results were obtained by Vogel and Skinner [10]. As b/a approaches unity, the frequencies approach those of an infinite strip of width $(a - b)$ having both edges clamped. Results for all nine combinations of edge conditions are available in Ref. [4].

One interesting observation is that frequency determinants for three of the annular plate cases are the same as for three other cases. For example, the frequency determinant for the plate with outer edge $(r = a)$ clamped and inner edge $(r = b)$ free is the same as that for the plate with outer edge free and inner edge clamped. The radius ratio, b/a, appears in the determinant. For $b/a < 1$ results for the first case are found. For $b/a > 1$ the second case is described.

For *sectorial* plates, exact solutions for frequencies and mode shapes are available only when the radial edges are simply supported. The logic follows that employed earlier for the sectorial membrane (see Fig. 5.11). That is, one takes the solution to the equation of motion as (6.42), except using (in general) non-integer values μ in place of n, and $\sin \mu\theta$ in place of $\cos n\theta$. Doing so satisfies simply supported edge conditions at $\theta = 0$ and $\theta = \alpha$ exactly if one chooses

$$\mu = \frac{n\pi}{\alpha} \quad (n = 1, 2, 3, \ldots) \tag{6.52}$$

as was done for the membrane. The boundary condition $w = 0$ at $\theta = 0$ and α is clearly then satisfied. The other boundary condition is

$$M_\theta = -D\left(v\frac{\partial^2 w}{\partial r^2} + \frac{1}{r}\frac{\partial w}{\partial r} + \frac{1}{r^2}\frac{\partial^2 w}{\partial \theta^2}\right) = 0 \tag{6.53}$$

at $\theta = 0$ and α. Because $w = 0$ along the edges, then $\partial w/\partial r = \partial^2 w/\partial r^2 = 0$. And it is clear that $\partial^2 w/\partial \theta^2$ is also zero if μ is chosen as in (6.52).

For $\alpha \leq 180°$ the regularity conditions at $r = 0$ require that the coefficients B_μ and D_μ of the solution be discarded, as was done above for complete circular plates. Applying the boundary conditions at $r = b$ yields frequency equation (6.48) for the sector with the clamped circular edge, and (6.51) for the simply supported circular edge, except that Bessel functions are now typically of non-integer order, μ, instead of n. The frequencies presented earlier in Tables 6.2 and 6.4 for circular plates may be regarded as special cases of sectorial plates, with $n = 1, 2, 3, 4, 5$ corresponding to $\alpha = 180°, 90°, 60°, 45°, 36°$, respectively.

When $\alpha \leq 180°$ singularities (infinite bending and twisting moments) do exist at the sharp re-entrant corner of plate at $r = 0$. For such sectorial plates it is incorrect to discard B_μ and D_μ. Rather, a careful limiting process must be undertaken which yields zero displacement and slopes at $r = 0$, but allows infinite second derivatives of w there. This yields relationships between B_μ and D_μ, and both coefficients are retained [11].

Sectorial plates not having both radial edges simply supported have no exact solutions. Frequencies and mode shapes must be determined by an approximate procedure, such as the Ritz method (see Sec. 6.5). Where bending moment (and stress) singularities exist at the sharp corner at $r = 0$ during the vibratory motion, these singularities must be accounted for in order to obtain accurate results [12, 13].

As in the membrane vibration problem (Chap. 5), exact solutions for plates in polar coordinates are also applicable to *annular sectorial plates*. Such plates are bounded by two circle arcs ($r = b$, $r = a$) and two radial lines ($\theta = 0$, $\theta = \alpha$), provided that the radial edges are simply supported. The solutions discussed above for annular plates are applicable, with non-integer μ in place of n, and $\sin \mu\theta$ in place of $\cos n\theta$ in (6.42). Applying two boundary conditions at $r = b$ and a yields the frequency determinant. Thus, the frequencies shown in Table 6.5 for $n = 1, 2, 3$ are also those of annular sectorial plates with $\alpha = 180°$, $90°$, $60°$, respectively.

6.5 Energy Functionals; Rayleigh and Ritz Methods

The potential energy of an infinitesimal volume $dV = dxdydz$ of a vibrating body which is due to the strain energy of deformation is

$$d(PE) = \frac{1}{2}(\sigma_x \epsilon_x + \sigma_y \epsilon_y + \tau_{xy} \gamma_{xy}) dxdydz \tag{6.54}$$

For a linearly elastic, isotropic material the stress–strain relations (6.10) may be inserted in (6.54) to give

$$d(PE) = \frac{E}{2(1-v^2)}\left[\epsilon_x^2 + \epsilon_y^2 + 2v\epsilon_x\epsilon_y + \frac{(1-v)}{2}\gamma_{xy}^2\right] dxdydz \tag{6.55}$$

Substituting the strain–curvature equations (6.8) into (6.55), and integrating over the thickness yields for the complete plate,

$$PE = \frac{1}{2}\int_A D\left\{(\kappa_x + \kappa_y)^2 - 2(1-v)\left[\kappa_x\kappa_y - \kappa_{xy}^2\right]\right\}dA \qquad (6.56)$$

where κ_x, κ_y, and κ_{xy}, are the curvatures and twist of the plate mid-surface, given by (6.13); A is the surface area of the plate ($dA = dxdy$); and D is as defined previously by (6.14). With D inside the integral in (6.56), plates having varying thickness or elastic modulus may be straightforwardly accommodated.

In polar coordinates the potential energy is found to be

$$PE = \frac{1}{2}\int_A D\left\{(\kappa_r + \kappa_\theta)^2 - 2(1-v)\left[\kappa_r\kappa_\theta - \kappa_{r\theta}^2\right]\right\}dA \qquad (6.57)$$

where κ_r, κ_θ, and $\kappa_{r\theta}$ are curvature and twist components given by (6.36).

The translational kinetic energy of a three-dimensional plate element is

$$d(KE) = \frac{1}{2}\rho dxdydz\left(\frac{\partial w}{\partial t}\right)^2 \qquad (6.58)$$

Consistent with the thin plate theory of Sec. 6.1, where the rotary inertia was neglected, the kinetic energy of the element due to rotations will also be neglected. Integrating through the plate thickness, the kinetic energy of the plate is

$$KE = \frac{1}{2}\int_A \rho h\left(\frac{\partial w}{\partial t}\right)^2 dA \qquad (6.59)$$

where ρ and h may vary (i.e., be functions of x and y).

The Rayleigh method may be employed as it was in previous chapters to obtain approximate free vibration frequencies. One assumes sinusoidal motion in time,

$$w(x,y,t) = W(x,y)\sin(\omega t + \phi) \qquad (6.60)$$

This yields maximum potential and kinetic energies, expressed in rectangular coordinates, during the vibratory motion as

$$PE_{\max} = \frac{1}{2}\int_A D\left\{\left(\nabla^2 W\right)^2 - 2(1-v)\left[\frac{\partial^2 W}{\partial x^2}\frac{\partial^2 W}{\partial y^2} - \left(\frac{\partial^2 W}{\partial x \partial y}\right)^2\right]\right\}dxdy \qquad (6.61)$$

$$KE_{max} = \frac{\omega^2}{2} \int_A \rho h W^2 dxdy \tag{6.62}$$

Using the Rayleigh method, one assumes a trial function $W(x,y)$ for the presumed free vibration mode, substitutes it into (6.61) and (6.62) (or the equivalent forms in polar coordinates), sets $PE_{max} = KE_{max}$, and solves for ω. How accurate the resulting estimate for ω is depends on how well one has chosen $W(x,y)$. If $W(x,y)$ should be the exact shape, then the exact value of ω results. If it is only approximate, then an approximate ω is determined. Because the approximate assumed shape imposes internal constraints on the deformation freedom of the plate, the approximate frequency is an upper bound to the exact value (Rayleigh's Principle). (The reader is referred to additional discussions of the Rayleigh method in Chaps. 2, 3, 4, and 5.)

Example 6.4 Use the Rayleigh method to determine an approximate value for the fundamental frequency of a clamped circular plate of radius "a."

Solution

The assumed mode shape must satisfy the geometric constraints at the boundaries. For this problem, they are

$$w(a,\theta) = \frac{\partial w}{\partial r}(a,\theta) = 0$$

One function that will do this is

$$W = (a^2 - r^2)^2$$

It is assumed that the fundamental mode shape is axisymmetric, i.e., it does not depend on θ. Moreover, if one plots this assumed shape, it does look reasonable for the first mode.

To demonstrate a point let us calculate PE_{max} as the sum of two parts. Using (6.36) in (6.57),

$$PE_{max} = \frac{D}{2}[I_1 - 2(1-v)I_2]$$

where $$I_1 \equiv \int_0^a \int_0^{2\pi} \left(\frac{\partial^2 W}{\partial r^2} + \frac{1}{r}\frac{\partial W}{\partial r} + \frac{1}{r^2}\frac{\partial^2 W}{\partial \theta^2} \right)^2 rdrd\theta$$

$$I_2 \equiv \int_0^a \int_0^{2\pi} \left(\frac{\partial^2 W}{\partial r^2} \left(\frac{1}{r}\frac{\partial W}{\partial r} + \frac{1}{r^2}\frac{\partial^2 W}{\partial \theta^2} \right) - \left[\frac{\partial}{\partial r} \left(\frac{1}{r}\frac{\partial W}{\partial \theta} \right) \right]^2 \right) rdrd\theta$$

Because the assumed W does not depend on θ, derivatives w.r.t. θ vanish in the above integrals. Substituting the assumed W, we find

$$I_1 = \frac{64}{3}\pi a^6, \quad I_2 = 0$$

whence $PE_{max} = \dfrac{32}{3}\pi a^6 D$.

Using (6.62) in polar coordinates, and substituting the assumed W,

$$KE_{max} = \frac{\rho h\omega^2}{2}\int_0^a\int_0^{2\pi}\left(a^8 - 4a^6r^2 + 6a^4r^4 - 4a^2r^6 + r^8\right)rdrd\theta = \frac{\pi\rho h\omega^2 a^{10}}{10}$$

Setting $PE_{max} = KE_{max}$ gives $\omega^2 a^2 \rho h/D = 320/3$, or $\omega a^2\sqrt{\rho h/D} = 10.33$. Comparing this with the exact value of 10.22 from Table 6.2, one finds that the approximate value of 10.33 is a rather close upper bound (1.1 percent error).

It is noted that I_2 vanished. This is desirable, for we know that $\omega a^2\sqrt{\rho h/D}$ should not depend on ν for this problem. More will be said about this later in this section.

A more simple function for W could have been chosen, which also satisfies both boundary conditions at $r = a$:

$$W = (a - r)^2$$

However, this function causes moment singularities at $r = 0$ and is therefore unacceptable, that is,

$$\frac{1}{r}\frac{\partial W}{\partial r} = 2\left(1 - \frac{a}{r}\right)$$

This gives infinite κ_θ at $r = 0$, according to (6.36), and infinite M_r and M_θ there, according to (6.37). Using this assumed displacement yields infinite PE_{max}, regardless of the amplitude of the motion. The corresponding frequency would be infinite.

It has been seen in previous chapters that the Ritz method is a straightforward procedure for obtaining improved upper bounds for the natural frequencies of vibrating systems using an energy approach. As described in Chap. 2, Rayleigh employed the method of assuming a single function for a mode shape, setting the maximum potential and kinetic energies during free vibration equal to each other, and solving for a corresponding approximate frequency. Whether he was the first person to do this is questionable. But since it appeared in his classical, excellent book [14] in 1877, it is attributed to him.

Ritz presented his method [15] three decades later, with no mention of Rayleigh. The method is considerably different than that of Rayleigh, both in logic and in application. There has been a tendency in the literature to call the Ritz method the "Rayleigh–Ritz" method. Indeed, the first author followed this trend when writing his own book on plate vibrations [4] more than four decades ago. However, he now believes that the two methods should be given their separate names, not only to recognize properly the very original thinking of Ritz, but also so that the Rayleigh method itself is not overlooked. This latter method is relatively simple, and therefore of considerable practical value.

As demonstrated earlier (see Secs. 2.13, 4.10, and 5.7), the Ritz method utilizes a *series* of admissible functions to represent the vibratory displacement of a two-dimensional continuous system. That is, $W(x,y)$ in (6.60) is written as

$$W(x,y)=\sum_{i=1}^{I}C_i\phi_i(x,y) \tag{6.63}$$

where each of the ϕ_i satisfies at least the *geometric* boundary conditions (the generalized force boundary conditions need not be satisfied, although they can be). For a plate the geometric B.C. are those imposed on displacements and slopes. The others involve normal bending moment (M_n) and edge reaction force (V_n). If the set of admissible function is mathematically *complete* [i.e., capable of representing any arbitrary displaced shape consistent with the geometric boundary conditions as $I \to \infty$ in (6.63)], then the Ritz procedure will converge monotonically to the exact frequencies as sufficient terms in (6.63) are taken. If not, then it will converge to frequencies which are upper bounds on the exact frequencies.

To use the Ritz method one formulates $PE_{\max}$ and $KE_{\max}$ in terms of C_i and φ_i of (6.63), and then minimizes the frequency w.r.t. C_i. That is, one lets the method determine C_i so as to obtain the best upper bounds for the frequencies. The frequency minimizing equations have been shown to be (see Sec. 2.13)

$$\frac{\partial}{\partial C_i}\left(PE_{\max}-\omega^2 KE^*_{\max}\right)=0 \quad (i=1,2,\ldots,I) \tag{6.64}$$

where $KE^*_{\max} = KE_{\max}/\omega^2$ is the integral over the plate area (6.62) without the frequency. Equation (6.64) yields I homogeneous, linear equations in C_i. Setting the determinant of the coefficient matrix equal to zero yields upper bound approximations to I frequencies of the plate.

In carrying out the operations involved with (6.64) it is typically better to take the partial derivatives w.r.t. C_i before doing the integrations over the plate area required by (6.61) and (6.62). Doing so will reduce considerably the amount of algebra required as the number of admissible functions used (I) is increased.

Computational labor with the Ritz method may also be reduced considerably by recognizing that the second term of PE_{max} in (6.61), i.e.,

$$\frac{\partial^2 W}{\partial x^2}\frac{\partial^2 W}{\partial y^2} - \left(\frac{\partial^2 W}{\partial x \partial y}\right)^2$$

vanishes when integrated over the area of either

1. A plate of arbitrary shape which is clamped along its boundary

or

2. A plate with straight edges which has $w = 0$ along its boundary (which includes the simply supported situation)

The first case was already observed in Example 6.4. The proofs of the two statements were made by Langhaar [16].

A well-known, simple, and straightforward procedure for generating sets of admissible functions which are mathematically complete has already been suggested for the membrane (Sec. 5.7). It may be generalized to the plate as follows. Choose $W(x,y)$ as

$$W(x,y) = G(x,y)^K \sum_{i=0}^{I}\sum_{j=0}^{J} C_{ij}x^i y^j \tag{6.65}$$

The function $G(x,y)$ is the equation of the plate boundary in implicit form. The various terms in it are taken to the powers $K = 0$, 1, and 2 for each segment of the plate boundary depending on whether it is free, simply supported, or clamped, respectively. For example, consider a rectangular plate having its edges $x = 0$ and $x = a$ clamped, $y = 0$ simply supported, and $y = b$ free, as depicted in Fig. 6.5. Then, $G(x,y) = x^2(x - a)^2 y$. This procedure is applicable to plates of arbitrary shape.

6.6 Approximate Solutions for Rectangular Plates

In Sec. 6.2, exact solutions were presented for the free vibration frequencies and mode shapes for the six cases of rectangular plates having two opposite edges simply supported, and the other two either clamped, simply supported, or free. This leaves 15 distinct cases, for which approximate solutions may be obtained.

There are numerous approximate methods which may be employed to solve plate vibration problems. The Rayleigh and Ritz methods have been emphasized in this book because of their

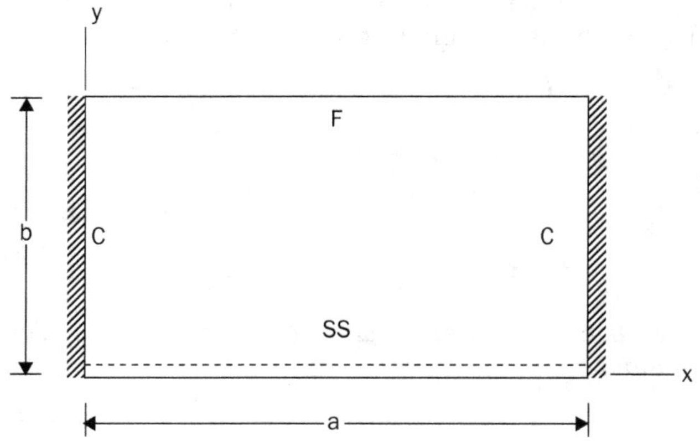

FIGURE 6.5 A C–SS–C–F rectangular plate.

historical significance, their relative ease in understanding in comparison with some of the other methods, their widespread usage in the published literature of free vibrations, and the capability of the Ritz method to obtain truly accurate frequencies and mode shapes. Some of the other approximate methods which have been used by plate vibration analysts include: Galerkin (which can be completely equivalent to the Ritz method), boundary collocation (also called "point matching"), generalized collocation, series (or superposition), and subdomain methods. Discussions and examples of applications of these methods are found in the book by Crandall [17] and in Ref. [18]. Other widely used methods include finite elements, finite differences, and boundary element methods.

Because of our understanding of the Ritz approach which has been developed and utilized in previous chapters here, it will now be used to obtain free vibration frequencies for the 15 remaining cases of rectangular plates having classical boundary conditions. The procedure to be followed is the one used by Ritz himself for the completely free plate [19], the only problem for which he published a solution using it.

Trial functions for the $\phi_i(x, y)$ in (6.63) are assumed which are the products of one-dimensional "beam functions." That is, (6.63) is rewritten as

$$W(x,y)=\sum_{p=1}^{P}\sum_{q=1}^{Q}A_{pq}X_p(x)Y_q(y) \tag{6.66}$$

where X_p and Y_q are eigenfunctions of vibrating beams in the two directions (x and y) which satisfy the corresponding beam boundary conditions at the ends, and A_{pq} are coefficients to be determined. For example, consider the C–SS–C–F rectangular plate shown in Fig. 6.5. In this case the $X_p(x)$ would be the C–C beam eigenfunctions (4.30b), and the $Y_q(y)$ would be the F–SS beam eigenfunctions (4.33b) of Chap. 4, with a and b used in place of ℓ for lengths.

The beam functions have certain advantages over algebraic polynomials. Most importantly, one would think that approximating the 2D behavior of a plate by functions in both directions which represent 1D behaviors, and which satisfy *all* the boundary conditions for the 1D problems, would reasonably represent the plate from a physical viewpoint. That is, using a single term $X_m Y_n$ of (6.66) may be regarded as a Rayleigh approximation (which would result from using the Rayleigh method with it) and the additional terms of (6.66) are corrections which improve the solution. Another advantage of the method is that the orthogonality of the beam functions, as well as their second derivatives, over the intervals of the integration considerably reduces the labor involved.

The Ritz method using beam functions as in (6.66) was employed in Ref. [3] to obtain frequencies for all 15 of the cases not having exact solutions. Upper limits $P = Q = 6$ were taken in (6.66), yielding 36 terms in each solution. Nondimensional frequencies were thus found as the roots of determinants of 36th order. In Ref. [3] the first nine frequencies thus obtained were presented for plates with $a/b = 0.4$, 2/3, 1, 1.5, and 2.5 for $\nu = 0.3$. One-term Rayleigh-type results were also given for comparison. A brief digest of these results is shown for square plates in Table 6.6, which is similar in form to Table 6.1. The plates are listed in order of decreasing fundamental frequencies, from all sides clamped to SS–SS–F–F. For the SS–F–F–F plate there is one zero frequency (rigid body rotation), which has been omitted. For the completely free case there are three zero frequencies (one translation, two rotations), which have also been omitted.

It is interesting to note that the C–C–C–C plate (Table 6.6) and the SS–SS–SS–SS plate (Table 6.1) both have degenerate 21 and 12 modes. Indeed, the nodal patterns for those plates are the same as those for square membranes (Fig. 5.6a) and superposition of them can cause diagonal nodal lines as for the membrane. For the F–F–F–F plate the 21 and 12 modes are rigid body rotations. However, the 31 and 13 modes do not exist in simple forms for the F–F–F–F plate. Rather, the second and third modes are composed of sums and differences of 31 and 13 terms, and have significantly different frequencies. Hence, they are not degenerate. The superimposed nodal patterns are essentially those of the 13 − 31 and 13 + 31 modes for the membrane, seen in Fig. 5.6b.

As a historical note it should be mentioned that the first significant work examining nodal patterns on rectangular plates of any kind was carried out experimentally by Chladni on free square plates and published in 1787 [20]. For experiments the completely

Edge conditions	Mode		
	1	2	3
C–C–C–C	35.99 (11)	73.41 (21)	73.41 (12)
C–C–C–SS	31.83 (11)	63.35 (12)	71.08 (21)
C–C–SS–SS	27.06 (11)	60.54 (21)	60.79 (12)
C–C–C–F	24.02 (11)	40.04 (12)	63.49 (21)
C–SS–C–F	23.46 (11)	35.61 (12)	63.13 (21)
C–F–C–F	22.27 (11)	26.53 (12)	43.66 (13)
C–C–SS–F	17.62 (11)	36.05 (12)	52.06 (21)
C–SS–SS–F	16.86 (11)	31.14 (12)	51.63 (21)
C–F–SS–F	15.28 (11)	20.67 (12)	39.78 (13)
C–C–F–F	6.942 (11)	24.03 (21)	26.68 (12)
C–SS–F–F	5.364 (11)	19.17 (12)	24.77 (21)
C–F–F–F	3.492 (11)	8.525 (12)	21.43 (21)
SS–SS–F–F	3.369 (11)	17.41 (12)	19.37 (21)
SS–F–F–F	6.648 (12)	15.02 (21)	25.49 (22)
F–F–F–F	13.49 (22)	19.79 (13–31)	24.43 (13+31)

Table 6.6 Nondimensional Frequencies $\omega a^2\sqrt{\rho h/D}$ for Square Plates (a/b = 1) Obtained by the Ritz Method (ν = 0.3)

free plate is the easiest to deal with, for it requires no edge fixtures. One simply supports the plate by threads or by points underneath, locating them at anticipated node lines so as to not affect the desired mode. Chladni did this, and strew sand on the plate surface. Exciting the plate by some simple means, such as a violin bow, one observes that as the sand particles are given sufficient excitation they break contact with the plate surface and collect along nodal lines, thus identifying them.

The addition of trial functions φ_i to the series of terms in (6.63) or (6.66) used with the Ritz method does not, in general, decrease *all* the previously found frequencies (i.e., yield improved upper bounds). As an example, consider the first two symmetric modes of the C–C–C–F square plate. If only a single term X_1Y_1 is used in (6.66), the Ritz method gives the result $\lambda_{11}^2 = 24.22$, where $\lambda^2 = \omega a^2\sqrt{\rho h/D}$ as in (6.31). This would also be the result, although somewhat more easily obtained, if the Rayleigh method were used. If the single term X_1Y_2 approximation to the second mode is used, the result is $\lambda_{12}^2 = 40.53$. If one uses both terms together in (6.70), the Ritz method yields $\lambda_{11}^2 =$ 24.07 and $\lambda_{12}^2 = 40.61$. Thus, while the two-term solution improves the fundamental frequency, it worsens the second. Accurate values of λ_{11}^2 and λ_{12}^2 obtained from the 36-term solution are seen in Table 6.6 to be 24.02 and 40.04, respectively. This behavior is discussed in more detail in Ref. [3].

The single term Rayleigh-type solutions supply reasonably accurate frequencies for some modes, but not for others. Table 6.7, taken from Ref. [3], lists the *average* percent differences between the frequencies from the 36-term Ritz solutions and the single term solutions for the first six modes of plates with $a/b = 0.4, 2/3, 1, 1.5$, and 2.5 (i.e., 30 values of λ^2 for each entry in Table 6.7), for $\nu = 0.3$. Also given is the *greatest* percent difference found among the 30 values, with the corresponding a/b and mode shape identified. It is seen that for the three cases having no free edge, the average and maximum differences between the two solutions is quite small, being less than 1 percent. The greatest differences occur for plates having two or more free edges. The worst case is the first torsional (12) frequency of the relatively slender ($a/b = 2.5$) cantilever plate (C–F–F–F), where the single term Rayleigh frequency is more than 24 percent too high. It should be noted that the maximum differences occur for the lower frequencies, most often for the second modes, as identified by the footnote to Table 6.7. In one case (C–SS–F–F) the maximum difference occurs for the fundamental mode. Simple formulas for free vibration frequencies of rectangular plates having any of the 21 possible

Edge conditions	Average percent difference	Mode with maximum difference		
		a/b	Mode	Percent difference
C-C-C-C	0.38	1	13-31	0.64
C-C-C-SS	0.41	1.5	21*	0.55
C-C-C-F	0.91	1	22	1.76
C-C-SS-SS	0.46	1	21*	0.76
C-C-SS-F	1.10	1	22	1.98
C-C-F-F	4.27	1	21*	10.24
C-SS-C-F	0.54	2.5	22	1.17
C-SS-SS-F	0.61	1	12*	1.25
C-SS-F-F	3.80	2.5	11†	14.77
C-F-C-F	0.58	2.5	12*	1.55
C-F-SS-F	0.77	2.5	12*	1.98
C-F-F-F	4.68	2.5	12*	24.36
SS-SS-F-F	2.93	1	12*	7.80
SS-F-F-F	3.41	2.5	12*	9.27
F-F-F-F	3.51	1	13-31*	13.06

*Second mode
†Fundamental mode

Table 6.7 Average and Maximum Percent Differences in $\lambda^2 = \omega a^2 \sqrt{\rho h / D}$ between 1- and 36-Term Solutions

combinations of simple edge conditions were presented by Warburton [21], obtained by the Rayleigh method using beam functions.

The primary reason that the single term Rayleigh solutions with beam functions are not very good when free edges are involved is that, while the clamped and simply supported boundary conditions are the same for beams and plates, for free edges they are not. The free edge beam conditions corresponding to zero bending moment and zero shearing force at $x = a$, for example, are $\partial^2 w/\partial x^2 = 0$ and $\partial^3 w/\partial x^3 = 0$, respectively (see Sec. 4.3). These differ from the plate free edge conditions by the terms $\nu\partial^2 w/\partial y^2$ and $(2 - \nu)\partial^3 w/\partial x\partial y^2$, respectively [see (6.12), (6.13), and (6.23a)]. Indeed, the beam functions impose unnecessary constraints on the free edges. Considering bending moment M_x to be zero along an edge $x = a$, for example, requires

$$\frac{\partial^2 w}{\partial x^2} + v\frac{\partial^2 w}{\partial y^2} = 0 \tag{6.67}$$

for all values of y. If $\partial^2 w/\partial x^2 = 0$ is enforced by the beam functions, then (6.67) implies that $\partial^2 w/\partial y^2 = 0$ for all y, that is, the edge can have no curvature in the y-direction. It therefore must remain straight, which is unrealistic and imposes a significant, frequency-raising constraint along the free edge. The beam functions impose a similar undesirable constraint on the higher derivative term $\partial^3 w/\partial x\partial y^2$ in (6.23a).

Displacement contour plots showing the mode shapes of completely free square plates are seen in Fig. 6.6. The plots show node lines (heavy lines) and other lines of constant displacement (light lines) for the first two modes of each of the four symmetry classes (SS, SA, AS, AA) of the modes. These plots show a limiting case of a shallow shell, where the Ritz method was used with algebraic polynomial displacement functions to solve the problem (see Sec. 7.3). The constant displacement lines are for values of $W/W_{\text{max}} = 0$ (node lines), 0.2, 0.4, 0.6, 0.8, and 1. From such "contour maps," one can see the mode shapes clearly and with more accuracy than from three-dimensional computer plots. It is interesting to observe that the degenerate SA and AS modes may be superimposed to obtain contour plots which have symmetry or antisymmetry about the two *diagonals* of the square, as do the other four contour plots shown (see Sec. 5.2 for discussion of degenerate mode superposition). For the six mode shapes of Fig. 6.6, the corresponding frequency parameters are $\omega a^2\sqrt{\rho h/D} = 13.47$ (AA-1), 19.60 (SS-1), 24.27 (SS-2), 34.80 (SA-1 and AS-1), 61.11 (SA-2 and AS-2), and 69.28 (AA-2). Sixteen polynomial admissible functions were used with each symmetry class of mode shape to obtain the frequencies, which have converged to four-digit exactitude. These values may be compared with the less accurate upper bounds shown in Table 6.6, obtained by using nine beam function products for each symmetry class.

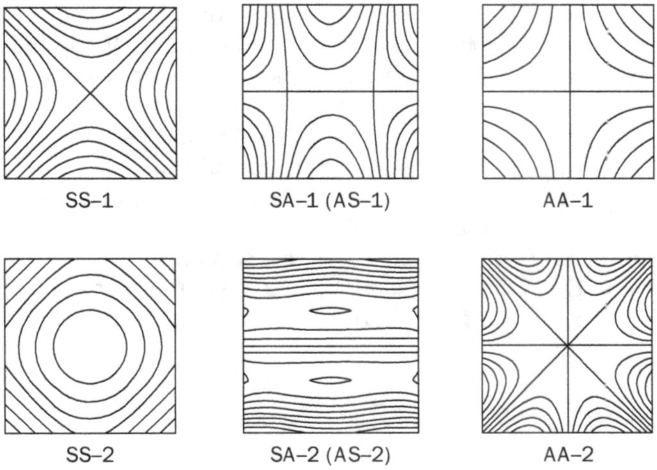

FIGURE 6.6 Mode Shapes for the first two modes of each of the four symmetry classes (SS, SA, AS, AA) of a completely free square plate.

Accurate frequencies for most of the 15 cases of rectangular plates not having exact free vibration solutions are also available in the book by Gorman [22]. He used a superposition method (also called the "series method" by others) based on exact solutions to the differential equation of motion, with boundary conditions being approximated. Considerable additional frequency data for rectangular plates may also be found in Ref. [4].

6.7 Other Free Vibration Problems for Plates According to Classical Plate Theory

In the previous sections of this chapter, plates which vibrate freely according to the classical equation of motion (6.24) have been studied for shapes having boundaries readily defined by rectangular and polar coordinates. Even so, the rectangular, circular, annular, and sectorial plates considered have been uncomplicated. For example, they could have had additional springs or masses attached, either at interior points or along the edges. Such problems can typically be handled straightforwardly by the Rayleigh or the Ritz methods. Or they could have had discontinuous boundary conditions, for example, a circular plate is clamped along one portion of its boundary and free along the remaining portion. The latter type of problem is more difficult to analyze, mainly because of the stress singularity at the boundary discontinuity.

But there are also innumerable other shapes of plates possible. Of particular practical importance are other shapes defined by straight line edges (parallelograms, trapezoids, other quadrilaterals, triangles, other polygonal shapes). Leissa's 1969 monograph [4] summarized frequency and mode shape data taken from approximately 70 publications which analyzed such problems. Hundreds of additional references of this type have appeared subsequently. There is also interest in other plate shapes having curvilinear boundaries (e.g., ellipses, segments of circles). Summarizing all the available data in an introductory textbook is out of the question. However, a few interesting cases will be looked at in this section.

An elliptical plate with semi-major and semi-minor axes a and b is depicted in Fig. 6.7. The equation of its boundary is

$$\frac{x^2}{a^2}+\frac{y^2}{b^2}=1 \tag{6.68}$$

An exact solution to the equation of motion (6.25) is achievable in elliptical coordinates (cf. [4], pp. 2–3). These coordinates are related to the rectangular xy-coordinates by

$$\begin{aligned} x &= c\cosh\xi\cos\eta \\ y &= c\sinh\xi\sin\eta \end{aligned} \tag{6.69}$$

The lines ξ = constant are ellipses, and the lines η = constant are hyperbolas which are everywhere orthogonal to the ellipses. The exact solution to (6.25), where ∇^4 is expressed in elliptical coordinates, is in terms of Mathieu functions (cf. [4], p. 3). However, applying the boundary conditions along the elliptical edge requires finding the roots of an infinite determinant to determine vibration frequencies.

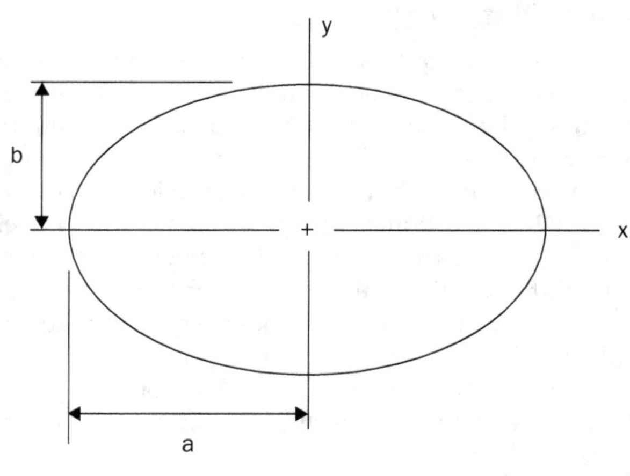

FIGURE 6.7 An elliptical plate.

a/b	$\omega a^2\sqrt{\rho h/D}$
1.00	10.2
1.25	8.38
2.00	6.88
3.00	6.32

TABLE 6.8 Fundamental Frequencies for Clamped Elliptical Plates

Successive truncation of the determinant yields frequencies to any desired degree of accuracy. This was the approach used by Shibaoka [23] to determine the fundamental frequencies for clamped elliptical plates. His results are shown in Table 6.8, with the special case of the circular plate ($a/b = 1$) added.

The Ritz method may be used straightforwardly to obtain accurate frequencies for elliptical plates. This was done by Leissa [24] for the simply supported case, using rectangular coordinates, for the fundamental mode. Using the approach suggested by (6.65) with nondimensional coordinates $\bar{x} = x/a$ and $\bar{y} = y/b$, a suitable trial function for the doubly symmetric fundamental mode is

$$W(x,y) = (\bar{x}^2 + \bar{y}^2 - 1)(C_{00} + C_{20}\bar{x}^2 + C_{02}\bar{y}^2) \tag{6.70}$$

In constructing the double integrals over the plate area for PE_{max} and KE_{max}, given by (6.61) and (6.62) converted to $\bar{x}$ and $\bar{y}$ coordinates, the limits of integration are taken from 0 to $\sqrt{1-\bar{x}^2}$ on $\bar{y}$ and from 0 to 1 on $\bar{x}$. These integrals may be evaluated exactly. Employing the frequency minimizing equations (6.64) results in a third-order determinant (which is given in [24] for arbitrary a/b and ν). Fundamental frequencies from this determinant are seen in Table 6.9. The two higher frequencies obtained from the third-order determinant are inaccurate for the corresponding two axisymmetric modes, and are not presented. The special case of $a/b = 1$ may be compared with exact data for $n = 0$ and $s = 0$ in Table 6.4. For $\nu = 0$ the three-term frequency of 4.447 is a close upper bound to the exact value of $\omega a^2\sqrt{\rho h/D} = 4.444$. For $\nu = 0.5$ the corresponding data are 5.219 and 5.213. As $a/b \to \infty$, the nondimensional frequency $\omega b^2\sqrt{\rho h/D}$ approaches that of an infinite strip of width $2b$ having its parallel edges simply supported. The exact value in this case is $\pi^2/4 = 2.467$, to four significant figures. Additional frequencies for clamped and simply supported elliptical plates were obtained by Young and Dickinson, using the Ritz method [25].

Table 6.10 lists frequency *ratios* for a free elliptical plate made of brass, as determined experimentally by Waller [26] for the aspect ratio $a/b = 1.98$. Figure 6.8 shows the corresponding nodal

	ν		
a/b	**0**	**0.25**	**0.50**
1.0	4.447	4.865	5.219
1.1	4.078	4.454	4.772
1.2	3.823	4.157	4.442
1.4	3.512	3.773	3.990
1.7	3.286	3.463	3.617
2.0	3.172	3.292	3.399
2.5	3.061	3.128	3.189
3.0	2.987	3.027	3.066
5.0	2.833	2.846	2.858
10.0	2.747	2.750	2.754
20.0	2.724	2.725	2.726

TABLE 6.9 Fundamental Frequencies $\omega a^2\sqrt{\rho h/D}$ for Simply Supported Elliptical Plates

	n				
s	**0**	**1**	**2**	**3**	**4**
0	–	–	1	2.58	4.7
1	–	1.77	3.27	5.68	8.29
2	4.25	6.57	9.43	12.6	–
3	10.6	14	–	–	–

TABLE 6.10 Experimentally Determined Relative Frequencies for a Free Elliptical Plate Having $a/b = 1.98$

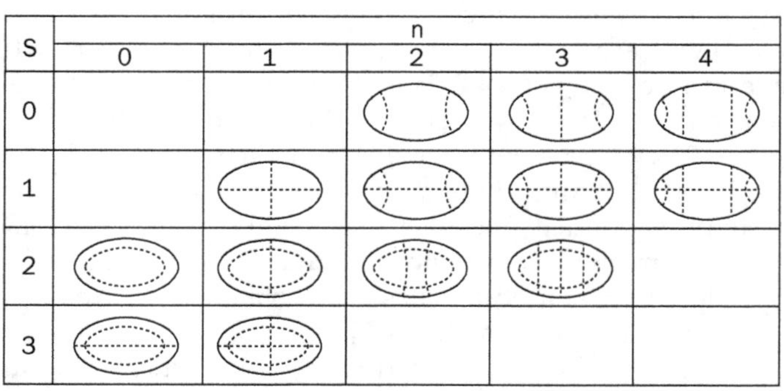

FIGURE 6.8 Experimental nodal lines for a free elliptical plate with $a/b = 1.98$.

patterns. The n and s designations are generalizations of the descriptors for circular plates.

Figure 6.9 shows the plan view of a trapezoidal plate. By varying the four dimensions a_1, a, b, and c one may obtain trapezoids of arbitrary shape. For example, if $c = b$, the shape has a symmetry axis; and if $c = 0$, one has a right-angled trapezoid. Another special, very important, case is when $a_1 = 0$. Then one has a triangular plate, for which arbitrary shapes may be obtained by varying a, b, and c. The free vibration frequencies and mode shapes of trapezoidal and triangular plates may be obtained straightforwardly by the Ritz method using trial functions in the form of (6.65). For completely free plates the analysis is simplified, for then $G = 1$ in (6.65). Jaber, Qatu, and Leissa [27, 28] made extensive studies of wide varieties of free triangular and trapezoidal plates by this approach.

Table 6.11 is a typical convergence study that one needs to carry out in order to be reasonably assured of achieving accurate frequencies. This is for a free, right triangular plate ($a_1 = c = 0$), with $a/b = 0.5$, and $\nu = 0.3$. The table is essentially a numerical experimentation to determine how many algebraic polynomial terms are needed in (6.65) to obtain a desired accuracy, and whether more terms should be taken in one direction than the other. Thus, for example, the 4×9 solution uses four terms in the x-direction and nine terms in the y-direction [$I = 3$, $J = 8$ in (6.65)] and yields a frequency determinant of order 36. Two sets of data are shown in

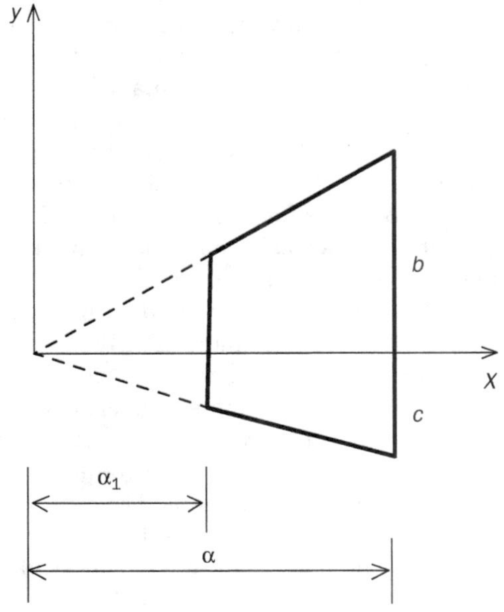

FIGURE 6.9 The plan view of a trapezoidal plate.

No. of Terms	Deter. size	Mode Number				
		1	2	3	4	5
6 × 6	36	6.4416	15.001	17.556	28.944	32.993
4 × 9	36	6.4289	15.015	17.332	28.979	32.232
7 × 7	49	6.4279	14.968	17.449	28.717	30.995
5 × 10	50	6.4279	14.979	17.300	28.709	30.773
8 × 8	64	6.4279	14.958	17.289	28.659	30.864
6 × 11	66	6.4278	14.959	17.288	28.674	30.762
Most Accurate		6.4278	14.958	17.288	28.659	30.762
Ref. [29]		6.429	15.00	17.33	28.94	32.23

TABLE 6.11 Convergence of $\omega a^2\sqrt{\rho h/D}$ for an Isosceles Right Triangular Plate (ν = 0.3).

Table 6.11. In the first set, an equal number of terms is taken in both directions ($I = J$). In the second set, more terms are taken in the y-direction than in the x-direction. Underlined frequencies are the most accurate ones (i.e., lowest upper bounds) for each of the six modes. Taking determinants of order 8 × 8 = 64 or 6 × 11 = 66, frequencies are seen to converge to at least four significant figures, and perhaps five, for the first three modes, although it was found that no better than three figure accuracy was obtained for the sixth mode. All calculations were made using double precision (16 significant figures). Using larger solutions (for example, 9 × 9) results in matrix "ill-conditioning," preventing accurate numerical results. The ill-conditioning problem can be effectively eliminated by using *orthogonal* polynomials (Legendre polynomials, in this case) in place of the ordinary polynomials in (6.65). However, this increases greatly the problem complexity and computational time required. Table 6.11 also shows frequencies obtained by Kim and Dickinson [29] using a similar procedure for right triangular plates.

Figure 6.10, taken from Ref. [27], shows nodal patterns of the first five mode shapes for free right triangular (c/b = 0) plates, with a/b = 0.5 and 1, for ν = 0.3. Corresponding nondimensional frequencies $\omega a^2(\rho h/D)^{1/2}$ are also given for each mode. For the isosceles triangle, one sees, as expected, that all modes are either symmetric or antisymmetric w.r.t. the diagonal symmetry axis present. Much more extensive frequencies and nodal patterns for triangular plates are available in Ref. [27]. Extensive, additional frequencies and nodal patterns for triangular plates were presented by Kim and Dickinson [30] for all combinations of simple edge conditions.

In analyzing polygonal plates by any method one should be aware of the bending stress singularities which may exist in sharp corners. They can affect the frequencies drastically, and need to be accounted for. An example is the parallelogram (or skew) plate ABCD

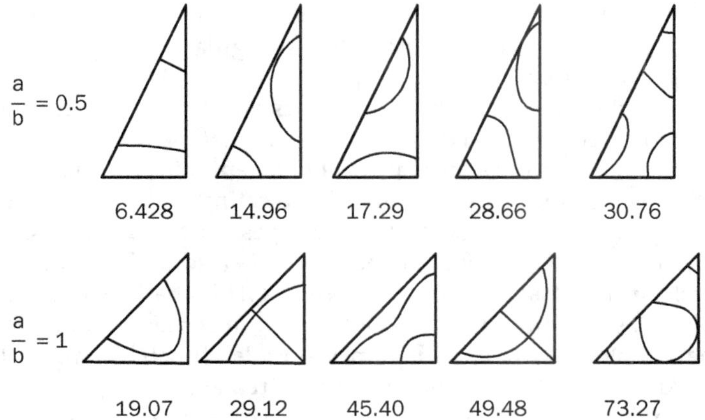

FIGURE 6.10 Nodal patterns and frequencies $\omega a^2\sqrt{\rho h/D}$ for free right triangular plates ($c/b = 0$, $\nu = 0.3$).

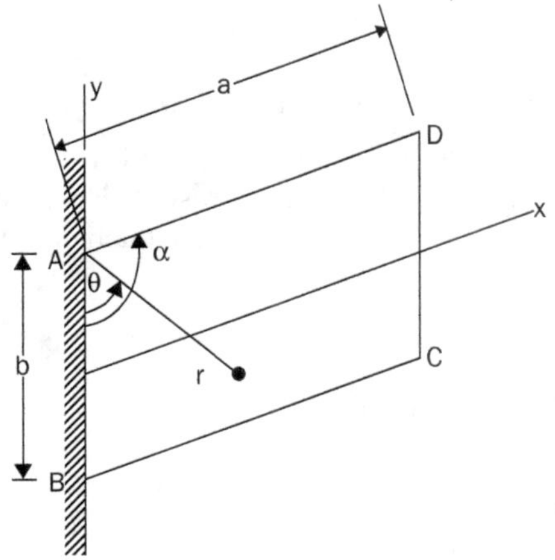

FIGURE 6.11 Cantilevered parallelogram (or skew) plate with bending stress singularity at point A.

shown in Fig. 6.11 which has one edge clamped (AB) and the other three free. A stress singularly exists in the corner A for all $\alpha > 90°$, the strength (importance) of the singularity increasing with α. Numerous researchers have attempted to obtain frequencies for such plates (a partial summary of their work is in [4], pp. 168–84), but reasonably accurate frequencies were found for only small skew angles ($90° < \alpha < 120°$). A later study with the Ritz method [31] has shown for such problems that for larger skew angles ($\alpha > 135°$) a relatively large number (64) of trial functions in the form of algebraic polynomials alone, used in (6.65), does not typically yield accurate frequencies, but the addition of a few singular, corner functions in the polar coordinates (r, θ) to account for the stress singularities improves the convergence greatly.

The difficulty described above occurs when the angle α of the clamped-free corner exceeds 90°, no matter what the actual shape of the plate is. For example, the plate could be trapezoidal or triangular. Similar singularities exist for other intersecting edges: for $\alpha > 90°$ with SS–SS edges; for $\alpha > 95°$ with SS–F edges; for $\alpha > 128°$ with C–SS edges; and for $\alpha > 180°$ with either C–C or F–F edges [32]. Thus, for example, if the parallelogram plate of Fig. 6.11 had all edges simply supported, stress singularities would result in corners A and C due to the vibratory motion, and proper singular functions would be required in an analysis to obtain accurate frequencies [33].

Many hundreds of research papers have been published which provide theoretical and/or experimental frequencies and mode shapes of plates which are described by (6.24) of classical theory. A large number of them are listed in Refs. [4, 34–36], which are reasonably comprehensible for the years preceding 1966, and the period 1973–1985.

6.8 Complicating Effects in Plate Vibrations

The analysis in this chapter has thus far been limited to plates which the equation of motion (6.24) may represent. However, plates may have other physical characteristics. A list of such characteristics is as follows:

1. Variable mass density
2. Variable material properties (E and ν)
3. Variable thickness
4. Anisotropic material
5. Laminated composite material
6. In-plane forces
7. Effects of surrounding media
8. Large amplitude (nonlinear) displacements
9. Shear deformation
10. Rotary inertia

These characteristics also exist for strings, bars, beams, and membranes and many of them have been discussed in the preceding chapters. For the plate they are typically more complicated, and these complications will be summarized briefly below, considering each effect by itself. Anisotropic and laminated composite plates will be discussed in Chap. 9.

Considerable variability in mass density can occur for some materials (e.g., rubber, Styrofoam), although usually not for metals. If the density (ρ) varies only with the in-plane coordinates (say x and y), then (6.24) still applies, with $\rho = \rho(x,y)$. The equation then has a variable coefficient, for which an exact solution is probably intractable (although possible in some cases). However, the Ritz method deals with such problems straightforwardly.

Materials with variable mass density will usually have significant variability in their material properties, as well, especially in E. However, variable material properties can also result independently as, for example, in a metal plate which is heated non-uniformly to high temperatures. Variable E and/or ν causes the flexural rigidity $D = Eh^3/12(1-\nu^2)$ to vary, and (6.24) is not appropriate. Variable thickness also affects the problem through D, as well as through the inertia term of the equation of motion. In substituting (6.12) into (6.15), treating D as a function of x and y, an equation much more complicated than (6.24) results. But the Ritz method is not significantly more complicated for such problems. Plates having variable density and/or material properties are called *nonhomogeneous* (or heterogeneous) plates.

A plate may be subjected to static, in-plane forces, in the same manner that a beam may have axial forces (see Sec. 4.11). For the plate subjected to static force resultants $T_x = \sigma_x h$, $T_y = \sigma_y h$, $T_{xy} = \tau_{xy} h$, the equation of motion (6.17) is generalized to (for free vibrations)

$$D\nabla^4 w + \rho h \frac{\partial^2 w}{\partial t^2} = T_x \frac{\partial^2 w}{\partial x^2} + 2T_{xy} \frac{\partial^2 w}{\partial x \partial y} + T_y \frac{\partial^2 w}{\partial y^2} \tag{6.71}$$

The in-plane forces enter the equations of motion in the same manner as they did for the membrane (see Sec. 5.1) and, indeed, if the plate bending stiffness vanishes ($D = 0$) the membrane equation (5.6) results. In general T_x, T_{xy}, and T_y are functions of x and y, and one must first solve a static, plane elasticity problem to determine those stress resultants throughout the plate. However, in such a case exact solutions of (6.71) are almost impossible. If $T_{xy} = 0$, and T_x and T_y are constants, then exact solutions of the form described in Sec. 6.2 are possible for rectangular plates having two opposite sides simply supported. If the in-plane forces are compressive (negative) in *any direction* anywhere within the plate, then buckling will occur at values of the loading which are large enough to cause zero frequencies. Frequencies may also be decreased, leading to buckling, by nonuniform, in-plane *tensile* forces on the boundary [37, 38].

The theory developed above for plates (as well as those used earlier for strings, bars, beams, and membranes) assumed that the vibratory motions take place in a vacuum. Of course, vibration experiments can be (and have been) conducted in the laboratory in a vacuum chamber. And important vibration problems do exist for structures in earth orbit, or in space, which are essentially in a vacuum. However, most practical vibratory situations take place in a surrounding medium, particularly air, and also water. In such cases the vibratory body has at least mass coupling with the medium. The theoretical problem involves solving two sets of equations of motion, one for the structural element, and the other for the surrounding medium (which may be infinite, or finite), and coupling them through the stresses at their interface. Indeed, in an early paper Lamb [39] did this for the first two modes of a clamped circular plate vibrating in an infinite expanse of water. He used the Rayleigh method, and assumed that the liquid was incompressible.

An interesting experimental and theoretical study was conducted by Lindholm et al. [40] on cantilever plates. Experimental frequencies were obtained for a variety of plates, both in air and water, and were compared with the theoretical values obtained by Young [41] and Barton [42] using the Ritz method with nine products of beam functions to approximate the modes. Some of their results are shown in Table 6.12. Others (for three additional modes, and for $a/b = 3$) are available in Ref. [4], as well as in Ref. [40]. All plates were made of 1080 cold-rolled steel. The results shown are for plates ranging from being moderately thick ($h/b = 0.1240$) to quite thin ($h/b = 0.0090$). The differences between the theoretical frequencies and the experimental ones in air are typically small, and are partly due to the inability to obtain perfect clamping experimentally. However, as expected, all frequencies in water are much lower.

"Large amplitude displacements" in a plate vibration problem refer to transverse displacements which are sufficiently large to cause significant stretching of the mid-surface during the motion. The phrase is somewhat misleading because the maximum displacement is typically much smaller than the in-plane dimensions of the plate. Indeed, it is usually on the order of the plate thickness. But such displacements cause significant stiffening of the system during transverse motion, which increases the free vibration frequencies. Strains of the mid-surface are

$$\epsilon_{x_0} = \frac{\partial u_0}{\partial x} + \frac{1}{2}\left(\frac{\partial w}{\partial x}\right)^2, \quad \epsilon_{y_0} = \frac{\partial v_0}{\partial x} + \frac{1}{2}\left(\frac{\partial w}{\partial y}\right)^2,$$

$$\gamma_{xy_0} = \frac{\partial v_0}{\partial x} + \frac{\partial u_0}{\partial y} + \frac{\partial w}{\partial x}\frac{\partial w}{\partial y} \tag{6.72}$$

With the introduction of the Airy stress function (ϕ) of plane elasticity, defined by

$$\sigma_{x_0} = \frac{\partial^2 \phi}{\partial y^2}, \quad \sigma_{y_0} = \frac{\partial^2 \phi}{\partial x^2}, \quad \tau_{xy_0} = -\frac{\partial^2 \phi}{\partial x \partial y} \tag{6.73}$$

		Nodal patterns								
		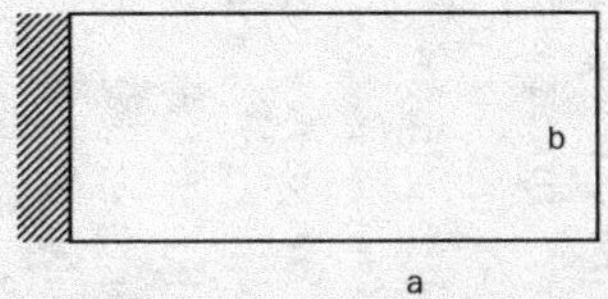			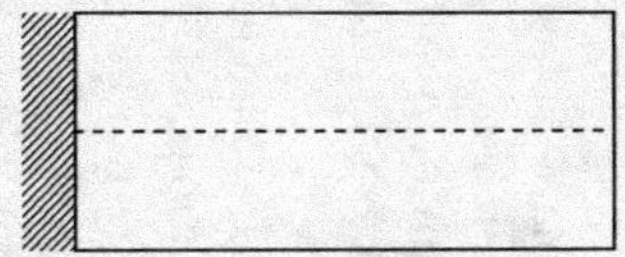			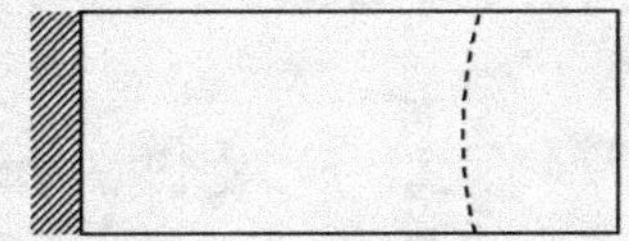		
		Theory	Experiment		Theory	Experiment		Theory	Experiment	
h/b	*a/b*	(vacuum)	Air	Water	(vacuum)	Air	Water	(vacuum)	Air	Water
0.1240	5	20.9	19.4	14.6	210	193	166	130	123	96
.0611	2	65.7	60.7	40.3	283	267	209	409	377	257
.0611	5	10.4	10.0	6.3	105	99.6	77.3	65.2	62. 3	40.1
.0238	1	99.5	96.3	51.4	243	241	154	610	591	355
.0238	2	24.7	24.2	12.1	106	108	67.5	154	151	80.0
.0238	3	10.9	10.8	5.1	68.1	67.7	41.6	68.2	66.9	33.3
.0131	½	223	214	106	342	339	189	1397	1339	739
.0131	1	55.6	52.9	23.3	136	129	68.7	341	326	158
.0131	2	13.8	12.9	5.1	59.3	58.2	29.8	85.9	80.8	34.4
.0090	½	159	147	63.5	244	228	110	998	920	452
.0090	1	39.7	37.9	14.6	97.1	95.3	44.2	243	236	102
.0090	2	9.86	9.3	3.12	42.4	42.0	18.8	61.4	57.8	21.1

TABLE 6.12 Frequencies (in Hertz) for C–F–F–F Rectangular Steel Plates in Vacuum, Air, and Water

a compatibility equation arising from (6.72) yields

$$\nabla^4 \phi = E\left[\left(\frac{\partial^2 w}{\partial x \partial y}\right)^2 - \frac{\partial^2 w}{\partial x^2}\frac{\partial^2 w}{\partial y^2}\right] \tag{6.74}$$

The equation of motion is given by (6.71), except that the in-plane stress resultants (T_x, etc.) are now caused by the vibratory stretching. Using (6.73), (6.71) becomes

$$D\nabla^4 w + \rho h \frac{\partial^2 w}{\partial t^2} = h\left(\frac{\partial^2 w}{\partial x^2}\frac{\partial^2 \phi}{\partial y^2} + \frac{\partial^2 w}{\partial y^2}\frac{\partial^2 \phi}{\partial x^2} - 2\frac{\partial^2 w}{\partial x \partial y}\frac{\partial^2 \phi}{\partial x \partial y}\right) \tag{6.75}$$

Equations (6.74) and (6.75) were derived for statically loaded plates by von Kármán [43]. They are both nonlinear due to the coupling terms on their right-hand sides. The fourth-order plane elasticity compatibility equation is thereby coupled with the fourth-order plate equation of motion, yielding an eighth-order system of nonlinear partial differential equations to be solved. No useful exact solutions have been known to be found.

As an example of the results one finds in the literature, Yamaki [44] applied the Galerkin method to the von Kármán equations (6.74) and (6.75) to study the effect of large displacements on the fundamental frequencies of circular plates. Two types of simply supported (Case I) and clamped (Case II) boundary conditions were considered. One type (*a*) had no radial in-plane restraint ($T_r = 0$), and the other (*b*) had complete radial in-plane restraint ($u_0 = 0$). The ratio of linear (small displacement) frequency to the nonlinear (large displacement) frequency is shown in Fig. 6.12 for varying ratio of maximum displacement (at the plate center) to the plate thickness. One sees that for displacement ratios of unity, significant increases in frequency occur for all four combinations of edge conditions, which is a "hard spring" behavior. However, if the plates are constrained radially [Cases I(b) and II(b)], the stiffness is increased considerably more than if they are not, and the frequencies are also increased considerably more.

An excellent book by Chia [45] presents results for the large amplitude vibrations of plates and laminated composites, including unsymmetric laminates.

The effects of shear deformation and rotary inertia on plates are similar to those which were found for beams (Sec. 4.12). That is, both of these effects cause the natural frequencies to decrease, and they are typically significant for the fundamental frequencies of thicker plates ($h/a > 0.1$, where a is the average in-plane dimension), or for the higher frequencies of thin plates ($h/a < 0.1$). Shear deformation effects were introduced into static plate theory by Reissner [46]. Subsequently, Mindlin [47] developed a more simple theory, which also considered

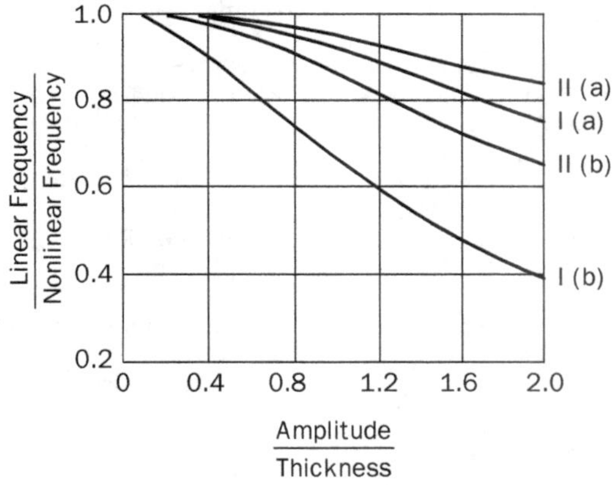

FIGURE 6.12 The ratio of linear frequency to the nonlinear frequency for two types of simply supported (Case I) and clamped (Case II) boundary conditions under various restraints.

dynamic effects, and this is the theory that is most often used. Instead of the normals to the mid-surface being required to remain normal, which results in (6.6), they are allowed to undergo independent rotations ψ_x and ψ_y in the x and y directions, respectively, yielding

$$u = -z\psi_x, \quad v = -z\psi_y \tag{6.76}$$

Thus, deformation is defined by three variables—w, ψ_x, ψ_y—each of which is a function of two space variables (x and y) and time, instead of the single displacement variable (w) which prescribes thin plate theory. Equations (6.76) are substituted into (6.7), and rotary inertia terms of the type seen on the R. H. S. of (6.3) (with $\partial w/\partial x$ replaced by ψ_x) are retained in the two moment equations of motion. This ultimately yields a set of three second-order differential equations which are coupled in w, ψ_x, and ψ_y [see Ref. [4], Eq. (12.72)]. The beam has only two second order, coupled equations, (4.127a,b), which makes the system fourth order and requires two conditions at each boundary. However, the system of plate equations is sixth order, requiring three boundary conditions at each edge.

The complicating effects discussed in this section have received enormous attention from researchers. Many of these are examined in review articles directed to this topic, prepared by Leissa [48–50].

References

1. S. Timoshenko and S. Woinowsky-Krieger, *Theory of Plates and Shells*, 2nd ed., McGraw-Hill Book Co., 1959.
2. W. Voigt, "Bemerkung zu dem Problem der transversalen Schwingungen rechteckiger Platten," *Nachr. Ges. Wiss.* (Göttingen), No. 6 (1893): 225–30.

3. A. W. Leissa, "The Free Vibration of Rectangular Plates," *J. Sound Vib.*, 31(1973): 257–93.
4. A. W. Leissa, *Vibration of Plates*, NASA SP-160, U.S. Govt. Printing Office, 353 pp. 1969, reprinted by The Acoustical Society of America, 1993.
5. J. Airey, "The Vibration of Circular Plates and Their Relation to Bessel Functions," *Proc. Phys. Soc. (London)*, 23(1911): 225–32.
6. H. Carrington, "The Frequencies of Vibration of Flat Circular Plates Fixed at the Circumference," *Phil. Mag.*, 50 (6)(1925): 1261–64.
7. V. S. Gontkevich, *Free Vibrations of Plates and Shells* (in Russian), Nauk. Dumka (Kiev), 1964, Translated by Lockheed Missiles and Space Co., Sunnyvale, Calif.
8. A. W. Leissa and Y. Narita, "Natural Frequencies of Simply Supported Circular Plates," *J. Sound Vib.*, 70 (2)(1980), 221–29.
9. K. Itao and S. H. Crandall, "Natural Modes and Natural Frequencies of Uniform, Circular, Free-Free Plates," *J. Appl. Mech.*, 46 (2)(1979): 448–59.
10. S. M. Vogel and D. W. Skinner, "Natural Frequencies of Transversely Vibrating Uniform Annular Plates," *J. Appl. Mech.*, 32 (1965): 926–31.
11. C. S. Huang, A. W. Leissa, and O.G. McGee, "Exact Analytical Solutions for the Vibrations of Sectorial Plates with Simply Supported Radial Edges," *J. Appl. Mech.*, 60 (1993): 478–83.
12. A. W. Leissa, O. G. McGee, and C. S. Huang, "Vibrations of Sectorial Plates having Corner Stress Singularities," *J. Appl. Mech.*, 60 (1993): 134–40.
13. O. G. McGee, C. S. Huang, and A. W. Leissa, "Vibrations of Completely Free Sectorial Plates," *J. Sound Vib.*, 164 (3)(1993): 565–69.
14. J. W. Strutt (Lord Rayleigh), *Theory of Sound*, vol. 1, The MacMillan Co., 1877; reprint, Dover Publications, 1945.
15. W. Ritz, "Über eine neue Methode zur Lösung gewisser Variations-probleme der mathematischen Physik," *Journal für die Reine und Angewandte Mathematik*, 1908.
16. H. L. Langhaar, *Energy Methods in Applied Mechanics*, John Wiley & Sons, 1962.
17. S. H. Crandall, *Engineering Analysis*, McGraw-Hill Book Co., 1956.
18. A. W. Leissa, W. E. Clausen, L. E. Hulbert, and A. T. Hopper, "A Comparison of Approximate Methods for the Solution of Plate Bending Problems," *AIAA J.*, 7(5)(1969): 920–28.
19. W. Ritz. "Theorie der Transversalschwingungen einer quadratischen Platte mit freien Rändern," *Annalen der Physik*, 28 (1909): 737–86, 1909.
20. E. F. F. Chladni, "Entdeckungen über die Theorie des Klanges," Leipzig, 1787.
21. G. B. Warburton, "The Vibration of Rectangular Plates," *Proc. Inst. Mech. Engr., Ser. A.*, 168 (12)(1954): 371–84.
22. D. J. Gorman, *Free Vibration Analysis of Rectangular Plates*, Elsevier, 1982.
23. Y. Shibaoka, "On the Transverse Vibration of an Elliptic Plate with Clamped Edges," *J. Phys. Soc. Japan*, 11 (7)(1956): 797–803.
24. A. W. Leissa. "Vibration of a Simply Supported Elliptical Plate," *J. Sound Vib.*, 6 (1)(1967): 145–48.
25. P. G. Young and S. M. Dickinson. "On the Free Vibration of Thin Isotropic and Rectangular Orthotropic Plates Involving Curved Boundaries," *J. Sound Vib.*, 165 (3)(1993): 511–26.
26. Mary Waller, "Vibrations of Free Elliptical Plates," *Proc. Phys. Soc. (London), Ser. B.*, 63 (1950): 451–55.
27. A. W. Leissa and N. A. Jaber, "Vibrations of Completely Free Triangular Plates," *Int. J. Mech. Sci.*, 34 (8)(1992): 605–16.
28. M. S. Qatu, N. A. Jaber, and A. W. Leissa, "Natural Frequencies for Completely Free Trapezoidal Plates," *J. Sound Vib.*, 167 (1)(1993): 183–91.
29. C. M. Kim and S. M. Dickinson, "The Free Flexural Vibration of Right Triangular Isotropic and Orthotropic Plates," *J. Sound Vib.*, 141 (1990): 291.
30. C. S. Kim and S. M. Dickinson, "The Free Flexural Vibration of Isotropic and Orthotropic General Triangular Shaped Plates," *J. Sound Vib.*, 152 (3)(1992): 383–403.

31. O. G. McGee, A. W. Leissa, and C. S. Huang, "Vibrations of Cantilevered Skew Plates with Comer Stress Singularities," *Int. J. Numer. Methods Eng.*, 35 (1992): 409–24.
32. M. L. Williams, "Surface Stress Singularities Resulting from Various Boundary Conditions in Angular Corners of Plates Under Bending," *Proc. U.S. Natl. Cong. of Applied Mech.*, 1951: 325–29.
33. C. S. Huang, O. G. McGee, A. W. Leissa, and J. W. Kim, "Accurate Vibration Analysis of Simply Supported Rhombic Plates by Considering Stress Singularities," *J. Vib. Acoust.*, 117 (1995): 245–251.
34. A. W. Leissa, "Recent Research in Plate Vibrations. 1973–1976: Classical Theory," *Shock Vib. Digest.*, 9 (10): 13–24.
35. A. W. Leissa, "Plate Vibration Research, 1976–1980: Classical Theory," *Shock Vib. Digest.*, 13 (9)(1981): 11–22.
36. A. W. Leissa, "Recent Studies in Plate Vibrations: 1981-85: Classical Theory," *Shock Vib. Digest.*, 19 (2)(1987): 11–18.
37. A. W. Leissa and E. F. Ayoub, "Vibration and buckling of a simply supported rectangular plate subjected to a pair of inplane concentrated forces," *J. Sound Vib.*, 127 (1)(1988): 155–171.
38. A. W. Leissa and E. F. Ayoub. "Free Vibrations and Tension Buckling of Circular Plates with Diametral Point Forces," *J. Appl. Mech.*, 57 (4)(1990): 995–99.
39. H. Lamb, "On the Vibrations of an Elastic Plate in Contact with Water," *Proc. Roy. Soc. (London), Ser. A.*, 98 (1920): 205–16.
40. H. Lindholm, D. D. Kana, W. Chu, and H. N. Abramson, "Elastic Vibration Characteristics of Cantilever Plates in Water," *J. Ship Res.*, 9 (1)(1965): 11–22.
41. D. Young. "Vibration of Rectangular Plates by the Ritz Method," *J. Appl. Mech.*, 17 (4)(1950). 448–53.
42. M. V. Barton, "Vibration of Rectangular and Skew Cantilever Plates," *J. Appl. Mech.*, 18 (1)(1951): 129–34.
43. Th. von Kármán, "Festigkeitsprobleme in Maschinenbau," *Encyclopadie der Mathematischen Wissenschaften*, 4 (4)(1910): 311–85.
44. N. Yamaki, "Influence of Large Amplitudes on Flexural Vibrations of Plates," *Zeitschrift für Angewandte Mathematik und Mechanik* (ZAMM), 41 (1961): 501–10.
45. C. Y. Chia, *Nonlinear Analysis of Plates*, McGraw-Hill Book Co., 1980.
46. E. Reissner, "The Effect of Transverse Shear Deformation on the Bending of Elastic Plates," *J. Appl. Mech.*, 12 (1945): A-69.
47. R. D. Mindlin, "Influence of Rotatory Inertia and Shear on Flexural Motions of Isotropic, Elastic Plates," *J. Appl. Mech.*, 18 (1)(1951): 31–38.
48. A. W. Leissa, "Recent Research in Plate Vibrations, 1973–1976: Complicating Effects," *Shock Vib. Digest.*, 10 (12)(1978): 21–35.
49. A. W. Leissa, "Plate Vibration Research, 1976–1980: Complicating Effects," *Shock Vib. Digest.*, 13 (10)(1981): 19–36.
50. A. W. Leissa, "Recent Studies in Plate Vibrations: 1981–85. Part II. Complicating Effects," *Shock Vib. Digest.*, 19 (3)(1987): 10–24.

Problems

1 A. Set up the frequency determinant for Example 6.2, expand it, and verify that the frequency equation given is correct.

B. Let $\nu = 0.3$ and $b/a = 2$. Evaluate the first (i.e., lowest) four values of λ.

C. One of the lower vibration modes has a nodal pattern having a single node line which is almost straight and parallel to the x-axis. On a plan view of the plate, plot the contour lines $W/W_{\max} = 0.2, 0.4, 0.6, 0.8, 1$ for that mode, where $W_{\max}$ is the value of W at the point having maximum vibratory displacement.

2 For an SS–SS–SS–F rectangular plate having $\nu = 0.3$, carry out an analysis to determine the range of mb/a for which the solution case $k^2 < \alpha^2$ is valid.

3 A. Determine the free vibration eigenfunctions for SS–C–SS–C rectangular plates.

B. Prove that the eigenfunctions are orthogonal over the plate area. (You may want to try a procedure that is similar to that which was employed for beams in Sec. 4.5.)

4 A rectangular ($a/b = 2$) SS–C–SS–C plate is vibrating freely in its fundamental mode with a maximum displacement, δ.

A. Determine the bending and twisting stresses σ_x, σ_y, and τ_{xy} as functions of x, y, and t.

B. Locate the points where σ_x, σ_y, and τ_{xy} are maxima. Determine these maximum values.

5 Consider a plate of width "a" and infinite length having two parallel, straight edges, as shown (Fig. 6.13). Let the two edges be clamped.

A. Determine its first three frequencies ($\omega a^2\sqrt{\rho h/D}$) of free vibration. Consider possible mode shapes $W(x,y)$ which do vary with y, and ones that do not.

B. Compare the frequencies of Part A with those given for square plates in Tables 6.1 and 6.6. Are they consistent? Why or why not?

C. Compare the fundamental frequency of Part A with those for elliptical plates given in Table 6.8. Are they consistent? Why or why not?

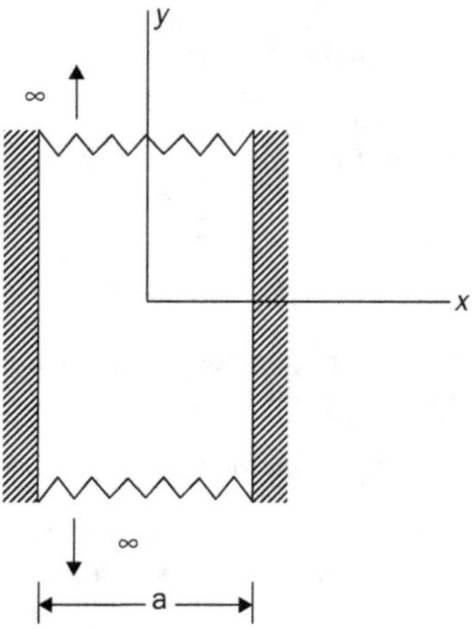

FIGURE 6.13 Problem 5.

6 A. Obtain the frequency equation for a free circular plate.

B. Let $\nu = 0.3$. Find the three lowest frequencies. Make three-dimensional sketches carefully of the corresponding mode shapes (no further calculations needed).

7 An annular plate has its outer edge clamped (Fig. 6.14). The inner edge is attached to a rigid circular disk of radius b and mass M, as shown. Formulate the frequency determinants from which *all* the natural frequencies of the system could be found. Express all elements of the determinants in terms of Bessel functions (not their derivatives) and the nondimensional parameters $\lambda(=\omega a^2\sqrt{\rho h/D})$, b/a and M/m_p, where m_p, is the mass of the annular plate without the rigid disk.

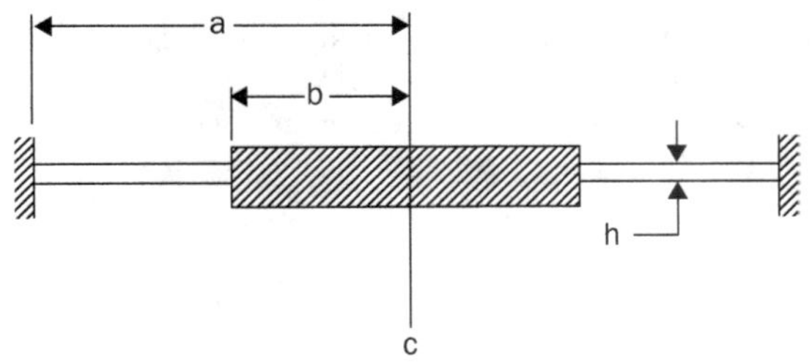

FIGURE 6.14 Problem 7.

8 A square plate of dimensions $a \times a$ has two *adjacent* edges simply supported and the other two free. Its material has a Poisson's ratio of 0.3.

A. Use the Rayleigh method with a displacement function expressed as a simple algebraic polynomial to find a first approximation to the fundamental natural frequency.

B. Add another polynomial term to the one used in Part A, and use the Ritz method to obtain a second approximation. (Ensure that the added polynomial term has the proper symmetry for the fundamental mode shape.)

C. Compare the two frequencies from above with that given in Table 6.6.

9 A circular plate has its boundary ($r = a$) completely free. However, it is supported transversely at its center by a spring of stiffness k. Let $\nu = 0.3$.

A. Use the Ritz method with an assumed displacement containing two independent polynomial trial functions to obtain an approximation to the lowest frequency of axisymmetric, free vibration. Express the nondimensional frequency $\omega a^2\sqrt{\rho h/D}$ in terms of a suitable nondimensional stiffness ratio (call it k^*) containing k. (Give some considerable thought to your choice of W before beginning this.)

B. Make a plot of $\omega a^2\sqrt{\rho h/D}$ versus k^* for $0 < k^* < \infty$. Use a logarithmic scale for the abscissa (k^*) with sufficient range to show clearly what is happening as k^* changes.

C. The limiting case as $k^* \to \infty$ is a rigid point support. Determine this value of $\omega a^2\sqrt{\rho h/D}$ from your result of Part A. Call it Ω^*. Determine the value of k^* for which $\omega a^2\sqrt{\rho h/D} = \Omega^*/2$. Plot the mode shape of the plate corresponding to this frequency.

10 A rectangular plate of dimensions a and b has all four edges simply supported. It is subjected to uniform static stresses σ_x and σ_y, at its edges ($\tau_{xy} = 0$).

A. Determine the free vibration frequency parameters ($\omega a^2\sqrt{\rho h/D}$) for the plate, as a function of $\sigma_x h a^2/D$ and the ratios σ_y/σ_x and a/b.

B. Let $\sigma_y/\sigma_x = -1$. Make a plot of the buckling stresses ($\sigma_x h a^2/D$) for which the frequencies become zero, versus b/a over the range $0 \le b/a \le 4$. Identify the *critical* (i.e., lowest) buckling stress values over this range (they vary with b/a).

CHAPTER 7

Shell Vibrations

Geometrically viewed, a shell is like a plate, except that it has curvature. Whereas a plate is flat, a shell is not. Nevertheless, like a plate it has one dimension, which we call its thickness (h), which is small compared to its other dimensions. The thickness need not be constant, but in many practical applications it is. And like a plate, deformation of a shell is characterized entirely by what happens at its midsurface and the normal to the midsurface. Thus, as in plate theory, shell theory represents the deformations of a three-dimensional body by equations which are mathematically two-dimensional. That is, only two independent space variables are needed to unequivocally define what is occurring at every point within the shell, instead of three.

For thin, isotropic plates, the in-plane natural frequencies are an order of magnitude higher than flexural frequencies. In addition, the differential equations of in-plane motion of isotropic plates are totally decoupled from those describing transverse motion. In other words, the membrane forces are decoupled from the bending and shear forces. This is not the same in a shell where curvature is present and the in-plane forces are coupled with the shear forces and bending and twisting moments. This leads to a set of equations for thin shells that have three displacement components as dependant variables: two in the tangential directions as well as the one in the transverse direction. The order of the resulting differential equations of motion is eight (instead of four for flat plates). The general shell equations can be specialized to those of plates (when curvature is set to zero) as well as those of membranes (when the flexural rigidity of the shell becomes very small).

Shells are among the most widely used structural and/or machine components. Most of the structural surfaces one encounters in real applications (e.g., airplane wing, fuselage, automotive exterior structures, and stamped brackets) are typically made of curved surfaces with a relatively small thickness (i.e., shells). In biological engineering, the human skull and some of the bones in the skeleton are shells. Conduits, pipes, tubes, and similar devices are all shell components. In addition, if one is interested in higher frequency content of a simplified thin-walled beam (e.g., a shaft), a shell theory will be needed. Frequently these high frequency modes in such structures are referred to as shell modes.

7.1 Introduction

Shells may have a great variety of curvatures. Examples include:

1. Circular cylindrical
2. Noncircular cylindrical
3. Conical
4. Spherical
5. Ellipsoidal (or spheroidal)
6. Paraboloidal
7. Hyperboloidal
8. Toroidal
9. Hyperbolic paraboloidal

That is, the shells are described by their middle surfaces, which may have any of the shapes listed. And this list is far from complete. For example, those listed as 1 and 3–8 are all shells of revolution. Surfaces of revolution are obtained by rotating a line segment about a polar axis, with each point on the segment generating a circle arc. If the line is straight, a conical surface is generated (Fig. 7.1). If it is also parallel to the polar axis, the special case of a circular cylinder results. If the line is curved, then other surfaces of revolution ensue. A circular line segment generates a spherical shell, a hyperbola segment generates a hyperboloidal shell, and so on. If the line segment is closed, a toroidal shell is generated. Figure 7.2 shows this for an elliptic toroidal surface. Clearly, innumerable other curves exist, both open and closed, which may generate surfaces of revolution. And there are still other surfaces which are not surfaces of revolution, such as the noncircular cylinder, the skewed cone, and the hyperbolic paraboloid. (Noncircular cylinders are obtained by rotating a line segment, called the "generator," which is straight and parallel to the polar axis, about the axis in a noncircular curve.)

Each of this great variety of possible shell curvatures has practical application. Circular cylinders are easily fabricated, and therefore widely used (e.g., tubes, ducts, cans, storage tanks, chimneys). Hyperboloidal shells are used for cooling towers. Other shells of revolution are used greatly for aircraft, missiles, spacecraft, and submarines. Hyperbolic paraboloidal shells are easily constructed, and are attractive, and are therefore used for some contemporary roofs of buildings. An enormous amount of research has taken place on shell vibrations. Reference [1] contains approximately 900 publications that deal with the free vibrations of shells.

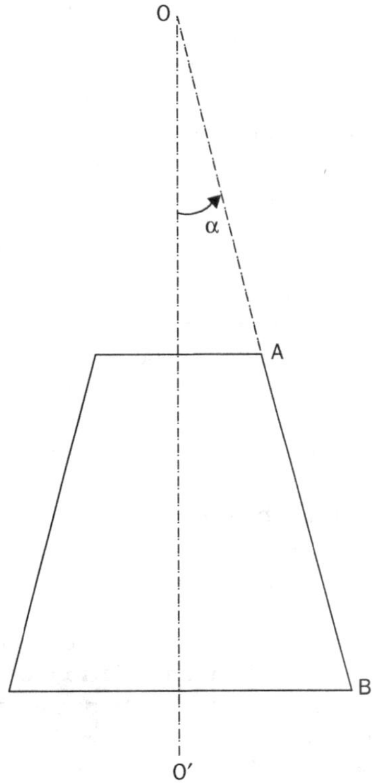

FIGURE 7.1 Conical surface generated by rotating the straight line segment AB about the polar axis O–O′.

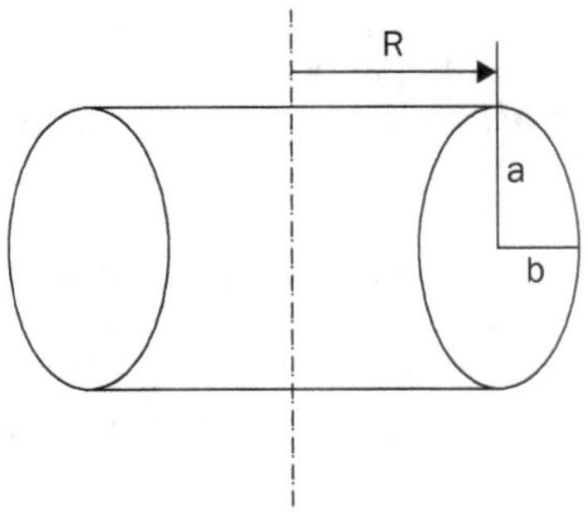

FIGURE 7.2 Elliptical toroidal surface.

In addition to having the added complexity of curvature, shells are more complicated than plates because their bending cannot, in general, be separated from their stretching. Thus, a "classical" bending theory of shells is governed by an eighth-order system of governing partial differential equations of motion, while the corresponding plate bending theory is only of the fourth order. This added complexity enters into the problem not only by means of more complex equations of motion, but through the boundary conditions as well. The classical bending theory of plates requires only two conditions to be specified along an edge, while a corresponding thin shell theory requires four specified conditions.

To complicate matters still further, whereas all academicians will agree on the form of the classical, fourth-order equation of motion for a plate (6.17), such agreement does not exist in shell theory. Numerous different shell theories have been derived and used for thin shells. Some of these are specialized, but quite limited and generally inaccurate. For example, "membrane theory" has been derived which considers only the stretching stiffness of a shell, and "inextensional theory" is available which includes only the shell bending stiffness. Both theories yield fourth-order systems of equations, which are much easier to deal with than the eighth order, complete (bending plus stretching) theories. However, both specialized theories have only narrow ranges of valid application for vibration problems, and should only be used by those who understand their severe limitations and can apply them properly

Not only is shell theory considerably more complicated than that for plates, but the study of vibrations is also more complicated for shells. For example, it is found that shell frequencies are more closely spaced and less easily identified, both theoretically and experimentally, than plate frequencies. Furthermore, the fundamental (lowest frequency) mode for a shell is generally not at all obvious, whereas for a plate it usually is. There are more parameters required to define the shell vibration problem. For example, consider a rectangular plate simply supported on all its edges. The complete frequency spectrum is determined by varying one parameter—the length-to-width ratio. For the cylindrically curved panel having the same edge conditions, however, three *additional* parameters can be independently varied—the thickness-to-radius ratio, the length-to-radius ratio, and Poisson's ratio (v).

This text will not attempt to present general theory for arbitrary thin shells. Readers who wish to learn more about such theory are referred to the excellent books by Kraus [2], Goldenveizer [3], Novozhilov [4], and Vlasov [5]. Various shell theories are derived and compared in the monograph by Leissa [1] (Chaps. 1 and 2, respectively). Flügge's books [6, 7] are particularly useful for shells of revolution. Although, with the exception of Refs. [1] and [6], these books typically present only static theories for shells, the inertia terms are easily added to the static equilibrium equations to obtain dynamic equations of motion, applicable to vibrations.

In this chapter the vibrations of two important types of shells will be examined, the shallow shell of arbitrary (but constant) curvature, and the circular cylindrical shell. The shallow shell is chosen because its theory is a clear generalization of that for the plate, which was considered in some detail in Chap. 6. Results will be shown only for free vibration frequencies and mode shapes. And none of the complicating effects mentioned for plates (see Sec. 6.8) will be considered here, although they certainly could be. That is, only constant thickness, isotropic, homogeneous, thin shells, vibrating in a vacuum, having no initial stress, will be analyzed.

7.2 Equations of Motion for Shallow Shells

The middle surface of a shallow shell of arbitrary curvature is depicted in Fig. 7.3. In terms of the rectangular coordinates shown there, its equation is

$$z = \frac{x^2}{2R_x} + \frac{xy}{R_{xy}} + \frac{y^2}{2R_y} \tag{7.1}$$

where R_x and R_y are radii of curvature in the x and y directions, respectively, as shown in Fig. 7.3, and R_{xy} is the corresponding coefficient describing the twist of the surface. This analysis will be limited to the case when R_x, R_{xy}, and R_y are constants; then (7.1) represents a quadratic surface. Figure 7.3 shows a shallow shell having boundaries which, when projected on the xy-plane (i.e., its planform), are rectangular. For analysis it is then usually convenient to choose the xy-coordinates to be parallel to the boundaries. However, if the x and y axes were rotated about the z-axis, it is

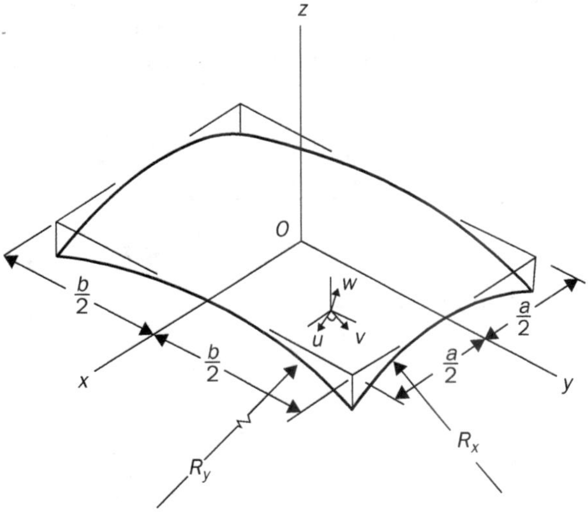

Figure 7.3 Shallow shell with rectangular planform.

possible to orient them so that $R_{xy} = \infty$ in (7.1). Then x and y are principal coordinates, and the new values of R_x and R_y are principal radii of curvature (taking on maximum and minimum values as second-order tensorial quantities, similar to stress and strain).

No precise definition of a "shallow" shell may be given in terms of its geometrical parameters. Rather, we will only say that a shell will be considered shallow if the theory presented here is reasonably accurate, for vibration analysis, when compared with deep shell theory. More will be said about this later. But, for the present, consider the shell segment shown in Fig. 7.4. It may be either circular cylindrical, with $R_x = R$, and $R_y = R_{xy} = \infty$ in (7.1), or spherical, with $R_x = R_y = R$ and $R_{xy} = \infty$ in (7.1). Then its shallowness may be measured in various ways; for example, the included angle (γ), the span-to-radius ratio (a/R), or the rise-to-span ratio (H/a). For vibration analysis we may usually consider the shallow shell theory to be applicable if γ is 60° or less.

Shallow shell theories were developed independently by Marguerre [8], Reissner [9], [10], and Vlasov [5]. A simplified derivation taken from Ref. [11], which is similar to that of Reissner, is presented below. It shows how the inplane bending and stretching effects of plate theory are combined in the shallow shell theory.

Consider an infinitesimal element of the shell. It will have stress resultants (forces per unit length) T_x, T_y, and T_{xy} tangent to its midsurface called "membrane forces" acting along its edges, as were shown for the membrane in Fig. 5.2. However, for the membrane they were initially applied, static forces. For the shell they are caused by deformation in the present representation, and are zero when the shell is not being deformed. In summing forces on the element in the z-direction, T_x, T_y, and T_{xy}, enter the equation as they did for the membrane in (5.1), except that the slope changes in traversing the element entail the initial curvatures (and twist), as well as the

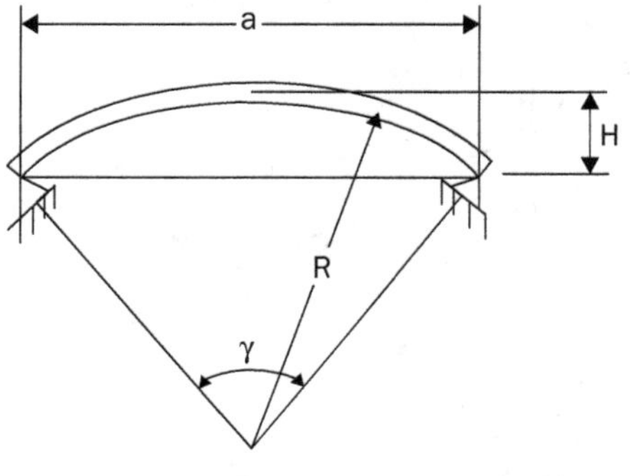

FIGURE 7.4 Circular cylindrical or spherical shallow shell.

curvature changes. Including the transverse shearing forces Q_x and Q_y, as they were in (6.2), the resulting equation of motion is

$$\frac{\partial Q_x}{\partial x}+\frac{\partial Q_y}{\partial y}+T_x\left(\frac{\partial^2 w}{\partial x^2}-\frac{1}{R_x}\right)+T_y\left(\frac{\partial^2 w}{\partial y^2}-\frac{1}{R_y}\right)$$
$$+2T_{xy}\left(\frac{\partial^2 w}{\partial x\partial y}-\frac{1}{R_{xy}}\right)+q=\rho h\frac{\partial^2 w}{\partial t^2} \quad (7.2)$$

Summing moments about axes parallel to the x and y coordinates yields the same Eqs. (6.4) and (6.5) as existed for the plate. Assuming the Kirchhoff hypothesis for the behavior of the normal to the shell midsurface as was done for thin plates (see Sec. 6.1), Eqs. (6.6)–(6.10) apply for the portion of the strain due to *bending*, and integrating the bending stresses through the shell thickness yields the same Eqs. (6.11)–(6.14) as for plates. Substituting (6.4), (6.5), (6.12), and (6.13) into (7.2) gives

$$D\nabla^4 w+\rho h\frac{\partial^2 w}{\partial t^2}=q+T_x\left(\frac{\partial^2 w}{\partial x^2}-\frac{1}{R_x}\right)$$
$$+T_y\left(\frac{\partial^2 w}{\partial y^2}-\frac{1}{R_y}\right)+2T_{xy}\left(\frac{\partial^2 w}{\partial x\partial y}-\frac{1}{R_{xy}}\right) \quad (7.3)$$

where $D = Eh^3/12(1-\nu^2)$ is the same flexural rigidity (6.14) used in plate analysis.

Equation (5.6) which was derived for the flat membrane may be regarded as a special case of (7.3) when the bending stiffness vanishes ($D = 0$), and when there is no initial curvature or twist ($1/R_x = 1/R_y = 1/R_{xy} = 0$). Similarly, (6.17) for the plate is a special case of (7.3) when $1/R_x = 1/R_y = 1/R_{xy} = 0$. If the plate had static, initial in-plane stressed resultants T_x, T_y, and T_{xy}, acting, then (6.77) arises. If not, then further simplification to (6.17) results.

Equation (7.3) may be regarded as primarily representing the bending taking place in the shell. The stretching enters through the membrane forces T_x, T_y, and T_{xy}, which are generated by deformation. If one sums forces in the x and y directions acting on a differential element, neglecting tangential inertia and tangential exciting forces, and considering the effects of the transverse shearing forces Q_x, and Q_y, to be small, (5.5) result. These are the classical equilibrium equations of plane elasticity. As in plane elasticity, it is convenient to introduce an Airy stress function, ϕ, defined by:

$$T_x=\frac{\partial^2\phi}{\partial y^2},\quad T_y=\frac{\partial^2\phi}{\partial x^2},\quad T_{xy}=-\frac{\partial^2\phi}{\partial x\partial y} \quad (7.4)$$

Then (5.5) are identically satisfied. However, as in plane elasticity, one must ensure compatibility of the strains due to stretching (i.e., the "membrane strains"). The membrane strains are

$$\epsilon_x = \frac{\partial u}{\partial x} + \frac{w}{R_x}, \quad \epsilon_y = \frac{\partial v}{\partial y} + \frac{w}{R_y}, \quad \gamma_{xy} = \frac{\partial v}{\partial x} + \frac{\partial u}{\partial y} + 2\frac{w}{R_{xy}} \tag{7.5}$$

where u and v are displacement components tangent to the shell midsurface, and w is normal, as shown in Fig. 7.3. A compatibility equation may be obtained by eliminating the displacements u and v among (7.5). The result is

$$\frac{\partial^2 \epsilon_x}{\partial y^2} + \frac{\partial^2 \epsilon_y}{\partial x^2} - \frac{\partial^2 \gamma_{xy}}{\partial x \partial y} = \frac{1}{R_x}\frac{\partial^2 w}{\partial y^2} + \frac{1}{R_y}\frac{\partial^2 w}{\partial x^2} - \frac{2}{R_{xy}}\frac{\partial^2 w}{\partial x \partial y} \tag{7.6}$$

When there is no curvature ($1/R_x = 1/R_y = 1/R_{xy} = 0$), it is seen that (7.6) becomes the well-known St. Venant compatibility equation of plane elasticity (cf. [1] in Chap. 5, p. 29) Substituting the stress–strain equations (6.10) into (7.6) as well as (7.4), we obtain:

$$\frac{1}{Eh}\nabla^4 \phi = \frac{1}{R_x}\frac{\partial^2 w}{\partial y^2} + \frac{1}{R_y}\frac{\partial^2 w}{\partial x^2} - \frac{2}{R_{xy}}\frac{\partial^2 w}{\partial x \partial y} \tag{7.7}$$

Finally, neglecting the curvature *changes* (e.g., $\partial^2 w/\partial^2 x$) in comparison with the initial curvatures (e.g., $1/R_x$), and substituting (7.4), the equation of motion (7.2) becomes

$$D\nabla^4 w + \rho h \frac{\partial^2 w}{\partial t^2} = q - \left(\frac{1}{R_x}\frac{\partial^2 \phi}{\partial y^2} + \frac{1}{R_y}\frac{\partial^2 \phi}{\partial x^2} - \frac{2}{R_{xy}}\frac{\partial^2 \phi}{\partial x \partial y} \right) \tag{7.8}$$

Equations (7.7) and (7.8) show clearly how the two fourth differential equations of plate bending and plane elasticity are coupled through the existence of curvature. Thus the shell is represented by an eighth-order system of differential equations. And solving a vibration problem requires satisfying four boundary conditions along each edge.

Equations (7.7) and (7.8) account for the possibility of a normal component of exciting pressure (q). Thus, they may be used for forced vibration problems. However, they do not permit tangential excitation components or consider tangential inertia. To include these, one may sum forces in three directions to obtain more general equations of motion. For the case of coordinates of principal curvature, the resulting equations written in terms of displacements in matrix form, are [12, 13].

$$\frac{Eh}{1-\nu^2}\begin{bmatrix}\mathcal{L}_{11} & \mathcal{L}_{12} & \mathcal{L}_{13}\\ \mathcal{L}_{12} & \mathcal{L}_{22} & \mathcal{L}_{23}\\ \mathcal{L}_{13} & \mathcal{L}_{23} & \mathcal{L}_{33}\end{bmatrix}\begin{bmatrix}u\\ v\\ w\end{bmatrix}+\rho h\begin{bmatrix}-\dfrac{\partial^2 u}{\partial t^2}\\ -\dfrac{\partial^2 v}{\partial t^2}\\ \dfrac{\partial^2 w}{\partial t^2}\end{bmatrix}=\begin{bmatrix}-p_x\\ -p_y\\ q\end{bmatrix} \tag{7.9}$$

where p_x and p_y, are the tangential components of exciting force (per unit surface area) acting in planes parallel to the xz- and yz-planes, respectively, and the L_{ij} are differential operators given by

$$\begin{aligned}
\mathcal{L}_{11} &= \frac{\partial^2}{\partial x^2}+\frac{(1-\nu)}{2}\frac{\partial^2}{\partial y^2}\\
\mathcal{L}_{22} &= \frac{(1-\nu)}{2}\frac{\partial^2}{\partial x^2}+\frac{\partial^2}{\partial y^2}\\
\mathcal{L}_{33} &= \frac{1}{R_x^2}+\frac{2\nu}{R_xR_y}+\frac{1}{R_y^2}+\frac{h}{12}\nabla^4\\
\mathcal{L}_{12} &= \frac{(1+\nu)}{2}\frac{\partial^2}{\partial x\partial y}\\
\mathcal{L}_{13} &= \left(\frac{1}{R_x}+\frac{\nu}{R_y}\right)\frac{\partial}{\partial x}\\
\mathcal{L}_{23} &= \left(\frac{\nu}{R_x}+\frac{1}{R_y}\right)\frac{\partial}{\partial y}
\end{aligned} \tag{7.10}$$

If one is using an energy approach (e.g., the Rayleigh or Ritz methods), the total potential energy is due to strain energy. In a deforming shell its $PE = PE_s + PE_b$, where PE_s arises from midsurface stretching and PE_b is due to bending (cf. [4]), where

$$PE_s=\frac{1}{2}\iint_A \frac{Eh}{(1-\nu^2)}\left\{(\epsilon_x+\epsilon_y)^2-2(1-\nu)\left[\epsilon_x\epsilon_y-\frac{\gamma_{xy}^2}{4}\right]\right\}dA \tag{7.11}$$

with ϵ_x, ϵ_y, and γ_{xy}, given by (7.5), and A is the area of the shell midsurface. The potential energy from bending is the same as (6.56) for a plate:

$$PE_b = \frac{1}{2}\iint_A D\left\{(\kappa_x + \kappa_y)^2 - 2(1-\nu)\left[\kappa_x\kappa_y - {\kappa_{xy}}^2\right]\right\}dA \tag{7.12}$$

where the curvature and twist components are those of (6.13). The last term in (7.11) would take the same form as that of (7.12) if the *tensorial* shear strain ($\epsilon_{xy} = \gamma_{xy}/2$) were used, instead of the engineering shear strain. The kinetic energy of the vibrating shell is

$$KE = \frac{1}{2}\iint \rho h\left\{\left(\frac{\partial u}{\partial t}\right)^2 + \left(\frac{\partial v}{\partial t}\right)^2 + \left(\frac{\partial w}{\partial t}\right)^2\right\}dA \tag{7.13}$$

It may be noted that the potential and kinetic energies given by (7.11)–(7.13) are written with E, h, ν, D, and ρ in the integrands. In these forms shells of variable thickness and/or nonhomogeneous material may be accommodated straightforwardly. In contrast, the equations of motion (7.7), (7.8), and (7.9) are for uniform thickness and homogeneous material.

7.3 Free Vibrations of Shallow Shells

Although (7.7), (7.8), and (7.9) are much more complicated than the plate equation, exact solutions for free vibrations are still possible. Moreover, the exact solutions can fit physically meaningful boundary conditions. For the shell these are "shear diaphragm" (also called "freely supported") conditions. They are a generalization of the plate "simply supported" boundary conditions.

Consider first (7.7) and (7.8), where the tangential inertia was neglected, take $1/R_{xy} = 0$, and assume solutions for free vibrations

$$w(x,y,t) = W(x,y)\sin\omega t$$

$$\phi(x,y,t) = \Phi(x,y)\sin\omega t \tag{7.14}$$

Assuming that the shell has a rectangular planform with dimensions $a \times b$, choose the origin of the x and y coordinates in one corner of the shell, so that its boundaries are $x = 0, a$ and $y = 0, b$. Suppose that all four edges of the shell are supported by shear diaphragms. Then assume

$$W(x,y) = W_{mn}\sin\alpha_m x\,\sin\beta_n y \tag{7.15}$$

where W_{mn} is an undetermined constant, $\alpha_m = m\pi/a$, $\beta_n = n\pi/b$, and m and n are integers. This is the same function as for the normal displacement of a vibrating rectangular plate having all edges simply supported (see Example 6.1). A stress function Φ which would then be appropriate to satisfy (7.7) and (7.8) is

$$\Phi(x,y) = \Phi_{mn} \sin \alpha_m x \, \sin \beta_n y \tag{7.16}$$

with Φ_{mn} a constant.

But before substituting these assumed solution forms into the governing equations, let us determine the boundary conditions that they yield. Consider, for example, the edges $x = 0$ and $x = a$. Clearly, $W(0,y) = W(a,y) = 0$. And using (6.12) and (6.13), it is seen that $M_x(0,y) = M_x(a,y) = 0$. Thus, the shell boundaries have no normal displacement, nor any normal bending moment, as did the corresponding simply supported plate. Now consider the other two boundary conditions along each edge. Substituting (7.16) into (7.14), and then into (7.4), shows that the membrane forces T_x and T_y are both zero at $x = 0$ and $x = a$, but that the membrane shearing force T_{xy} is not zero. From the stress–strain equations (6.10) it follows then that $\epsilon_x = \epsilon_y = 0$ along the boundaries. Substituting ϵ_y and w into the second of (7.5) one finds that $v = 0$ along $x = 0$ and $x = a$. These four boundary conditions along an edge are the shear diaphragm conditions. To summarize, for the shallow shell having rectangular boundaries $x = 0,a$ and $y = 0,b$ supported by shear diaphragms the boundary conditions being satisfied are

$$\text{At } x = 0 \text{ and } x = a: \quad w = M_x = v = T_x = 0 \tag{7.17a}$$

$$\text{At } y = 0 \text{ and } y = b: \quad w = M_y = u = T_y = 0 \tag{7.17b}$$

Now let us examine the *physical* meaning of the shear diaphragm boundary conditions. Figure 7.5 shows a section taken through a shallow shell supported at $x = 0$ and a by shear diaphragms. The shear diaphragms may be regarded as thin, flat plates. Each plate is rigidly attached to the shell, normal to its boundary, and also rigidly held at its other end. Because the plate is thin, its transverse bending stiffness is small. Hence, the T_x and M_x it can cause at the shell boundary are small. But the plate is very stiff in its own plane. Hence, it prevents shell displacements at the boundary in this plane ($w = v = 0$). Moreover, in the dynamic problem of shell vibration, the inertia effect of the plate is small because it contributes significant inertia only in the tangential (u) direction. It should be noted that if the shear diaphragms were in the z-direction, instead of normal to the shell, the boundary condition $w = 0$ would not apply. Instead, a linear combination of u and w would be zero.

The shear diaphragm boundary conditions can also be approximated by a piano hinge which constrains the v and w

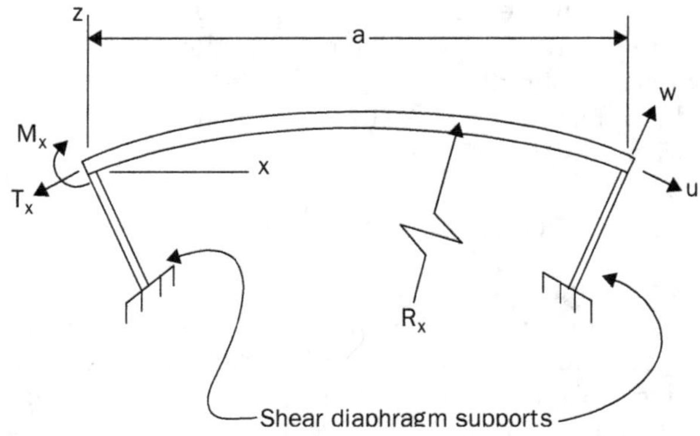

FIGURE 7.5 A shallow shell supported by shear diaphragms.

displacement components at the shell boundary, but not u (i.e., a movable hinge). However, constraining a shell boundary between two knife-edges is different. If the knife-edges are compressed, then $u = v = w = 0$. If not, then $T_x = T_{xy} = w = 0$. Both types of constraint cause free vibration frequencies which are significantly different than those of shear diaphragm supports.

Returning to the solution of the vibration problem, substituting (7.14), (7.15) and (7.16) into (7.7) and (7.8) yields

$$\begin{bmatrix} D(\alpha_m^2+\beta_n^2)^2 - \rho h\omega_{mn}^2 & -\left(\dfrac{\alpha_m^2}{R_y}+\dfrac{\beta_n^2}{R_x}\right) \\ \left(\dfrac{\alpha_m^2}{R_y}+\dfrac{\beta_n^2}{R_x}\right) & \dfrac{1}{Eh}(\alpha_m^2+\beta_n^2)^2 \end{bmatrix} \begin{bmatrix} A_{mn} \\ B_{mn} \end{bmatrix} = \begin{bmatrix} 0 \\ 0 \end{bmatrix} \tag{7.18}$$

For a nontrivial solution, setting the determinant of the coefficient matrix equal to zero permits one to obtain the free vibration frequencies from

$$\omega_{mn}^2\rho h(\alpha_m^2+\beta_n^2)^2 = Eh\left(\frac{\alpha_m^2}{R_y}+\frac{\beta_n^2}{R_x}\right) + D(\alpha_m^2+\beta_n^2)^4 \tag{7.19}$$

Converting (7.19) to nondimensional form gives

$$\Omega_{mn}^2 = \frac{(1-\nu^2)\left[(R_x/R_y)(R_x\alpha_m)^2+(R_x\beta_n)^2\right]^2 + (h^2/12R_x^2)\left[(R_x\alpha_m)^2+(R_x\beta_n)^2\right]^4}{\left[(R_x\alpha_m)^2+(R_x\beta_n)^2\right]^2} \tag{7.20}$$

where Ω_{mn}^2 is a nondimensional frequency parameter given by

$$\Omega_{mn} = \omega_{mn} R_x \sqrt{\frac{\rho(1-\nu^2)}{E}} \tag{7.21}$$

It should be noted in (7.20) that $R_x\alpha = m\pi R_x/a$, which involves the shallowness ratio R_x/a, and then $R_x\beta = n\pi R_x/b = n\pi(R_x/a)(b/a)$, which involves also the aspect ratio (b/a) of the shell planform. Because m and n take on all integer values of m and n, a doubly infinite set of free vibrations frequencies is given by (7.20).

If the frequencies are to be determined from (7.9), then displacement components may be assumed as

$$\begin{aligned} u(x,y,t) &= U_{mn} \cos \alpha_m x \sin \beta_n y \sin \omega t \\ v(x,y,t) &= V_{mn} \sin \alpha_m x \cos \beta_n y \sin \omega t \\ w(x,y,t) &= W_{mn} \sin \alpha_m x \sin \beta_n y \sin \omega t \end{aligned} \tag{7.22}$$

Substituting these into (6.12), (6.13), (7.5), and (6.10) shows that the shear diaphragm boundary conditions (7.16) are exactly satisfied. Then substituting (7.22) into (7.9) yields for free vibrations:

$$\begin{bmatrix} \alpha_m^2 + \frac{1-\nu}{2}\beta_n^2 - \frac{\rho(1-\nu^2)}{E}\omega_{mn}^2 & \frac{(1+\nu)}{2}\alpha_m\beta_n & -\left(\frac{1}{R_x}+\frac{\nu}{R_y}\right)\alpha_m \\ \frac{(1+\nu)}{2}\alpha_m\beta_n & \frac{(1-\nu)}{2}\alpha_m^2 + \beta_n^2 - \frac{\rho(1-\nu^2)}{E}\omega_{mn}^2 & -\left(\frac{\nu}{R_x}+\frac{1}{R_y}\right)\beta_n \\ -\left(\frac{1}{R_x}+\frac{\nu}{R_y}\right)\alpha_m & -\left(\frac{\nu}{R_x}+\frac{1}{R_y}\right)\beta_n & \frac{1}{R_x^2}+\frac{2\nu}{R_xR_y}+\frac{1}{R_y^2}+\frac{h}{12}(\alpha_m^2+\beta_n^2) - \frac{\rho(1-\nu^2)}{E}\omega_{mn}^2 \end{bmatrix} \begin{bmatrix} U_{mn} \\ V_{mn} \\ W_{mn} \end{bmatrix} = \begin{bmatrix} 0 \\ 0 \\ 0 \end{bmatrix} \tag{7.23}$$

Setting the determinant of the coefficient matrix of (7.23) equal to zero yields a cubic equation in ω_{mn}^2, which has three roots for each set of m and n (see [13]) in contrast with the previous solution (7.20) arising from (7.18) which had only one root. The three roots will typically consist of one value much smaller than the other two, corresponding to a mode shape which is predominantly (but not entirely) bending (small U_{mn}/W_{mn} and V_{mn}/W_{mn}), and two higher values corresponding to mode shapes which are predominantly stretching. The mode shapes are determined for each value of ω_{mn} the usual manner, by substituting ω_{mn} back into (7.23) and solving for U_{mn}/W_{mn} and V_{mn}/W_{mn}. Equations (7.18) and (7.20) yield only single frequency ω_{mn} for each m and n because the tangential inertia terms were discarded.

In discussing nodal patterns and mode shapes of free vibration for shells, it is typical to describe them in terms of the normal (w) component of displacement because in the lower frequency modes this is usually the largest component. Thus, a "node line" is a line on the shell surface (more precisely, at its middle surface) where w is zero everywhere. However, along such lines u and v are typically not zero. They are merely small in comparison with the maximum values of w in the mode shape. The mode shapes arising from (7.22) and (7.23) may be called doubly symmetric for m and n both odd, and there are three additional symmetry classes in the mode shapes.

Table 7.1 lists nondimensional frequencies Ω for shallow shells with square planforms ($a/b = 1$), obtained from both (7.20) and (7.23). Three important types of shell curvature are considered: spherical ($R_x/R_y = 1$), circular cylindrical ($R_x/R_y = 0$), and hyperbolic paraboloidal ($R_x/R_y = -1$). These three curvatures are depicted in Fig. 7.6.

		n		
R_x/R_y	M	1	3	5
1	1	0.9481 (0.9546)	0.9691 (0.9704)	1.5098 (1.0604)
	3	0.9691 (0.9704)	1.0056 (1.0064)	1.1294 (1.1298)
	5	1.0579 (1.0604)	1.1294 (1.1298)	1.3045 (1.3049)
0	1	0.4746 (0.4783)	0.8756 (0.8768)	1.0274 (1.0275)
	3	0.2018 (0.2020)	0.5743 (0.5747)	0.9264 (0.9266)
	5	0.4643 (0.4644)	0.6558 (0.6560)	1.0098 (1.0101)
−1	1	0.0351 (0.0356)	0.7825 (0.7836)	0.9945 (0.9948)
	3	0.7825 (0.7836)	0.3200 (0.3205)	0.7531 (0.7537)
	5	0.9945 (0.9948)	0.7531 (0.7537)	0.8898 (0.8904)

TABLE 7.1 Nondimensional Frequencies $\Omega = \omega R_x \sqrt{\rho(1-\nu^2)/E}$, Including and Neglecting (Values in Parentheses) Tangential Inertia, for the Doubly Symmetric Modes of Shallow Shells with Square Planforms ($a/b = 1$) Having Shear Diaphragm Edge Supports; $h/R_x = 0.001$, $a/R_x = 0.4$, $\nu = 0.3$

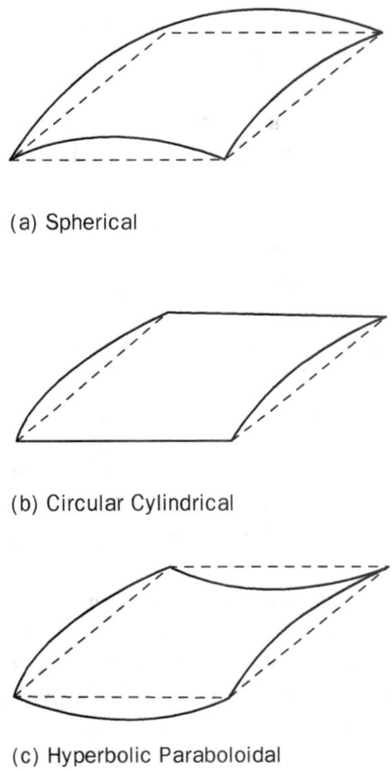

FIGURE 7.6 Spherical ($R_x/R_y = 1$), circular cylinder ($R_x/R_y = 0$), and hyperbolic paraboloidal ($R_x/R_y = -1$) shallow shells.

Results are shown for quite thin shells ($h/R_x = 0.001$), which are rather shallow ($a/R_x = 0.4$, corresponding to $\gamma = 23.1°$ in Fig. 7.4), and a material similar to most commonly used metals ($\nu = 0.3$). Data is given only for the doubly symmetric modes (m and n odd).

Most importantly, in Table 7.1 one notes that the lowest frequencies for shells with spherical curvature are much higher than those of the corresponding modes of circular cylindrical shells, and that the ones for hyperbolic paraboloids are typically lower than for either of the other two. For example, consider the mode having $m = n = 1$. Then $\omega_s/\omega_c = 1.99$ and $\omega_s/\omega_h = 27.01$, where the subscripts s, c, and h identify spherical, cylindrical, and hyperbolic paraboloidal curvatures. These frequency differences are due to the relative stiffnesses of the shells; the spherical shell (with the shear diaphragm boundary conditions) is stiffer than the other two, whereas the hyperbolic paraboloid is the most flexible. It is also interesting to note that the fundamental (i.e., lowest) frequency for the cylindrical shell is not for $m = n = 1$. Two frequency values are given in Table 7.1 for each m and n. The first one is the lowest root arising from (7.23), the second (in parentheses) is from (7.20), which neglects tangential inertia. It is seen that neglecting tangential inertia results in frequencies which are only slightly higher, the percent differences decreasing as the mode numbers (m and n) increase.

Table 7.1 also demonstrates how closely spaced the frequencies can be for a shell. Consider the spherical shell ($R_x/R_y = 1$). For the first four doubly symmetric modes shown in the table, the highest (1.0056) is only 6.1 percent higher than the lowest (0.9481). And there are five other modes not shown, corresponding to $(m, n) = (1, 2)$, (2, 1), (2, 2), (2, 3), (3, 2), which have intermediate frequencies. Thus, the first nine frequencies fall within 6.1 percent of each other. Of course, several of the modes are degenerate ($\omega_{mn} = \omega_{nm}$). However, if the symmetry were altered in a very small way (say, by adding a very small point mass to the shell), the frequencies would change only slightly, and all nine modes would be distinct (i.e., no degeneracies).

Another extremely important characteristic of vibrating shells will now be demonstrated. Frequencies of shallow shells will be compared with those of plates having the same thickness and planform area. To make the distinction clear, let ω_s be the shell frequency, and ω_p be a plate frequency. From (7.19)

$$\frac{\rho h\omega_s^2}{D} = \frac{Eh}{D}\frac{[(\alpha^2/R_y)+(\beta^2/R_x)]^2}{(\alpha^2+\beta^2)^2} + (\alpha^2+\beta^2)^2 \tag{7.24}$$

The plate frequency is obtained as a limiting case of (7.24), as R_x and R_y approach infinity, giving

$$\frac{\rho h\omega_p^2}{D} = (\alpha^2+\beta^2)^2 \tag{7.25}$$

This is the frequency which was found in the preceding chapter for the rectangular plate having all edges simply supported (see Example 6.1). For any given mode (m, n) the ratio of the frequencies squared is obtained by dividing (7.24) by (7.25), resulting in

$$\left(\frac{\omega_s}{\omega_p}\right)^2 = 1 + \frac{Eh}{D}\frac{[(\alpha^2/R_y)+(\beta^2/R_x)]^2}{(\alpha^2+\beta^2)^4} \tag{7.26}$$

One may regard the second term on the R.H.S. of (7.26) as a factor which changes the frequency if the plate is given curvature. This term is always positive (except for certain modes of shells having negative R_x/R_y, when it can be zero), indicating that the shell frequency is always greater than the corresponding plate frequency. The next question is: How much greater?

To reduce the number of parameters which may be varied in (7.26), let us take a shallow shell having a square planform boundary (a = b), and consider only the mode shape having one half-sine wave in each direction ($m = n = 1$). This is often the fundamental (lowest frequency) mode, but not always, as we have seen in Table 7.1. Setting $a = b$, and $m = n = 1$, and also choosing $\nu = 0.3$, specifies (7.26) as

$$\left(\frac{\omega_s}{\omega_p}\right)^2 = 1 + 7.0065 \times 10^{-3}\left(\frac{R_x}{h}\right)^2\left(\frac{a}{R_x}\right)^4\left(\frac{R_x}{R_y}+1\right)^2 \tag{7.27}$$

The frequency ratio is thus a function of three nondimensional parameters—thickness ratio (R_x/h), shallowness parameter (a/R_x) and curvature ratio (R_x/R_y).

For the hyperbolic paraboloidal curvature ($R_x/R_y = -1$), (7.27) gives $\omega_s/\omega_p = 1$. This result is surprising for it says that for such shells *with shear diaphragm edge supports*, fundamental frequencies are the same as those of plates with the same planform. However, this statement applies only to the very special configuration being discussed here. It is not valid for other edge conditions or other boundary shapes. Even more remarkably, if the tangential inertia is considered, the fundamental frequency of the hyperbolic paraboloid shell is slightly lower than that of the plate, as seen from the data in Table 7.1.

Variation of ω_s/ω_p, with the shallowness parameter as obtained from (7.27) is described in Table 7.2 for spherical ($R_x/R_y = 1$) shells. From (7.27) it is clear that cylindrical shells ($R_x/R_y = 0$) have frequency ratio values that are one-fourth of those in Table 7.2. All three forms of shallowness indicators described earlier in Sec. 7.2 are shown in Table 7.2. The first one, a/R_x is that used in (7.27). The second one, γ, is the angle included by the circular arc, determined from $\gamma = 2$ arc sin $(a/2R_x)$. The third one, H/a, is the *rise* of the shell (H) divided by its span. The geometric parameters a, R_x, γ, and H are shown in Fig. 7.4 which is drawn for the case $a/R_x = 0.60$, $\gamma = 34.9°$, and $H/a = 0.076$. This case is clearly a shallow shell, but less shallow than for the smaller a/R_x in Table 7.2.

Consider first the data in Table 7.2 for the shell of moderate thickness ($h/R_x = 0.01$). It is seen that for $a/R_x = 1$, the shell frequency is 16.77 times that of the corresponding plate. It is true that this

Shallowness indicators			ω_s/ω_p for:	
a/R_x	γ (deg)	H/a	$h/R_x = 0.01$	$h/R_x = 0.001$
1.00	60.0	0.134	16.77	167.41
0.80	47.2	.104	10.76	107.15
0.60	34.9	.076	6.11	60.28
0.40	23.1	.051	2.86	26.80
0.20	11.5	.025	1.20	6.77
0.10	5.7	.013	1.01	1.95
0.05	2.9	.006	1.00	1.08

TABLE 7.2 Ratio of Shell Frequency (ω_s) to Plate Frequency (ω_p) for Spherical Shallow Shells Having Square Boundaries Which Are Supported by Shear Diaphragms, Varying with Shallowness (a/R_x), for $m = n = 1$ and $\nu = 0.3$

shell pushes the limits of validity for shallow shell theory, and that if tangential inertia were included the ratio would be slightly smaller. For the clearly shallow configuration with $a/R_x = 0.60$ (Fig. 7.4), $\omega_s/\omega_p = 6.11$. The thin shells ($h/R_x = 0.001$) show an even greater contrast, having ω_s/ω_p on the order of 100 for the deeper cases. For the extremely shallow shell with $a/R_x = 0.10$, one computes $H/h = (H/a)(a/R_x)(R_x/h) = (0.013)(0.1)(1000) = 1.3$, that is, its rise is only 1.3 times its thickness. Nevertheless, its frequency is almost twice (1.95) that of the plate. Thus, one reaches the important conclusion that (at least for most curvatures) frequencies of shallow shells supported by shear diaphragms for modes with $m = n = 1$ are typically much higher than those of simply supported plates having the same planforms. This frequency increase is due to the much greater stiffness of a shell, especially with spherical curvature. For higher modes, the frequency ratio is typically less. Because the straight *nodal lines duplicate the conditions which exist at shear diaphragm supports*, $m = n = 2$, for example, corresponds to a square planform having a/R_x only half as large as that for $m = n = 1$.

As for rectangular plates, it is also possible to obtain exact solutions for the free vibration frequencies and mode shapes of shallow shells of rectangular planform, supported on two opposite edges by shear diaphragms, with arbitrary boundary conditions on the other two edges (e.g., clamped, immovable hinges, shear diaphragms, free). However, the procedure is a long and tedious one. If tangential inertia is neglected, then the solution forms (7.15) and (7.16) may be generalized to

$$W(x,y) = W_m(y)\sin\alpha_m x, \quad \phi(x,y) = \phi_m(y)\sin\alpha_m x \tag{7.28}$$

where, as before, $\alpha = m\pi/a$ ($m = 1, 2, \ldots$). It may be shown that (7.28) satisfy the shear diaphragm B.C. at $x = 0$ and a exactly. Substituting (7.28) into (7.7) and (7.8) yields two fourth-order ordinary differential equations in W_m and Φ_m. Solving this eighth-order system of equations is rather complex, but it can be done exactly. This yields W_m and Φ_m with eight undetermined constants of integration. Then, in general, one applies four B.C. at each of the edges $y = 0$ and $y = b$ to obtain an eighth-order frequency determinant. The eigenvalues of this determinant, corresponding to the nondimensional frequency parameters, may be found as accurately as desired.

If tangential inertia is included one may proceed similarly by assuming, in place of (7.22),

$$\begin{aligned} u(x,y,t) &= U_m(y) \cos \alpha_m x \sin\omega t \\ v(x,y,t) &= V_m(y) \sin \alpha_m x \sin \omega t \\ w(x,y,t) &= W_m(y) \sin \alpha_m x \sin \omega t \end{aligned} \tag{7.29}$$

Substituting (7.29) into (7.9), another eighth-order system of ordinary differential equations in U, V, and W, is generated. It also has an exact solution to which the eight remaining B.C. may be applied.

Other exact solutions may be found by interchanging sine and cosine in (7.15), (7.16), (7.22), (7.28), and (7.29). This yields boundary conditions which are antitheses of the shear diaphragm B.C. given by (7.17). For example, along $x = 0$ and $x = a$,

$$Q_x = \frac{\partial w}{\partial x} = T_{xy} = u = 0 \tag{7.30}$$

The resulting frequencies will be identical to those of shear diaphragm supports. But the conditions expressed by (7.30) are physically unrealizable, or even capable of being approximated reasonably.

For shallow shells *not* having at least two opposite edges supported by shear diaphragms, one may obtain accurate frequencies and mode shapes by various approximate methods (e.g., Ritz, finite elements, finite differences). Leissa and Narita [14] used the Ritz method to analyze completely free shallow shells of rectangular planform. Assuming sinusoidal motions in time for the displacement components in free vibrations,

$$u = U(x,y) \sin \omega t, \quad v = V(x,y) \sin \omega t, \quad w = W(x,y) \sin \omega t \tag{7.31}$$

they may be chosen as algebraic polynomials having no constraints imposed by the boundaries:

$$U(x,y)=\sum_{i=0}^{I}\sum_{j=0}^{J}A_{ij}x^i y^j$$

$$V(x,y)=\sum_{k=0}^{K}\sum_{l=0}^{L}B_{kl}x^k y^l$$

$$W(x,y)=\sum_{m=0}^{M}\sum_{n=0}^{N}C_{mn}x^m y^n \tag{7.32}$$

Using (7.32) in the energy equations (7.11)–(7.13), and writing the Ritz frequency minimizing equations similar to (6.64) for each of the coefficients A_{ij}, B_{kl}, and C_{mn}, yields the frequency determinant. It was shown in Ref. [14] that reasonably accurate, upper bound frequencies are obtained by choosing the upper limits $I, J, \ldots, N$ of the summations in (7.32) at 3, giving third-degree variations in both x and y for all three displacement components, and frequency determinants of order 48.

Table 7.3 displays nondimensional frequency parameters $\omega a^2\sqrt{\rho h/D}$ for completely free shallow shells of square planform ($a/b = 1$), moderate thickness ($a/h = 100$) and $\nu = 0.3$. Results are given for the first two nonzero frequencies of each symmetry class for three types of curvature ($R_x/R_y = 0, 1, -1$) and for three shallowness ratios ($a/R_x = 0, 0.2, 0.5$). The values for $a/R_x = 0$ (i.e., $R_x \to \infty$) are those for a free square plate. Symmetry classes are identified as either symmetric (S) or antisymmetric (A) with respect to axes through the center of the shell, parallel to the x and y axes, respectively. It can be seen that the frequencies of the shallow shells are greater than those of the flat plate for all eight modes with one exception—the first doubly antisymmetric (AA-1) mode of the hyperbolic paraboloidal shell ($R_x/R_y = -1$). Because tangential inertia is included when retaining all three velocity components in the kinetic energy (7.13), this frequency is slightly less than that of the plate. Moreover, the fundamental mode is the AA-1 for each type of curvature and shallowness ratio, with very little difference in frequency from that of the plate (13.468). The second mode is doubly symmetric in each case. For this mode, the spherical curvature ($R_x/R_y = 1$) changes the frequency only slightly from that of the plate, whereas the hyperbolic paraboloid frequencies show the largest increases. This is in contrast with results shown earlier for the shells supported by shear diaphragms (Table 7.1) where the SS-1 frequencies were largest for the spherical curvature and smallest for the hyperbolic paraboloid. Thus, the stiffening influence of shell curvature depends strongly on the edge conditions and modes being considered.

R_x/R_y	a/R_x	SS		SA		AS		AA	
		1	2	1	2	1	2	1	2
–	0	19.60	24.27	34.80	61.11	34.80	61.11	13.47	69.28
0	0.2	21.90	38.47	34.85	75.30	37.64	61.15	13.48	70.95
	0.5	22.07	54.33	34.87	98.22	48.71	61.33	13.51	72.48
1	0.2	19.76	42.35	35.88	73.89	35.88	73.89	13.52	69.60
	0.5	20.00	49.62	36.86	87.72	36.86	87.72	13.58	70.72
−1	0.2	24.74	52.57	36.96	77.06	36.96	77.06	13.46	77.65
	0.5	25.70	64.26	38.92	103.77	38.92	103.77	13.42	79.40

TABLE 7.3 Frequency Parameters $\omega a^2 \sqrt{\rho h / D}$ of Completely Free Shallow Shells Having Square Planform ($a/b = 1$, $a/h = 100$, $\nu = 0.3$)

Figures 7.7, 7.8, and 7.9 are contour plots of normal displacement which describe the mode shapes corresponding to the frequencies for the cylindrical, spherical, and hyperbolic paraboloidal shallow shells, respectively, of Table 7.3 having moderate depth ($a/R_x = 0.5$), taken from Ref. [14]. They may be compared with the contour plots for the completely free square plate, shown previously in Fig. 6.6. As before, the contour lines are for values of $W/W_{max} = 0$ (node lines), 0.2, 0.4, 0.6, 0.8, and 1. One can imagine, for example, how the contour lines for the SS-1 and SS-2 modes for the cylindrical shell (Fig. 7.7) evolve from those of the plate as the curvature increases. The SA-1 and AS-1 modes are similar in appearance to the degenerate modes for the plate, but the two shell frequencies are much different from each other, reflecting the differences in shell stiffness parallel to and perpendicular to the curvature. For $R_x/R_y = 1$ and -1 (Figs. 7.8 and 7.9) the SA-1 and AS-1 modes remain degenerate because the curvature has the same magnitude in both directions.

A similar procedure was followed [15] to obtain accurate frequencies and mode shape contour plots for shallow shells of rectangular planform fixed in the four corners. In that case the displacement functions (7.32) may still be used, along with supplementary conditions that $U = V = W$ in each of the four corners, enforced by using Lagrange multipliers. Shallow shell theory was also used to make extensive studies of the free vibrations of stationary and rotating turbomachinery blades [16–18], with particular emphasis on the cantilevered shell.

Extensions to laminated composite materials were also made [19,20]. Some early research on vibrations of shallow spherical shells having circular boundaries was carried out by Reissner and Johnson [21,22].

FIGURE 7.7 Mode shapes for completely free, circular cylindrical, shallow shells ($R_x/R_y = 0$, $a/R_x = 0.5$).

FIGURE 7.8 Mode shapes for completely free, spherical, shallow shells ($R_x/R_y = 1$, $a/R_x = 0.5$).

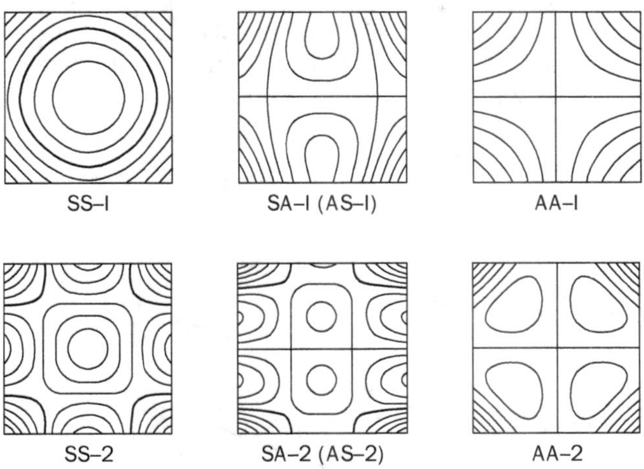

FIGURE 7.9 Mode shapes for completely free, hyperbolic paraboloidal, shallow shells ($R_x/R_y = -1$, $a/R_x = 0.5$).

7.4 Equations of Motion for Circular Cylindrical Shells

We turn now to the more general circular cylindrical shell which may be deep (say, $\gamma > 90°$ in Fig. 7.4) and open, or it may be closed ($\gamma = 360°$ in Fig. 7.4). Such shells have vast practical application, and yet the deep cylindrical shell theory is not as complicated as it is for other curvatures.

Following procedures similar to those of Sec. 7.2, taking an infinitesimal element of the cylindrical shell, summing forces and moments, involving the Kirchhoff hypothesis, using strain-displacement and stress–strain equations, integrating the force resultants and moment resultants through the shell thickness, and finally combining all equations, one can obtain a set of three equations of motion. Various forms of equations may arise, depending on how certain terms are neglected in comparison with others, but proper equations of motion for thin, circular cylindrical shells will take the form (see Ref. 7.1, pp. 31–34):

$$[\mathcal{L}]\{u_i\} = \{0\} \tag{7.33}$$

where $\{u_i\}$ is the vector of displacement components

$$\{u_i\} = \begin{Bmatrix} u \\ v \\ w \end{Bmatrix} \tag{7.34}$$

the orthogonal displacement components u, v, and w being taken in the longitudinal, circumferential, and radial directions, respectively, as shown in Fig. 7.10, and $\mathcal{L}$ is a matrix differential operator. This operator may be written as the sum of operators

$$[\mathcal{L}] = [\mathcal{L}_{D-M}] + k[\mathcal{L}_{MOD}] \tag{7.35}$$

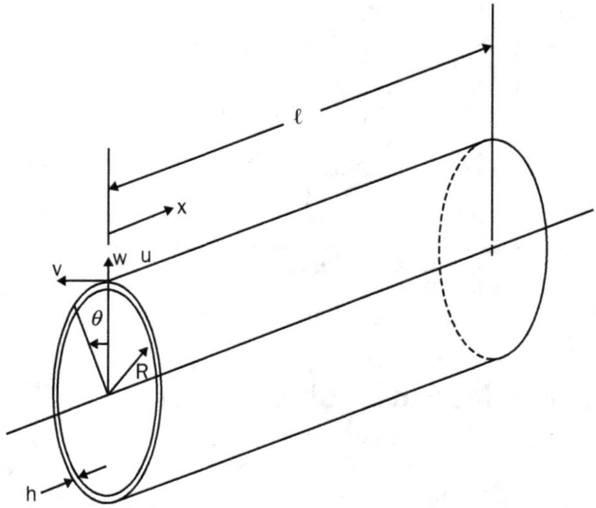

Figure 7.10 Closed circular cylindrical shell and coordinate system.

where $[\mathcal{L}_{D-M}]$ is the differential operator according to the theory developed by Donnell [23,24] and Mushtari [25,26], $[\mathcal{L}_{MOD}]$ is a "modifying" operator which alters the Donnell–Mushtari operator to yield another shell theory, and k is the nondimensional thickness parameter defined as

$$k = \frac{h^2}{12R^2} \tag{7.36}$$

with h being the constant shell thickness, and R being the radius to the shell midsurface (Fig. 7.10).

The Donnell–Mushtari (D–M) operator is found to take the form

$$[\mathcal{L}_{D-M}] = \begin{bmatrix} \left[\frac{\partial^2}{\partial s^2} + \frac{(1-v)}{2}\frac{\partial^2}{\partial \theta^2} - \rho\frac{(1-v^2)R^2}{E}\frac{\partial^2}{\partial t^2}\right] & \frac{1+v}{2}\frac{\partial^2}{\partial s \partial \theta} & v\frac{\partial}{\partial s} \\ \frac{1+v}{2}\frac{\partial^2}{\partial s \partial \theta} & \left[\frac{(1-v)}{2}\frac{\partial^2}{\partial s^2} + \frac{\partial^2}{\partial \theta^2} - \rho\frac{(1-v^2)R^2}{E}\frac{\partial^2}{\partial t^2}\right] & \frac{\partial}{\partial \theta} \\ v\frac{\partial}{\partial s} & \frac{\partial}{\partial \theta} & 1 + k\nabla^4 + \rho\frac{(1-v^2)R^2}{E}\frac{\partial^2}{\partial t^2} \end{bmatrix} \tag{7.37}$$

where $s = x/R$, and $\nabla^4 = \nabla^2\nabla^2$, where

$$\nabla^2 = \frac{\partial^2}{\partial s^2} + \frac{\partial^2}{\partial \theta^2} \tag{7.38}$$

Setting $R_x = R$, and $1/R_y = 1/R_{xy} = 0$ in the shallow shell equations of (7.9) and (7.10), it is seen that they are exactly the same as those of the D–M theory. Moreover, looking closely at the differential operator as it is presented in (7.9), it is clear that all terms in it arise from the membrane stiffness [terms multiplied by $Eh/(1 - v^2)$] except for the $(h^2/12)\,\nabla^4 w$ term arising from L_{33}. This term results from the bending stiffness [note that $Eh/(1 - v^2)$ times $h^2/12$ is $Eh^3/12(1 - v^2) = D$, the flexural rigidity].

Modifying differential operators can take different forms, depending on the simplifying assumptions made in deriving the shell theory, as described above. However, it should be noted from (7.35) that they only represent differences in the *bending* stiffness of the cylindrical shell. They do not affect the stretching stiffness. One widely used cylindrical shell theory was derived by Arnold and Warburton [27,28]. Although derived in a different manner, the same equations for circular cylindrical shells arise from the theories of Goldenveizer [3] and Novozhilov [4]. The corresponding modifying operator is

$$[\mathcal{L}_{MOD}] = \begin{bmatrix} 0 & 0 & 0 \\ 0 & 2(1-v)\dfrac{\partial^2}{\partial s^2} + \dfrac{\partial^2}{\partial \theta^2} & -(2-v)\dfrac{\partial^3}{\partial s^2 \partial \theta} - \dfrac{\partial^3}{\partial \theta^3} \\ 0 & -(2-v)\dfrac{\partial^3}{\partial s^2 \partial \theta} - \dfrac{\partial^3}{\partial \theta^3} & 0 \end{bmatrix} \tag{7.39}$$

Another popular cylindrical shell theory results from independent derivations by Flügge [6,7], and by Byrne [29], Lur'ye [30], and Biezeno and Grammell [31]. The modifying differential operator for it is

$$[\mathcal{L}_{MOD}] = \begin{bmatrix} \dfrac{(1-v)}{2}\dfrac{\partial^2}{\partial \theta^2} & 0 & -\dfrac{\partial^3}{\partial s^3} + \dfrac{(1-v)}{2}\dfrac{\partial^3}{\partial s \partial \theta^2} \\ 0 & \dfrac{3(1-v)}{2}\dfrac{\partial^2}{\partial s^2} & -\dfrac{(3-v)}{2}\dfrac{\partial^3}{\partial s^2 \partial \theta} \\ -\dfrac{\partial^3}{\partial s^3} + \dfrac{(1-v)}{2}\dfrac{\partial^3}{\partial s \partial \theta^2} & -\dfrac{(3-v)}{2}\dfrac{\partial^3}{\partial s^2 \partial \theta} & 1 + 2\dfrac{\partial^2}{\partial \theta^2} \end{bmatrix} \tag{7.40}$$

It can be seen that the two modifying operators (7.39) and (7.40) are considerably different from each other. Seven other modifying operators derived by others for circular cylindrical shells may be found in Ref. [1, pp. 33–34]. And there are still others, derived by other researchers. An obvious question is: Can such differing equations represent the same problem properly? This will be partly answered in the next section.

7.5 Solution for Deep or Closed Circular Cylindrical Shells

Exact solutions for the free vibration frequencies of closed circular cylindrical shells are straightforwardly obtainable if the ends $x = 0$

and $x = \ell$ (Fig. 7.10) are both supported by shear diaphragms. The boundary conditions are those given earlier by (7.17a). These conditions are closely approximated in physical application by means of rigidly attaching a thin, flat, circular cover plate at each end of the shell.

The exact solutions are taken as

$$
\begin{aligned}
u &= A \cos \alpha_m x \cos n\theta \sin \omega t \\
v &= B \sin \alpha_m x \sin n\theta \sin \omega t \\
w &= C \sin \alpha_m x \cos n\theta \sin \omega t
\end{aligned}
\tag{7.41}
$$

where $\alpha_m = m\pi/\ell$ $(m = 1, 2, \ldots)$, and A, B, C are undetermined coefficients. These displacement functions satisfy the shear diaphragm boundary conditions at $x = 0$ and ℓ exactly. For the present, n is also taken in integer values $(n = 0, 1, 2, \ldots)$. These integer values are required for the periodicity of the displacements circumferentially, e.g., $u(x, \theta) = u(x, \theta + 2\pi)$.

Substituting, for example, the displacement components (7.41) into the D–M equations of motion, determined by (7.37) yields the following set of homogeneous equations in A, B, and C:

$$
\begin{bmatrix}
\left[-\lambda^2 - \dfrac{(1-\nu)}{2}n^2 + \rho\dfrac{(1-\nu^2)R^2\omega^2}{E}\right] & \dfrac{(1+\nu)}{2}\lambda n & \nu\lambda \\
\dfrac{(1+\nu)}{2}\lambda n & \left[-\dfrac{(1-\upsilon)}{2}\lambda^2 - n^2 + \rho\dfrac{(1-\nu^2)R^2\omega^2}{E}\right] & -n \\
-\upsilon\lambda & n & \left[1 + k(\lambda^2 + n^2)^2 - \rho\dfrac{(1-\nu^2)R^2\omega^2}{E}\right]
\end{bmatrix}
\begin{bmatrix} A \\ B \\ C \end{bmatrix}
=
\begin{bmatrix} 0 \\ 0 \\ 0 \end{bmatrix}
\tag{7.42}
$$

where $\lambda = m\pi R/\ell$. For a nontrivial solution one sets the determinant of the coefficient matrix of (7.42) equal to zero. Expanding this determinant gives a cubic equation in the frequencies squared. In terms of the nondimensional frequency parameter

$$\Omega = \omega R\sqrt{\frac{\rho(1-\nu^2)}{E}} \tag{7.43}$$

this frequency equation from the D–M theory is

$$\Omega^6 - K_2\Omega^4 + K_1\Omega^2 - K_0 = 0 \tag{7.44}$$

where

$$K_2 = 1 + \frac{1}{2}(3-\nu)(n^2+\lambda^2) + k(n^2+\lambda^2)^2$$

$$K_1 = \frac{1}{2}(1-\nu)\left[(3+2\nu)\lambda^2 + n^2 + (n^2+\lambda^2)^2 + \frac{(3-\upsilon)}{(1-\upsilon)}k(n^2+\lambda^2)^3\right]$$

$$K_0 = \frac{1}{2}(1-\nu)\left[(1-\nu^2)\lambda^4 + k(n^2+\lambda^2)^4\right] \tag{7.45}$$

and again $\lambda = m\pi R/\ell$. Thus, for each m and n there are three free vibration frequencies, corresponding to the three roots of (7.44). Mode shapes are obtained by substituting each Ω, in turn, back into (7.42) to determine corresponding eigenvectors.

If another shell theory is used, the K_i are modified by adding terms $k\,\Delta K_i$ to each of (7.45). For example, for the Goldenveizer–Novozhilov (also Arnold–Warburton) theory, the ΔK_i are [1]

$$\Delta K_2 = 2(1-\nu)\lambda^2 + n^2$$

$$\Delta K_1 = 2(1-\nu)\lambda^2 + n^2 + 2(1-\nu)\lambda^4 - (2-\nu)\lambda^2 n^2 - \frac{1}{2}(3+\nu)n^4$$

$$\Delta K_0 = \frac{1}{2}(1-\nu)\left[4(1-\nu)\lambda^4 + 4\lambda^2 n^2 + n^4 - 2(4-\nu^2)\lambda^4 n^2 - 8\lambda^2 n^4 - 2n^6\right] \tag{7.46}$$

Because the ΔK_i are each multiplied by k before adding them to the K_i, in (7.45), one might think that the difference is trivial. For example, if $h/R = 1/500$, $k = 3.333 \times 10^{-7}$. Even if $h/R = 1/20$, which is as thick a shell as one may typically use thin shell theory accurately, $k = 2.083 \times 10^{-4}$, which is still very small. Nevertheless, these small values of k can make significant differences in frequencies.

A study was made in Ref. [1] comparing the frequencies obtained from 12 shell theories with an exact solution from three-dimensional theory. This showed that the frequencies from the D–M theory are inaccurate for small circumferential wave numbers ($n = 0, 1, 2, 3, 4$), especially for thicker long shells in modes having small longitudinal wave numbers ($\ell/mR > 4$). (More will be said about this later in this chapter.) Most of the shell theories, such as those determined by (7.39) or (7.40), agreed closely with three-dimensional theory. However, others did not. Membrane theory, obtained by deleting all terms in (7.35) containing k, was shown to be highly unreliable.

A comprehensive vibration study of the circular cylindrical shell supported at both ends by shear diaphragms was made earlier by Forsberg [32–34] using the Donnell and Flügge theories. In Fig. 7.11, taken from Ref. [34], the frequency parameter Ω is plotted as a function of the length-to-radius ratio, ℓ/mR, for numbers of circumferential waves n varying between 0 and 28 for a relatively thin shell ($R/h = 500$) according to the accurate Flügge theory. It is

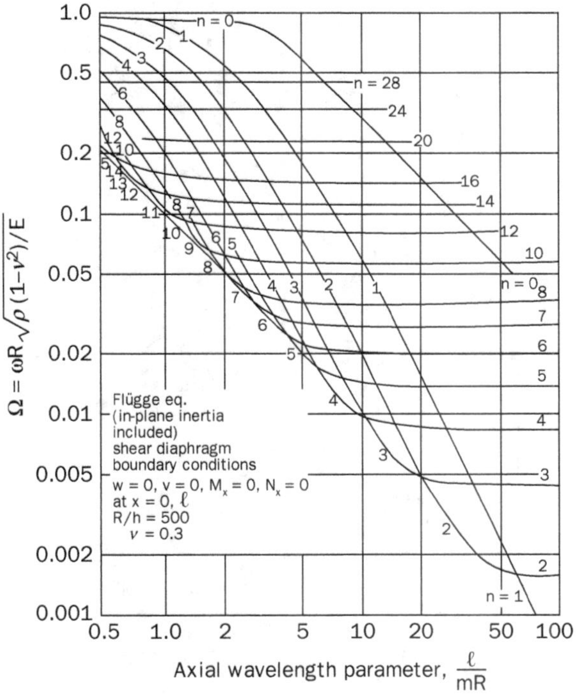

FIGURE 7.11 Variation of Ω with ℓ/mR and n ($R/h = 500$).

obvious from Fig. 7.11 that, for any number of circumferential waves, the lowest frequency always occurs for $m = 1$. But the fundamental frequency for the shell depends strongly on the length-to-radius ratio ℓ/R. For example, for $R/h = 500$ and $\ell/R = 2$, Fig 7.11 shows that the fundamental frequency occurs for $m = 1$ and $n = 8$. That is this mode shape involves eight circumferential waves. However, there are more than 90 modes, with values of m up to 6, and n up to 24, having frequencies less than that for the simple mode shape $m = 1$, $n = 2$ [34].

Some of the mode shapes obtained for various m and n are depicted in Fig. 7.12. The case $n = 0$ is for axisymmetric (shown) and torsional (not shown) modes. The case $n = 1$ is particularly interesting for it describes the shell bending overall as if it were a simply supported beam (with the shear diaphragms preserving the circular cross-section at both ends). Higher n correspond to other flexural modes. It is interesting to see in Fig. 7.11 that for the beam bending mode ($n = 1$) to be the fundamental shell frequency for this very thin shell ($R/h = 500$) requires an extremely long cylinder ($\ell/R > 70$).

Figure 7.13 (from [34]) shows the *fundamental* frequencies Ω for various ℓ/R and R/h ratios. These are the *envelopes* of the complete

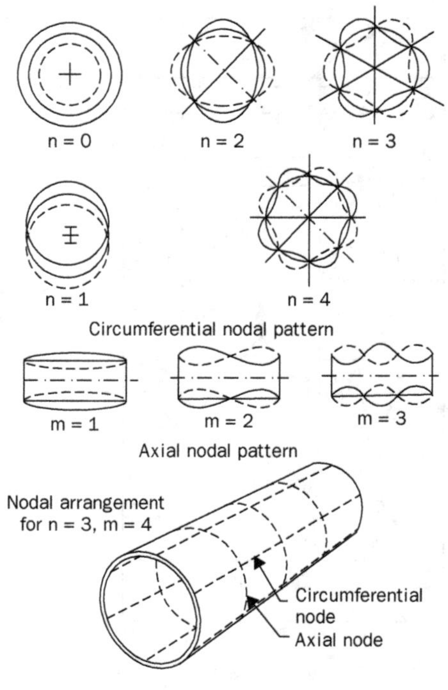

FIGURE 7.12 Some mode shapes for circular cylindrical shells supported at both ends by shear diaphragms.

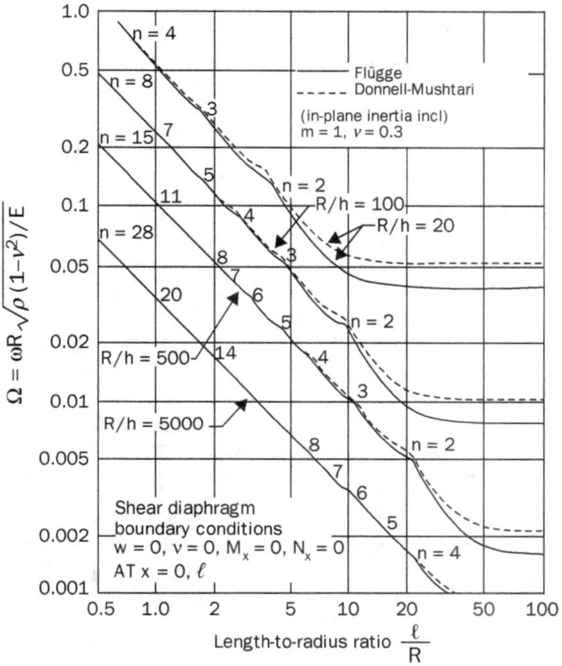

FIGURE 7.13 Fundamental frequency parameters Ω for various ℓ/mR and R/h parameters.

frequency spectrum. For example, the envelope of Fig. 7.11 appears as a single curve (containing numerous cusps) in Fig. 7.13. One also sees how the Donnell–Mushtari theory deviates from the accurate Flügge theory for large ℓ/R, especially for the thicker shells.

Although the axisymmetric modes ($n = 0$) vibration frequencies are never among the lowest ones for a cylindrical shell (e.g., Fig. 7.11), they are worthy of special attention. Looking at the matrix differential operators for the various theories, (7.37), (7.39), and (7.40), it is seen that the second of the three equations of motion becomes uncoupled and yields a purely torsional mode shape [to obtain this case, one interchanges sin $n\theta$ and cos $n\theta$ in (7.41)]. This is seen clearly for the D–M theory by inserting $n = 0$ in (7.42). The other two modes are combinations of radial (w) and axial (u) displacements, uncoupled from the torsion (v). Figure 7.14 (from [35]) shows the frequencies for all three modes, for a broad range of R/h, using the Flügge theory. It is seen that the lowest frequency can correspond to either a radial or torsional mode, depending on ℓ/mR, but never an axial mode. In the vicinity of $\ell/mR = 3$ there is strong coupling between the radial and axial modes. Away from this "veering" region these modes are essentially uncoupled.

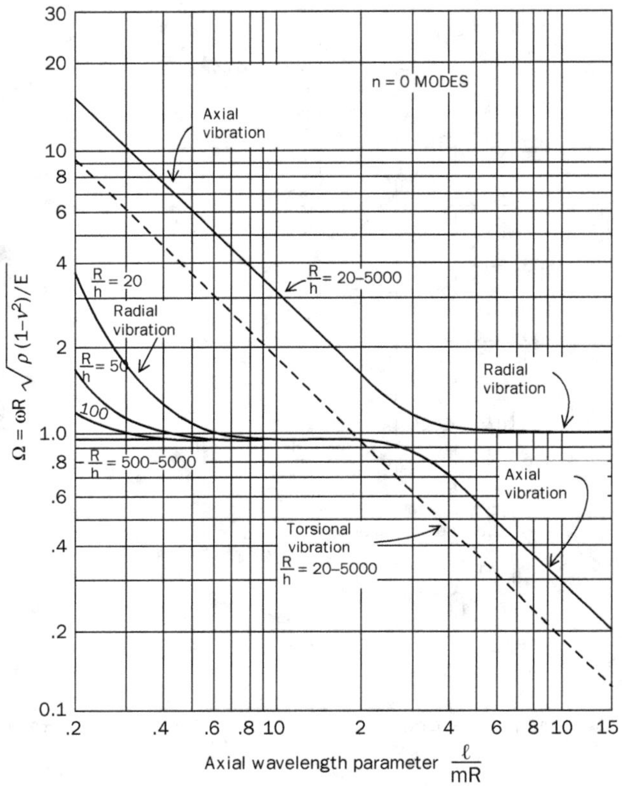

FIGURE 7.14 Axisymmetric (n = 0) frequency parameters; Flügge theory, v = 0.3.

Figure 7.15 (from [35]) is an interesting set of curves showing how the frequency varies with n for different lengths of shells, for the case when $R/h = 100$. (Because Ω contains R, one considers R to be fixed and varies ℓ to obtain different length ratios. Then Ω is a measure of how ω varies.) Here one sees clearly and dramatically how, for a given shell length and a fixed m, the frequency varies with the circumferential wave number (n). For example, for $\ell/R = 10$, the minimum frequency is seen to be at value of n slightly greater than 2. This is the fundamental frequency of the shell. For $\ell/R = 1$ it is in the vicinity of $n = 7$.

In studying Fig. 7.15 one should wonder how smooth, continuous curves can be drawn for varying n, when n can only take on integer values for closed shells. And yet the Ω shown for non-integer n are correct eigenvalues of the cubic equation (7.44) in Ω^2. The answer is that the roots of (7.44) for non-integer n are solutions for another class of problems—the open shell—for which the closed shell may be considered a special case. The open shell has two straight-line boundaries parallel to the shell generators,

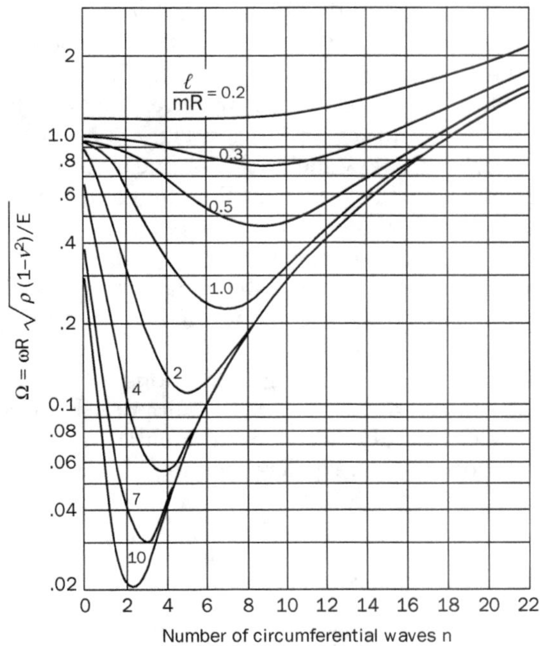

FIGURE 7.15 Variation of Ω with n; Flügge, $R/h = 100$, $\nu = 0.3$.

say $\theta = \pm\gamma/2$ or, $\theta = 0$ and γ, in addition to the two cylindrical boundaries, $x = 0$ and ℓ. Thus Fig. 7.4 may represent a cross-section of the open shell, except that now the shell may be deep (unlimited γ). Indeed, γ may even exceed 360°, which corresponds to a thin shell with overlapped longitudinal boundaries.

In the case of the open shell, the analytical solution and numerical results discussed earlier in this section all apply to shell panels which have all four edges supported by shear diaphragms. Along the straight line edges the corresponding boundary conditions are

$$w = M_\theta = u = T_\theta = 0 \tag{7.47}$$

These are satisfied exactly along the edges $\theta = \pm\gamma/2$ by the displacement functions (7.41), provided that $n = \gamma/\pi$, where γ is in radians. For $\gamma = 180°, 120°, 90°, 60°, \ldots$ (i.e., $\pi, 2\pi/3, \pi/2, \pi/3, \ldots$ radians), one obtains the integer values of n (1, 2, 3, 4,...) for which we have already seen frequencies in Figs. 7.11–7.13 and 7.15. Frequencies for open shells with other R/h and ℓ/mR may be obtained by finding roots of the frequency equations (7.44)–(7.46) using $n = \gamma/\pi$ as described. The complete frequency spectrum for the open shell requires one to use $n = p\gamma/\pi$, where $p = 1, 2, 3, \ldots$ describes the number of circumferential waves present.

Understanding now that the closed cylindrical shell solutions are also applicable to open shells, the reason for the deviation of the D–M theory from the accurate Flügge theory in Fig. 7.13 becomes clear. It was shown earlier that the D–M theory corresponds to the shallow shell theory. Since the curves for $n = 4$ in Fig. 7.13 correspond to $\gamma = 45°$, which is clearly a shallow shell (Fig. 7.4), one may expect the D–M theory to agree reasonably well with the Flügge theory for all R/h and ℓ/R. For $n = 3$ the agreement is worse. For $n = 2$ ($\gamma = 90°$) the disagreement is large for the longer shells.

Extensive frequency and mode shape data for cylindrical shells having their circular boundaries supported by shear diaphragms may be found in Ref. [1].

For cylindrical shells having conditions at their circular boundaries *other than* shear diaphragm supports a more general solution to the eighth-order system of differential equations may be found. The solutions (7.41) are generalized to

$$u = Ae^{\lambda s} \cos n\theta \cos \omega t$$

$$v = Be^{\lambda s} \sin n\theta \cos \omega t$$

$$w = Ce^{\lambda s} \cos n\theta \cos \omega t \tag{7.48}$$

where $s = x/R$. Substituting (7.48) into the D–M equations of motion, for example, as determined by (7.37), leads to the same set of equations of motion given by Eq. (7.42), except that λ^2 is replaced by $-\lambda^2$ in the diagonal elements, and λ is replaced by $-\lambda$ in the first column of the coefficient matrix. For a nontrivial solution, the determinant of the coefficient matrix is set equal to zero, which results in an algebraic equation of the fourth degree in λ^2:

$$\lambda^8 + g_6\lambda^6 + g_4\lambda^4 + g_2\lambda^2 + g_0 = 0 \tag{7.49}$$

where the g_i involve Ω^2, as defined by (7.43), the thickness ratio parameter $k = h^2/12R^2$, and also ν and n.

An exact procedure is available in mathematics for finding roots for algebraic equations of degree as high as four (but no higher). One finds it in many handbooks (cf. [36], p. 9). However, the procedure is quite complicated. With a computer, it is simple matter to use a reliable root-finding program, which may follow an iterative logic. The roots typically take the forms

$$\lambda = \pm\lambda_1,\ \pm i\lambda_2,\ \pm(\lambda_3 + i\lambda_4) \tag{7.50}$$

($i = \sqrt{-1}$). These may be used for solutions for u, v, w where the λ_1 and λ_2 generate exponential (or hyperbolic) and trigonometric functions of s, respectively, and $\lambda_3 + i\lambda_4$, give rise to products of exponential (or hyperbolic) functions and trigonometric functions. For each root the ratios A/C and B/C may be found by returning to the original matrix equation in A, B, and C. The general solutions for u, v, and w are then expressible in terms of eight independent, real constants of integration, which is to be expected from an eighth-order system of differential equations. More details of these procedures may be found in Ref. [1, p. 84].

Having first solved one eigenvalue problem to determine the λ, A/C, and B/C for the solution forms (7.48), it is then necessary to solve another eigenvalue problem to determine the free vibration frequencies (eigenvalues) and mode shapes (eigenfunctions). However, this is straightforward. Four boundary conditions are applied at each end of the shell. This yields, in general, eight homogeneous, simultaneous equations in terms of the unknown integration constants, which is the second eigenvalue problem. If the shell has a plane of symmetry parallel to its circular boundaries (say, both ends clamped, or both ends free), then if the solution forms use the hyperbolic functions rather than the exponential functions one may take advantage of the symmetry by placing the origin of the x-coordinate at the symmetry plane, separating the symmetric and antisymmetric vibration modes, and finding frequencies from fourth-order eigenvalue determinants, rather than eighth-order ones.

Figure 7.16, taken from Ref. [34], shows the characteristics of a shell clamped at both ends ($u = v = w = \partial w/\partial x = 0$) vibrating in a mode shape having eight circumferential waves ($n = 8$) and a single longitudinal wave, which is one of its lowest frequency modes. The shell is thin ($R/h = 500$), and is as long as its diameter ($\ell/R = 2$). Displacement components are normalized with respect to the maximum value of $w\,(W_{\max})$. It is seen that the tangential (circumferential) component is an order of magnitude smaller than $w(V_{\max}/W_{\max} = -0.12)$, and that the longitudinal component is yet another order of magnitude smaller. Nondimensional values of axial force (N_x), axial moment (M_x), hoop (circumferential) force (N_θ), and hoop (circumferential) moment (M_θ) are also shown. What is particularly interesting about the latter three curves is the "edge effect" seen there. For the shell of Fig. 7.16 this is a zone of length approximately 0.1ℓ at each end, where the forces and moments change rapidly, before becoming smooth functions. This edge effect is not seen in rods, beams, membranes, and plates, but is common in shells.

The Ritz method may also be applied straightforwardly to solve free vibration problems for circular cylindrical shells. Accurate frequencies and mode shapes may be obtained relatively easily, but accurate modal forces and moments are more difficult to achieve, not only because they involve differentiating the displacement mode shapes, but because of the rapidly changing edge zone described above.

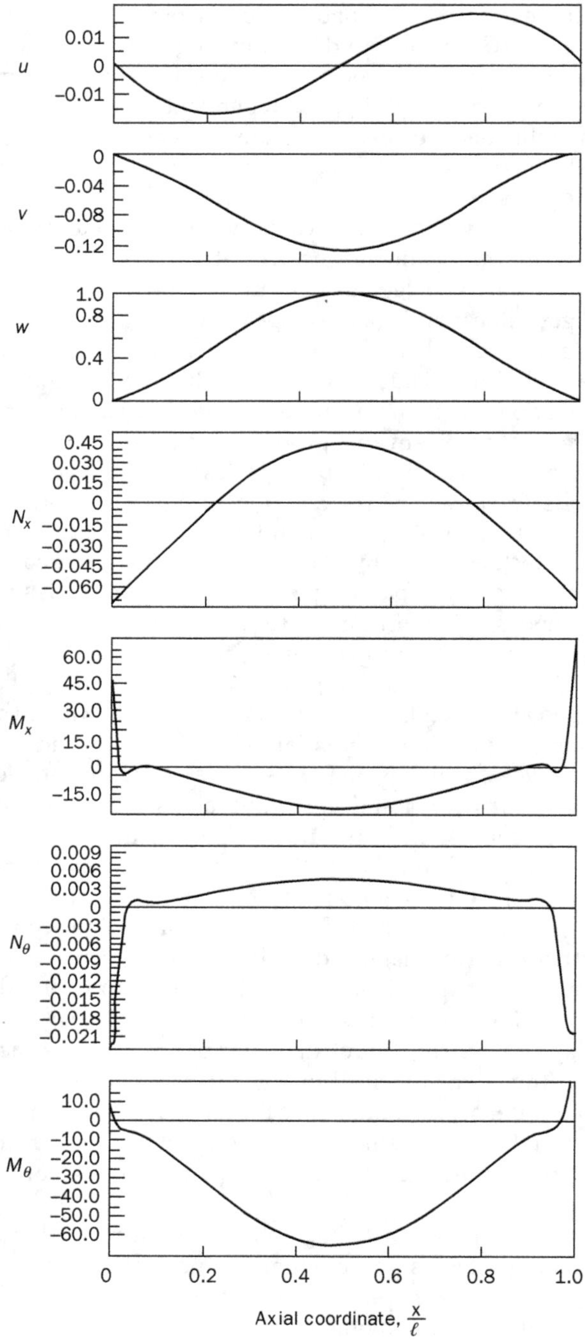

FIGURE 7.16 Modal characteristics for a clamped-clamped shell; $R/h = 500$, $\ell/R = 2$, $m = 1$, $n = 8$.

Extensive frequency and mode shape data for circular cylindrical shells with a variety of boundary conditions is available in Ref. [1]. This data is both theoretical and experimental, the theoretical results being obtained using exact and other approaches.

Another comparison between shallow shell theory and deep shell theory for circular cylindrical shells (in addition to that made earlier in this section for shear diaphragm supports) is available for cantilevered, open shells [37]. In this case one circular edge is clamped, and the other three edges are free.

References

1. A. W. Leissa, *Vibration of Shells,* NASA SP-288, U.S. Govt. Printing Office, 1973, 428 pp. reprinted by The Acoustical Society of America, 1993.
2. H. Kraus, *Thin Elastic Shells,* John Wiley and Sons, 1967.
3. A. L. Goldenveizer, *Theory of Thin Shells* (translation from the Russia edition), Pergamon Press, 1961.
4. V. V. Novozhilov, *The Theory of Thin Elastic Shells* (translation from the Russian edition), P. Noordhoff Ltd., 1964.
5. V. Z. Vlasov, *General Theory of Shells and its Applications in Engineering* (translation from the Russian edition of 1949), NASA TT F-99, 1964.
6. W. Flügge, *Statik und Dynamik der Schalen,* Julius Springer (Berlin), 1934.
7. W. Flügge, *Stresses in Shells,* Springer-Verlag (Berlin), 1962.
8. K. Marguerre, "Zur Theorie der gekrümmten Platte grosser Formänderung," *Proc. Fifth Inter. Congr. Appl. Mech.,* 1938: 93–101.
9. E. Reissner, "Stresses and small displacements of shallow spherical shells," *J. Math. Phys.,* Pt. I, 25 (1946): 80–85; Pt. II, 26 (1947): 273–300.
10. E. Reissner, "On transverse vibrations of thin shallow elastic shells," *Quart. Appl. Math.,* 13 (1955): 169.
11. F. W. Niedenfuhr, A. W. Leissa, and M. J. Gaitens, "A method of analysis for shallow shells of revolution supported elastically on concentric rings," *Developments in Theoretical and Applied Mechanics,* II (1965): 47–67.
12. S. A. Ambartsumyan, *Theory of Anisotropic Shells* (translation from the 1961 Russian edition), NASA TT F-118, 1964.
13. A. W. Leissa and A. S. Kadi, "Curvature effects on shallow shell vibrations," *J. Sound Vib.,* 16 (2)(1971): 173–87.
14. A. W. Leissa and Y. Narita, "Vibrations of completely free shallow shells of rectangular planform," *J. Sound Vib.,* 96 (2)(1984): 207–18.
15. Y. Narita and A. W. Leissa, "Vibrations of corner point supported shallow shells of rectangular planform," *Earthquake Eng. Struct. Dyn.,* 12 (1984): 651–61.
16. A. W. Leissa, J. K. Lee, and A. J. Wang, "Vibrations of cantilevered shallow cylindrical shells of rectangular planform," *J. Sound Vib.,* 78 (3)(1981): 311–328.
17. A. W. Leissa, J. K. Lee, and A. J. Wang, "Vibrations of cantilevered doubly curved shallow shells," *Int. J. Solids and Struct.,* 19 (5)(1983): 411–24.
18. A. W. Leissa, J. K. Lee, and A. J. Wang, "Rotating blade vibration analysis using shells," *J. Eng. for Power.,* 104 (2)(1982): 296–302.
19. A. W. Leissa and M. S. Qatu, "Equations of elastic deformation for laminated composite shallow shells," *J. Appl. Mech.* 58 (1)(1991): 181–88.
20. M. S. Qatu and A. W. Leissa, "Free vibrations of completely free doubly curved laminated composite shallow shells," *J. Sound Vib.* 151 (1)(1991): 9–29.
21. E. Reissner, "On transverse vibrations of thin shallow elastic shells," *Quart. Appl. Math.,* 13 (2)(1955): 169–76.

22. M. W. Johnson and E. Reissner, "On transverse vibrations of shallow spherical shells," *Quart. Appl. Math.*, 15 (4)(1958): 367–80.
23. L. H. Donnell, "Stability of Thin Walled Tubes Under Torsion," NACA Report No. 479, 1933.
24. L. H. Donnell, "A discussion of thin shell theory," *Proc. Fifth Intern. Congr. Appl. Mech.*, 1938.
25. Kh. M. Mushtari, "On the stability of cylindrical shells subjected to torsion" (in Russian), *Trudy, Kaz, Avais, In-ta.*, 2 (1938).
26. Kh. M. Mushtari, "Certain generalizations of the theory of thin shells" (in Russian), *Izv. Fiz. Mat. Ob-va. Pri. Kaz. Un-te.*, 11 (8)(1938).
27. R. N. Arnold and G. B. Warburton, "Flexural vibrations of the walls of thin cylindrical shells having freely supported ends," *Proc. Roy. Soc. (London)*, Ser. A., 197 (1949): 238–56.
28. R. N. Arnold and G. B. Warburton, "The flexural vibrations of thin cylinders," *Inst. Mech. Engineers (Ser. A.)* Proc., 167 (1953): 62–80.
29. R. Byrne, "Theory of small deformations of a thin elastic shell," Seminar Reports in Math., Univ. of Calif. Pub. in Math., N.S., 2 (1)(1944): 103–52.
30. A. I. Lur'ye, "General theory of elastic shells" (in Ukranian), *Prik. Mat. Mekh.*, 4 (1)(1940): 7–34.
31. C. B. Biezeno and R. Grammel, *Engineering Dynamics*, Vol. II, Blackie and Son, Ltd., 1956.
32. K. Forsberg, "Influence of boundary conditions on the modal characteristics of thin cylindrical shells," *AIAA J.*, 2 (12)(1964): 2150–57.
33. K. Forsberg, "Axisymmetric and beam-type vibrations of thin cylindrical shells," Paper 66-447, AIAA 4th Aerospace Sciences Meeting (Los Angeles, Calif.), 1966.
34. K. Forsberg, "A Review of Analytical Methods Used to Determine the Modal Characteristics of Cylindrical Shells," NASA CR- 613, 1966.
35. W. F. Bozich, "The Vibration and Buckling Characteristics of Cylindrical Shells Under Axial Load and External Pressure," Air Force Flight Dynamics Lab, Wright-Patterson AFB, Report No. AFFDL-TR-67-28, 1967.
36. R. S. Burington, *Handbook of Mathematical Tables and Formulas*, 3rd ed., Handbook Publishers, Inc., Sandusky, Ohio, 1949.
37. J. K. Lee, A. W. Leissa, and A. J. Wang, "Vibrations of cantilevered circular cylindrical shells: shallow versus deep shell theory," *Inter. J. Mech. Sci.*, 25 (5) (1983): 361–83.

Problems

1 Make a complete and thorough derivation of the necessary equations which result in (7.7) and (7.8). Explain carefully all the terms used and all steps taken.

2 Make a complete and thorough derivation of the necessary equations which result in (7.9) and (7.10). Explain carefully all the terms used and all steps taken.

3 A. Neglecting tangential inertia, (7.19) was derived. Determine completely the corresponding eigenfunctions for the three displacement components.

B. One may also neglect tangential inertia by omitting two of the inertia terms in (7.9). Assuming solutions as in (7.21), determine whether or not the resulting frequencies are the same as given by (7.19). If not, try to explain why.

4 A circular cylindrical shallow shell has all four edges supported by shear diaphragms. It has a rectangular planform ($a/b = 2$), is quite thin ($h/R_x = 0.001$) and has considerable depth ($a/R_x = 0.8$). Let $\nu = 0.3$.

A. Including tangential inertia, calculate the first ten frequency parameters $\Omega_{mn} = \omega_{mn}\sqrt{\rho(1-\nu^2)/E}$, listing them in order of ascending frequency.

B. For each of the ten modes of Part A, calculate the amplitude ratios U_{mn}/W_{mn} and V_{mn}/W_{mn}.

C. Determine the two higher frequencies for the same m and n as the fundamental frequency, and calculate the amplitude ratio corresponding to it.

D. Sketch carefully the mode shape for the fundamental frequency.

5 It was shown in Sec. 7.3 that hyperbolic paraboloidal shallow shells having square planforms ($a/b = 1$) have fundamental frequencies the same as that for simply supported plates, if tangential inertia is neglected. Investigate what happens if:

A. other modes are considered.

B. other planforms are considered.

6 Consider the axisymmetric ($n = 0$) free vibration modes of closed circular cylindrical shells supported at both ends by shear diaphragms:

A. Write down the displacement functions u, v, and w that yield all possible modes.

B. Obtain formulas which express the frequency parameters Ω in terms of ℓ/mR, k, and ν corresponding to the Flügge theory.

C. Let $\nu = 0.3$, confirm the accuracy of the formulas of Part B by obtaining frequencies for $\ell/mR = 1, 3$, and 10, and comparing them with Fig. 7.15.

D. Determine the amplitude ratios A/C and B/C of the displacements corresponding to all frequencies found in Part C. Compare those results with Fig. 7.15. Make complete sketches for mode shapes for $\ell/mR = 1$.

7 A closed circular cylindrical shell has both ends supported by shear diaphragms.

A. Determine the frequency determinants for both Flügge theory and membrane theory.

B. Let $\nu = 0.3$. On a single graph plot Ω versus ℓ/mR for both theories for modes having two circumferential waves ($n = 2$). Use the range $0.1 \le \ell/mR \le 10$. Discuss the accuracy of the membrane theory for this mode.

8 A circular cylindrical shell panel has all four edges supported by shear diaphragms. Its opening angle is γ (see Fig. 7.4), $\ell/R = 2$, and $R/h = 100$. Let $\nu = 0.3$ for the material.

A. Obtain the frequency determinant, according to the Flügge theory.

B. Plot curves on a single graph showing the nondimensional frequency $\Omega = \omega R\sqrt{\rho(1-\upsilon^2)/E}$ versus γ for $0 \le \gamma \le 360°$. Plot sufficient curves so that at least the four lowest frequencies may be seen for each γ. Compare this graph carefully with the information available in Fig. 7.15.

C. Suppose $\gamma = 360°$. By means of a sketch show carefully what problem this corresponds to. How does it differ physically from an ordinary closed shell? Are the frequencies for $\gamma = 360°$ the same as $n = 0$ for a closed shell? If so, why? If not, why not?

CHAPTER 8

Vibrations of Three-Dimensional Bodies

The vibrations studies in the preceding chapters involved modeling physical bodies mathematically as one-dimensional (strings, bars, beams) or two-dimensional (membranes, plates, shells). During the reading of these earlier chapters, one should logically wonder whether or not three-dimensional analyses could not have been carried out and, if so, why the lower dimensional approaches were really needed. The present chapter should answer both questions to a reasonable degree.

Inasmuch as problems amenable to rectangular coordinates and their analysis are generally the least complicated, and best understood by most, these were the first ones treated in Chaps. 5, 6, and 7. Subsequently, configurations best suited to polar coordinates were taken up. In a similar way, in the present chapter the rectangular parallelepiped is first studied, using rectangular coordinates. Then bodies suitable to cylindrical coordinates are taken up.

Three-dimensional equations of motion are therefore first derived in rectangular coordinates. Some exact solutions of them are seen to exist, but such solutions are shown to usually apply only to unrealistic boundary conditions on parallelepiped faces. Instead, the Ritz method, which has been applied extensively to 1D and 2D problems throughout the earlier chapters, is laid out here, and is used again to obtain reasonably accurate frequencies for some cantilevered rectangular parallelepipeds.

In a similar way the 3D equations of motion in cylindrical coordinates are taken up, and solutions examined. For bodies of revolution (e.g., cylinders, cones, spheres) solutions which separate out the circumferential coordinate (θ) are available, thus, reducing the 3D problem to a set of significantly easier 2D problems. But even then, it is shown that limited solutions exist, which also satisfy the usual boundary conditions of interest.

Therefore, the Ritz method is again taken up, and used to find free vibration frequencies of solid and hollow cylinders which may have free or fixed surfaces. For some of the cylinders, the accurate 3D frequencies are compared with ones calculated by rod, beam, and plate theories. A brief mention of more general Ritz solutions for other bodies of revolution (cones, rings, spheres, hemispheres, paraboloids), both solid and hollow (e.g., shells) is also made. Finally, analysis of twisted and skewed, cantilevered parallelepipeds is described.

All work in this chapter, unlike others, involves only free vibrations with no damping. Generalizations to consider damping and/or exciting forces could be made, as were done in the early chapters.

8.1 Equations of Motion in Rectangular Coordinates

As we have done in previous chapters, let us begin by considering the free body diagram of a typical 3D element taken from within a vibrating body. From this the equations of motion may be derived, expressed in terms of the internal force intensities (stresses).

A typical element of infinitesimal dimensions *dx*, *dy*, and *dz* is depicted in Fig. 8.1. Each face of the element has one component of normal stress (σ) and two components of shear stress (τ) acting on it. The positive senses of the stresses are those conventionally used in the theory of elasticity. That is, normal stresses are positive in tension, and positive shear stresses go in the positive *x*, *y*, and *z* directions on the positive faces (i.e., those having the largest coordinates of *x*, *y*, and *z*). The normal stresses in the *x*, *y*, and *z* directions are σ_x, σ_y, and σ_z, respectively. A typical shear stress τ_{ij} is identified by its subscripts, where *i* indicates the face *normal* to the

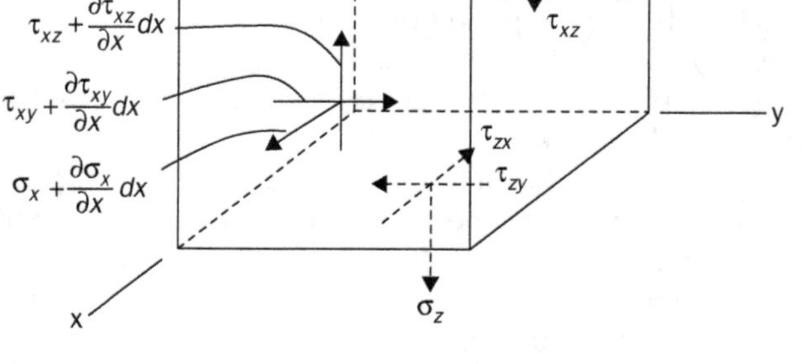

FIGURE 8.1 Three-dimensional element in rectangular coordinates, showing direction of positive stress components.

i direction, and j denotes the direction of the stress vector. Figure 8.1 shows the complete stress components acting on three of the six faces, with directions of positive stress as indicated. The stresses on the positive faces are incrementally different than those of the opposite (negative) faces in a typical situation, as shown by the changes in σ_x, τ_{xz}, and τ_{xy} in Fig. 8.1.

Let us also admit the possibility of body *force* components p_x, p_y, and p_z (force/volume) acting, caused by electric, magnetic, gravitational, or other fields. Then, summing forces in the x-direction yields

$$-\sigma_x dydz + \left(\sigma_x + \frac{\partial \sigma_x}{\partial x}dx\right)dydz - \tau_{yx}dxdz + \left(\tau_{yx} + \frac{\partial \tau_{yx}}{\partial y}dy\right)dxdz$$
$$-\tau_{zx}dxdy + \left(\tau_{zx} + \frac{\partial \tau_{zx}}{\partial z}dz\right)dxdy + p_x dxdydz = (\rho dxdydz)\frac{\partial^2 u}{\partial t^2} \tag{8.1}$$

Here the mass density (ρ) is per unit volume, and the components of displacement in the x, y and z directions are u, v, and w, respectively. Cancelling terms, and dividing through (8.1) by the volume $dxdydz$ yields one equation of motion. Following this procedure for the y and z directions, as well, gives the other two. The resulting complete set is

$$\frac{\partial \sigma_x}{\partial x} + \frac{\partial \tau_{yx}}{\partial y} + \frac{\partial \tau_{zx}}{\partial z} + p_x = \rho\frac{\partial^2 u}{\partial t^2}$$
$$\frac{\partial \tau_{xy}}{\partial x} + \frac{\partial \sigma_y}{\partial y} + \frac{\partial \tau_{zy}}{\partial z} + p_y = \rho\frac{\partial^2 v}{\partial t^2}$$
$$\frac{\partial \tau_{xz}}{\partial x} + \frac{\partial \tau_{yz}}{\partial y} + \frac{\partial \sigma_z}{\partial z} + p_z = \rho\frac{\partial^2 w}{\partial t^2} \tag{8.2}$$

Consider next the angular motions of the element of Fig. 8.1. Let the angular rotations (in radians) about the x, y, and z axes be θ_x, θ_y, and θ_z, respectively. Let us begin by summing moments about an axis parallel to y (call it y') passing through the center of the element. The normal stresses have no moment about this axis. Of the various shear stress components acting on the six faces, only those involving τ_{xz} and τ_{zx} have moments. There is also the possibility of body *couples* m_x, m_y, and m_z (moment/volume) acting, caused by external fields. Then

$$-\tau_{xz}dydz\left(\frac{dx}{2}\right) - \left(\tau_{xz} + \frac{\partial \tau_{xz}}{\partial x}dx\right)dydz\left(\frac{dx}{2}\right) + \tau_{zx}dxdy\left(\frac{dz}{2}\right)$$
$$+\left(\tau_{zx} + \frac{\partial \tau_{zx}}{\partial z}dz\right)dxdy\left(\frac{dz}{2}\right) + m_y dxdydz = dI_{y'}\frac{\partial^2 \theta_y}{\partial t^2} \tag{8.3}$$

where $dI_{y'}$ is the mass moment of inertia of the element about y'. It is

$$dI_{y'} = \rho\left[\frac{dx(dz)^3}{12} + \frac{(dx)^3 dz}{12}\right]dy \tag{8.4}$$

Substituting (8.4) into (8.3) and dividing through the equation by the volume $dxdydz$, one sees that $dI_{y'}$ is a higher order term. Discarding it, and the other two higher order terms, yields the moment equation of motion. Following this procedure, as well, for moments about axes parallel to the x and y directions yields the following set of equations:

$$-\tau_{zy} + \tau_{yz} + m_x = 0$$

$$-\tau_{xz} + \tau_{zx} + m_y = 0$$

$$-\tau_{yx} + \tau_{xy} + m_z = 0 \tag{8.5}$$

Significant body couples exist in only very special cases. Discarding them, (8.5) will become

$$\tau_{yx} = \tau_{xy}, \quad \tau_{zy} = \tau_{yz}, \quad \tau_{xz} = \tau_{zx} \tag{8.6}$$

Then, in the case of *free* vibrations, (8.2) may be written as

$$\frac{\partial \sigma_x}{\partial x} + \frac{\partial \tau_{xy}}{\partial y} + \frac{\partial \tau_{xz}}{\partial z} = \rho\frac{\partial^2 u}{\partial t^2}$$

$$\frac{\partial \tau_{xy}}{\partial x} + \frac{\partial \sigma_y}{\partial y} + \frac{\partial \tau_{yz}}{\partial z} = \rho\frac{\partial^2 v}{\partial t^2}$$

$$\frac{\partial \tau_{xz}}{\partial x} + \frac{\partial \tau_{yz}}{\partial y} + \frac{\partial \sigma_z}{\partial z} = \rho\frac{\partial^2 w}{\partial t^2} \tag{8.7}$$

As we proceeded in other chapters, it now becomes necessary to relate the stresses in (8.7) to the displacements. This is accomplished by using the strain–displacement and stress–strain equations of classical 3D elasticity (cf. [1] pp. 5–10). Assuming small strains, they are related to the displacements by

$$\epsilon_x = \frac{\partial u}{\partial x}, \quad \epsilon_y = \frac{\partial v}{\partial y}, \quad \epsilon_z = \frac{\partial w}{\partial z} \tag{8.8}$$

$$\gamma_{xy} = \frac{\partial v}{\partial x} + \frac{\partial u}{\partial y}, \quad \gamma_{yz} = \frac{\partial w}{\partial y} + \frac{\partial v}{\partial z}, \quad \gamma_{xz} = \frac{\partial w}{\partial x} + \frac{\partial u}{\partial z}$$

where γ_{xy}, γ_{yz}, and γ_{xz} are the *engineering* (not tensorial) shear strains.

For an *anisotropic* material in its most general case the linear, stress–strain relationships involve 21 independent elastic coefficients, each of which must be given in order that the problem is completely defined (cf. [2] p. 10). For an *orthotropic* material in its most general case, the number reduces to nine ([2], p. 19). In both cases the equations of motion (8.7) in terms of displacements become exceedingly complicated. However, for an *isotropic* material there are only two independent coefficients, and the strains may be expressed in terms of the stresses by

$$\epsilon_x = \frac{1}{E}[\sigma_x - \nu(\sigma_y + \sigma_z)]$$

$$\epsilon_y = \frac{1}{E}[\sigma_y - \nu(\sigma_x + \sigma_z)]$$

$$\epsilon_z = \frac{1}{E}[\sigma_z - \nu(\sigma_x + \sigma_y)] \tag{8.9}$$

and

$$\tau_{xy} = \frac{\gamma_{xy}}{G}, \quad \tau_{yz} = \frac{\gamma_{yz}}{G}, \quad \tau_{xz} = \frac{\gamma_{xz}}{G} \tag{8.10}$$

where the shear modulus (G) is related to Young's modulus (E) and Poisson's ratio (ν) by

$$G = \frac{E}{2(1+\nu)} \tag{8.11}$$

The inverse relationships for (8.9) are

$$\sigma_x = \lambda e + 2G\epsilon_x$$

$$\sigma_y = \lambda e + 2G\epsilon_y$$

$$\sigma_z = \lambda e + 2G\epsilon_z \tag{8.12}$$

where $e \equiv \epsilon_x + \epsilon_y + \epsilon_z$ is the trace of the strain tensor, and λ is the Lamé parameter,

$$\lambda = \frac{\nu E}{(1+\nu)(1-2\nu)} \tag{8.13}$$

Substituting (8.12) and (8.13) into (8.7) and assuming that the material is homogeneous, so that the elastic coefficients are all constants, results in the following three equations of motion expressed in terms of the displacements:

$$(\lambda+G)\frac{\partial e}{\partial x}+G\nabla^2 u=\rho\frac{\partial^2 u}{\partial t^2}$$

$$(\lambda+G)\frac{\partial e}{\partial y}+G\nabla^2 v=\rho\frac{\partial^2 v}{\partial t^2}$$

$$(\lambda+G)\frac{\partial e}{\partial z}+G\nabla^2 w=\rho\frac{\partial^2 w}{\partial t^2} \tag{8.14}$$

where ∇^2 is now the three-dimensional Laplacian operator

$$\nabla^2=\frac{\partial^2}{\partial x^2}+\frac{\partial^2}{\partial y^2}+\frac{\partial^2}{\partial z^2}$$

Thus, (8.14) are a set of sixth-order differential equations.

There are only three types of simple boundary conditions which may exist clearly on any surface of the body from a physical point of view:

1. Fixed surfaces (all three displacements components are zero)
2. Sliding surfaces (zero normal displacement, both shear stresses are zero)
3. Free surfaces (all three stress components—one normal, two shear—are zero)

Other boundary conditions may exist from a *mathematical* point of view. They involve combinations of stresses and displacements on any face. But, physically, they are dubious, to say the least. Some of these will be seen in the next section.

8.2 Exact Solutions in Rectangular Coordinates

One form of exact solution to (8.14) is

$$u(x,y,z,t)=A\sin\alpha x\cos\beta y\cos\gamma z\sin\omega t$$

$$v(x,y,z,t)=B\cos\alpha x\sin\beta y\cos\gamma z\sin\omega t$$

$$w(x,y,z,t)=C\cos\alpha x\cos\beta y\sin\gamma z\sin\omega t \tag{8.15}$$

If one chooses the origin of the coordinate system in one corner of a rectangular parallelepiped having dimensions $a \times b \times c$, as shown in Fig. 8.2, and also sets

$$\alpha = \frac{\ell \pi}{a}, \quad \beta = \frac{m\pi}{b}, \quad \gamma = \frac{n\pi}{c} \tag{8.16}$$

where ℓ, m, and n are integers, then (8.15) shows that the *normal* components of displacements are zero along the faces $x = 0,a$; $y = 0,b$; $z = 0,c$. Moreover, substituting (8.15) and (8.16) into (8.8) and (8.12) shows that both tangential (shear) stress components are also zero along these faces. Then, substituting (8.15) and (8.16) into the equations of motion (8.14) yields a third-order frequency determinant. Expanding it results in a cubic frequency equation in ω^2, the squares of the free vibration frequencies. Thus for each set of ℓ, m, and n there are three frequencies.

Let us examine the *physical* meanings of the boundary conditions described above. These conditions could be approximated by enclosing a block of elastic material by a much more rigid outer material, with well-lubricated contact surfaces between the two bodies. However, an initial compressive prestress would be required at all surfaces, so that they remain in contact at all times during the vibration. This would be more difficult to achieve. Thus, the solution given by (8.15) and (8.16) has little practical value.

A second type of solution to (8.14) would interchange sine and cosine in (8.15) still using (8.16). This would correspond to a parallelepiped which now has all its faces restrained tangentially, but with no normal constraint (i.e., $\sigma_n = 0$). This situation is even more difficult to approximate physically than that for the previous type of solution.

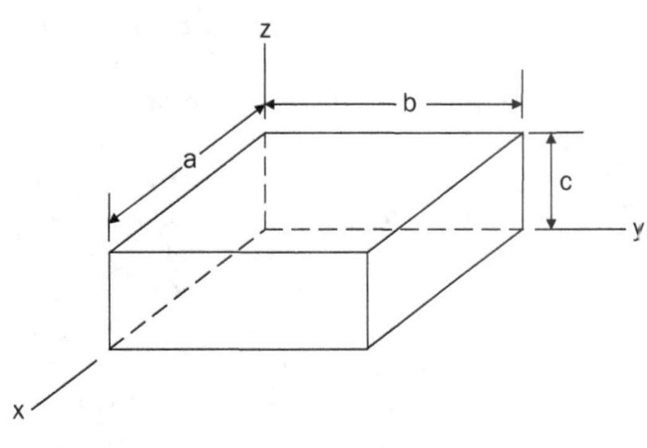

FIGURE 8.2 A rectangular parallelepiped, with a coordinate system useful for the solution (8.15) and (8.16).

A more general type of solution to (8.14) may be assumed in the form

$$u(x,y,z,t) = U(z)\sin\alpha x\cos\beta y\sin\omega t$$

$$v(x,y,z,t) = V(z)\cos\alpha x\sin\beta y\sin\omega t$$

$$w(x,y,z,t) = W(z)\cos\alpha x\cos\beta y\sin\omega t \tag{8.17}$$

Substituting (8.17) into the equations of motion (8.14) yields a set of three coupled, ordinary differential equations in U, V, and W. Each equation is of order two. Thus, the differential order of the set is six. Solving this set yields U, V, and W as functions of z, with each having two constants of integration. Applying three boundary conditions to the faces $z = 0,c$ (see Fig. 8.2) results in a sixth-order determinant for the natural frequencies.

Following the laborious process described in the preceding paragraph, one can prescribe the faces $z = 0,c$ to be both stress-free, and then regard the parallelepiped of Fig. 8.2 to be a "plate" of thickness c. Some analysts have followed this route, using the alternative form of (8.17) wherein sine and cosine are interchanged as described earlier in this section. They then claim that the resulting solution is a 3D free vibration solution for a rectangular plate. But the resulting displacements and stresses along the faces $x = 0,a$ and $y = 0,b$ do not duplicate realistic physical conditions for simply supported plates.

8.3 Approximate Solutions for Rectangular Parallelepipeds

For rectangular parallelepipeds (Fig. 8.2) with faces which are either completely fixed or free, no solutions for free vibration frequencies and mode shapes are available which satisfy exactly both the equations of motion and the boundary conditions. The analyst should use an approximate method which, as refinement in the solution is pursued, would converge to the exact solution.

One approach to the problem was set forth by Fromme and Leissa [3] in 1970. They used the "method of associated periodicity" to develop a solution procedure for the completely free parallelepiped. With this method, solution functions of the type given by (8.15) and (8.16) were used, with extra terms added to permit relaxation of the normal displacement constraints at the boundaries. However, computers available at that time were inadequate to permit obtaining accurate 3D frequencies, and numerical results were obtained for cases having one length infinite. This has the effect of reducing the mathematical dimensionality of the problem to two. Later, Hutchinson and Zillmer [4] used a similar series (or superposition) method to treat the same problem, and obtained numerical results for parallelepipeds having all lengths finite.

The Ritz method may also be used in a straightforward manner to achieve accurate frequencies and mode shapes for rectangular parallelepipeds. As we have seen in previous chapters, this method involves the potential and kinetic energies of the vibrating body.

The potential energy (*PE*) of a 3D elastic body undergoing free vibration with no active external forces being applied is its strain energy of deformation. This may be written in terms of its internal stresses and strains as

$$PE = \frac{1}{2}\int_V (\sigma_x \epsilon_x + \sigma_y \epsilon_y + \sigma_z \epsilon_z + \tau_{xy}\gamma_{xy} + \tau_{yz}\gamma_{yz} + \tau_{xz}\gamma_{xz})dV \tag{8.18}$$

where the integral is taken over the volume (V) of the body. Substituting (8.10)–(8.13) into (8.18) generates this functional in terms of the three displacement components (u, v, w) and, in the case of an isotropic body, any two of the three elastic coefficients E, G, and ν. If the body is also homogeneous, then E, G, and ν are constants.

The kinetic energy is simply

$$KE = \frac{1}{2}\int_V \rho(\dot{u}^2 + \dot{v}^2 + \dot{w}^2)dV \tag{8.19}$$

where, the dots indicate single derivatives in time. For a homogeneous body, the mass density (ρ) is a constant.

For free vibrations one may choose

$$u(x,y,z,t) = U(x,y,z)\sin\omega t \tag{8.20}$$

and similarly for v and w. This yields, for example, the maximum kinetic energy during a cycle of motion as

$$KE_{\text{max}} = \frac{\rho\omega^2}{2}\int_0^a\int_0^b\int_0^c (U^2 + V^2 + W^2)dxdydz \tag{8.21}$$

Similarly, one obtains PE_{max} as the integral of a quadratic form in the first derivatives of U, V, and W.

The functions U, V, and W must be *admissible*. That is, they must satisfy the *geometric* boundary conditions of the problem exactly. For faces which are either fixed or free this may be accomplished by choosing

$$U = x^{e_1} y^{e_2} z^{e_3} (x-a)^{e_4} (y-b)^{e_5} (z-c)^{e_6} \sum_{i=0}^{I} \sum_{j=0}^{J} \sum_{k=0}^{K} A_{ijk} x^i y^j z^k$$

$$V = x^{e_7} y^{e_8} z^{e_9} (x-a)^{e_{10}} (y-b)^{e_{11}} (z-c)^{e_{12}} \sum_{\ell=0}^{L} \sum_{m=0}^{M} \sum_{n=0}^{N} B_{\ell mn} x^\ell y^m z^n$$

$$W = x^{e_{13}} y^{e_{14}} z^{e_{15}} (x-a)^{e_{16}} (y-b)^{e_{17}} (z-c)^{e_{18}} \sum_{p=0}^{P} \sum_{q=0}^{Q} \sum_{r=0}^{R} C_{pqr} x^p y^q z^r \qquad (8.22)$$

In the above, the exponents $e_1, \ldots, e_{18}$ are either zero or one—zero if a face is free, and one if it is fixed. This guarantees that the conditions of zero displacement at a fixed face are satisfied. Moreover, these constraints on the polynomials are minimal, and the resulting sets of displacement functions are mathematically complete. That is, as the upper limits ($I, J, \ldots, R$) of the summations are increased, the body is free to assume an arbitrary deformed shape, and the frequencies will approach the exact values monotonically from above. The polynomials (8.22) are substituted into $PE_{max} - KE_{max}$, and the quantity is minimized w.r.t. the coefficients A_{ijk}, $B_{\ell mn}$, and C_{pqr} to determine the eigenvalues (frequencies) and eigenfunctions in the usual manner (cf. Secs. 2.13 and 4.10).

This procedure was followed by Leissa and Zhang [5] to analyze the 3D vibrations of a *cantilevered* rectangular parallelepiped. Because of the two-fold symmetry present in this situation, the coordinate origin in Fig. 8.2 is moved to the center of the face $x = 0$, so that the xy- and xz-planes are planes of symmetry. Then $e_1 = e_7 = e_{13} = 1$ in (8.22), with all other $e_i = 0$. There are four symmetry classes in the free vibration modes, which may be readily identified by the shape of the axial (U) displacement. Mode shapes are either symmetric (S) or antisymmetric (A) with respect to the xy- and xz-planes. The V and W components will then also have appropriate symmetries. The four symmetry classes of mode shapes may be described as

1. Doubly symmetric (SS) modes; e.g., axial extension
2. Symmetric–antisymmetric (SA) modes; e.g., bending about the y-axis
3. Antisymmetric–symmetric (AS) modes; e.g., bending about the z-axis
4. Doubly antisymmetric (AA) modes; e.g., torsion about the x-axis, with cross-sectional warping

Taking advantage of the symmetry reduces the sizes of the eigenvalue determinants to be evaluated by a factor of four. The exponents $i, j, \ldots, r$ in (8.22) are chosen to reflect the symmetry. Whereas all powers of x are retained for all modes, the SS-modes, for example, take i, k, n, and q as even integers; and m and r as odd integers.

Table 8.1 is a convergence study for the frequencies of a cube ($a = b = c$) having $\nu = 0.3$. Here the results for the nondimensional frequency parameter $\omega a(\rho/E)^{1/2}$ are arranged in increasing order of determinant size, and the first five frequencies are presented for each symmetry class. "Polynomial type" in the second column identifies the number of terms taken in the x, y, and z directions for all three displacement components. It is seen that convergence is monotonic from above as polynomial terms are added in the *same direction*. Table 8.1 also shows that it is generally desirable for this problem to take more terms in the x direction than in y or z. Thus, for examples, the $4 \times 2 \times 2$ results are all better than those from the $2 \times 4 \times 4$ solutions, even though the latter require evaluation of determinants twice as large. Although computer storage limitations at the time of this work (1982) prohibited use of determinants of order greater than 120, the frequencies for determinants of order $(4 \times 3 \times 3) \times 3 = 108$ are seen to be reasonably well convergent.

Numerical results were presented in Ref. [5] for the series of cantilevered rectangular parallelepipeds shown in Fig. 8.3, with Poisson's ratio of 0.3. Based on the convergence study described above, $4 \times 3 \times 3$ solutions were used for all results, yielding 108 modal degrees of freedom for each symmetry class of modes. These results are interesting, and will be described below.

The first five nondimensional frequencies $\omega a(\rho/E)^{1/2}$ for SS modes are listed in Table 8.2. The lowest frequency for each configuration is predominantly axial extension, and elementary rod theory (see Chap. 3) gives the classical result $\omega a(\rho/E)^{1/2} = \pi/2 = 1.5708$, independent of the ratios a/b and b/c. The second value, according to that theory, is $3\pi/2 = 4.7124$. A modified one-dimensional theory was suggested by Love (see [6], p. 428) which takes into account the *lateral* inertia of a rod undergoing axial extension, this inertia being caused by Poisson effects. However, comparing the two sets of 1D frequencies with the much more accurate 3D frequencies listed in Table 8.2 shows that the modified theory is only more accurate than the classical theory for the two very short configurations, D and E.

The first five SS mode shapes for the *cube* (configuration A) are partially described by the contour plots of Fig. 8.4. Here the vibratory displacements U, V, and W occurring at the free end face (x = a) are plotted, normalized w.r.t. the maximum displacement component, corresponding to the frequencies listed in Table 8.2. Contour lines are separated by increments of 0.1. Thus, the first mode, which is predominantly axial extension, has U_{max} occurring in the four free

Symmetry class	Polynomial type	Deter. size	Mode number				
			1	2	3	4	5
SS	4 × 2 × 2	48	1.6022	2.6492	2.9460	3.2553	3.3668
	6 × 2 × 2	72	1.6005	2.6488	2.9421	3.2538	3.3642
	3 × 3 × 3	81	1.6049	2.5866	2.9400	3.0595	3.2919
	2 × 4 × 4	96	1.6483	2.6366	3.0623	3.1931	3.3360
	4 × 3 × 3	108	1.6000	2.5812	2.9154	3.0541	3.2795
SA	4 × 2 × 2	48	0.67402	1.7766	2.8762	3.1486	3.4072
and	6 × 2 × 2	72	.67316	1.7754	2.8721	3.1312	3.4034
AS	3 × 3 × 3	81	.67460	1.7994	2.7659	3.1582	3.3269
	2 × 4 × 4	96	.69742	1.8869	2.8010	3.4332	3.5205
	4 × 3 × 3	108	.67087	1.7695	2.7562	3.0797	3.2985
AA	4 × 2 × 2	48	0.91391	2.1933	2.7343	2.7751	4.0001
	6 × 2 × 2	72	.91369	2.1923	2.7231	2.7747	3.9734
	3 × 3 × 3	81	.91194	2.1830	2.7570	2.8010	3.9911
	2 × 4 × 4	96	.92578	2.2091	2.8106	3.1979	4.2501
	4 × 3 × 3	108	.90930	2.1801	2.7011	2.7480	3.9088

TABLE 8.1 Convergence Study for a Cantilevered Cube; Values of $\omega a(\rho/E)^{1/2}$, $\nu = 0.3$

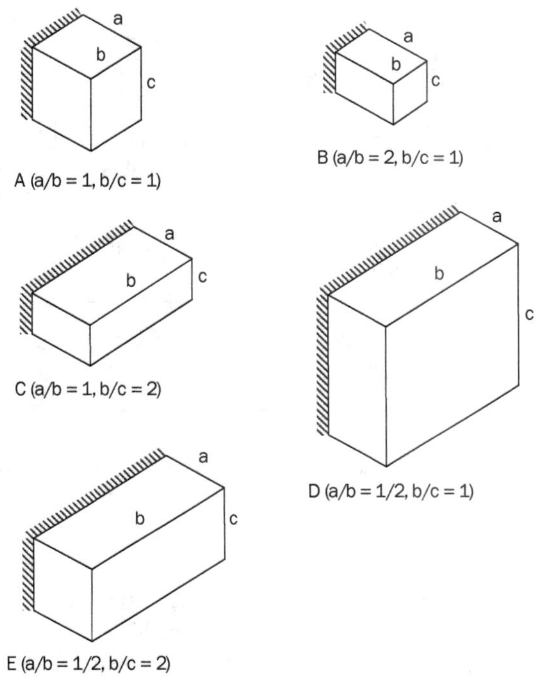

FIGURE 8.3 Cantilevered parallelepiped configurations analyzed.

Configuration	a/b	b/c	Mode number				
			1	2	3	4	5
A	1	1	1.600 1.543*	2.581 ...	2.915 ...	3.054 ...	3.280 ...
B	2	1	1.594 1.564*	4.581	5.065	5.491 ...	5.861 ...
C	1	2	1.596 1.553*	2.797 ...	3.199 ...	4.361 ...	4.746 ...
D	1/2	1	1.467 1.466*	1.562 ...	1.797 ...	1.845 ...	2.011 ...
E	1/2	2	1.533 1.503*	1.684 ...	2.034 ...	2.815 ...	2.895 ...
Elementary rod theory			1.571	...	...	...	...

*Modified one-dimensional theory.

TABLE 8.2 Frequency Parameters $\omega a(\rho/E)^{1/2}$ for SS Modes of Cantilevered Parallelepipeds, $\nu = 0.3$

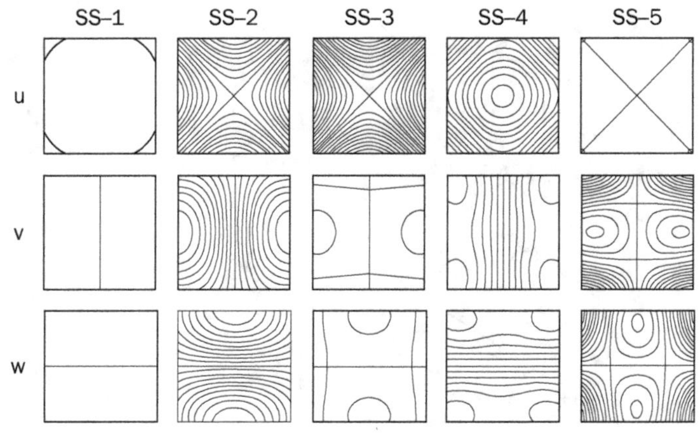

Figure 8.4 Displacement contours on the end face of a cantilevered cube (SS modes).

corners, with contours of $U/U_{max} = 0.9$ located inside of the corners. No other contour lines appear, so $U/U_{max} > 0.8$ for all other end points. No non-zero contour lines are seen for V and W for this mode, so V_{max}/U_{max} and W_{max}/U_{max} are less than 0.1 everywhere.

In the special case of the cube the SS symmetry class of U displacements divides further into two subclasses, which are either symmetric or antisymmetric with respect to the diagonals. Examples of the latter subclass are modes 2, 3, and 5 in Fig. 8.4. Mode 2 is comprised more strongly of transverse displacements V and W, with U/V_{max} (or U/W_{max}) only slightly exceeding 0.8. Mode 3 consists predominantly of diagonally antisymmetric axial warping at the free end; however, examination of other cross-sections show the v and w displacements becoming dominant with decreasing x. Mode 4 is mainly a form of symmetric axial warping, with the corners moving in opposite directions than the center. The interior node lines separating the corners from the center portion are nearly straight. Mode 5 consists almost purely of V and W displacements.

While the contour plots of Fig. 8.4 require some thought in order to interpret them, they provide definitive information about the mode shapes. If further sets of plots were made for, say, $x/a = 0.2$, 0.4, 0.6, and 0.8, the understanding would be complete. These plots are in contrast with 3D plots commonly made with computational packages, which may be pleasing at a first glance, but typically cannot describe completely what is happening, especially for the higher modes. Figure 8.4 also clearly illustrates that *there are no nodal lines or nodal surfaces in 3D modes, except for symmetry axes* ($y = z = 0$). That is, a nodal point must have all three of its displacement components zero, and this only occurs occasionally along symmetry axes (cf. modes 2, 3, and 5 in Fig. 8.4).

Nondimensional frequencies for the SA and AS modes of the cantilevered cube are given in Table 8.3. Comparison of frequencies for the first mode is made among elementary beam theory (Sec. 4.3), Timoshenko beam theory (Sec. 4.12), and the present 3D theory. The lowest frequency SA and AS modes for each configuration are predominantly bending about the y and z axes, respectively, with additional shear deformation, anticlastic bending, and cross-sectional warping present in the 3D modes. A shear deformation coefficient $k = 5/6$ was used for the Timoshenko beam calculations. Table 8.3 shows that the lowest elementary beam frequencies are significantly in error for configurations B and C (14.4 and 13.4 percent differences,

Configu-ration	a/b	b/c	SA or AS	Mode number 1	2	3	4	5
A	1	1	Both	0.6709	1.770	2.756	3.080	3.299
				0.6493[a]	1.733[a]	...	...	...
				1.1050[b]	...	...	...	...
B	2	1	Both	0.4437	1.671	3.724	4.628	5.432
				0.4321[a]	1.632[a]	3.407a	...	...
				0.5075[b]	...	...	...	...
C	1	2	SA	0.4473	1.664	2.278	3.394	3.782
				0.4321[a]	1.532[a]	...	...	3.407[a]
				0.5075[b]	...	...	...	...
			AS	0.6674	1.774	3.068	4.258	4.773
				0.6493[a]	1.733[a]	...	...	...
				1.1050[b]	...	...	...	...
D	1/2	1	Both	0.8300	1.532	1.765	1.879	2.319
				0.7996[a]	...	...	...	...
				2.0300[b]	...	...	...	...
E	1/2	2	SA	0.6748	1.354	1.807	2.312	2.798
				0.6493[a]	...	1.733[a]	...	...
				1.1050[b]	...	...	...	...
			AS	0.8271	1.691	2.313	2.827	2.938
				0.8000[a]	1.654[a]	...	...	...
				2.2100[b]	...	...	...	...

[a] Thick beam theory, including shear deformation and rotary inertia.
[b] Elementary beam theory.

TABLE 8.3 Frequency Parameters $\omega a(\rho/E)^{1/2}$ for SA and AS Modes of Cantilevered Parallelepipeds, $\nu = 0.3$

respectively, from the 3D frequencies), and for the other configurations the theory cannot be meaningfully applied. The Timoshenko beam frequencies are seen to be reasonably good approximations to the 3D frequencies (3.2, 2.6, 3.4, 2.7, 3.7, 2.7, 3.3 percent differences for the seven modes shown), irrespective of the very large "beam" depths in some cases (b/a or $c/a = 2$). Thus, the addition of shear deformation (and rotary inertia) effects appears to compensate well, even for very deep beams. The other effects (e.g., anticlastic bending, cross-sectional warping, thickness-stretch) appear to be minor, at least for the first frequencies. The data of Table 8.3 also show that the Timoshenko beam theory tends to overcorrect the elementary beam theory; that is, it decreases the frequencies, but decreases them somewhat too much. This overcorrection has been observed in many other comparisons of beam theory with 3D theory.

Table 8.4 gives the first five frequencies of doubly antisymmetric modes for each of the five configurations. The lowest frequency mode for each configuration is torsional, and a doubly antisymmetric U displacement results from cross-sectional warping. The third mode of the cube, and the second and fourth modes of configuration B, were also identified from modal displacement plots as being

Configuration	a/b	b/c	Mode number 1	2	3	4	5
A	1	1	0.9093	2.180	2.701	2.748	3.909
			0.9507[a]	...	2.927[a]	...	...
			0.6726[b]	...	2.015[b]	...	...
B	2	1	0.9040	2.719	4.179	4.886	4.944
			0.9213[a]	2.784[a]	...	4.705[a]	...
			0.6716[b]	2.015[b]	...	3.358[b]	...
C	1	2	0.7883	2.220	3.439	3.797	4.467
			0.8837[a]	...	...	...	...
			0.5420[b]	...	...	...	...
D	1/2	1	0.9164	1.355	1.906	2.309	2.523
			1.0174[a]	...	...	...	...
			0.6716[b]	...	...	...	...
E	1/2	2	0.8213	1.731	2.142	2.619	2.839
			1.0682[a]	...	...	...	...
			0.5420[b]	...	...	...	...

[a] Modified one-dimensional torsion theory, including warping constraint.
[b] Elementary torsion theory.

Table 8.4 Frequency Parameters $\omega a(\rho/E)^{1/2}$ for AA Modes of Cantilevered Parallelepiped, s, $\nu = 0.3$

predominantly torsional. Comparison values for torsional frequencies are given in Table 8.4 from two types of 1D analysis: (1) the classical torsion theory (Sec. 3.2) and (2) a modified theory which accounts for the warping constraint at the clamped boundary, presented by Leissa and Ewing [7]. It is seen that the elementary torsion theory gives highly inaccurate results for all the configurations. The frequencies from the improved theory are much closer to the 3D frequencies, which demonstrate the importance of considering torsional warping constraint when it exists (that is, for non-circular cross-sections), especially for short bars.

Plots of U, V, and W for the first five SA and AA modes of the cube (configuration A), similar to those shown in Fig. 8.4 for the SS modes, are also available in Ref. [5]. For the SA modes it was found that mode 2 is predominantly thickness-shear; mode 3 is bending in the transverse (y) direction, called "chordwise bending" in two-dimensional plate analysis; and modes 4 and 5 are more complex. The AA modes consist mainly of warping (U) displacements for modes 2 and 4; whereas mode 5 is a complex one, with strong coupling among all the displacements.

Taken altogether, Tables 8.2–8.4 also show the *density* of the eigenvalues from the 3D spectrum, as may be compared with 1D or 2D analyses. For example, the data of these tables show that for the cube (configuration A) there are 12 frequencies falling in the range $2.7 < \omega a(\rho/E)^{1/2} < 3.3$ (including both SA and AS modes, which are degenerate). But, of course, the various types of 1D and 2D analyses provide only approximate frequencies for restricted portions of the complete spectrum.

The effect of varying Poisson's ratio was also studied in Ref. [5]. Results are shown for the cantilevered cube in Table 8.5 for $\nu = 0$, 0.3, and 0.5, which encompasses the complete range of possible ν for an

Symmetry class	ν	Mode number				
		1	2	3	4	5
SS	0	1.5708	2.856	2.856	3.295	3.295
	0.3	1.6000	2.581	2.915	3.054	3.280
	0.5	1.6887	2.421	2.764	3.153	3.274
SA and AS	0	0.6914	1.885	2.803	3.415	3.522
	0.3	0.6709	1.769	2.756	3.080	3.299
	0.5	0.7020	1.723	2.711	3.004	3.177
AA	0	1.0336	2.416	2.980	3.068	4.275
	0.3	0.9093	2.180	2.701	2.748	3.909
	0.5	0.8598	2.076	2.553	2.631	3.745

TABLE 8.5 Effect of Poisson's Ratio on $\omega a(\rho/E)^{1/2}$ for a Cantilevered Cube ($a/b = b/c = 1$)

isotropic material. The following conclusions about the frequencies may be reached:

1. Some are greatly affected by changes in ν, others are not.
2. Some are increased by increasing ν, others are decreased, and still others do not change monotonically.
3. Some become degenerate, at least for the SS modes, when $\nu = 0$.

In trying to determine why frequencies change as they do with changing ν, one must realize that the Poisson effects change the inertia, as well as the stiffness and that the coupling between types of modes in the same symmetry class changes.

Finally, it should be mentioned that the fixed boundary condition on one face ($x = 0$) of a cantilevered parallelepiped is rather idealized. It could be approached experimentally by bonding a parallelepiped of relatively small Young's modulus (E) to a wall with large E (e.g., a wooden block attached to a steel wall). If the moduli are equal, or nearly so, the wall can only provide an *elastic* constraint, which causes much different frequencies, especially the lowest ones. Moreover, not only is the stiffness of the system affected by similar E, but also the inertia. That is, vibratory motion within the wall adds inertia to the system.

8.4 Exact Solutions in Cylindrical Coordinates

For bodies of circular cylindrical shape it is convenient to solve the 3D free vibration problem in terms of cylindrical coordinates r, θ, z, as shown in Fig. 8.5. In cylindrical coordinates the equations of motion (8.14) become ([8], p. 414):

$$L_{11}u + L_{12}v + L_{13}w = \delta^2 \frac{\partial^2 u}{\partial t^2}$$

$$L_{21}u + L_{22}v + L_{23}w = \delta^2 \frac{\partial^2 v}{\partial t^2}$$

$$L_{31}u + L_{32}v + L_{33}w = \delta^2 \frac{\partial^2 w}{\partial t^2} \tag{8.23}$$

where u, v, and w now are the displacement components in the r, θ, and z directions, respectively, $\delta^2 \equiv 2v\rho/\lambda$, and λ is defined by (8.13). The L_{ij} differential operators are

$$L_{11} = (1-2\nu)\left(\frac{\partial^2}{\partial r^2} + \frac{1}{r}\frac{\partial}{\partial r} + \frac{1}{r^2}\frac{\partial^2}{\partial \theta^2}\right) + 2(1-\nu)\frac{\partial^2}{\partial z^2} \tag{8.24a}$$

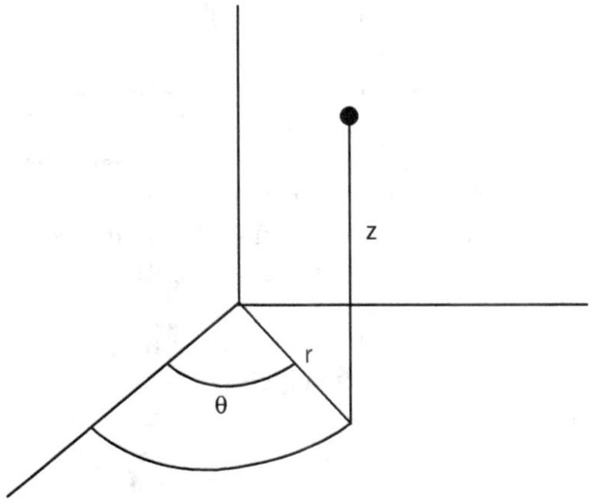

FIGURE 8.5 Cylindrical coordinates.

$$L_{12} = L_{21} = \frac{1}{r}\frac{\partial^2}{\partial\theta\partial z} \tag{8.24b}$$

$$L_{13} = \frac{\partial^2}{\partial r\,\partial z} + \frac{1}{r}\frac{\partial}{\partial z} \tag{8.24c}$$

$$L_{22} = (1-\nu)\left(\frac{\partial^2}{\partial r^2} + \frac{1}{r}\frac{\partial}{\partial r} - \frac{1}{r^2} + \frac{\partial^2}{\partial z^2}\right) + 2(1-\nu)\frac{1}{r^2}\frac{\partial^2}{\partial\theta^2} \tag{8.24d}$$

$$L_{23} = \frac{1}{r}\frac{\partial^2}{\partial r\,\partial\theta} + (3-4\nu)\frac{1}{r^2}\frac{\partial}{\partial\theta} \tag{8.24e}$$

$$L_{31} = \frac{\partial^2}{\partial r\,\partial z} \tag{8.24f}$$

$$L_{32} = \frac{1}{r}\frac{\partial^2}{\partial r\,\partial\theta} - (3-4\nu)\frac{1}{r^2}\frac{\partial}{\partial\theta} \tag{8.24g}$$

$$L_{33} = 2(1-\nu)\left(\frac{\partial^2}{\partial r^2} + \frac{1}{r}\frac{\partial}{\partial r} - \frac{1}{r^2}\right) + (1-2\nu)\left(\frac{1}{r^2}\frac{\partial^2}{\partial\theta^2} + \frac{\partial^2}{\partial z^2}\right) \tag{8.24h}$$

A solution to (8.23) will be sought in the following form:

$$u(r,\theta,z,t) = U(r,z)\cos n\theta \sin\omega t$$

$$v(r,\theta,z,t) = V(r,z)\sin n\theta \sin \omega t$$

$$w(r,\theta,z,t) = W(r,z)\cos n\theta \sin \omega t \tag{8.25}$$

where $n = 0, 1, 2, \ldots$. This form guarantees the periodicity of each of the displacement components with θ; for example, $u(r,\theta + 2\pi,z,t) = u(r,\theta,z,t)$. When $n = 0$ one obtains the *axisymmetric* modes, which involve only u and w (for example, longitudinal and/or radial extension).

A complementary set of functions may also be used for (8.25), replacing $\cos n\theta$ by $\sin n\theta$, and conversely. This gives the same vibratory mode shapes rotated by $90/n$ degrees in θ, and the same frequencies, except for $n = 0$. For $n = 0$ the complementary set yields the torsional modes which involve only v. Thus, for the circular cross-section there is no warping of the cross-section during torsional vibration (which may also be proven easily by geometrical symmetry arguments).

Because of the simplicity of the displacements for the torsional modes of circular cylindrical bars, *meaningful exact* solutions of the 3D dynamic equations of elasticity are possible for these modes. Taking $u = w = 0$, and v as

$$v = V(r,z)\sin \omega t \tag{8.26}$$

Two of the three equations of motion in cylindrical coordinates (those obtained from summing forces in the radial and longitudinal directions) are satisfied exactly. The remaining one is

$$\frac{\partial^2 V}{\partial r^2} + \frac{1}{r}\frac{\partial V}{\partial r} - \frac{V}{r^2} + \frac{\partial^2 V}{\partial z^2} = \frac{\rho}{G}\frac{\partial^2 V}{\partial t^2} \tag{8.27}$$

where $G = E/2(1 + v)$ is the shear modulus of the material. Assuming

$$V(r,z) = R(r)Z(z) \tag{8.28}$$

and substituting it into (8.27), the dependent variables in (8.28) are separated, giving

$$\frac{d^2 R}{dr^2} + \frac{1}{r}\frac{dR}{dr} + \left(\beta^2 - \frac{1}{r^2}\right)R = 0 \tag{8.29a}$$

$$\frac{d^2 Z}{dz^2} + \alpha^2 Z = 0 \tag{8.29b}$$

where

$$\beta^2 = \left(\frac{\rho}{G}\right)\omega^2 - \alpha^2 \tag{8.30}$$

Solving (8.29b) and applying free–free boundary conditions at the two ends:

$$\tau_{\theta z} = G\frac{\partial v}{\partial z} = 0 \quad \text{at} \quad z = \pm\frac{L}{2} \tag{8.31}$$

results in

$$\alpha L = \begin{cases} 2m\pi, & \text{for symmetric modes} \\ (2m-1)\pi, & \text{for antisymmetric modes} \end{cases} \tag{8.32}$$

where $m = 0, 1, 2, \ldots$ for symmetric modes and $m = 1, 2, 3, \ldots$ for antisymmetric modes. The corresponding mode shape functions are

$$Z(z) = \begin{cases} \cos 2m\pi z/L, & \text{for symmetric modes} \\ \sin(2m-1)\pi z/L, & \text{for antisymmetric modes} \end{cases} \tag{8.33}$$

The boundary conditions $\sigma_z = \tau_{rz} = 0$ at the free ends are identically satisfied by the assumed displacements.

Equation (8.29a) is Bessel's equation with a solution

$$R(r) = AJ_1(\beta r) + BY_1(\beta r) \tag{8.34}$$

where J_1 and Y_1 are Bessel functions of order one of the first and second kinds, respectively, and A and B are arbitrary coefficients. Satisfying the regularity condition that v is finite at $r = 0$, it is determined that $B = 0$. Applying the free surface boundary condition

$$\tau_{r\theta} = G\left(\frac{\partial v}{\partial r} - \frac{v}{r}\right) = 0 \qquad \text{at} \quad r = R \tag{8.35}$$

yields

$$(\beta R)J_0(\beta R) = 2J_1(\beta r) \tag{8.36}$$

The boundary conditions $\sigma_r = \tau_{rz} = 0$ on the free cylindrical surface are exactly satisfied by the assumed displacements. Using (8.30) the nondimensional torsional frequencies are given by

$$\omega R\sqrt{\frac{\rho}{G}} = \sqrt{(\beta R)^2 + (\alpha R)^2} \tag{8.37}$$

where α is given by (8.32) and (βR) are the roots of (8.36). These roots have been tabulated ([9], p. 414).

A special case arises for the root $\beta R = 0$. In this case the differential equation (8.29a) reduces to an Euler form having the solution

$$R(r) = C_1 r + \frac{C_2}{r} \tag{8.38}$$

Applying the regularity conditions at $r = 0$ requires that $C_2 = 0$. Hence, for these mode shapes, v varies linearly with r, and as given by (8.33) in z. These are in agreement with the strength of materials assumptions made in deriving the elementary, 1D torsion theory found in Chap. 3. That is, the elementary, 1D torsional vibration solutions are a special case of the complete spectrum of the exact 3D torsional modes and frequencies. With $\beta R = 0$, (8.37) yields $\omega R\sqrt{\rho/G} = \alpha R$, with α given by (8.32). The other modes have the tangential displacement (v) varying as $J_1(\beta r)$, with the core of the rod rotating in one direction, and one or more radial layers rotating in opposite directions during vibratory motion. These are higher frequency modes.

When the rod has one end ($z = 0$) fixed and the other end ($z = L$) free, applying the boundary conditions at the ends yields

$$\alpha L = \left(\frac{2m-1}{2}\right)\pi \quad (m = 1,2,3,\ldots) \tag{8.39}$$

with a mode shape function

$$Z(z) = \sin \alpha z \tag{8.40}$$

Again, there are exact 3D solutions for the special case when $\beta r = 0$, agreeing with the 1D elementary theory, and more general solutions with frequencies given by (8.37) and mode shape function $R(r) = J_1(\beta r)$. It is also clear that fixed–free torsional frequencies of the rod of length L are the same as the antisymmetric mode frequencies of a free–free rod of length $2L$.

For a hollow cylinder having both inner and outer cylindrical surfaces free, the boundary conditions which must still be satisfied are $\tau_{r\theta} = 0$ at $r = R_i$ and $r = R_0$. In this case both terms of (8.34) are retained, and applying the two conditions yields a second-order frequency determinant

$$J_2(\beta R_0)Y_2(K\beta R_0) = Y_2(\beta R_0)J_2(K\beta R_0) \tag{8.41}$$

where $K = R_i/R_0$.

Numerical results for the torsional frequencies of solid and hollow circular cylinders are givens in Secs. 8.5 and 8.6.

Above in this section the solution to the equations of motion (8.23) has been presented thoroughly for $n = 0$ for the *torsional* modes only, where the longitudinal (u) and radial (w) displacement components vanish. The other $n = 0$ modes involve both u and w, coupled, where the circumferential displacement (v) vanishes. For these modes, substituting (8.25) into (8.23) yields two, second-order, partial differential equations coupled in u and w. Solutions to these equations may be found which apply straightforwardly to certain *special* boundary conditions.

For $n \neq 0$, using (8.25) does not uncouple the equations of motion (8.23), and solutions become more difficult. Then, substituting (8.24) and (8.25) into (8.23), one obtains a set of three partial differential equations in the remaining two independent variables r and z. Solutions to them can be found which satisfy certain specialized and limited boundary conditions. But for problems of importance such as, for example, a completely free cylinder or a cylinder which is free except for being fixed at one end, any solution will involve infinite series of terms and infinite frequency determinants.

Nevertheless, the fact that solutions to (8.23) exist in the form of (8.25) establishes one important fact: Mathematically, (8.25) reduces the three dimensional problem to a set of two dimensional problems—one for each value of n. Physically, this means that the cylinder can have a separate set of free vibration modes for each Fourier component, n.

8.5 Approximate Solutions for Solid Cylinders

It was shown in the preceding section that, although exact solutions to the equation of motion expressed in cylindrical coordinates do exist, a complete set of physically meaningful boundary conditions cannot be satisfied, at least not with single-term solutions. Consequently, 3D frequencies for circular cylinders of finite length must be sought for by approximate methods. Among such methods which have been used successfully are: Ritz, series (or superposition) and finite elements. Accurate frequencies for cylinders are easier to obtain than those for parallelepipeds because of the periodicity of displacements in the circumferential (θ) direction. This permits the analyst to separate one coordinate out of the solution, leaving a problem which is mathematically only 2D for each circumferential wave number.

Let us describe how the 3D frequencies and mode shapes of circular cylinders having either fixed or free boundaries may be determined by the Ritz method. Displacement components are again assumed as (8.25), where $n = 0$ provides both the axisymmetric and torsional modes, as discussed in Sec. 8.4. The functions U, V, and W are taken as power series.

$$U(r,z) = f_1(r,z)\sum^{I}\sum^{J} A_{ij} r^i z^j$$

$$V(r,z) = f_2(r,z)\sum^{K}\sum^{L} B_{kl} r^k z^l$$

$$W(r,z) = f_3(r,z)\sum^{P}\sum^{Q} C_{pq} r^p z^q \tag{8.42}$$

where the f_i depend on the geometric boundary conditions to be enforced. For example, in terms of the coordinates shown in Fig. 8.5:

1. $f_1 = f_2 = f_3 = 1$ for all boundaries free
2. $f_1 = f_2 = f_3 = z$ for one end fixed, all other boundaries free
3. $f_1 = f_2 = f_3 = z(z-L)(r-R)$ for all boundaries fixed, where $r = R$ is the cylindrical surface, and $z = 0, L$ are the ends of the cylinder

Because the engineering strains are related to the cylindrical components of displacement by (cf. [1])

$$\epsilon_r = \frac{\partial u}{\sigma r}, \quad \epsilon_\theta = \frac{1}{r}\frac{\partial v}{\partial \theta} + \frac{u}{r}, \quad \epsilon_z = \frac{\partial w}{\partial z}$$

$$\gamma_{r\theta} = \frac{\partial v}{\partial r} + \frac{1}{r}\frac{\partial u}{\partial \theta} - \frac{v}{r}, \quad \gamma_{rz} = \frac{\partial w}{\partial r} + \frac{\partial u}{\partial z}, \quad \gamma_{z\theta} = \frac{\partial v}{\partial z} + \frac{1}{r}\frac{\partial w}{\partial \theta} \tag{8.43}$$

and the stresses are linearly related to the strains, it is clear that care must be exercised in choosing the lower limits for i, k, and p in (8.42) so as to avoid unacceptable strain and stress singularities at $r = 0$, in the case of a solid cylinder. This is accomplished by taking:

1. For the axisymmetric modes ($n = 0$): $i = 1, 2, 3, \ldots$ and $p = 0, 1, 2, \ldots$.
2. For the torsional modes ($n = 0$): $k = 1, 2, 3, \ldots$.
3. For the flexural modes ($n = 1$): $i, k, p = 1, 2, 3, \ldots$ and terms $A_{00} + A_{01}z$ and $B_{00} + B_{01}z$ added to U and V, respectively. These are rigid body translation and rotation terms needed for completeness of the functions for all boundary conditions.
4. For general modes ($n \geq 2$): $i, k, p = 1, 2, 3, \ldots$.

In the special case for $n = 1$, where the terms $U = A_{00} + A_{01}z$ and $V = B_{00} + B_{01}z$ are included, rigid body motions require that $V = -U$. Hence the singularities which they would otherwise cause in the strains ϵ_θ and $\gamma_{r\theta}$ in (8.43) are eliminated.

Substituting (8.42), (8.43) and the stress-strain relations into $V_{max} - T_{max}$, and minimizing the frequency w.r.t. the coefficients A_{ij}, B_{kl}, and C_{pq} (see Sec. 8.3), yields a frequency determinant for each circumferential wave number (n).

Extensive numerical results were presented by Leissa and So [10,11] for circular cylinders using the Ritz method as described above. Table 8.6 shows the convergence of nondimensional frequency parameters $\omega R\sqrt{\rho/G}$ for representative modes (for $n = 1$) of a completely free cylinder having a length-to-diameter ratio, $L/D = 1$.

TR	*TZ*	*DET*	1	2	3	4	5
2	2	12	2.20851	3.00282	3.81261	4.84408	6.11667
2	4	24	2.09061	2.95247	3.47701	4.24351	5.30966
2	6	36	2.09060	2.95258	3.47781	4.24325	5.28632
2	8	48	2.09060	2.95224	3.45998	3.48509	4.24327
3	2	18	2.09656	2.94999	3.75652	4.79013	5.40423
3	4	36	1.99716	2.88066	3.45393	4.16805	4.85705
3	6	54	1.99715	2.88064	3.45375	4.16767	4.85131
3	8	72	1.99715	2.88064	3.45375	4.16760	4.85069
4	2	24	2.09245	2.86580	3.67701	4.74594	5.31706
4	4	48	1.99390	2.80175	3.40084	4.11744	4.73661
4	6	72	1.99389	2.80174	3.40069	4.11702	4.73014
4	8	96	1.99389	2.80174	3.40069	4.11702	4.73010
5	2	30	2.09239	2.85671	3.67607	4.74565	5.27452
5	4	60	1.99381	2.80162	3.40047	4.11715	4.71564
5	6	90	1.99381	2.80161	3.40032	4.11672	4.70970
5	8	120	1.99381	2.80161	3.40032	4.11672	4.70970
6	2	36	2.09238	2.86561	3.67564	4.74559	5.27348
6	4	72	1.99381	2.80153	3.40029	4.11707	4.71444
6	6	108	1.99381	2.80152	3.40015	4.11664	4.70853
7	2	42	2.09238	2.86561	3.67563	4.74558	5.27317
7	4	84	1.99381	2.80153	3.40029	4.11707	4.71427
7	6	126	1.99380	2.80152	3.40015	4.11664	4.70840
8	2	48	2.09238	2.86561	3.67563	4.74558	5.27316
8	4	96	1.99381	2.80153	3.40028	4.11707	4.71425
8	6	144	1.99380	2.80152	3.40015	4.11664	4.70839

Table 8.6 Convergence of Frequencies $\omega R\sqrt{\rho/G}$ for a Cylinder; the Five Lowest Modes with $n = 1$ Which Are Symmetric in z, with $L/D = 1$ and $\nu = 0.3$, Where Boundary Conditions Are Free–Free and Rigid Body Mode Frequencies Are Excluded

For these boundary conditions there is a symmetry plane in the middle of the cylinder, the coordinate origin is located there, and the ends of the cylinder are at $z = \pm L/2$. This permits one to separate modes which are symmetric w.r.t. the midplane from those that are antisymmetric, and reduces the size of frequency determinants needed by a factor of two. The data appearing in Table 8.6 is for symmetric modes. The symbols TR and TZ indicate the total number of polynomial terms used in the r or z directions, respectively, in the displacement functions (8.42), giving rise to determinant orders (DET) of $2 \times TR \times TZ$ to be evaluated for the eigenvalues. The higher values of TR and TZ were limited by numerical ill-conditioning which typically arises with the ordinary polynomials. The underlined data correspond to the lowest (i.e., closest upper bound) frequency obtained for that mode with the smallest determinant size needed. It is observed in Table 8.6 that the first two frequencies here converged to six significant figures, but the next three have lesser accuracy.

Similar convergence trends were seen for the antisymmetric modes of $n = 1$, and for both symmetry classes with $n = 0$ and 2 [10]. More slender rods ($L/D = 5$) were found to require more terms in z than in r. But *determinant sizes* requires to obtain the same degree of convergence as for $L/D = 1$ were approximately the same.

Accurate (i.e., well converged) frequencies obtained by the 3D Ritz analysis are given in Table 8.7 for the axisymmetric ($n = 0$) modes of completely free rods having various L/D ratios (3, 5, 10, 20) and $\nu = 0.3$. The lower frequency modes are predominantly longitudinal vibration. Therefore, in Table 8.7 comparison is made with 1D frequencies calculated by elementary rod theory (see Chap. 3) or 1DE, and also with the modified 1D theory (1DM) which accounts for

L/D	Theory	1	2	3	4	5
				Modes *symmetric* in z		
3	3D	0.8388	2.3264	2.9449	3.1815	3.6843
	1DE	0.8443	2.5328	4.2214	5.9099	7.5985
	%	(0.6)	(8.9)	(43.3)	(85.8)	(106.2)
	1DM	0.8391	2.4029	3.6905	4.6656	5.3739
	%	(0.0)	(3.3)	(25.3)	(46.6)	(45.9)
5	3D	0.5054	1.4846	2.3265	2.8831	2.9574
	1DE	0.5066	1.5197	2.5328	3.5460	4.5591
	%	(0.2)	(2.4)	(8.9)	(23.0)	(54.2)
	1DM	0.5054	1.4902	2.4029	3.2135	3.9098
	%	(0.0)	(0.4)	(3.3)	(11.5)	(32.2)

TABLE 8.7 Comparison of Frequencies $\omega R\sqrt{\rho/G}$ from 3D ($n = 0$), 1D Elementary (1DE), and 1D Modified (1DM) Rod Theories for *Completely Free* Cylinders ($\nu = 0.3$)

L/D	Theory	1	2	3	4	5
10	3D	0.2531	0.7559	1.2470	1.7143	2.1385
	1DE	0.2533	0.7598	1.2664	1.7730	2.2796
	%	(0.1)	(0.5)	(1.6)	(3.4)	(6.6)
	1DM	0.2531	0.7561	1.2492	1.7266	2.1835
	%	(0.0)	(0.0)	(0.2)	(0.7)	(2.1)
20	3D	0.1266	0.3794	0.6310	0.8802	1.1259
	1DE	0.1266	0.3799	0.6332	0.8865	1.1398
	%	(0.0)	(0.1)	(0.4)	(0.7)	(1.2)
	1DM	0.1266	0.3795	0.6310	0.8805	1.1272
	%	(0.0)	(0.0)	(0.0)	(0.0)	(0.1)
Modes *antisymmetric* in z						
3	3D	1.6387	2.8112	2.9541	3.4977	3.6679
	1DE	1.6886	3.3771	5.0657	6.7542	8.4428
	%	(3.0)	(20.1)	(71.5)	(93.1)	(130.2)
	1DM	1.6484	3.0862	4.2153	5.0490	5.6490
	%	(0.6)	(9.8)	(42.7)	(44.3)	(54.0)
5	3D	1.0035	1.9332	2.6432	2.9451	3.1076
	1DE	1.0131	2.0263	3.0394	4.0525	5.0657
	%	(1.0)	(4.8)	(15.0)	(37.6)	(63.0)
	1DM	1.0043	1.9579	2.8221	3.5760	4.2153
	%	(0.1)	(1.3)	(6.8)	(21.4)	(35.6)
10	3D	0.5054	1.0035	1.4846	1.9333	2.3284
	1DE	0.5066	1.0131	1.5197	2.0263	2.5328
	%	(0.2)	(1.0)	(2.4)	(4.8)	(8.8)
	1DM	0.5054	1.0043	1.4902	1.9579	2.4029
	%	(0.0)	(0.1)	(0.4)	(1.3)	(3.2)
20	3D	0.2531	0.5054	0.7559	1.0035	1.2478
	1DE	0.2533	0.5066	0.7598	1.0131	1.2664
	%	(0.1)	(0.2)	(0.5)	(1.0)	(1.5)
	1DM	0.2531	0.5054	0.7561	1.0043	1.2492
	%	(0.0)	(0.0)	(0.0)	(0.1)	(0.1)

Values in parentheses indicate the percent differences between 1D and 3D results.

TABLE 8.7 Comparison of Frequencies $\omega R\sqrt{\rho/G}$ from 3D (n = 0), 1D Elementary (1DE), and 1D Modified (1DM) Rod Theories for *Completely Free* Cylinders (ν = 0.3) (*Continued*)

lateral displacements due to Poisson effects (see Sec. 8.3). Percents by which the 1D frequencies differ from the 3D frequencies are also given, in parentheses, in the table.

In Table 8.7, it is seen that the 3D analysis always yields frequencies lower than those of the 1D analysis. However, for the lower frequencies of slender bars, the differences are negligible. For $L/D = 3$, the 1D elementary theory is seen to be reasonably accurate for only the fundamental frequency, where is the 1D modified theory is reasonably accurate for the first two frequencies (symmetric and antisymmetric modes). The tenth frequency (fifth antisymmetric mode) shows 1D values differing by 130.2 percent and 54 percent from the 3D value. Thus, while the modified 1D theory is significant improvement over the elementary 1D theory, it becomes highly inaccurate for the higher longitudinal frequencies, especially when the rod is relatively short.

One must also realize that some of the 3D modes for $n = 0$ consist of a central core moving longitudinally in one direction while the outer part of the rod moves in the opposite direction. These are typically higher frequency modes, especially for larger L/D; but for $L/D = 3$, at least two of the frequencies listed in Table 8.7 correspond to such modes. Comparison with the 1D frequencies is therefore unrealistic for such modes.

Frequencies of *fixed–free* cylinders were also determined in Ref. [10]. For these boundary conditions there is no symmetry plane. The first five frequencies for $\nu = 0.3$ and $L/D = 3, 5, 10, 20$ are given in Table 8.8. Again, comparisons are made with the two 1D theories. Interestingly, for this case the 1D theories are seen to yield frequencies which are slightly too low for the fundamental mode, but too high for the higher mode frequencies of the less slender (small L/D) configurations.

It is also interesting to compare the 3D frequencies of Table 8.8 with those of the symmetric modes for the free–free cylinders in Table 8.7. The latter modes are similar to those of the fixed–free cylinder in that $v = w = 0$ at the symmetry plane. However, the radial displacement component (u) is not zero there. The 1D symmetric mode frequencies for the free–free cylinder are exactly twice those for the fixed–free one (because the former has a length $L/2$, while the latter is L). However, if one takes the five symmetric mode frequencies for $L/D = 3$ listed in Table 8.7, they are found to be only 1.972, 1.846, 1.451, 1.205, and 1.2514 times those of Table 8.8. (The highest two frequencies may correspond to different types of modes.) Thus, the radial constraint at the completely fixed face is very significant, especially for short (small L/D) cylinders, and the higher frequencies. It was also found that convergence of the fixed–free 3D solutions was much slower than the free–free ones due to the stress singularities that exist in the fixed corner.

In applying the 3D Ritz method to the circular cylinder, all the torsional modes described in Sec. 8.4 were found to any degree of accuracy desired. Convergence studies similar to that of Table 8.6 showed that, as expected, only one polynomial term in the radial direction is required ($TR = 1$) to obtain accurate frequencies for the modes of the elementary torsion type, although many more terms in r are needed for the other modes to represent $R(r) = J_1(\beta r)$. Similarly,

L/D	Theory	1	2	3	4	5
3	3D	0.4254	1.2601	2.0301	2.6407	2.9441
	1DE	0.4221	1.2664	2.1107	2.9550	3.7993
	%	(−0.8)	(0.5)	(4.0)	(11.9)	(29.0)
	1DM	0.4215	1.2492	2.0337	2.7542	3.3984
	%	(−0.9)	(−0.9)	(0.2)	(4.3)	(15.4)
5	3D	0.2546	0.7606	1.2553	1.7274	2.1629
	1DE	0.2533	0.7598	1.2664	1.7730	2.2796
	%	(−0.5)	(−0.1)	(0.9)	(2.6)	(5.4)
	1DM	0.2531	0.7561	1.2492	1.7266	2.1835
	%	(−0.6)	(−0.6)	(−0.5)	(0.0)	(0.9)
10	3D	0.1270	0.3807	0.6332	0.8834	1.1306
	1DE	0.1266	0.3799	0.6332	0.8865	1.1398
	%	(−0.3)	(−0.2)	(0.0)	(0.3)	(0.8)
	1DM	0.1266	0.3795	0.6310	0.8805	1.1272
	%	(−0.3)	(−0.3)	(−0.3)	(−0.3)	(−0.3)
20	3D	0.06346	0.1903	0.3170	0.4435	0.5697
	1DE	0.06332	0.1900	0.3166	0.4432	0.5699
	%	(−0.2)	(−0.2)	(−0.1)	(−0.1)	(0.0)
	1DM	0.06332	0.1899	0.3163	0.4425	0.5683
	%	(−0.2)	(−0.2)	(−0.2)	(−0.2)	(−0.2)

Values in parentheses indicate the percent differences between 1D and 3D results.

TABLE 8.8 Comparison of Frequencies $\omega R\sqrt{\rho/G}$ from 3D (n = 0), 1D Elementary, and 1D Modified Rod Theories for *Fixed–Free* Cylinders (ν = 0.3)

more polynomial terms in z are needed to represent the mode shape functions in z given by (8.33) or (8.37). However, convergence to six significant figures was easily achieved for the first five frequencies for arbitrary L/D and both types of end conditions.

When $n = 1$ in (8.26) the lowest frequencies which arise correspond to beam bending (flexural) modes. That is, the circular cylinder behaves in these modes as a beam of length L bending about its centerline. Elementary (Euler–Bernoulli) and improved (Timoshenko) 1D beam theories were developed and applied in Chap. 4. It will now be interesting to compare the frequencies arising from these analytical models with accurate ones from the 3D theory.

Such comparisons are made in Table 8.9 for free–free beams, and in Table 8.10 for fixed–free (i.e., cantilever) beams, both tables taken from Ref. [10]. The corresponding frequency equations and resulting eigenvalues for the elementary theory are given in Chap. 4 (see Table 4.1). Frequency equations for the Timoshenko beams are rather complicated. They are available in Ref. [12].

L/D	Theory	1	2	3	4	5
				Modes *symmetric* in *z*		
3	3D	0.4049	1.3779	2.1250	2.5971	2.8058
	1DE	0.5011	2.7077	6.6862	12.4330	19.9481
	%	(23.7)	(96.5)	(214.6)	(378.7)	(611.0)
	1DI	0.4037	1.3674	2.1509	2.6833	3.2696
	%	(−0.3)	(−0.8)	(1.2)	(3.3)	(16.5)
5	3D	0.1646	0.6818	1.2889	1.8404	2.1146
	1DE	0.1804	0.9748	2.4070	4.4759	7.1813
	%	(9.6)	(43.0)	(86.8)	(143.2)	(239.6)
	1DI	0.1643	0.6771	1.2762	1.8338	2.1533
	%	(−0.1)	(−0.7)	(−1.0)	(−0.4)	(1.8)
10	3D	0.04397	0.2156	0.4675	0.7585	1.0670
	1DE	0.04509	0.2437	0.6018	1.1190	1.7953
	%	(2.6)	(13.0)	(28.7)	(47.5)	(68.3)
	1DI	0.04395	0.2149	0.4648	0.7520	1.0546
	%	(0.0)	(−0.3)	(−0.6)	(−0.9)	(−1.2)
20	3D	0.01120	0.05885	0.1389	0.2439	0.3673
	1DE	0.01127	0.06092	0.1504	0.2797	0.4488
	%	(0.7)	(3.5)	(8.3)	(14.7)	(22.2)
	1DI	0.01120	0.05880	0.1386	0.2430	0.3651
	%	(0.0)	(−0.1)	(−0.2)	(−0.4)	(−0.6)
				Modes *antisymmetric* in *z*		
3	3D	0.8735	1.7670	2.1002	2.5561	2.8788
	1DE	1.3812	4.4759	9.3386	15.9695	24.3687
	%	(58.1)	(153.3)	(344.6)	(524.8)	(746.5)
	1DI	0.8674	1.7680	2.1397	2.6699	3.2423
	%	(−0.7)	(0.1)	(1.9)	(4.5)	(12.6)
5	3D	0.3986	0.9833	1.5818	1.9013	2.1149
	1DE	0.4972	1.6113	3.3619	5.7490	8.7727
	%	(24.7)	(63.9)	(112.5)	(202.4)	(314.8)
	1DI	0.3969	0.9746	1.5675	1.9372	2.1363
	%	(−0.4)	(−0.9)	(−0.9)	(1.9)	(1.0)
10	3D	0.1162	0.3346	0.6099	0.9110	1.2228
	1DE	0.1243	0.4028	0.8405	1.4373	2.1932
	%	(7.0)	(20.4)	(37.8)	(57.8)	(79.4)
	1DI	0.1160	0.3332	0.6054	0.9022	1.2078
	%	(−0.2)	(−0.4)	(−0.7)	(−1.0)	(−1.2)

TABLE 8.9 Comparison of Frequencies $\omega R\sqrt{\rho/G}$ from 3D (n = 1), 1D Elementary (Euler–Bernoulli, 1DE) Beam, and 1D Improved (Timoshenko, 1DI) Beam Theories for *Completely Free* Cylinders (ν = 0.3)

L/D	Theory	1	2	3	4	5
20	3D	0.03052	0.09527	0.1887	0.3036	0.4338
	1DE	0.03108	0.1007	0.2101	0.3593	0.5483
	%	(1.8)	(5.7)	(11.3)	(18.4)	(26.4)
	1DI	0.03050	0.09513	0.1882	0.3022	0.4311
	%	(−0.0)	(−0.1)	(−0.3)	(−0.4)	(−0.6)

Values in parentheses indicate the percent differences between 1D and 3D results.

TABLE 8.9 Comparison of Frequencies $\omega R\sqrt{\rho/G}$ from 3D (n = 1), 1D Elementary (Euler–Bernoulli, 1DE) Beam, and 1D Improved (Timoshenko, 1DI) Beam Theories for *Completely Free* Cylinders (ν = 0.3) (*Continued*)

L/D	Theory	1	2	3	4	5
3	3D	0.07517	0.3643	0.8186	1.2961	1.7729
	1DE	0.07874	0.4935	1.3817	2.7076	4.4759
	%	(4.8)	(35.4)	(68.8)	(108.9)	(152.5)
	1DI	0.07425	0.3606	0.8095	1.2800	1.7550
	%	(−1.2)	(−1.0)	(−1.1)	(−1.2)	(−1.0)
5	3D	0.02797	0.1562	0.3828	0.6512	0.9432
	1DE	0.02835	0.1776	0.4974	0.9747	1.6113
	%	(1.4)	(13.7)	(29.9)	(49.7)	(70.8)
	1DI	0.02772	0.1549	0.3794	0.6445	0.9315
	%	(−0.9)	(−0.8)	(−0.9)	(−1.0)	(−1.2)
10	3D	0.007089	0.04300	0.1149	0.2121	0.3288
	1DE	0.007087	0.04441	0.1244	0.2437	0.4028
	%	(0.0)	(3.3)	(8.2)	(14.9)	(22.5)
	1DI	0.007047	0.04275	0.1142	0.2106	0.3256
	%	(−0.6)	(−0.6)	(−0.6)	(−0.7)	(−1.0)
20	3D	0.001778	0.01105	0.03054	0.05876	0.09527
	1DE	0.001772	0.01110	0.03109	0.06092	0.1007
	%	(−0.4)	(0.5)	(1.8)	(3.7)	(5.7)
	1DI	0.001769	0.01099	0.03038	0.05844	0.09442
	%	(−0.5)	(−0.5)	(−0.5)	(−0.6)	(−0.9)
40	3D	0.0004449	0.002782	0.007764	0.01514	0.02498
	1DE	0.0004429	0.002776	0.007772	0.01523	0.02518
	%	(−0.4)	(−0.2)	(0.1)	(0.6)	(0.8)
	1DI	0.0004428	0.002769	0.007726	0.01507	0.02475
	%	(−0.5)	(−0.5)	(−0.5)	(−0.5)	(−0.9)

Values in parentheses indicate the percent differences between 1D and 3D results.

TABLE 8.10 Comparison of Frequencies $\omega R\sqrt{\rho/G}$ from 3D (n = 1), 1D Elementary (Euler–Bernoulli, IDE) Beam, and 1D Improved (Timoshenko, IDI) Beam Theories, for *Fixed–Free* Cylinders (ν = 0.3)

For free–free cylinders (Table 8.9) it is seen that Timoshenko theory is reasonably accurate in predicating the first five frequencies for each symmetry class (i.e., the first ten frequencies overall) for $L/D \geq 5$. As expected, the elementary beam theory frequencies are significantly in error for all modes when $L/D = 10$ (2.6 percent–79.4 percent errors from the first to the tenth frequencies). For the thickest configuration considered ($L/D = 3$), the Timoshenko theory is reasonably accurate for only the first two frequencies of each symmetry class. It is also interesting to note that, while the elementary theory always gives frequencies which are too high because shear deformation and rotary inertia effects are ignored, the Timoshenko theory often overcorrects, yielding frequencies which are somewhat too low.

Similar results are shown for fixed–free (cantilevered) cylinders in Table 8.10. Here, only the first five frequencies for each L/D are given (as compared with the first ten in Table 8.9). Although 3D frequencies are listed with at least four significant figures, accuracy to only three figures can be guaranteed for all of the data in Table 8.10, because of the slower convergence obtained for the fixed–free conditions. Nevertheless, it is interesting to note that elementary 1D theory predicts the lower frequencies significantly better in the fixed–free condition than when both ends are free. Conversely, the Timoshenko theory is less accurate for fixed–free ends than for free–free ones.

To summarize the comparisons made above between frequencies from 1D theories and those from the accurate 3D theory, suppose one can accept frequencies which are accurate to within 1 percent of the correct values. Then one finds, for example, from the tabular data for a free–free rod, with a length-to-diameter ratio (L/D) of 5 that only the first two longitudinal mode frequencies are satisfactorily determined by elementary 1D theory, and that the improved theory gives only one more. None of the bending frequencies are acceptable from the 1D elementary theory, but the Timoshenko beam theory is much better, providing seven admissible frequencies. For the torsional modes, the 1D theory is exact (for circular cross-sections only) as long as the rotations of elements in a given cross-section are all in the same direction. Higher frequencies which involve counter-rotation within a cross-section are missed entirely. However, for $L/D = 5$, one finds that the first ten torsional frequencies are those of elementary theory.

Frequencies from the 3D Ritz analysis presented above have been limited to $n = 0$ and 1, so that they be compared directly with the results from 1D analyses. Further 3D results will now be given to determine how the frequencies for other values of n relate. These data are taken from Ref. [11].

Tables 8.11 and 8.12 display the frequencies of completely free cylinders for modes which are symmetric and antisymmetric, respectively, w.r.t. their midplanes, for $L/D = 1, 1.5, 2, 3$, and 5, and for $n = 0, 1, 2, 3$. Poisson's ratio is taken as 0.3.

		L/D				
n	**Modes**	**1**	**1.5**	**2**	**3**	**5**
0[a]	1	5 2.326	3 1.639	3 1.247	3 0.839	4 0.505
	2	10 3.067	14 2.922	18 2.920	15 2.326	12 1.485
	3	20 3.989	3.500	20 3.028	2.945	2.326
	4	4.896	4.042	3.824	3.181	2.883
	5	5.593	4.814	4.185	3.684	2.957
0[t]	1	12 3.142	5 2.094	5 1.571	5 1.047	5 0.628
	2	5.136	4.189	3.142	10 2.094	10 1.257
	3	6.020	5.136	4.712	3.142	16 1.885
	4	6.283	5.546	5.136	4.200	2.520
	5	8.115	6.283	5.370	5.136	3.335
1	1	3 1.994	2 1.176	1 0.774	1 0.405	1 0.165
	2	8 2.802	9 2.533	6 2.091	6 1.378	6 0.682
	3	16 3.400	13 2.888	15 2.780	12 2.125	11 1.289
	4	4.117	20 3.341	16 2.827	20 2.597	15 1.850
	5	4.708	3.732	3.158	2.806	2.136
2	1	6 2.338	7 2.163	7 2.118	13 2.128	2.131
	2	7 2.488	8 2.350	11 2.348	17 2.348	2.324
	3	4.070	3.361	17 2.858	18 2.437	2.348
	4	4.457	4.073	3.602	2.990	2.436
	5	5.135	4.362	4.078	3.411	2.736
3	1	15 3.336	18 3.245	3.248	3.253	3.253
	2	18 3.631	3.618	3.615	3.612	3.597
	3	4.948	4.202	3.835	3.634	3.614
	4	5.802	4.906	4.524	3.945	3.636
	5	6.074	5.557	4.865	4.397	3.789

a — axisymmetric, t — torsional.

TABLE 8.11 Frequencies $\omega R\sqrt{\rho/G}$ for *Symmetric* Modes of *Free–Free* Cylinders with $\nu = 0.3$

Exactitude of all data in Tables 8.11 and 8.12 to the four significant figures shown has been verified by convergence studies. Underlined numerals 1, 2, … , 20 preceding frequency values are used to identify the first 20 frequencies, in order, of the complete spectrum (symmetric and antisymmetric modes) for each length-to-diameter ratio.

It is interesting to note in studying Tables 8.11 and 8.12 together that the fundamental frequency corresponds to a symmetric bending mode (n = 1) for L/D = 2, 3, and 5, but for L/D = 1 and 1.5 it is an antisymmetric torsional mode. In spite of this, although the torsional

n	Modes	*L/D* = 1		1.5		2		3		5	
0[a]	1	9	2.872	12	2.810	10	2.326	8	1.639	9	1.004
	2	17	3.541	16	2.993	19	2.959		2.811	18	1.933
	3		4.707		4.026		3.519		2.954		2.643
	4		5.481		4.160		3.737		3.498		2.945
	5		6.226		4.866		4.261		3.668		3.114
0[t]	1	1	1.571	1	1.047	2	0.785	2	0.524	2	0.314
	2		4.712	17	3.142	12	2.356	7	1.571	7	0.942
	3		5.370		5.236		3.927		2.618	13	1.571
	4		6.970		5.241		5.195		3.665		2.199
	5		7.854		6.020		5.498		4.738		2.843
1	1	4	2.151	4	1.774	4	1.431	4	0.873	3	0.399
	2	11	3.081	10	2.590	8	2.127	9	1.767	8	0.983
	3	19	3.964	15	2.956	14	2.727	11	2.100	14	1.582
	4		4.577		3.686		3.160	19	2.556	17	1.901
	5		4.969		4.147		3.402		2.879	19	2.116
2	1	2	1.978	6	2.107	9	2.146	14	2.135	20	2.131
	2	14	3.192	11	2.693	13	2.436	16	2.334		2.334
	3		4.038		3.407		3.177		2.680		2.355
	4		4.704		4.079		3.512		3.146		2.565
	5		5.397		4.661		4.092		3.367		2.946
3	1	13	3.190	19	3.260		3.257		3.252		3.253
	2		4.139		3.734		3.631		3.604		3.599
	3		4.826		4.410		4.170		3.748		3.612
	4		5.789		4.889		4.402		4.181		3.696
	5		6.440		5.511		4.936		4.296		3.925

a — axisymmetric, t — torsional.

TABLE 8.12 Frequencies in $\omega R\sqrt{\rho/G}$ for *Antisymmetric* Modes of *Free–Free* Cylinders with $\nu = 0.3$

frequency $\omega R\sqrt{\rho/G} = 1.571$ $(= \pi/2)$ is the fundamental one for $L/D = 1$, the second torsional frequency of 3.142 $(= \pi)$ does not appear until the 12th mode. For the more slender cylinders (e.g., $L/D = 5$), more flexural ($n = 1$) modes are included among the first 20, as expected. Whereas 3 of the first 20 frequencies for $L/D = 1$ are for $n = 3$, none are included for $L/D \geq 2$. The lowest ranking of longitudinal modes ($n = 0$) in Tables 8.11 and 8.12 is seen to be the third frequencies for $L/D = 1.5$, 2, and 3.

It is seen in Tables 8.11 and 8.12 that for $L/D = 1$, the second mode is antisymmetric with $n = 2$. This corresponds to a plate-like mode.

For $L/D = 0.5$, one finds that the first two modes are both for $n = 2$, the first being an antisymmetric mode ($\omega R\sqrt{\rho/G} = 1.591$) and the second being a symmetric mode ($\omega R\sqrt{\rho/G} = 2.345$).

Extensions of Tables 8.11 and 8.12 for higher circumferential wave numbers ($n = 4 - 9$) are available in the dissertation by So [13]. These are higher frequency modes.

The effect of Poisson's ratio on the frequencies of free–free cylinders is shown in considerable detail in Ref. [11] for $L/D = 1$. The frequencies are found to *decrease* with increasing ν over the entire range $0 \le \nu \le 0.5$ (except for the torsional frequencies, which are unaffected).

		L/D				
n	**Modes**	**1**	**1.5**	**2**	**3**	**5**
0[a]	1	3 1.286	3 0.855	3 0.640	4 0.425	4 0.255
	2	10 2.960	10 2.400	7 1.859	7 1.260	8 0.761
	3	3.169	2.947	2.783	2.030	1.255
	4	4.182	3.307	2.951	2.641	1.727
	5	4.297	3.747	3.346	2.944	2.163
0[t]	1	2 0.785	2 0.524	2 0.393	2 0.262	3 0.157
	2	6 2.356	5 1.571	5 1.178	5 0.785	6 0.471
	3	3.927	2.618	9 1.963	9 1.309	9 0.785
	4	5.195	3.666	2.749	1.833	1.100
	5	5.498	4.725	3.544	2.363	1.418
1	1	1 0.506	1 0.264	1 0.159	1 0.075	1 0.028
	2	4 1.444	4 0.930	4 0.650	3 0.364	2 0.156
	3	8 2.588	6 1.904	6 1.383	6 0.819	5 0.383
	4	9 2.854	8 2.290	8 1.928	8 1.296	7 0.652
	5	3.346	2.772	2.368	10 1.774	10 0.945
2	1	5 2.162	7 2.137	10 2.132	2.131	2.132
	2	7 2.518	9 2.381	2.350	2.341	2.326
	3	3.449	2.765	2.503	2.356	2.331
	4	3.760	3.229	2.847	2.475	2.346
	5	4.347	3.652	3.292	2.702	2.378
3	1	3.258	3.253	3.253	3.253	3.260
	2	3.698	3.629	3.614	3.609	3.597
	3	4.289	3.814	3.674	3.610	3.600
	4	4.651	4.190	3.878	3.663	3.613
	5	5.350	4.544	4.168	3.786	3.619

a — axisymmetric, t — torsional.

Table 8.13 Frequencies in $\omega R\sqrt{\rho/G}$ for *Fixed–Free* Cylinders with $\nu = 0.3$

Frequencies of fixed–free cylinders (i.e., all displacement components zero on one end) are given in Table 8.13 for $\nu = 0.3$. Because of the stress singularities which occur in the corner of the fixed end, the Ritz solutions converge more slowly. Nevertheless, convergence studies [13] have demonstrated that the frequencies in Table 8.13 do represent the exact values to at least three significant figure accuracy. Again, the order of the modes for each L/D is indicated by underlined numerals preceding the frequency values, with the first ten being identified.

A comprehensive and excellent 3D study of the free vibrations of completely free cylinders was conducted by Hutchinson [14–16] using infinite series solutions of the equation of motion. He showed plots of $\omega R_0\sqrt{\rho/G}$ versus L/D with $\nu = 0.344$, for $n = 0$, 1, 2, and 3 over the range $0 < L/D \leq 2$. The plotted frequencies agreed closely with extensive experimental data presented earlier by McMahon [17]. Some amazingly accurate theoretical frequencies were also obtained by Pickett [18], using another method, in 1945 before electronic digital computers were available.

8.6 Approximate Solutions for Hollow Cylinders

The Ritz method described in Sec. 8.5 may also be used straightforwardly to obtain accurate free vibrations frequencies of hollow circular cylinders. A representative hollow circular cylinder of inner diameter $D_i(= 2R_i)$, outer diameter $D_0(= 2R_0)$, and length L is shown in Fig. 8.6. The most straightforward procedure is to choose the coordinate origin along the symmetry axis ($r = 0$), and assume displacement components as in (8.25) and (8.42). In this case no regularity conditions are required at the origin, and all the indices $(i, \ldots, q)$ in the summation of (8.50) not only can begin with zero, but *must* (to provide sets of functions which are mathematically complete).

Besides the *global* coordinate system above, one may also use a *local* coordinate system which has the same θ and z coordinates, but $r = 0$ is then at the middle of the cylinder wall; that is, the curvilinear surfaces are at $r = \pm H/2$, where $H = (D_0 - D_i)/2$ is the wall thickness. This local coordinate system has certain computational advantages over the global system.

Both coordinate systems were used by So and Leissa [19] to obtain extensive, accurate frequencies for hollow cylinders. Table 8.14 is a representation of numerous convergence studies made by So [13], using both coordinate systems, for various geometrical parameters $(L/D_0,\ D_i/D_0)$ and the first few circumferential wave numbers (n). It was found that frequencies obtained using both coordinate systems were identical, thus establishing the completeness of the functions (8.42) in terms of the local coordinates. However, numerical

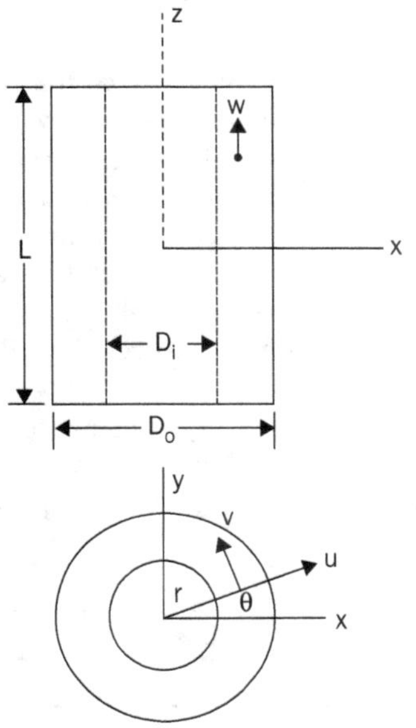

FIGURE 8.6 Hollow circular cylinder with displacement components and dimensions.

TR	*TZ*	*DET*	1	2	3	4	5
2	2	12	2.06394	2.58242	3.12189	4.72849	8.00733
2	4	24	1.92317	2.50001	3.09439	4.05925	4.83453
2	6	36	1.92312	2.49992	3.09438	4.05779	4.78566
2	8	48	1.92312	2.49992	3.09438	4.05779	4.78565
4	2	24	2.03605	2.54974	3.10594	4.71112	7.29447
4	4	48	1.89353	2.41800	3.07966	4.03210	4.57431
4	6	72	1.89345	2.48088	3.07964	4.02915	4.53028
4	8	96	1.89345	2.48088	3.07964	4.02915	4.53025
6	2	36	2.03586	2.54950	3.10571	4.71097	7.28659
6	4	72	1.89327	2.48076	3.07946	4.03134	4.56575
6	6	108	<u>1.89320</u>	2.48066	<u>3.07944</u>	4.02836	4.52259
6	8	144	1.89320	2.48066	3.07944	<u>4.02835</u>	4.52257
6	10	180	1.89320	<u>2.48065</u>	3.07944	4.02835	4.52257
8	2	48	2.03586	2.54950	3.10571	4.71097	7.28656
8	4	96	1.89327	2.48076	3.07946	4.03133	4.56554
8	6	144	1.89320	2.48066	3.07944	4.02835	4.52255
8	8	192	1.89320	2.48065	3.07944	4.02835	<u>4.52253</u>

TABLE 8.14 Convergence of Frequencies $\omega R_0\sqrt{\rho/G}$ for the First Five Symmetric Modes of $n = 1$ for a Completely Free Hollow Cylinder with $L/D_0 = 1$, $D_i/D_0 = 0.5$, and $\nu = 0.3$

ill-conditioning was found to be significantly less with the local coordinates.

Table 8.14 describes the convergence of the first five symmetric mode frequencies of a completely free, hollow cylinder ($L/D_0 = 1$, $D_i/D_0 = 0.5$, $\nu = 0.3$) for modes having $n = 1$. These modes are predominantly flexural or thickness-shear in character. For these modes, the order of determinant (DET) is related to TR and TZ by $3 \times TR \times TZ$, where again TR and TZ are the number of terms of (8.42) used in the r and z directions, respectively. The first five frequencies have converged to six significant figures by $TR \times TZ = 8 \times 8$ (or determinant size $DET = 192$).

Tables 8.15–8.17 (taken from [19]) present accurate values of the frequencies $\omega R_0 \sqrt{\rho/G}$ for the longitudinally symmetric (S) and antisymmetric (A) modes of completely free, hollow cylinders having

D_i/D_0		0.1		0.5		0.9	
n	**s**	**S**	**A**	**S**	**A**	**S**	**A**
	1	3.3188	1.4331	2.2335	1.3882	1.6979	1.6481
	2	8.1611	4.4905	9.9566	8.3210	5.9050	12.6939
	3	11.3529	7.4321	11.3428	9.1270	13.2223	25.9346
	4	11.5429	9.6202	12.2074	14.1327	20.2580	27.1552
0^t	1	5.1423	7.8540	6.8138	7.8540	15.7080	7.8540
	2	8.4574	9.3877	12.8555	10.3978	31.4159	23.5618
	3	11.7385	11.5418	15.7080	15.0649	31.4821	32.4470
	4	15.0444	14.1237	17.1222	20.6016	35.1832	39.2699
1	1	2.7475	2.7173	2.8055	1.9431	2.3932	1.6880
	2	6.0311	5.6433	7.3717	8.0388	5.9533	8.0835
	3	6.8793	7.6190	9.8675	8.5342	13.1191	12.7261
	4	10.2331	8.2382	11.3864	8.9449	15.9063	23.5117
2	1	2.2099	0.8909	0.9490	0.6907	0.1382	0.2769
	2	4.1485	4.0638	4.1774	3.1233	3.7708	1.9152
	3	6.8392	7.0180	8.6297	8.3999	6.0965	8.7070
	4	8.4934	8.7818	9.7212	8.7928	12.8965	12.8223
3	1	3.5938	1.8593	2.2486	1.6805	0.3883	0.8203
	2	5.7879	5.3505	5.7173	4.4498	5.3064	2.3822
	3	8.7332	8.1441	9.4408	8.8076	6.3300	9.5976
	4	10.0785	9.7189	10.2698	8.9864	12.6858	12.9815

Note: a — axisymmetric, t — torsional.

TABLE 8.15 Frequencies $\omega R_0 \sqrt{\rho/G}$ for Completely Free, Hollow Circular Cylinders Having $L/D_0 = 0.2$ and $\nu = 0.3$

D_i/D_0		0.1		0.5		0.9	
n	**s**	**S**	**A**	**S**	**A**	**S**	**A**
0[a]	1	2.3167	2.8291	2.0425	2.1513	1.6466	1.6906
	2	3.0122	3.5237	2.3047	3.0914	1.7071	1.8231
	3	3.8609	4.7566	2.8930	5.2855	2.1679	2.7816
	4	4.8592	5.4372	4.3810	5.6419	2.7410	4.6575
0[t]	1	3.1416	1.5708	3.1416	1.5708	3.1416	1.5708
	2	5.1424	4.7124	6.2832	4.7124	6.2832	4.7124
	3	6.0261	5.3769	6.8138	6.9926	9.4248	7.8540
	4	6.2832	6.9750	7.5032	7.8540	12.5664	10.9956
1	1	1.9922	2.1117	1.8932	1.6041	1.4845	1.2940
	2	2.8110	3.0755	2.4807	2.9311	2.1573	1.7692
	3	3.3468	3.9120	3.0794	3.2380	2.1649	2.6629
	4	4.1065	4.5739	4.0284	4.6748	2.7668	2.8037
2	1	2.2130	1.9128	0.9700	1.0451	0.1427	0.1852
	2	2.4456	3.1381	1.9351	2.4981	1.0898	1.5868
	3	4.0290	4.0040	3.2929	3.4401	2.1737	1.9709
	4	4.3618	4.6540	4.3187	4.2654	2.6689	2.8791
3	1	3.3346	3.1876	2.2868	2.3475	0.3999	0.4662
	2	3.6254	4.1351	2.8033	3.4771	0.9644	1.5825
	3	4.9457	4.8242	4.1991	4.2579	2.2500	2.8468
	4	5.7999	5.7877	5.3248	5.2834	3.3062	3.0486

Note: a — axisymmetric, t — torsional.

TABLE 8.16 Frequencies $\omega R_0\sqrt{\rho/G}$ for Completely Free, Hollow Circular Cylinders Having $L/D_0 = 1$ and $\nu = 0.3$

length-to-outside-diameter ratios (L/D_0) of 0.2, 1, and 5, respectively. The former configuration (Table 8.15) is a thick circular plate, whereas the latter (Table 8.17) is a thick circular tube. In each table, data is given for three diameter ratios (D_i/D_0 = 0.1, 0.5, 0.9) and four circumferential wave numbers (n = 0, 1, 2, 3). Poisson's ratio is taken as 0.3. Two cases are shown for n = 0: axisymmetric and torsional modes. Exactitude of all data to at least four significant figures shown was verified by convergence studies.

Based on these tables, the first 10 modes for each cylinder are identified in sequence in Table 8.18. In studying Table 8.18, one notes that, for example, the fundamental frequency for L/D_0 = 1 corresponds to an antisymmetric torsional mode for D_i/D_0 = 0.1, but for D_i/D_0 = 0.5 and 0.9 it is a symmetric mode of n = 2. For D_i/D_0 = 0.9, the configurations are circular cylindrical shells. It is generally known that for such shapes the higher circumferential wave num-

D_i/D_0		0.1		0.5		0.9	
n	s	S	A	S	A	S	A
0^a	1	0.5054	1.0034	0.5051	0.9998	0.5044	0.9904
	2	1.4838	1.9301	1.4621	1.8329	1.3949	1.5885
	3	2.3168	2.6214	2.0434	2.1436	1.6468	1.6708
	4	2.8444	2.8963	2.1781	2.2101	1.6844	1.6924
0^t	1	0.6283	0.3142	0.6283	0.3142	0.6283	0.3142
	2	1.2566	0.9425	1.2566	0.9425	1.2566	0.9425
	3	1.8850	1.5708	1.8850	1.5708	1.8850	1.5708
	4	2.5133	2.1991	2.5133	2.1991	2.5133	2.1991
1	1	0.1651	0.3990	0.1776	0.4096	0.2018	0.4344
	2	0.6811	0.9807	0.6726	0.9344	0.6740	0.8756
	3	1.2835	1.5718	1.1842	1.3469	1.0469	1.0874
	4	1.8224	1.8668	1.5020	1.5037	1.2258	1.2524
2	1	2.0599	2.0596	0.9804	0.9831	0.1439	0.1458
	2	2.2112	2.2131	1.0063	1.0552	0.1708	0.2526
	3	2.2216	2.2529	1.1441	1.2777	0.3824	0.5330
	4	2.3466	2.4843	1.4512	1.6555	0.6875	0.8369
3	1	3.2503	3.2503	2.3007	2.3016	0.4032	0.4055
	2	3.5919	3.5941	2.3269	2.3580	0.4179	0.4449
	3	3.6081	3.6062	2.4072	2.4780	0.4935	0.5655
	4	3.6313	3.6915	2.5729	2.6927	0.6580	0.7650

Note: a — axisymmetric, t — torsional.

TABLE 8.17 Frequencies $\omega R_0\sqrt{\rho/G}$ for Completely Free, Hollow Circular Cylinders Having $L/D_0 = 5$ and $\nu = 0.3$.

bers ($n > 2$) become more important in the frequency spectrum (cf. [8]). This is verified in Table 8.18. Conversely, for completely free annular plates it is known that the fundamental mode is in bending with $n = 2$ (cf. [20]). This is seen in Table 8.18 for $L/D_0 = 0.2$ and $D_i/D_0 = 0.1$ and 0.5.

For the longer cylinders ($L/D_0 = 1$ and 5) which are also thick ($D_i/D_0 = 0.1$ and 0.5) it is seen in Table 8.18 that the *antisymmetric* torsional modes ($n = 0^t$) are particularly important. For $L/D_0 = 1$ and $D_i/D_0 = 0.1$, the fundamental frequency is in this mode. Table 8.16 shows that this fundamental frequency is $\omega R_0\sqrt{\rho/G} = \pi/2(=1.5708)$. Interestingly, the second antisymmetric torsional frequency is $3\pi/2(=4.7124)$. Similarly, the first and fourth symmetric torsional frequencies are π and 2π. For all these modes the core of the cylinder rotates in the same direction as its outer elements. For the other torsional frequencies (e.g., $\omega R_0\sqrt{\rho/G} = 5.1424, 5.3769, 6.026, 6.950$)

L/D_0	D_i/D_0	Sequence									
		1	2	3	4	5	6	7	8	9	10
0.2	0.1	$\underline{2,1}$	$\underline{0^a,1}$	$\underline{3,1}$	2,1	$\underline{1,1}$	1,1	$\underline{4,1}$	$0^a,1$	3,1	$\underline{5,1}$
	0.5	$\underline{2,1}$	2,1	$\underline{0^a,1}$	$\underline{3,1}$	$\underline{1,1}$	$0^a,1$	3,1	$\underline{4,1}$	1,1	$\underline{2,2}$
	0.9	2,1	$\underline{2,1}$	3,1	4,1	$\underline{3,1}$	5,1	$\underline{4,1}$	$\underline{0^a,1}$	$\underline{1,1}$	$0^a,1$
1	0.1	$\underline{0^t,1}$	$\underline{2,1}$	1,1	$\underline{1,1}$	2,1	$0^a,1$	2,2	1,2	$\underline{0^a,1}$	0,2
	0.5	2,1	$\underline{2,1}$	$\underline{0^t,1}$	$\underline{1,1}$	1,1	2,2	$0^a,1$	$\underline{0^a,1}$	3,1	
	0.9	2,1	$\underline{2,1}$	3,1	$\underline{3,1}$	4,1	$\underline{4,1}$	3,2	2,2	4,2	
5	0.1	1,1	$\underline{0^t,1}$	$\underline{1,1}$	$0^a,1$	$0^t,1$	1,2	$\underline{0^t,2}$	$\underline{1,2}$	$\underline{0^a,1}$	$0^t,2$
	0.5	1,1	$\underline{0^t,1}$	$\underline{1,1}$	$0^a,1$	$0^t,1$	1,2	$\underline{1,2}$	$\underline{0^t,2}$	2,1	$\underline{2,1}$
	0.9	2,1	$\underline{2,1}$	2,2	1,1	$\underline{2,2}$	$\underline{0^t,1}$	3,1	$\underline{3,1}$	3,2	$\underline{1,1}$

Notes: a — axisymmetric, t — torsional.
Underlined numerals are for antisymmetric modes.

TABLE 8.18 Modes (n,s) Corresponding to the First 10 Frequencies for the Hollow Circular Cylinders of Tables 8.15–8.17 ($\nu = 0.3$)

the core rotates in a direction opposite to that of the outer elements, with a nodal surface (of non moving points) occurring between them.

The length-to-outside diameter ratio (L/D_0) of 1 represents a typical 3D hollow cylinder. From Table 8.16, it is noted that as D_i/D_0 becomes smaller (i.e., the cylinder becomes thicker), the frequencies not only increase on the whole, but becomes dense in a saturated manner.

Three-dimensional vibration analyses of completely free, hollow circular cylinders were also conducted by several others. Particularly notable is the work of Hutchinson and El-Azhari [21,22], who used a series method to obtain extensive, reasonably accurate frequencies. Gladwell and Vijay [23] applied the finite element method. Singhal and Williams [24] obtained excellent experimental results.

Hutchinson and El-Azhari [22] chose four geometrical shapes $(L/D_0, D_i/D_0) = (0.2, 0.1), (0.2, 0.5), (0.5, 0.1)$, and $(0.5, 0.5)$, and compared their 3D series solutions with 2D frequencies from Mindlin's thick plate theory. Tables 8.19 and 8.20 compare the frequencies for thick annular plates having $L/D_0 = 0.2$ and 0.5, respectively, obtained by the Ritz method [19] and those presented in Ref. [22].

It is observed from Table 8.19 that most of the frequencies of Hutchinson and El-Azhari [22] agree well with the Ritz ones in an upper bound manner. However, there are some data which are too high or too low. The Ritz data presented are accurately converged to the exact values. The large disagreements are underlined. From

D_I/D_O	n	Method	s			
			1	2	3	4
0.1	0	R	1.433	4.491	7.432	9.620
		S	1.440	4.498	7.439	9.387
		M	1.433	4.484	7.454	9.606
	1	R	2.717	5.643	7.619	8.238
		S	2.724	5.649	7.622	8.238
		M	2.717	5.637	7.725	8.498
	2	R	0.891	4.064	7.018	8.782
		S	0.896	4.071	7.025	8.782
		M	0.892	4.058	7.021	8.944
	3	R	1.859	5.350	8.144	9.719
		S	1.864	5.358	8.149	9.719
		M	1.864	5.343	8.176	9.866
0.5	0	R	1.388	8.321	9.127	14.133
		S	1.398	8.327	9.128	10.398
		M	1.388	8.324	9.370	10.593
	1	R	1.943	8.039	8.534	8.945
		S	1.950	8.040	8.539	8.946
		M	1.951	8.189	8.659	9.162
	2	R	0.691	3.123	8.400	8.793
		S	–	3.127	8.404	8.794
		M	–	3.142	8.461	8.964
	3	R	1.680	4.450	8.808	8.986
		S	1.682	4.453	8.990	10.234
		M	1.684	4.475	8.899	9.076

TABLE 8.19 Comparison of Frequencies $\omega R_O\sqrt{\rho/G}$ for Completely Free, Hollow Cylinders with L/D_O = 0.2 and ν = 0.3 by the 3D Ritz Approach (R) [19], Hutchinson's 3D Series (S) [22], and Mindlin's 2D Plate Theory (M) [22]

Tables 8.19 and 8.20 it appears that, on the whole, the Mindlin thick plate theory gives reasonably accurate results, even for the very thick plates (L/D_O = 0.5), at least for the lower frequencies.

8.7 Other Three-Dimensional Bodies

In Secs. 8.4–8.6 earlier in this chapter it was shown that the immensity of a 3D analysis of bodies having polar symmetry can be reduced greatly by assuming trigonometric functions of the polar angle (θ) for the three displacements components (u, v, w). This results in a set

D_i/D_0	n	Method	s			
			1	2	3	4
0.1	0	R	2.363	4.758	6.201	7.911
		S	2.364	4.758	6.026	6.204
		M	2.361	4.928	6.080	6.465
	1	R	2.986	4.027	4.628	6.279
		S	2.986	4.027	4.628	6.023
		M	3.016	4.147	4.715	6.444
	2	R	1.551	3.999	5.198	5.763
		S	1.552	4.000	5.198	5.756
		M	1.555	4.039	5.341	5.913
	3	R	2.804	4.959	6.474	6.988
		S	2.805	4.959	6.474	6.992
		M	2.823	4.983	6.759	7.306
0.5	0	R	1.984	5.772	8.258	9.084
		S	1.985	5.774	7.503	8.259
		M	1.985	6.720	7.547	10.010
	1	R	1.999	3.930	5.839	7.706
		S	2.000	3.930	5.841	6.401
		M	2.005	4.064	6.583	8.207
	2	R	1.039	2.846	5.172	6.157
		S	1.040	2.846	5.173	6.159
		M	1.040	2.860	5.399	6.730
	3	R	2.320	3.946	6.392	6.805
		S	2.321	3.946	6.392	6.806
		M	2.324	3.971	6.749	7.311

TABLE 8.20 Comparison of Frequencies $\omega R_0\sqrt{\rho/G}$ for Completely Free, Hollow Cylinders with L/D_0 = 0.5 and ν = 0.3 by the 3D Ritz Approach (R) [19], Hutchinson's 3D Series (S) [22], and Mindlin's 2D Plate Theory (M) [22]

of 2D problems, one for each Fourier component (n) of interest. But even with this simplification, physical problems having important boundary conditions, such as fixed or free, typically have no exact solutions. Approximate methods, such as that of Ritz must then be used to obtain accurate free vibration frequencies and corresponding mode shapes. But the size of eigenvalue (frequency) determinants needed are much smaller than what would be required without considering polar symmetry.

In this manner accurate frequencies for solid and hollow cylinders were calculated and presented earlier. In similar ways analysis of other bodies having polar symmetry may be undertaken. This was done by Kang and Leissa, using the Ritz method, to publish the first known results from 3D analysis for a considerable variety of bodies of interest, some of which will now be described briefly. Extensive tables of accurate natural frequencies for each configuration are given in the published journal articles listed below.

The thick conical shell [25] was examined utilizing the cross-section and cylindrical coordinates (r, θ) depicted in Fig. 8.7. The energy functionals contained volume elements which were hollow cylinders of thickness *dr*. The resulting volume integrations are relatively simple for the cross-section in Fig. 8.7. But the edge at $r = R$ is seen to be different from that usually associated with shells. Frequencies were given for moderately thick and thicker shells ($h/R = 0.1, 0.3$) and various cone angles (α). In a similar manner solid cones with and without an axial circular hole (Fig. 8.8) were treated [26]. For this situation it was convenient for integration to measure r from the hole boundary (Fig. 8.8). A solid cone was then achieved by having an extremely small ($a/h_b = 10^{-5}$) hole radius.

Three-dimensionally, a circular ring is a toroid. Various rings having isosceles trapezoidal cross-sections (Fig. 8.9) were examined [27] by means of a 3D Ritz analysis. Accurate frequencies were found for some interesting cross-sections (Fig. 8.10), which include the square as a special case. Isosceles triangular cross-sections were also treated by setting $h_t = 0$ (Fig. 8.9). Circular rings with elliptical cross-sections (Fig. 8.11) were analyzed, carrying out the r–z integrations over the cross-sectional areas; integration on θ is duplicated for

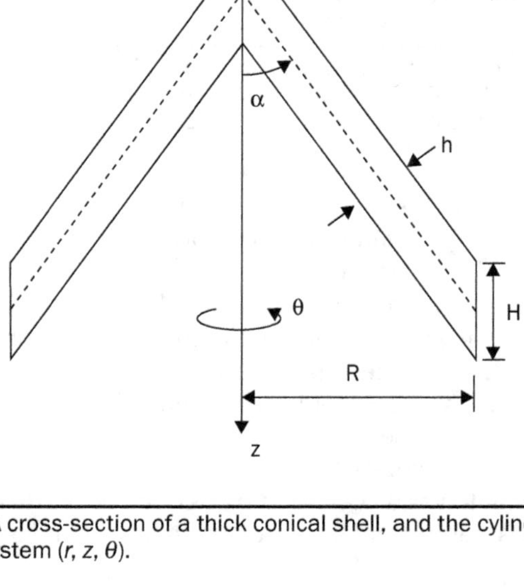

FIGURE 8.7 A cross-section of a thick conical shell, and the cylindrical coordinate system (r, z, θ).

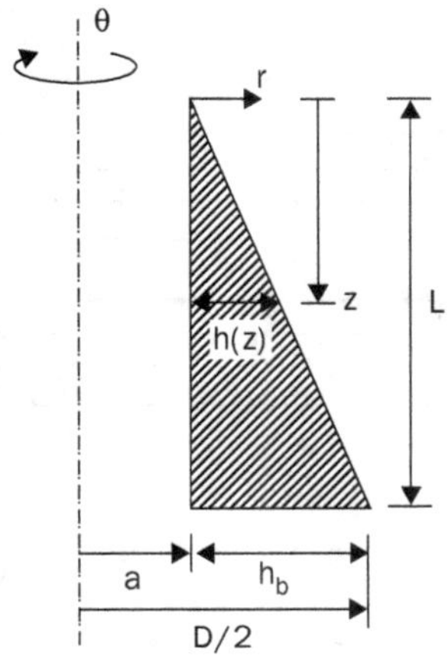

FIGURE 8.8 Right triangular cross-section of a circular cylindrical body of revolution with linearly varying wall thickness and coordinate system (r, θ, z).

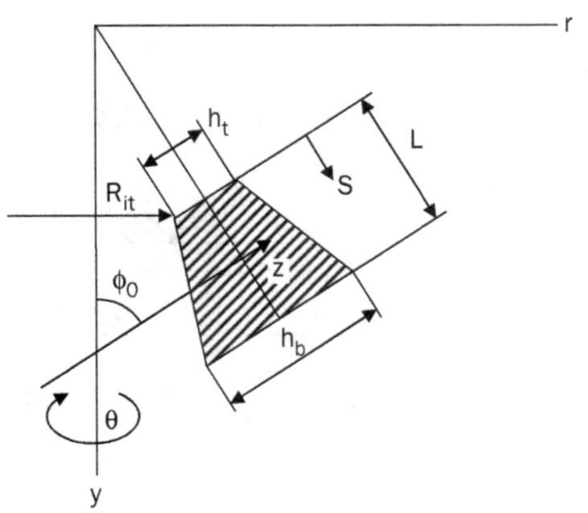

FIGURE 8.9 A circular ring with isosceles trapezoidal cross-section and the coordinate system (s, z, θ).

all integrals, and could therefore be ignored [28]. Frequencies for circular cross-sections, an obvious special case, were also reported.

In Ref. [29] a general, 3D method of analysis was presented for determining free vibration frequencies and mode shapes of hollow bodies of revolution (i.e., thick shells), not limited to straight-line generators or constant thickness. The middle surface of the shell may have an arbitrary curvature, and the wall thickness may vary arbitrarily. A meridian line of arbitrary shape that generates the shell midsurface when rotated is shown in Fig. 8.12. This generating curve

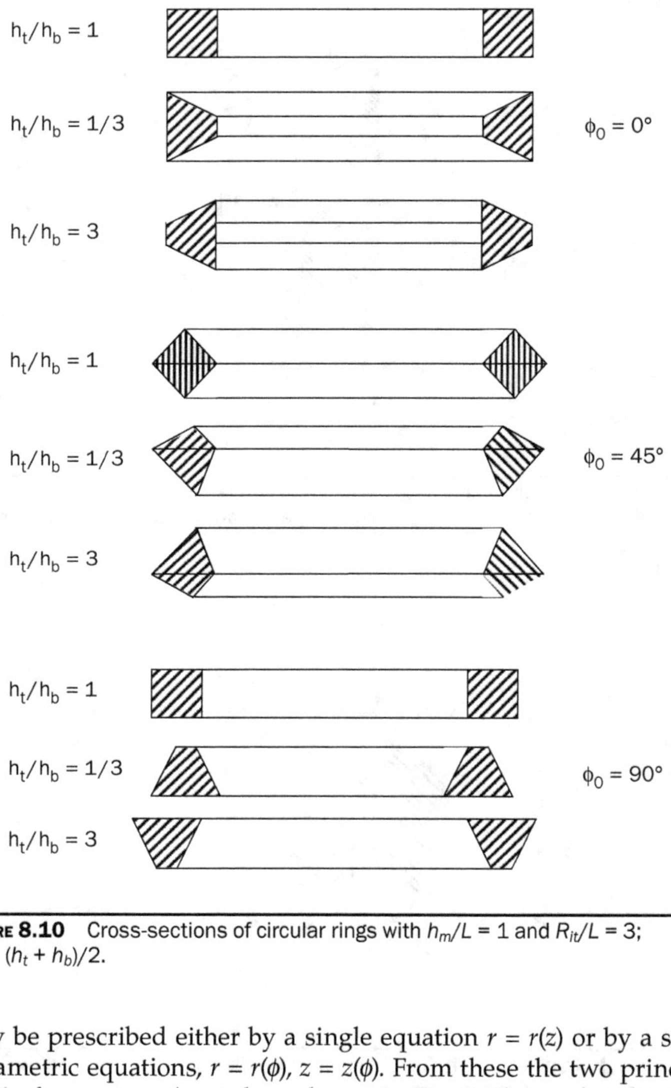

FIGURE 8.10 Cross-sections of circular rings with $h_m/L = 1$ and $R_{it}/L = 3$; $h_m = (h_t + h_b)/2$.

may be prescribed either by a single equation $r = r(z)$ or by a set of parametric equations, $r = r(\phi)$, $z = z(\phi)$. From these the two principal radii of curvature (ρ_1 and ρ_2, shown in Fig. 8.12) may be obtained. This was carried out for many shells of classical curvature [29]. Utilizing tensor analysis, the 3D strains were expressed in terms of the displacement components, and from these the energy functional needed for the Ritz method were derived [29].

The general 3D method described above was used to obtain extensive, accurate frequencies for several different types of thick shells of importance: spherical [30], hemispherical [31], paraboloidal [32,33], and hyperboloidal [34]. A typical hyperboloidal shell, which could simulate a cooling tower, is seen in Fig. 8.13.

Three-dimensional bodies which do not have rotational symmetry typically require very large frequency determinants for

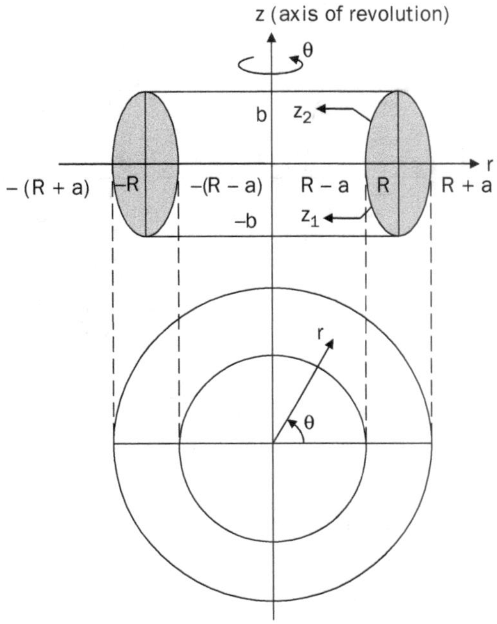

FIGURE 8.11 Cross-section of a thick, complete, circular ring having an elliptical cross-section, its planform and the circular cylindrical coordinate system (r, θ, z).

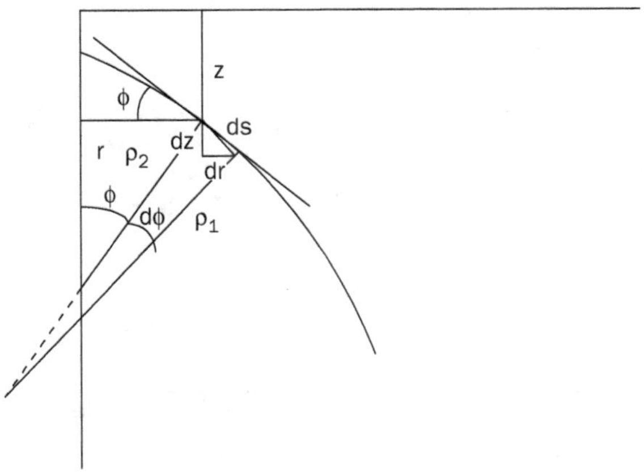

FIGURE 8.12 Meridian of middle surface of shell of revolution.

accurate (i.e., well-converged) results. The twisted, cantilevered parallelepiped (Fig. 8.14) was taken up by Leissa and Jacob [35] using the Ritz method. The three displacement components (u, v, w) were

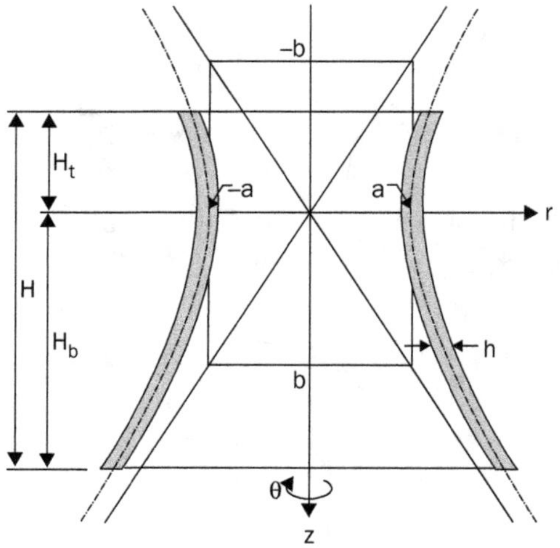

FIGURE 8.13 Hyperboloidal shell of revolution with cylindrical coordinate system (r, θ, z).

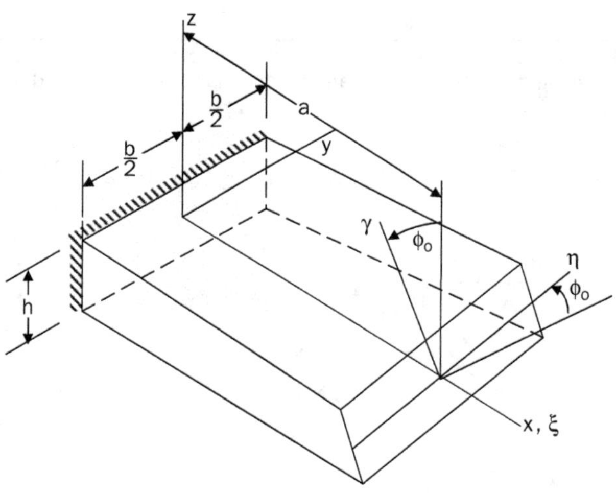

FIGURE 8.14 Coordinates and dimensions for a twisted cantilevered parallelepiped.

assumed to be algebraic polynomials, as in Sec. 8.3 previously. But some of the symmetry present in regular parallelepipeds is lost, which causes larger frequency determinants to be needed. Accurate (i.e., well-converged) results required using one of the few supercomputers in existence of that time.

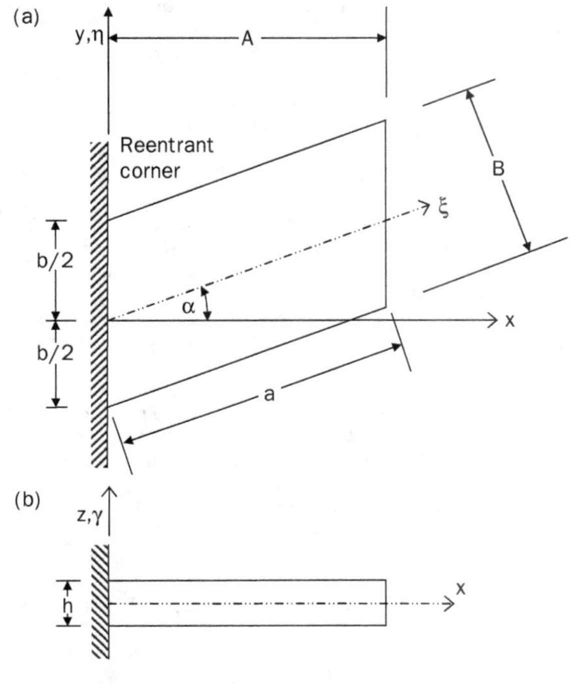

Figure 8.15 Skewed cantilevered parallelepiped: (a) plan view; (b) side view.

In a similar manner frequencies were found for the 3D skewed cantilever parallelepiped (Fig. 8.15), using algebraic polynomials in the x–y–z coordinates to represent the three displacement components [36], and also using a supercomputer for the calculations. Determinant orders as large as 288 were required to obtained reasonably accurate (three digit) convergence for skew angles (α) as large as 45°. Calculations were also made using versions of a good, 3D finite element computer program (MSC/NASTRAN®), requiring determinants as large as 2520. Moreover, reasonably accurate frequencies could thus be found for α not exceeding 30°. The strength of the stress singularities in the re-entrant corner (Fig. 8.15) increases as the skew angle (α) increases, which makes computational convergence slower.

References

1. S. P. Timoshenko and J. N. Goodier, *Theory of Elasticity*, 3rd ed. McGraw-Hill Book Co., 1934.
2. S. G. Lekhnitskii, *Theory of Elasticity of an Anisotropic Elastic Body*, Holden-Day, 1963, translated from a 1950 publication in Russian.
3. J. A. Fromme and A. W. Leissa, "Free vibration of the rectangular parallelepiped," *J. Acoust. Soc. of Amer.*, 48 (1970): 290–98.

4. J. R. Hutchinson and S. D. Zillmer, "Vibrations of a free rectangular parallelepiped," *J. Appl. Mech.*, 50 (1983): 123–30.
5. A. W. Leissa and Z. Zhang, "On the three-dimensional vibrations of the cantilevered rectangular parallelepiped," *J. Acoust. Soc. Am.*, 73 (1983): 2013–21.
6. A. E. H. Love, *A Treatise on the Mathematical Theory of Elasticity*, 4th ed. MacMillan, 1927, reprinted by Dover Publications, 1944.
7. A. W. Leissa and M. S. Ewing, "Comparison of beam and shell theories for the vibrations of thin turbomachinery blades," *J. Eng. Power*, 105 (1983): 383–92.
8. A. W. Leissa, *Vibration of Shells*, NASA SP-288, U.S. Govt. Printing Office, 1973, 428 pp. Reprinted by the Acoustical Society of America, 1993.
9. M. Abramowitz and I. A. Stegun, *Handbook of Mathematical Functions.* U.S. Govt. Printing Office, Washington, DC, 1964.
10. A. W. Leissa and J. So, "Comparisons of vibration frequencies for rods and beams from one-dimensional and three-dimensional analyses," *J. Acoust. Soc. Am.*, 98 (1995): 2122–35.
11. A. W. Leissa and J. So, "Accurate vibration frequencies of circular cylinders from three-dimensional analysis," *J. Acoust. Soc. Amer.*, 98 (1995): 2136–41.
12. T. C. Huang, "The effect of rotatory inertia and of shear deformation on the frequency and normal mode equations of uniform beams with simple end conditions," *J. Appl. Mech.*, 28 (1961): 579–84.
13. J. So, "Three dimensional vibration analysis of elastic bodies of revolution," Ph.D. Dissertation, The Ohio State University, 1993.
14. J. R. Hutchinson, "Axisymmetric vibrations of a free finite-length rod," *J. Acoust. Soc. Am.*, 51 (1972): 233–40.
15. J. R. Hutchinson, "Vibration of solid cylinders," *J. Appl. Mech.*, 47 (1980): 901–7.
16. J. R. Hutchinson, "Transverse vibrations of beams, exact versus approximate solutions," *J. Appl. Mech.*, 48 (1981): 923–28.
17. G. W. McMahon, "Experimental study of the vibrations of solid, isotropic elastic cylinders," *J. Acoust. Soc. Am.*, 36 (1964): 85–92.
18. G. Pickett, "Flexural vibration of unrestrained cylinders and disks," *J. Appl. Phys.*, 16 (1945): 820–31.
19. J. So and A. W. Leissa, "Free vibrations of thick hollow circular cylinders from three-dimensional analysis," *J. Vibr. Acoust.*, 119 (1997): 89–95.
20. A. W. Leissa, *Vibration of Plates*, NASA SP-160, U.S. Govt. Printing Office, 1969, 353 pp. Reprinted by The Acoustical Society of America, 1993.
21. J. R. Hutchinson and S. A. El-Azhari, "Vibrations of free hollow circular cylinders," *J. Appl. Mech.*, 53 (1986): 641–46.
22. J. R. Hutchinson and S. A. El-Azhari, "On the vibration of thick annular plates," Refined Dynamical Theories of Beams, Plates and Shells and Their Applications, *Proceedings of Euromech-Colloquium 219* (1986): 102–11.
23. G. M. L. Gladwell and D. K. Vijay, "Natural frequencies of free finite length circular cylinders," *J. Sound Vib.*, 42 (1975): 387–97.
24. R. K. Singal and K. Williams, "A theoretical and experimental study of vibrations of thick circular shells and rings," *ASME J. Vib. Acoust. Stress Rel. Des.*, 110 (1988): 533–37.
25. J. H. Kang and A. W. Leissa, "Three-dimensional vibration analysis of thick, complete conical shells," *J. Appl. Mech.*, 71 (2004): 502–7.
26. J. H. Kang and A. W. Leissa, "Three-dimensional vibration analysis of solid cones with and without an axial circular cylindrical hole," *Int. J. Solids Struct.*, 41 (2004): 3735–46.
27. J. H. Kang and A. W. Leissa, "Three-dimensional vibrations of thick, circular rings with isosceles trapezoidal and triangular cross-sections," *J. Vib. Acoust.*, 122 (2000): 132–39.
28. J. H. Kang and A. W. Leissa, "Natural frequencies of thick, complete circular rings with an elliptical or circular cross-section from a three-dimensional theory," *Arch. Appl. Mech.*, 75 (2006): 425–39.
29. J. H. Kang and A. W. Leissa, "Three-dimensional vibration analysis of thick shells of revolution," *J. Eng. Mech.*, 125 (1999): 1365–71.

30. J. H. Kang and A. W. Leissa, "Three-dimensional vibrations of thick spherical shell segments with variable thickness, *Int. J. Solids Struct.*, 37 (2000): 4811–23.
31. J. H. Kang and A. W. Leissa, "Three-dimensional vibration analysis of solid and hollow hemispheres having variable thickness with and without axial conical holes," *J. Vibr. Cont.*, 10 (2004): 199–214.
32. J. H. Kang and A. W. Leissa, "Three-dimensional vibration analysis of paraboloidal shells," *Japan Soc. Mech. Eng. Int. J.*, 45 (2002): 2–7.
33. J. H. Kang and A. W. Leissa, "Free vibration analysis of complete paraboloidal shells of revolution with variable thickness and solid paraboloids from a three-dimensional theory," *Comput. Struct.*, 83 (2005): 2594–608.
34. J. H. Kang and A. W. Leissa, "Three-dimensional vibration analysis of thick hyperboloidal shells of revolution," *J. Sound Vib.*, 282 (2005): 277–96.
35. A. W. Leissa and K. I. Jacob, "Three dimensional vibrations of twisted cantilevered parallelepipeds," *J. Appl. Mech.*, 53 (1986): 614–18.
36. O. G. Mc Gee and A. W. Leissa, "Three-dimensional free vibrations of thick skewed cantilever plates," *J. Sound Vib.*, 144 (1991): 305–22.

Problems

1 A. Substitute (8.17) into the equations of motion (8.14). Solve the resulting sixth order set of ordinary differential equations in $U(z)$, which will contain six constants of integration.

B. Locate the coordinate origin so that the two faces of a rectangular parallelepiped are at $z = \pm c/2$, and they are both completely free. Develop two determinants of third order, one for symmetric free vibration modes, and the other for antisymmetric ones in z of the body.

2 Derive the 3D equations of motion (8.23) and (8.24) in cylindrical coordinates. Follow the same logic that was used to arrive at Eqs. (8.14), except that all relationships used will now be in cylindrical coordinates. (Hint: The strain–displacement relationships are also derived in some theory of elasticity textbooks.)

3 A. Determine the characteristic (frequency) equations for the torsional modes of a solid circular cylinder having its cylindrical surface ($r = R$) fixed, and both ends ($z = \pm L/2$) free.

B. Let $L/2R = 1$. Calculate the first six nondimensional frequencies $\omega R\sqrt{\rho/G}$. Obtain the corresponding eigenfunctions.

C. Determine whether interior nodal *surfaces* (i.e., surfaces where $u = v = w = 0$) exist for any of the mode shapes corresponding to the frequencies found in Part B. If any exist, determine their locations.

4 Determine solutions $U(r,z)$ and $W(r,z)$ to (8.23), using (8.25), for $n = 0$. Describe carefully at least two sets of boundary conditions to which these solutions can be applied.

5 Assume that a three-dimensional body has material damping (Sec. 3.6). Derive a generalization of the equations of motion (8.14) which includes the material damping.

6 A solid circular cylinder has a radius R and a length L, and is completely free. Use the information and results available in earlier chapters to determine its *first* natural frequency for each of the following modes:

A. Axial extension

B. Torsion

C. Beam bending (elementary theory)

D. Beam bending (Timoshenko theory)

Compare these results with those of 3D theory in Tables 8.11 and 8.12.

CHAPTER 9

Vibrations of Composite Continuous Systems

The use of composite materials as structural components in many engineering applications has been expanding rapidly in the past four decades. As a result, composite materials and structures constitute a considerable portion of today's airplanes, missiles, submarines, and sport equipment. In addition, they are finding interesting applications in the areas of automotive engineering, construction materials, biomedical equipment, and others.

The first attribute of composite materials that has attracted engineers to use them in many applications was probably their light-weight. Light-weight structures and materials were considered closely by the aerospace and defense industries early on. In today's engineering environments, light-weight structures are being considered closely in many fields because of their positive impacts on efficiency, noise and vibrations, as well as other attributes. In addition to offering light-weight, composite materials are achieving both high strength and high stiffness which made them competitive when compared with common metallic materials. Moreover, composite materials offer more parameters for designers to optimize or tailor their design to certain applications. They can also be friendly to specific environments like the ability to embed special layers or elements that achieve certain requirements including thermal or electric conductivity. Also, composites made of environmental-friendly materials (e.g., biodegradable materials) are receiving particular attention for many interesting applications.

There are several types of composite materials. In general, composite materials are those materials made of more than one type of materials. Traditional examples of reinforced concrete, thermoplastics reinforced with short fibers, honeycomb composites, sandwich panels are all composite materials. Some of the most common types of composites are those made of layers, each of these layers is composed of long fibers embedded in a resin. Using various approaches (e.g., mixture theory), each layer can be represented at a macro-scale as an orthotropic material. These laminated composites are widely used in the aerospace industry and are receiving attention in other industries like automotive and medical equipment. These will be the focus of this chapter.

Advances in the manufacturing of composite materials and structures made them affordable for many engineering applications. These attributes of higher strength and stiffness to weight ratios as well as their affordability led to their extensive use. Laminated materials can exist in modern applications as beam, curved beam, plate, and shell components.

Vibration of composite structures has received considerable attention by researchers. A recent book on the subject [1] listed hundred of publications on the dynamics of composite shells alone.

In this chapter, we will review some of the fundamental equations used for straight beams, curved beams, plates, and shells. In addition, we will present solutions for the vibration problem of these components with various boundary conditions. We will present exact solutions as well as numerical ones using mainly the Ritz method. The focus here will be on free vibration analysis.

Unless stated otherwise, we will make the following assumptions:

1. The fibers are parallel to the upper and lower surfaces of their layer within the shell, plate, or beam.
2. The fibers do not follow straight lines for shells or curved beams. Instead, the curvature of these fibers follow that of the shell (or curved beam) maintaining the same distance from the upper and lower surfaces.
3. The angle between the fibers in one layer to those in another layer remains constant.
4. The fibers are evenly distributed in each layer.

Structural elements can have various types of orthotropy depending on the fiber orientations. The fibers may follow rectangular orthotropy when one finds a rectangular coordinate system for each layer where the fibers in that layer are parallel to one coordinate and perpendicular to the other coordinates. Other types of orthotropy including circular or polar orthotropy exist either naturally (e.g., wood) or synthetically. In the latter, for each layer there exists a polar coordinate system where the fibers are either in the radial or tangential directions. This consideration will introduce a fundamental concern in the formulation of the problem which is no longer a function of the boundaries or the geometrical shape only. The material and its orthotropy must be considered in the formulation of the equations at a fundamental level. A challenge will rise when a circular plates, for example, is made out of materials having rectangular orthotropy (or vice versa). The analyst then faces the challenge of whether to formulate the problem using rectangular or polar coordinates. Such problems are seldom treated analytically.

It should be noted here that composite materials are generally more flexible in shear than metallic material. Thus, shear deformation can be a major contributor to the overall deformation of the structural element at hand. Its impact is noticed in elements that are thinner than those of metallic materials. Thus, shear deformation (and rotary inertia) will receive more attention in this chapter than previous chapters.

9.1 Differential Equation of a Laminated Body in Rectangular Coordinates

As typically done in mechanics and experienced in the previous chapters, we will first relate the external forces to internal stresses (or stress resultants) through force and moment equilibrium of a differential element. Second, the deformation at any point of the engineering body is related to engineering strains or "average" deformation. Once the state of stress and that of strain are determined independently, the constitutive (i.e., stress–strain) relations can be used to complete the set of equations needed to solve the problem. As will be seen, it is in the constitutive equations, that composites materials are significantly different from metallic or isotropic materials.

In this section, we will cover the fundamental equations that govern the mechanics of elastic laminated bodies in *rectangular* coordinates. It is assumed here, unless stated otherwise, that the materials are perfectly elastic. The only displacement within the body treated here is related to its deformation. Furthermore, only small deformations (and rotations) are treated. The deformation of the body is broken into components u, v, and w parallel to the x, y, and z coordinates; respectively. The strain displacement relations can be written as

$$\epsilon_x = \frac{\partial u}{\partial x}, \quad \epsilon_y = \frac{\partial v}{\partial y}, \quad \epsilon_z = \frac{\partial w}{\partial z}$$

$$\gamma_{xy} = \frac{\partial v}{\partial x} + \frac{\partial u}{\partial y}, \quad \gamma_{xz} = \frac{\partial w}{\partial x} + \frac{\partial u}{\partial z}, \quad \gamma_{yz} = \frac{\partial w}{\partial y} + \frac{\partial v}{\partial z} \tag{9.1}$$

In order to derive stress–strain relations, consider a laminated composite thin structure constructed from very thin layers of composite material laminae. The materials of each lamina consist of parallel, continuous fibers of one material (e.g., glass, boron, carbon, graphite) embedded in a matrix material (e.g., epoxy resin). The matrix material has the primary purpose of transferring shear stress between the fibers as needed.

On a macroscopic level, each layer will be regarded as being homogeneous and orthotropic. However, the fibers of a typical layer may not be parallel to the coordinates in which the equations are expressed. This yields anisotropy at the macro-level of the laminate. For an orthotropic layer, the stress–strain relations can be presented in terms of the layers fiber directions (or coordinates 1 and 2 in Fig. 9.1) in 3D as

$$\begin{bmatrix} \sigma_1 \\ \sigma_2 \\ \sigma_3 \\ \sigma_{23} \\ \sigma_{13} \\ \sigma_{12} \end{bmatrix} = \begin{bmatrix} Q_{11} & Q_{12} & Q_{13} & 0 & 0 & 0 \\ Q_{12} & Q_{22} & Q_{23} & 0 & 0 & 0 \\ Q_{13} & Q_{23} & Q_{33} & 0 & 0 & 0 \\ 0 & 0 & 0 & Q_{44} & 0 & 0 \\ 0 & 0 & 0 & 0 & Q_{55} & 0 \\ 0 & 0 & 0 & 0 & 0 & Q_{66} \end{bmatrix} \begin{bmatrix} \epsilon_1 \\ \epsilon_2 \\ \epsilon_3 \\ \gamma_{23} \\ \gamma_{13} \\ \gamma_{12} \end{bmatrix} \tag{9.2}$$

Notice that we described the fiber coordinates of the laminate as 1 and 2, where direction 1 is parallel to the fibers and 2 is perpendicular to them. The material constants Q_{ij} are defined in terms of the material properties of the orthotropic laminate.

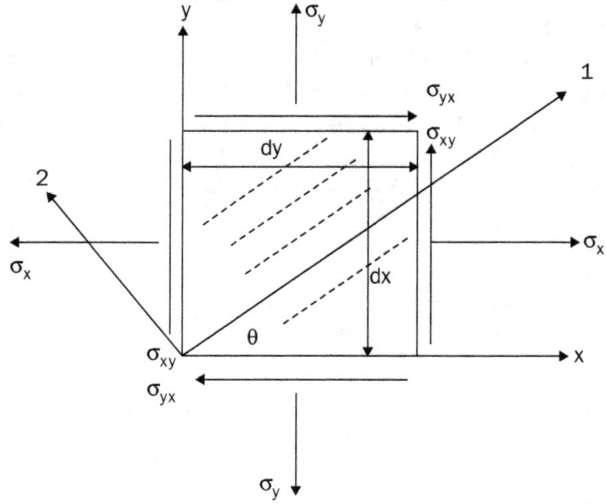

FIGURE 9.1 Coordinate systems of fiber reinforced materials.

$$
\begin{aligned}
&Q_{11} = E_{11}\frac{1-\nu_{23}\nu_{32}}{\Delta}, \quad Q_{22} = E_{22}\frac{1-\nu_{31}\nu_{13}}{\Delta} \\
&Q_{33} = E_{33}\frac{1-\nu_{12}\nu_{21}}{\Delta}, \quad Q_{44} = G_{23}, \quad Q_{55} = G_{13}, \quad Q_{66} = G_{12} \\
&Q_{12} = E_{11}\frac{\nu_{21}+\nu_{31}\nu_{23}}{\Delta} = E_{22}\frac{\nu_{12}+\nu_{32}\nu_{13}}{\Delta} \\
&Q_{13} = E_{11}\frac{\nu_{31}+\nu_{21}\nu_{32}}{\Delta} = E_{22}\frac{\nu_{13}+\nu_{12}\nu_{23}}{\Delta} \\
&Q_{23} = E_{22}\frac{\nu_{32}+\nu_{12}\nu_{31}}{\Delta} = E_{33}\frac{\nu_{23}+\nu_{21}\nu_{13}}{\Delta} \\
&\Delta = 1-\nu_{12}\nu_{21}-\nu_{23}\nu_{32}-\nu_{31}\nu_{13}-2\nu_{21}\nu_{32}\nu_{13}
\end{aligned}
\tag{9.3}
$$

where E_{11}, E_{22}, and E_{33} are modulii of elasticity in the 1, 2 and 3 directions respectively; G_{12}, G_{23}, and G_{13} are modulii of rigidity and ν_{ij} (i,j = 1, 2, 3, $i \neq j$) are Poisson's ratios. It should be noted that the Poisson's ratios are governed by the equation $\nu_{ij}/E_{ii} = \nu_{ji}/E_{jj}$. There are only nine independent material properties for *each* layer. These are E_{11}, E_{22}, E_{33}, G_{12}, G_{23}, G_{13}, ν_{12}, ν_{23}, and ν_{13}.

Consider the stress element shown in Fig. 9.1. The orientation of the fibers makes the angle θ with the rectangular coordinates x and y. The transformation of stresses (and strains) from the 1, 2 coordinates (Fig. 9.1) to the x and y coordinates can be performed by using the transformation matrix T. This matrix is

$$
T = \begin{bmatrix}
m^2 & n^2 & 0 & 0 & 0 & 2mn \\
n^2 & m^2 & 0 & 0 & 0 & -2mn \\
0 & 0 & 1 & 0 & 0 & 0 \\
0 & 0 & 0 & m & -n & 0 \\
0 & 0 & 0 & n & m & 0 \\
-mn & mn & 0 & 0 & 0 & (m^2-n^2)
\end{bmatrix}
\tag{9.4}
$$

where $m = \cos(\theta)$ and $n = \sin(\theta)$. Note that the inverse of the transformation matrix T can be found by replacing θ with $-\theta$. The transformation from fiber coordinates 1 and 2 to global coordinates x and y can be done now as follows:

$$
\begin{bmatrix} \sigma_x \\ \sigma_y \\ \sigma_z \\ \sigma_{yz} \\ \sigma_{xz} \\ \sigma_{xy} \end{bmatrix} = T^{-1} \begin{bmatrix} \sigma_1 \\ \sigma_2 \\ \sigma_3 \\ \sigma_{23} \\ \sigma_{13} \\ \sigma_{12} \end{bmatrix}, \quad \text{and} \quad
\begin{bmatrix} \epsilon_x \\ \epsilon_y \\ \epsilon_z \\ \epsilon_{yz} \\ \epsilon_{xz} \\ \epsilon_{xy} \end{bmatrix} = T^{-1} \begin{bmatrix} \epsilon_1 \\ \epsilon_2 \\ \epsilon_3 \\ \gamma_{23} \\ \gamma_{13} \\ \gamma_{12} \end{bmatrix}
\tag{9.5}
$$

The stress–strain relationship for a typical nth lamina (typically called monoclinic) becomes:

$$\begin{bmatrix} \sigma_x \\ \sigma_y \\ \sigma_z \\ \sigma_{yz} \\ \sigma_{xz} \\ \sigma_{xy} \end{bmatrix} = \begin{bmatrix} \bar{Q}_{11} & \bar{Q}_{12} & \bar{Q}_{13} & 0 & 0 & \bar{Q}_{16} \\ \bar{Q}_{12} & \bar{Q}_{22} & \bar{Q}_{23} & 0 & 0 & \bar{Q}_{26} \\ \bar{Q}_{13} & \bar{Q}_{23} & \bar{Q}_{33} & 0 & 0 & \bar{Q}_{36} \\ 0 & 0 & 0 & \bar{Q}_{44} & \bar{Q}_{45} & 0 \\ 0 & 0 & 0 & \bar{Q}_{45} & \bar{Q}_{55} & 0 \\ \bar{Q}_{16} & \bar{Q}_{26} & \bar{Q}_{36} & 0 & 0 & \bar{Q}_{66} \end{bmatrix} \begin{bmatrix} \epsilon_x \\ \epsilon_y \\ \epsilon_z \\ \gamma_{yz} \\ \gamma_{xz} \\ \gamma_{xy} \end{bmatrix} \tag{9.6}$$

where, as discussed earlier, σ_x, σ_y, and σ_z are normal stress components; σ_{xz}, σ_{yz}, and σ_{xy} are shear stress components; ϵ_x, ϵ_y, and ϵ_z are normal strain components and γ_{xz}, γ_{yz}, and γ_{xy} are the engineering shear strains. The positive directions of the stresses are shown in Fig. 9.1. The constants $\bar{Q}_{ij}$ are the elastic stiffness coefficients, which are found from

$$[\bar{Q}] = [T]^{-1}[Q][T] \tag{9.7}$$

Performing the matrix multiplication in the above equation, the stiffness coefficients $\bar{Q}_{ij}$ are

$$\begin{aligned}
\bar{Q}_{11} &= Q_{11}m^4 + 2(Q_{12} + 2Q_{66})m^2n^2 + Q_{22}n^4 \\
\bar{Q}_{12} &= (Q_{11} + Q_{22} - 4Q_{66})m^2n^2 + Q_{12}(m^4 + n^4) \\
\bar{Q}_{13} &= Q_{13}m^2 + Q_{23}n^2 \\
\bar{Q}_{16} &= -mn^3Q_{22} + m^3nQ_{11} - mn(m^2 - n^2)(Q_{12} + 2Q_{66}) \\
\bar{Q}_{22} &= Q_{11}n^4 + 2(Q_{12} + 2Q_{66})m^2n^2 + Q_{22}m^4 \\
\bar{Q}_{23} &= Q_{23}m^2 + Q_{13}n^2 \\
\bar{Q}_{33} &= Q_{33} \\
\bar{Q}_{26} &= mn^3Q_{11} - m^3nQ_{22} - mn(m^2 - n^2)(Q_{12} + 2Q_{66}) \\
\bar{Q}_{36} &= (Q_{13} - Q_{23})mn \\
\bar{Q}_{66} &= (Q_{11} + Q_{22} - 2Q_{12})m^2n^2 + Q_{66}(m^2 - n^2)^2 \\
\bar{Q}_{44} &= Q_{44}m^2 + Q_{55}n^2 \\
\bar{Q}_{55} &= Q_{55}m^2 + Q_{44}n^2 \\
\bar{Q}_{45} &= (Q_{55} - Q_{44})mn
\end{aligned} \tag{9.8}$$

The equations of motion can be derived directly from Newton's second law, where the sum of forces is equal to the mass multiplied by the acceleration in each direction. This is done in Chap. 8 and (8.2) offer the equilibrium equations. The free vibration equations for a 3D body are described earlier as well in (8.7).

The boundary terms have to be obtained by physical arguments. For the boundaries with z = constant, the boundary conditions that can be applied are

$$\begin{aligned} \sigma_{0z} - \sigma_z &= 0 \quad \text{or} \quad w = w^0 \\ \sigma_{0xz} - \sigma_{xz} &= 0 \quad \text{or} \quad u = u^0 \\ \sigma_{0yz} - \sigma_{yz} &= 0 \quad \text{or} \quad v = v^0 \end{aligned} \tag{9.9}$$

where σ_{0z}, σ_{0xz}, and σ_{0yz} are surface tractions and u°, v°, and w° are specified displacement functions at z = constant. Similarly for the boundaries x = constant and y = constant, a 3D element has six surfaces. With three equations describing boundary conditions at each surface, a total of 18 equations can be obtained.

It should be mentioned that the above equations are developed for single-layered bodies. For multiple layered bodies (Fig. 9.2), the subject of this chapter, both displacements and stresses have to be continuous going from the layer k to the next layer $k + 1$ in a laminate having N number of layers. The following conditions should be met to insure that there are no free internal surfaces (i.e., delamination) between the layers.

$$\begin{aligned} u(x,y,z=h_k)\ ]_{\text{layer } k} &= u(x,y,z=h_k)\ ]_{\text{layer } k+1} \\ v(x,y,z=h_k)\ ]_{\text{layer } k} &= v(x,y,z=h_k)\ ]_{\text{layer } k+1} \\ w(x,y,z=h_k)\ ]_{\text{layer } k} &= w(x,y,z=h_k)\ ]_{\text{layer } k+1} \\ \sigma_z(x,y,z=h_k)\ ]_{\text{layer } k} &= \sigma_z(x,y,z=h_k)\ ]_{\text{layer } k+1} \\ \sigma_{xz}(x,y,z=h_k)\ ]_{\text{layer } k} &= \sigma_{xz}(x,y,z=h_k)\ ]_{\text{layer } k+1} \\ \sigma_{yz}(x,y,z=h_k)\ ]_{\text{layer } k} &= \sigma_{yz}(x,y,z=h_k)\ ]_{\text{layer } k+1} \\ \text{For } k &= 1, \ldots, N-1 \end{aligned} \tag{9.10}$$

The above equations will be used and specialized for various types of structural elements. In particular, we will focus on beams, curved beams, plates, and shells.

For thin bodies, where the normal strain ϵ_z is negligible, the stress–strain equations for an element of material in the *kth lamina* may be written in the conventional form used for laminated plates [1] as:

$$\begin{bmatrix} \sigma_x \\ \sigma_y \\ \tau_{xy} \end{bmatrix}_k = \begin{bmatrix} \bar{Q}_{11} & \bar{Q}_{12} & \bar{Q}_{16} \\ \bar{Q}_{12} & \bar{Q}_{22} & \bar{Q}_{26} \\ \bar{Q}_{16} & \bar{Q}_{26} & \bar{Q}_{66} \end{bmatrix}_k \begin{bmatrix} \epsilon_x \\ \epsilon_y \\ \gamma_{xy} \end{bmatrix}_k \tag{9.11}$$

Note that in the above equations transverse shear stress relation with transverse shear strain is decoupled from the rest of the equations and can be written independently as a 2 × 2 matrix. Such matrix will involve the coefficients $\bar{Q}_{44}$, $\bar{Q}_{45}$, and $\bar{Q}_{55}$. If transverse shear strains are neglected, such equation will not be needed.

Although the layers are orthotropic in their material coordinates, the above equations are the same as those of material with general anisotropy. This is a result of the lamination at an angle other than 0° or 90° with respect to the global (x,y) coordinates. In other words, the $\bar{Q}_{16}$ and $\bar{Q}_{26}$ coefficients result from a coordinate transformation and arise only if the plate coordinates (x,y) are neither parallel nor perpendicular to the fibers. Correspondingly, if the plate coordinates are parallel or perpendicular to the fibers, then the terms $\bar{Q}_{16}$ and $\bar{Q}_{26}$ are zeros.

The stresses and stress couples over the thickness of the thin body (h) are integrated to obtain the force and moment resultants. Doing so results in equal inplane shear force resultants (i.e., $N_{xy} = N_{yx}$), and twisting moments (i.e., $M_{xy} = M_{yx}$).

Substituting (9.1) and (9.11) into equations (6.11) for stress resultants and integrating stresses over the cross-section for force resultants, we obtain

$$\begin{bmatrix} N_x \\ N_y \\ N_{xy} \\ -- \\ M_x \\ M_y \\ M_{xy} \end{bmatrix} = \left[\begin{array}{ccc|ccc} A_{11} & A_{12} & A_{16} & B_{11} & B_{12} & B_{16} \\ A_{12} & A_{22} & A_{26} & B_{12} & B_{22} & B_{26} \\ A_{16} & A_{26} & A_{66} & B_{16} & B_{26} & B_{66} \\ \hline B_{11} & B_{12} & B_{16} & D_{11} & D_{12} & D_{16} \\ B_{12} & B_{22} & B_{26} & D_{12} & D_{22} & D_{26} \\ B_{16} & B_{26} & B_{66} & D_{16} & D_{26} & D_{66} \end{array}\right] \begin{bmatrix} \epsilon_{0x} \\ \epsilon_{0y} \\ \gamma_{0xy} \\ -- \\ \kappa_x \\ \kappa_y \\ 2\kappa_{xy} \end{bmatrix} \tag{9.12}$$

where A_{ij}, B_{ij}, and D_{ij} are the stiffness coefficients arising from the piecewise integration over the thickness of the thin body:

$$A_{ij} = \sum_{k=1}^{N} \bar{Q}_{ij}^{(k)} \left(z_k - z_{k-1}\right)$$

$$B_{ij} = \frac{1}{2}\sum_{k=1}^{N} \bar{Q}_{ij}^{(k)} \left(z_k^2 - z_{k-1}^2\right)$$

$$D_{ij} = \frac{1}{3}\sum_{k=1}^{N} \bar{Q}_{ij}^{(k)} \left(z_k^3 - z_{k-1}^3\right) \tag{9.13}$$

and where z_k is the distance from the plate midsurface to the surface of the kth layer having the furthest z-coordinate (Fig. 9.2). The vector on the right side describes middle surface strains and curvature changes (6.13). Note here that the curvatures in (9.12) are defined as $\kappa_x = -\partial^2 w/\partial x^2$, $\kappa_y = -\partial^2 w/\partial y^2$, and $\kappa_{xy} = -\partial^2 w/\partial x \partial y$.

In addition, the above equations are general for any laminate. More specialized equations can be obtained for special types of laminates. For example, for cross ply laminates made of 0° and 90° orientations, all the stiffness parameters A_{16}, A_{26}, B_{16}, B_{26}, D_{16}, and D_{26} are zeros. In addition, if the cross-ply laminate is symmetric about its middle surface, then, the rest of the B_{ij} parameters will be zeros. If the plies (angle or cross-plies) are symmetrically laminated, all the B_{ij} vanish resulting in decoupling the inplane motion from transverse motion. This, as will be found later, will have major impact on the analysis performed.

Example 9.1 A laminated thin body is made of graphite/epoxy material with material properties as $E_{11} = 138$ GPa, $E_{21} = 8.96$ GPa, $G_{12} = 7.1$ GPa, and $\nu_{12} = 0.30$, (or $E_{11}/E_{21} = 15.4$, $G_{12}/E_{21} = 0.79$) and is constructed of

A. Single Layer 0° ply
B. Symmetric cross-ply $[0, 90]_s$ laminate
C. Unsymmetric cross-ply [0, 0, 0, 90] laminate
D. Symmetric $[30, 60]_s$ laminate
E. Antisymmetric angle-ply [30, 60, −60, −30]
F. Unsymmetric angle-ply laminate [30, 45, 60, 75]

Assume a thickness of one unit. Determine all the stiffness parameters A_{ij}, B_{ij}, and D_{ij} for each of the laminates above.

Solution

A. Consider the single layer

We will assume $\nu_{32} = \nu_{23} = \nu_{31} = \nu_{13} = 0$. Using (9.3) one gets

$$\nu_{21} = \frac{\nu_{12}E_{22}}{E_{11}} = \frac{0.3(8.96)}{138} = 0.01948$$

$$\Delta = 1 - \nu_{12}\nu_{21} = 0.9942$$

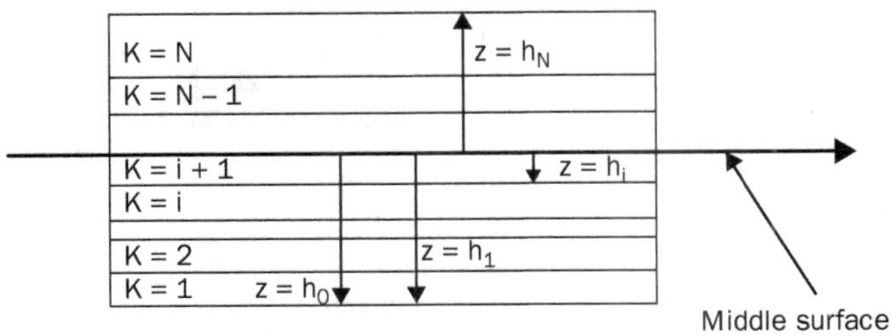

Figure 9.2 Nomenclature for stacking sequence.

which yields (in a nondimensional form)

$$Q_{11} = E_{11} = \frac{15.4\,E_{22}}{\Delta} = 15.49\,E_{22}$$

$$Q_{22} = \frac{E_{22}}{\Delta} = 1.006\,E_{22}$$

$$Q_{12} = 0.01959\,E_{11} = 0.3018\,E_{22}$$

$$Q_{66} = G_{12} = 0.79\,E_{22}$$

It should be noted here that E_1 has been used to replace E_{11} and similarly for E_2. Also, note that $Q_{16} = Q_{26} = 0$ for the single layer. The other parameters are irrelevant here.

Noting that for a 0° layer $m = 1$, and $n = 0$, and using (9.8), the $[\bar{Q}]$ matrix as described in (9.11) for the single layer laminate can be calculated.

From the $[\bar{Q}]$ matrix and using (9.13) we can calculate the A matrix (in a nondimensional form)

$$A_{ij}/E_{22}h = \begin{bmatrix} 15.49050 & 0.30176 & 0 \\ 0.30176 & 1.00588 & 0 \\ 0 & 0 & 0.79000 \end{bmatrix}$$

Consider here B_{11},

$$B_{11} = \frac{1}{2}\left\{\bar{Q}_{11}^{(1)}\left((0.5)^2 - (-0.5)^2\right)\right\} = 0$$

Similarly, one finds that the B matrix is a zero matrix. Consider D_{11};

$$D_{11} = \frac{1}{3}\left\{\bar{Q}_{11}^{(1)}\left((0.5)^3 - (-0.5)^3\right)\right\} = \frac{\bar{Q}_{11}^{(1)}}{12}$$

Similarly, the D matrix is calculated (in a nondimensional form)

$$D_{ij}/E_{22}h^3 = \begin{bmatrix} 1.29088 & 0.02515 & 0 \\ 0.02515 & 0.08382 & 0 \\ 0 & 0 & 0.06583 \end{bmatrix}$$

B. Symmetric cross-ply $[0, 90]_s$ laminate

$$A_{ij}/E_{22}h = \begin{bmatrix} 8.24820 & 0.30176 & 0 \\ 0.30176 & 8.24820 & 0 \\ 0 & 0 & 0.79000 \end{bmatrix}$$

$$B_{ij}/E_{22}h^2 = \begin{bmatrix} 0 & 0 & 0 \\ 0 & 0 & 0 \\ 0 & 0 & 0 \end{bmatrix}$$

$$D_{ij}/E_{22}h^3 = \begin{bmatrix} 1.14000 & 0.02515 & 0 \\ 0.02515 & 0.23471 & 0 \\ 0 & 0 & 0.06853 \end{bmatrix}$$

C. Unsymmetric cross-ply [0, 0, 0, 90] laminate

$$A_{ij}/E_{22}h = \begin{bmatrix} 11.86940 & 0.30176 & 0 \\ 0.30176 & 4.62704 & 0 \\ 0 & 0 & 0.79000 \end{bmatrix}$$

$$B_{ij}/E_{22}h^2 = \begin{bmatrix} -1.35794 & 0 & 0 \\ 0 & 1.35794 & 0 \\ 0 & 0 & 0 \end{bmatrix}$$

$$D_{ij}/E_{22}h^3 = \begin{bmatrix} 0.76279 & 0.02515 & 0 \\ 0.02515 & 0.23471 & 0 \\ 0 & 0 & 0.06853 \end{bmatrix}$$

D. Symmetric $[30, 60]_s$

$$A_{ij}/E_{22}h = \begin{bmatrix} 5.86079 & 2.68918 & 3.13602 \\ 2.68918 & 5.86079 & 3.13602 \\ 3.13602 & 3.13602 & 3.17742 \end{bmatrix}$$

$$B_{ij}/E_{22}h^2 = \begin{bmatrix} 0 & 0 & 0 \\ 0 & 0 & 0 \\ 0 & 0 & 0 \end{bmatrix}$$

$$D_{ij}/E_{22}h^3 = \begin{bmatrix} 0.71472 & 0.22410 & 0.34748 \\ 0.22410 & 0.26208 & 0.17519 \\ 0.34748 & 0.17519 & 0.26479 \end{bmatrix}$$

E. Antisymmetric angle-ply [30, 60, −60, −30]

$$A_{ij}/E_{22}h = \begin{bmatrix} 5.86079 & 2.68918 & 0 \\ 2.68918 & 5.86079 & 0 \\ 0 & 0 & 3.17742 \end{bmatrix}$$

$$B_{ij}/E_{22}h^2 = \begin{bmatrix} 0 & 0 & -0.95630 \\ 0 & 0 & -0.61171 \\ -0.95630 & -0.61171 & 0 \end{bmatrix}$$

$$D_{ij}/E_{22}h^3 = \begin{bmatrix} 0.71472 & 0.22410 & 0 \\ 0.22410 & 0.26208 & 0 \\ 0 & 0 & 0.26479 \end{bmatrix}$$

F. Unsymmetric angle-ply laminate [30, 45, 60, 75]

$$A_{ij}/E_{22}h = \begin{bmatrix} 4.49173 & 2.49023 & 2.58135 \\ 2.49023 & 7.62775 & 3.27054 \\ 2.58135 & 3.27054 & 2.97846 \end{bmatrix}$$

$$B_{ij}/E_{22}h^2 = \begin{bmatrix} -0.86657 & -0.17408 & -0.44094 \\ -0.17408 & 1.21473 & 0.16210 \\ -0.44094 & 0.16210 & -0.17408 \end{bmatrix}$$

$$D_{ij}/E_{22}h^3 = \begin{bmatrix} 0.42678 & 0.17022 & 0.20836 \\ 0.17022 & 0.65779 & 0.22272 \\ 0.20836 & 0.22272 & 0.21090 \end{bmatrix}$$

General observations can be made from Example 9.1 that will impact the analysis performed on these composite structures. These are

1. For symmetric cross-ply laminates (all $B_{ij} = 0$, $A_{16} = A_{26} = D_{16} = D_{26} = 0$).
2. For all cross-ply laminates ($A_{16} = A_{26} = B_{16} = B_{26} = D_{16} = D_{26} = 0$).
3. For all symmetric laminates, the B_{ij} terms are zeros.
4. For anti-symmetric angle-ply laminates ($B_{11} = B_{12} = B_{22} = B_{66} = 0$; $A_{16} = A_{26} = D_{16} = D_{26} = 0$).
5. For the generally laminated body (asymmetric angle-ply), all stiffness matrices can be fully populated.

9.2 Laminated Beams

A classical beam theory (CBT) will be developed here to treat thin beams. Applying the traditional assumptions for thin beams (normals to the beam midsurface remain straight and normal, both rotary inertial and shear deformation are neglected), strains and curvature changes at the middle surface are:

$$\epsilon_0 = \frac{\partial u_0}{\partial x}, \qquad \kappa = -\frac{\partial^2 w_0}{\partial x^2} \tag{9.14}$$

where u_0, and w_0 are displacements in x and z directions, respectively. ϵ_0 is middle surface strain and κ is the curvature change. Normal strain at any point would be

$$\epsilon = \epsilon_0 + z\kappa \tag{9.15}$$

Force and moment resultants are calculated by integrating stresses over the cross-sectional area in a manner similar to (4.168), substituting the stress strain relations found earlier for a laminated body (9.11) and ignoring the strains in the y and z directions yields

$$\begin{bmatrix} N \\ M \end{bmatrix} = \begin{bmatrix} A_{11} & B_{11} \\ B_{11} & D_{11} \end{bmatrix} \begin{bmatrix} \epsilon_0 \\ \kappa \end{bmatrix} \tag{9.16}$$

where

$$A_{11} = \sum_{k=1}^{N} b\overline{Q}_{11}^{k}\left[(h_k - h_{k-1})\right]$$

$$B_{11} = \sum_{k=1}^{N} b\overline{Q}_{11}^{k} \frac{\left(h_k^2 - h_{k-1}^2\right)}{2}$$

$$D_{11} = \sum_{k=1}^{N} b\overline{Q}_{11}^{k} \frac{\left(h_k^3 - h_{k-1}^3\right)}{3} \tag{9.17}$$

Note here that the above definitions are different from those used for general laminate analysis in the literature. The width in the above terms is included in the definitions of these terms, while it is customary to leave this term out in general laminate analysis (9.13). In addition, the above formulation does not account for any bending–twisting coupling in the beam. Its applicability for angle-ply laminates should be tested.

Figure 9.3 shows a free body diagram of a differential beam element that can be used to derive the equations of motion. When this figure is compared with that of Fig. 4.2, it should be noted here that we included both bending and axial forces in the differential element for laminated composites. This is necessary because of the possible coupling that composites have between axial and bending deformations. Applying equilibrium equations in a manner similar to that done previously in Chap. 4 yields the equation of motion. The equations of motion for a laminated beam can be written as

$$\frac{\partial N}{\partial x} + p_x = I_1 \frac{\partial^2 u}{\partial t^2}$$

$$\frac{\partial^2 M}{\partial x^2} + p_z = I_1 \frac{\partial^2 w}{\partial t^2} \tag{9.18a}$$

where

$$I_1 = \sum_{k=1}^{N} b\rho^{(k)} \left(h_k - h_{k-1}\right) \tag{9.18b}$$

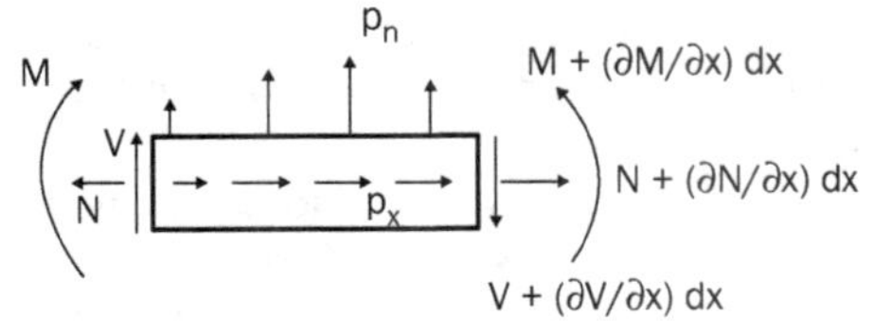

Figure 9.3 A differential element for a laminated beam.

Note here that the second equation in (9.18a) resulted after substituting the summation of moments equation into that of the summation of forces in the vertical direction equation [as was done earlier in (4.170)]. Inserting the strain and curvature relations in the force and moment resultants equations and using those in the equations of motion one can express the equations of motion in terms of displacements as

$$\begin{bmatrix} L_{11} & L_{12} \\ L_{21} & L_{22} \end{bmatrix} \begin{bmatrix} u_0 \\ w_0 \end{bmatrix} + \begin{bmatrix} -I_1 & 0 \\ 0 & I_1 \end{bmatrix} \frac{\partial^2}{\partial t^2} \begin{bmatrix} u_0 \\ w_0 \end{bmatrix} + \begin{bmatrix} p_x \\ -p_n \end{bmatrix} = \begin{bmatrix} 0 \\ 0 \end{bmatrix} \tag{9.18c}$$

where

$$L_{11} = A_{11} \frac{\partial^2}{\partial x^2}$$
$$L_{22} = D_{11} \frac{\partial^4}{\partial x^4}$$

and

$$L_{12} = L_{21} = -B_{11} \frac{\partial^3}{\partial x^3}$$

For symmetrically laminated beams, all terms containing B_{11} vanish. Note that even when the beam is symmetrically laminated, the coupling between in-plane and transverse displacements (as was found earlier for isotropic beams) vanishes and the problem reduces to transverse vibration of a beam and a longitudinal vibration of a rod.

On each boundary, one must specify three conditions:

$$\begin{aligned} u_0 &= 0 \quad \text{or} \quad N = 0 \\ w_0 &= 0 \quad \text{or} \quad Q = 0 \\ \frac{\partial w_0}{\partial x} &= 0 \quad \text{or} \quad M = 0 \end{aligned} \tag{9.18d}$$

Boundaries may also be elastically constrained, with the constraints being represented as translational and rotational springs at the beam edges.

Simply supported boundary condition is considered here. The following condition is used for such boundaries

$$w_0 = N_x = M_x = 0 \qquad \text{on } x = -a/2, a/2 \tag{9.19}$$

The above equations would be satisfied if

$$[u,w] = \sum_{m=1}^{M} [A_m \sin(\alpha_m x), C_m \cos(\alpha_m x)] \sin(\omega t) \tag{9.20}$$

where $\alpha_m = \frac{m\pi}{a}$ and a is the beam length and m is an odd number.

In a dynamic analysis, the external forces can be expanded in a Fourier series in x

$$[p_x, p_z] = \sum_{m=1}^{M} \left[p_{xm} \sin(\alpha_m x), p_{zm} \cos(\alpha_m x) \right] \tag{9.21}$$

Substituting these equations in the equations of motion, we have the characteristic equation

$$\begin{bmatrix} C_{11} & C_{12} \\ C_{21} & C_{22} \end{bmatrix} \begin{bmatrix} A_m \\ C_m \end{bmatrix} - \omega^2 \begin{bmatrix} I_1 & 0 \\ 0 & I_1 \end{bmatrix} \begin{bmatrix} A_m \\ C_m \end{bmatrix} - \begin{bmatrix} p_{xm} \\ p_{zm} \end{bmatrix} = 0 \tag{9.22}$$

where

$$C_{11} = \alpha_m^2 A_{11}$$

$$C_{22} = \alpha_m^4 D_{11}$$

$$C_{21} = C_{12} = \alpha_m^3 B_{11} \tag{9.23}$$

The nontrivial solution for natural frequencies in a free vibration analysis can be found by setting the determinant of the characteristic equation matrix to zero.

One should note here that if the laminate is *symmetric*, the B_{11} term vanishes and the bending frequencies are totally decoupled from axial ones. This yields a frequency parameter equation similar to that of (4.24). As a result, the following formula for the natural frequencies (of a symmetrically laminated simply supported composite beam) can be applied:

$$\omega_n = \left(\frac{n\pi}{\ell} \right)^2 \sqrt{\frac{D_{11}}{I_1 h}} \tag{9.24}$$

where I_1 is density as defined in (9.18b), ℓ is length, h is the thickness of the beam, and D_{11} is defined in (9.17).

As mentioned earlier, D_{11} does not quite capture all the couplings that exist in a laminate. That is why other formulations for D_{11} are proposed here. Vinson and Sierakowski [2] developed an equivalent flexural stiffness parameter for the laminate, which includes coupling. They proposed the following formula for equivalent modulus of elasticity of the each layer in the laminate

$$\frac{1}{E_x^k}=\frac{\cos^4(\theta_k)}{(E_{11})_k}+\left(\frac{1}{(G_{12})_k}-\frac{2\nu_{12}}{(E_{11})_k}\right)\cos^2(\theta_k)\sin^2(\theta_k)+\frac{\sin^4(\theta_k)}{(E_{22})_k} \tag{9.25}$$

An equivalent value for A_{11}, B_{11}, and D_{11} can then be

$$(A_{11})_{V-S}=\sum_{k=1}^{N} bE_x^k\left(h_k-h_{k-1}\right)$$

$$(B_{11})_{V-S}=\sum_{k=1}^{N} bE_x^k\frac{\left(h_k^2-h_{k-1}^2\right)}{2}$$

$$(D_{11})_{V-S}=\sum_{k=1}^{N} bE_x^k\frac{\left(h_k^3-h_{k-1}^3\right)}{3} \tag{9.26}$$

An alternative procedure proposes first finding the flexibility matrix [the inverse of the *ABD* matrix (9.12)]. We will call this the *J* matrix. It is derived for the whole laminate. An equivalent value to D_{11} can be calculated as [3]

$$(D_{11})_f=\frac{b}{J_{44}},\quad \text{where } J=[ABD]^{-1} \tag{9.27}$$

where J_{44} is the term in 4th row and 4th column of the inverse of the *ABD* matrix of the laminate (*J* matrix). Table 9.1 shows a comparison of the various stiffness parameters proposed in (9.12), and by Vinson and Sierakowski in (9.26). It is interesting to see that the Vinson–Sierakowski expression and those of (9.12) yield close results for all cross-ply laminates (symmetric and asymmetric) but they are significantly different for angle ply laminates.

The strain energy stored in a beam during elastic deformation is

$$PE=\frac{1}{2}\int_v(\sigma\epsilon)dV=\frac{1}{2}\int(N\epsilon_0+M\kappa)dx \tag{9.28}$$

Laminate	A_{11}/E_2bh		B_{11}/E_2bh^2		D_{11}/E_2bh^3	
	A_{11}	$(A_{11})_{V\text{-}S}$	B_{11}	$(B_{11})_{V\text{-}S}$	D_{11}	$(D_{11})_{V\text{-}S}$
$[0]_4$	15.4923	15.4018	0	0	1.2910	1.2835
$[0/90]_s$	8.2491	8.2009	0	0	1.1401	1.1335
$[0_2, 90_2]$	8.2491	8.2009	1.8108	1.8002	0.6874	0.6834
$[45]_4$	5.0678	1.7483	0	0	0.4223	0.1457
$[30_2/60_2]$	5.8632	2.1511	0.9054	0.2236	0.4886	0.1793

Table 9.1 A Comparison of Various Stiffness Parameters for Laminated Composite Beams with $E_1/E_2 = 15.4$, $G_{12}/E_2 = 0.79$, $\nu_{12} = 0.3$

where V is the volume. Writing the strain energy functional for the k*th* lamina, and summing for n number of laminates yields

$$PE = \frac{1}{2}\int_L \left[A_{11}(\epsilon_0)^2 + 2B_{11}\epsilon_0\kappa + D_{11}\kappa^2 \right] dx \tag{9.29}$$

Substituting the strain–displacement and curvature–displacement equations 4.2 into 4.13 yields the strain energy functional in terms of the displacements

$$PE = \frac{1}{2}\int_L \left(A_{11}\left(\frac{\partial u_0}{\partial x}\right)^2 + D_{11}\left(-\frac{\partial^2 w_0}{\partial x^2}\right)^2 + 2B_{11}\left(\frac{\partial u_0}{\partial x}\right)\left(-\frac{\partial^2 w_0}{\partial x^2}\right)\right) dx \tag{9.30}$$

It should be noted here that the coupling term B_{11} actually reduces the potential energy in the system, thus yielding lower natural frequencies when compared with symmetrically laminated beams. Using the distributed external force components p_x in the tangential direction, and p_z in the normal direction, the work done by the external forces as the beam displaces is

$$W = \int_L (p_x u_0 + p_z w_0) dx \tag{9.31}$$

The kinetic energy for each lamina is

$$KE_k = \frac{1}{2} b\rho^{(k)} \int_{L_k} \int_{z_{k-1}}^{z_k} \left[\left(\frac{\partial u_0}{\partial t}\right)^2 + \left(\frac{\partial w_0}{\partial t}\right)^2 \right] dx \tag{9.32}$$

where $\rho^{(k)}$ is the lamina density per unit volume, and t is time. The kinetic energy of the entire beam is (neglecting rotary inertia terms):

$$KE = \sum_{k=1}^{N} T_k = \frac{I_1}{2}\int\left[\left(\frac{\partial u_0}{\partial t}\right)^2 + \left(\frac{\partial w_0}{\partial t}\right)^2\right]dx \tag{9.33}$$

where I_1 is the average mass density of the beam per unit length as defined in (9.18b).

The above energy expressions can be used in an energy-based analysis like finite element or Ritz analyses.

9.3 Laminated Thick Beams

A shear deformation beam theory (SDBT) will be developed here to treat thick beams. It is well known that composite materials, in general, offer a stiffness that is more flexible in shear when compared with isotropic materials [1–3]. Also, composite beams used in general applications tend to be thicker than metallic ones (e.g., sandwich composites). For these reasons, it becomes necessary to include shear deformation in a reliable laminated beam theory. This was done earlier for isotropic beams (Sec. 4.12).

Strains and curvature changes at the middle surface are [1,4]:

$$\epsilon_0 = \frac{\partial u_0}{\partial x}, \quad \kappa = \frac{\partial \psi}{\partial x}, \quad \gamma = \frac{\partial w_0}{\partial x} + \psi \tag{9.34}$$

where ϵ_0 is middle surface strain, γ is the shear strain at the neutral axis, and ψ is the rotation of a line element perpendicular to the original direction (see Sec. 4.12).

Force and moment resultants as well as shear forces are calculated using

$$\begin{bmatrix} N \\ M \\ Q \end{bmatrix} = \begin{bmatrix} A_{11} & B_{11} & 0 \\ B_{11} & D_{11} & 0 \\ 0 & 0 & A_{55} \end{bmatrix} \begin{bmatrix} \epsilon_0 \\ \kappa \\ \gamma \end{bmatrix} \tag{9.35}$$

In order to find the shear force, it is assumed that the shear stresses are distributed parabolically across the cross-section in a manner similar to that for isotropic beams. In reality, there are discontinuities in shear stresses as one goes from one layer to the next. It is, however, found that the above assumption is accurate in assessing shear deformation. Implementing this assumption, the A_{55} term above becomes [1, 2].

$$A_{55} = \frac{5}{4}\sum_{k=1}^{N} b\overline{Q}_{55}^{k}\left[(h_k - h_{k-1}) - \frac{4}{3h^2}\left(h_k^3 - h_{k-1}^3\right)\right] \tag{9.36}$$

where $\overline{Q}_{55}^{k}$ is defined in (9.8). The equations of motion considering rotary inertia and shear deformation would then be [1,4]

$$\frac{\partial N}{\partial x} + p_x = I_1\frac{\partial^2 u}{\partial t^2} + I_2\frac{\partial^2 \psi}{\partial t^2}$$

$$\frac{\partial Q}{\partial x} + p_z = I_1\frac{\partial^2 w}{\partial t^2}$$

$$\frac{\partial M}{\partial x} - Q = I_2\frac{\partial^2 u}{\partial t^2} + I_3\frac{\partial^2 \psi}{\partial t^2} \tag{9.37}$$

where

$$(I_1, I_2, I_3) = \sum_{k=1}^{N} b\rho^{(k)}\left((h_k - h_{k-1}), \frac{1}{2}\left(h_k^2 - h_{k-1}^2\right), \frac{1}{3}\left(h_k^3 - h_{k-1}^3\right)\right) \tag{9.38}$$

The term (I_2) is of particular interest. It may be referred to as the coupling inertia and it only appears if the material density is not symmetric about the middle surface, which is rarely the case. This term does not exist for symmetrically laminated plates and even unsymmetrically laminated plates if the material used is the same, regardless of the fiber orientation.

Further mathematical manipulation enables one to express the equations of motion in terms of displacements. Multiplying the second equation in (9.37) of the equations of motion through by –1 and substituting previous strain– and curvature–displacement equations (9.34) and relations (9.35) and (9.36) into (9.37), the equations of motion may be expressed in terms of the midsurface displacements and slope in matrix form (for free vibrations) as:

$$\begin{bmatrix} L_{11} & L_{12} & L_{13} \\ L_{21} & L_{22} & L_{23} \\ L_{31} & L_{32} & L_{33} \end{bmatrix}\begin{bmatrix} u_0 \\ w_0 \\ \psi \end{bmatrix} - \begin{bmatrix} I_1 & 0 & I_2 \\ 0 & -I_1 & 0 \\ I_2 & 0 & I_3 \end{bmatrix}\frac{\partial^2}{\partial t^2}\begin{bmatrix} u_0 \\ w_0 \\ \psi \end{bmatrix} = \begin{bmatrix} 0 \\ 0 \\ 0 \end{bmatrix} \tag{9.39}$$

where

$$L_{11} = A_{11}\frac{\partial^2}{\partial x^2}$$

$$L_{12} = L_{21} = 0$$

$$L_{13} = L_{31} = B_{11}\frac{\partial^2}{\partial x^2}$$

$$L_{22} = -A_{55}\frac{\partial^2}{\partial x^2}$$

$$L_{23} = L_{32} = -A_{55}\frac{\partial}{\partial x}$$

$$L_{33} = D_{11}\frac{\partial^2}{\partial x^2} - A_{55}$$

For symmetrically laminated beams, all terms containing B_{11} vanish causing coupling between extensional and bending deformation to disappear.

Consider a simply supported beam where the following boundary conditions are used

$$w_0 = N_x = \frac{\partial \psi}{\partial x} = 0 \quad \text{on } x = -\frac{a}{2}, \frac{a}{2} \tag{9.40}$$

The above equations of motion as well as boundary conditions would be satisfied if

$$[u_0, w_0, \psi] = \sum_{m=1}^{m} [A_m \sin(\alpha_m x), C_m \cos(\alpha_m x), B_m \sin(\alpha_m x)] \sin(\omega t) \tag{9.41}$$

where $\alpha_m = m\pi/a$ and $m = 1,3,...$ and coordinates are at the center of the beam. Substituting these equations in the equations of motion we have the following equation for free vibration of simply supported thick laminated beams

$$\begin{bmatrix} C_{11} & C_{12} & C_{13} \\ C_{21} & C_{22} & C_{23} \\ C_{31} & C_{32} & C_{33} \end{bmatrix} \begin{bmatrix} A_m \\ C_m \\ B_m \end{bmatrix} + \omega^2 \begin{bmatrix} I_1 & 0 & I_2 \\ 0 & I_1 & 0 \\ I_2 & 0 & I_3 \end{bmatrix} \begin{bmatrix} A_m \\ C_m \\ B_m \end{bmatrix} = \begin{bmatrix} 0 \\ 0 \\ 0 \end{bmatrix} \tag{9.42}$$

where

$$C_{11} = -\alpha_m^2 A_{11}$$

$$C_{22} = -\alpha_m^2 A_{55}$$

$$C_{33} = -\alpha_m^2 D_{11} - A_{55}$$

$$C_{21} = C_{12} = 0$$

$$C_{31} = C_{13} = -\alpha_m^2 B_{11}$$

$$C_{23} = C_{32} = A_{55}\alpha_m$$

Previous formulation uses B_{11} which does not account for any coupling other than the bending–stretching coupling terms. Other couplings including bending–twisting, stretching–twisting as well as other stretching bending terms are not accounted for. To overcome this problem, equations in (9.26) are proposed.

Tables 9.2 and 9.3 show results obtained using both the CBT and SDBT with various coefficients obtained directly from the *ABD* matrix (9.12) and those obtained using expressions (9.26).

	CBT		SDBT		FEM
n	**S_{11},(9.13)**	**$(S_{11})_{VS}$**	**$(S_{11})^*$, (9.13)**	**$(S_{11})_{VS}$**	**3D**
0/0/0/0					
1	1.575	1.571	1.501	1.497	1.492
2	6.302	6.283	5.318	5.307	5.250
3	14.179	14.137	10.270	10.254	10.08
4	25.206	25.133	15.615	15.597	15.25
5	39.385	39.270	21.040	21.022	20.49
0/90/90/0					
1	1.480	1.476	1.418	1.414	1.415
2	5.922	5.905	5.082	5.071	5.046
3	13.324	13.285	9.927	9.910	9.80
4	23.688	23.618	15.225	15.206	14.96
5	37.012	36.904	20.644	20.624	20.21
0/0/90/90					
1	0.746	0.744	0.738	0.736	0.735
2	2.979	2.970	2.857	2.849	2.813
3	6.683	6.664	6.114	6.099	5.936
4	11.831	11.797	10.212	10.190	9.775
5	18.387	18.333	14.880	14.852	14.067

$(S_{11})_{VS}$ refers to $(A_{11})_{VS}$, $(B_{11})_{VS}$, and $(D_{11})_{VS}$.

TABLE 9.2 Natural Frequencies ($\omega a^2 / 2\pi\sqrt{I_1 / E_{11} I}$) of Simply Supported Cross-Ply Beams with Various Theories (a/h = 20, b/h = 2, Graphite/Epoxy, E_{11}/E_{22} = 15.4, G_{12}/E_{22} = 0.79, ν_{12} = 0.3)

n	CBT		SDBT		FEM
	S_{11}	$(S_{11})_{VS}$	S_{11}	$(S_{11})_{VS}$	3D
45/45/45/45					
1	0.9010	0.5292	0.8859	0.5257	0.5690
2	3.6042	2.1169	3.3792	2.0631	2.053
3	8.1094	4.7630	7.0920	4.5048	4.534
4	14.4167	8.4676	11.6064	7.7049	7.821
5	22.5261	13.2307	16.5932	11.5098	11.85
30/30/60/60					
1	0.8186	0.5476	0.8075	0.5439	0.5958
2	3.2715	2.1894	3.1048	2.1327	2.165
3	7.3499	4.9227	6.5842	4.6513	4.770
4	13.0388	8.7430	10.8916	7.9437	8.132
5	20.3167	13.6434	15.7248	11.8466	12.15

$(S_{11})_{VS}$ refers to $(A_{11})_{VS}$, $(B_{11})_{VS}$, and $(D_{11})_{VS}$.

TABLE 9.3 Natural Frequencies ($\omega a^2/2\pi\sqrt{I_1/E_{11}I}$) of Simply Supported Beams with Various Theories (a/h = 20, b/h = 2, Graphite/Epoxy, E_{11}/E_{22} = 15.4, G_{12}/E_{22} = 0.79, ν_{12} = 0.3)

In order to find which expressions deliver most accurate results, a procedure similar to that of Chap. 8 is developed. Instead of using the Ritz method, a finite element model is built using commercial software and 3D elements are used. Convergence studies are conducted for a typical problem to verify that the finite element results are converging.

It is interesting to see that indeed, using the expressions A_{11}, B_{11}, and D_{11} as defined in (9.13) with a CBT will yield relatively accurate frequencies for the *fundamental* frequencies of all *cross-ply* laminates. For higher frequencies, a SDBT is needed and still the stiffness parameters obtained using (9.13) will yield relatively accurate results for cross-ply. For angle-ply laminates (Table 9.3), it is clear that expressions (9.13) are not accurate and expressions (9.26) are relatively accurate, although they seem to offer more flexibility (i.e., lower frequencies) that 3D finite element results.

For angle ply laminates, it is clear that an expression similar to that of Vinson and Sierakowski (9.26) or the flexibility matrix (9.27) is needed with a CBT to be able to predict the lower frequency parameters and with a shear deformation beam theory to be able to provide accurate prediction for higher frequency parameters. It is interesting that for the fifth mode, where the length of the half sine wave to thickness ratio of 4, the SDBT yielded results that are relatively accurate when compared with those obtained with 3D finite elements.

The strain energy stored in a beam during elastic deformation is

$$PE = \frac{1}{2}\int\left(N\epsilon_0 + M\frac{\partial \psi}{\partial x} + Q\gamma\right)dx \tag{9.43}$$

Writing the strain energy functional for the kth lamina, and summing for n number of layers yields

$$PE = \frac{1}{2}\int_L\left[A_{11}(\epsilon_0)^2 + B_{11}\epsilon_0\kappa + D_{11}\kappa^2 + A_{55}\gamma^2\right]dx \tag{9.44}$$

Substituting the strain displacement relations into the above energy expression yields the strain energy functional in terms of the slope and displacements

$$PE = \frac{1}{2}\int_L\left[A_{11}\left(\frac{\partial u_0}{\partial x}\right)^2 + D_{11}\left(\frac{\partial \psi}{\partial x}\right)^2 + 2B_{11}\left(\frac{\partial u_0}{\partial x}\right)\frac{\partial \psi}{\partial x} + A_{55}\left(\frac{\partial w_0}{\partial x} + \psi\right)^2\right]dx \tag{9.45}$$

A similar term to that of (9.31) can be developed for work done by external forces. Finding the kinetic energy for each layer and then summing for all layers yield

$$KE = \sum_{k=1}^{N} KE_k = \frac{1}{2}\int_L\left[I_1\left(\frac{\partial u_0}{\partial t}\right)^2 + I_1\left(\frac{\partial w_0}{\partial t}\right)^2 + 2I_2\left(\frac{\partial \psi}{\partial t}\frac{\partial u_0}{\partial t}\right) + I_3\left(\frac{\partial \psi}{\partial t}\right)^2\right]dx \tag{9.46}$$

The above energy functionals can be used in energy-based analyses similar to the Ritz method.

9.4 Beams with Tubular Cross-Sections

Composite beams with tubular cross-sections are receiving attention in engineering applications where weight reduction is a priority. They are currently being used in automotive applications as driveline shafts. Figure 9.4 shows a laminated beam with a tubular cross-

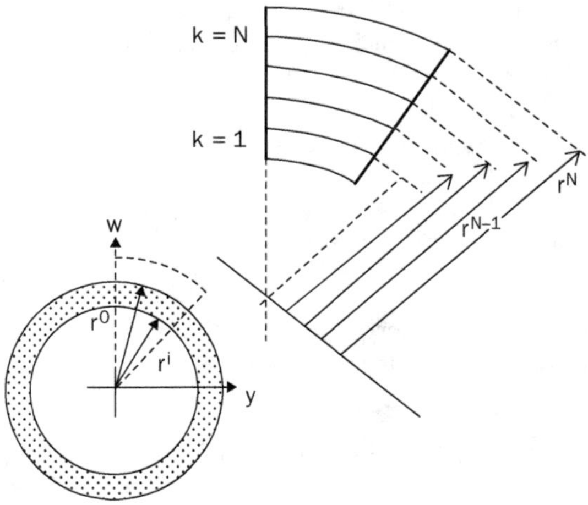

FIGURE 9.4 A tubular cross-section of composite beam made up of *N*-layer laminate.

section. For such a beam with symmetric lamination, the following parameters are to be used

$$A_{11} = \pi \sum_{k=1}^{N} \overline{Q}_{11}^{k} \left[\left(r_k^2 - r_{k-1}^2 \right) \right] \tag{9.47}$$

$$D_{11} = \frac{\pi}{4} \sum_{k=1}^{N} \overline{Q}_{11}^{k} \left[\left(r_k^4 - r_{k-1}^4 \right) \right] \tag{9.48}$$

Table 9.4 lists natural frequencies obtained for a laminate that is mainly a cross-ply by various analyses. The results show that most of the models can predict the fundamental natural frequency for this beam. Also the Euler–Bernoulli beam theory used here is among the better models as compared with the experimental data provided in Ref. [5] and predicts the natural frequency with a 4.88 percent difference. It should be noted here that the model used by Iqbal and Qatu [9] considered only cross-ply laminates and the problem at hand has limited angle ply layers.

It should also be noted here that the advanced models that use shell theory predicted the behavior fairly well. These models are also accurate for predicting high frequency modes (often referred to as shell modes) as described in Chap. 7 for isotropic shells. The results show, however, that the analyses presented here is accurate for bending modes of such beams.

The effect of ply orientation on reduction of stiffness and consequently natural frequency is presented in Table 9.5. These

Author	Method used	Frequency (Hz)
Zinberg and Symonds [5,7]	Measured experimentally	100
Kim and Bert [6]	Sanders shell theory	97.87
	Donnell shallow shell theory	106.65
Bert and Kim [7]	Bresse–Timoshenko beam theory	96.47
Chang et al. [8]	Continuum based Timoshenko beam	96.03
Iqbal and Qatu [9]	Finite element analysis	95.4
	Euler–Bernoulli beam theory	102.47
Present study	CBT	96.12
	SDBT	94.71

TABLE 9.4 Tubular Beam Fundamental Natural Frequencies (Hz) by Different Authors. Material is Boron/Epoxy (E_{11} = 211 GPa, E_{22} = 24 GPa, G_{12} = G_{13} = G_{23} = 6.9 GPa, ν = 0.36, Density = 1967 kg/m^3), Length = 2470 mm, Mean Diameter = 126.9 mm, Thickness = 1.321 mm. (90, 45, −45, 0, 0, 0, 0, 0, 0, 90) Laminate (From Inner to Outer)

	Lamination angle						
Theory	**0**	**15**	**30**	**45**	**60**	**75**	**90**
Sanders Shell [6]	92.12	72.75	50.13	39.77	35.33	33.67	33.28
Bernoulli-Euler [6]	107.08	89.88	71.15	52.85	38.20	31.42	30.22
Bresse-Timoshenko [6]	101.20	86.82	69.95	52.38	37.97	32.90	30.05
Present FEM analysis	100.28	68.80	45.51	35.90	31.96	30.57	30.27
Present CBT approach*	108.42	71.12	46.05	36.15	32.17	30.78	30.50
Present SDBT approach*	104.43	70.50	45.91	36.06	32.09	30.70	30.36

* $(A_{11})_{VS}$, $(B_{11})_{VS}$, and $(D_{11})_{VS}$ are used.

TABLE 9.5 Effect of Lamination Angle on Fundamental Natural Frequencies of Tubular Graphite-Epoxy Beams. Graphite/Epoxy Materials, (E_{11} = 139 GPa, E_{22} = 11 GPa, G_{12} = G_{13} = 6.05 GPa, G_{23} = 3.78 GPa, ν = 0.313, density = 1478 Kg/m^3)

results are for the first natural frequency of a graphite epoxy tubular beam with the same geometry of the previous one from Ref. [6]. Results of the present CBT, SDBT, and FEM using shell elements are presented.

Iqbal and Qatu [9] analyzed laminated composite shafts for automotive engineering applications when they have intermediate supports. They considered single and multi-segmented shafts with one and two supports in their analyses and studied their impacts on the natural frequencies.

9.5 Laminated Thin Curved Beams

A classical beam theory (CBT) will be developed here to treat thin curved beams. A laminated curved beam (Fig. 9.5) is characterized by its middle surface, which is defined by the polar coordinate α (Figs. 9.5 and 4.24), where $\alpha = R\theta$ as defined in (4.165).

Middle surface strain and curvature change are the same as those developed earlier in Sec. 4.13 for isotropic beams and described in (4.166). The strain at an arbitrary point can be found from (4.167).

Normal strain at an arbitrary point can be found from the equation $\epsilon = (\epsilon_0 + z\kappa)/(1 + z/R)$ as described in Ref. [1]. Note the presence of the $(1 + z/R)$ term in the denominator of this equation. The term is negligible when the beam is thin [1]. Subsequently, the force and moment stress resultants for curved thin beams can be found from (9.16) and (9.17) found earlier for straight beams.

The equations of motion may be obtained by taking a differential element of a beam (Fig. 4.25) and requiring the sum of the external and internal forces in the α and z direction and the sum of the external and internal moments in the out-of-plane direction to be zero. The equations of motion are the same as those derived earlier for isotropic beams (4.169), except that the term (ρA) there should be replaced by I_1 defined earlier in this chapter (9.18b). Further mathematical manipulation in a free vibration analysis will yield

$$\begin{bmatrix} L_{11} & L_{12} \\ L_{21} & L_{22} \end{bmatrix} \begin{bmatrix} u_0 \\ w_0 \end{bmatrix} + \begin{bmatrix} -I_1 & 0 \\ 0 & I_1 \end{bmatrix} \frac{\partial^2}{\partial t^2} \begin{bmatrix} u_0 \\ w_0 \end{bmatrix} = \begin{bmatrix} 0 \\ 0 \end{bmatrix} \tag{9.49}$$

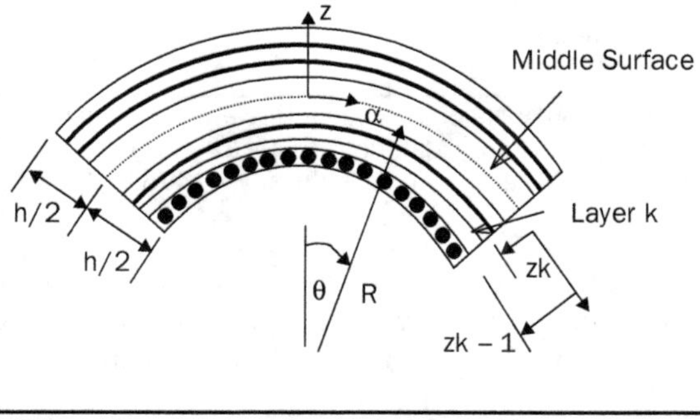

FIGURE 9.5 Parameters used in a laminated curved beam.

where

$$L_{11} = A_{11}\frac{\partial^2}{\partial\alpha^2} + 2\frac{B_{11}}{R}\frac{\partial^2}{\partial\alpha^2} + \frac{D_{11}}{R^2}\frac{\partial^2}{\partial\alpha^2}$$

$$L_{22} = D_{11}\frac{\partial^4}{\partial\alpha^4} - 2\frac{B_{11}}{R}\frac{\partial^2}{\partial\alpha^2} + \frac{A_{11}}{R^2}$$

$$L_{12} = L_{21} = -B_{11}\frac{\partial^3}{\partial\alpha^3} + \frac{A_{11}}{R}\frac{\partial}{\partial\alpha} - \frac{D_{11}}{R}\frac{\partial^3}{\partial\alpha^3} + \frac{B_{11}}{R^2}\frac{\partial}{\partial\alpha}p$$

For symmetrically laminated beams, all terms containing B_{11} vanish. Note that even when the curved beam is symmetrically laminated, there exist coupling between in-plane and transverse displacements (as was found earlier for isotropic beams).

Substituting the strain–displacement and curvature–displacement equations for curved beams (4.166) into (9.29) yields the strain energy functional in terms of the midsurface displacements

$$PE = \frac{1}{2}\int_L \left(A_{11}\left(\frac{\partial u_0}{\partial\alpha} + \frac{w_0}{R}\right)^2 + D_{11}\left(-\frac{\partial^2 w_0}{\partial\alpha^2} + \frac{1}{R}\frac{\partial u_0}{\partial\alpha}\right)^2 + 2B_{11}\left(\frac{\partial u_0}{\partial\alpha} + \frac{w_0}{R}\right)\left(-\frac{\partial^2 w_0}{\partial\alpha^2} + \frac{1}{R}\frac{\partial u_0}{\partial\alpha}\right)\right) d\alpha \tag{9.50}$$

Finding the kinetic energy for each lamina and summing the energy expressions for each lamina together yields an energy expression similar to (9.33).

Boundary conditions can be obtained through a variational formulation [1]. On each boundary, one must specify three conditions:

$$\begin{aligned} u_0 = 0 \quad &\text{or} \quad N + \frac{M}{R} = 0 \\ w_0 = 0 \quad &\text{or} \quad Q = 0 \\ \frac{\partial w_0}{\partial\alpha} = 0 \quad &\text{or} \quad M = 0 \end{aligned} \tag{9.51a}$$

Note that there is an additional term (M/R) in the first boundary condition. This term does not exist for straight beams. Boundaries may also be elastically constrained, with the constraints being represented as translational and rotational springs at the beam edges. In such cases, the boundary conditions are generalized to

$$k_u u_0 + \left(N + \frac{M}{R}\right) = 0$$
$$k_w w_0 + Q = 0$$
$$k_\psi \frac{\partial w_0}{\partial \alpha} - M = 0 \tag{9.51b}$$

at the edge $\alpha = -a/2$, where k_u and k_w are translational spring stiffnesses and $k\psi$ is the rotational spring stiffness. The signs of N, M, and Q in the above equations change at the edge $\alpha = a/2$.

The simple support boundary conditions can take two forms for curved beams at one boundary, namely S1 and S2 (4.176a). Similarly for the free (4.176b) and clamped (4.176c) boundaries.

The above equations constitute a complete and consistent set of equations for the analysis of laminated composite thin curved beams. It should be noted again here that the A_{11}, B_{11}, and D_{11} should be examined closely for angle-ply composites. Alternative expressions should be considered for beams that are not cross-ply as described in (9.26) and (9.27).

Simply supported curved beams (with S1 boundary condition) will be studied in this section. For such beams, straightforward exact solution can be used (similar to that used in Chap. 4 for curved isotropic beams).

The S1 boundary conditions are exactly satisfied at both ends ($\alpha = -a/2$ and $\alpha = a/2$) for thin beams by choosing displacement functions similar to those of (4.177). The external forces (important in static or dynamic analyses) can be expanded in a Fourier series in α as in (4.178), if a static analysis or a forced vibration analysis are needed. Substituting these equations into (9.49) yields for free vibration analysis

$$\begin{bmatrix} C_{11} & C_{12} \\ C_{21} & C_{22} \end{bmatrix} \begin{bmatrix} A_m \\ C_m \end{bmatrix} - \omega^2 \begin{bmatrix} I_1 & 0 \\ 0 & I_1 \end{bmatrix} \begin{bmatrix} A_m \\ C_m \end{bmatrix} = \begin{bmatrix} 0 \\ 0 \end{bmatrix} \tag{9.52}$$

where

$$C_{11} = \alpha_m{}^2 \left[A_{11} + \frac{2B_{11}}{R} + \frac{D_{11}}{R^2} \right]$$
$$C_{22} = D_{11}\alpha_m{}^4 + 2\left(\frac{B_{11}}{R}\right)\alpha_m^2 + \left(\frac{A_{11}}{R^2}\right)$$

and

$$C_{21} = C_{12} = B_{11}\alpha_m{}^3 + \left[\left(\frac{A_{11}}{R}\right) + \left(\frac{B_{11}}{R^2}\right)\right]\alpha_m + \left(\frac{D_{11}}{R}\right)\alpha_m{}^3$$

The solution $u_0 = \cos(\alpha_m \alpha) \sin(\omega t)$; and $w_0 = \sin(\alpha_m \alpha) \sin(\omega t)$ is another solution for laminated curved beams having vertical hinge supports (i.e., $u_0 = \partial w_0/\partial \alpha = Q = 0$). This will yield a frequency determinant which is the same as that of simply supported beams, and consequently and interestingly, the same natural frequencies.

Table 9.6 shows the first four natural frequencies obtained for simply supported deep thin beams. The nondimensional frequency parameter $\Omega = \omega a^2 \sqrt{12\rho_1 / E_{11}h^2}$ $(= \omega a^2 \sqrt{I_1 / E_{11} I})$, where A is the cross-sectional area and I is the moment of inertia $(bh^3/12)$. Two orthotropy ratios and an unsymmetric (or asymmetric) laminate [0, 90] is used. It is observed there that the effect of curvature on the natural frequencies is the highest for the fundamental frequency and it is less for higher frequencies.

Although exact solutions can be obtained for curved thin beams with other boundaries, they are mathematically complex and not attractive. The Ritz method will be used here in a general but simple form which can be used for arbitrary boundary conditions. This is done by using a procedure similar to that of Chap. 4 and summarized in equations (4.180) and (4.181).

Table 9.7 shows results obtained using the Ritz method with algebraic polynomials. The table shows results obtained for the S2 and C2 boundary conditions. Note here that the frequencies are much higher than those for the S1 case, due to the axial end constraints. Interestingly, this is the case even without curvature because of the

R/ℓ	m			
	1	2	3	4
$E_{11}/E_{22} = 15$				
–	4.7037	18.810	42.320	75.222
10	4.6936	18.795	42.295	75.181
5	4.6707	18.767	42.255	75.131
2	4.5176	18.593	42.049	74.881
1	4.0115	18.030	41.431	74.188
$E_{11}/E_{22} = 40$				
–	4.0072	16.024	36.040	64.061
10	3.9938	16.011	36.017	64.021
5	3.9758	15.980	35.985	63.976
2	3.8432	15.831	35.805	63.757
1	3.4153	15.349	35.271	63.156

TABLE 9.6 Exact Frequency Parameters $\Omega = \omega a^2 \sqrt{12\rho / E_{11}h^2}$ for Simply Supported (S1) [0°, 90°] Laminated Curved Thin Beams, $\ell/h = 100$

R/ℓ	m							
	1	2	3	4	1	2	3	4
	S2 Boundary conditions				C2 Boundary conditions			
–	6.4291	18.807	44.533	75.166	10.661	29.386	57.601	95.567
100	7.9180	18.806	44.781	75.160	10.660	29.385	57.599	95.576
20	14.722	18.796	46.079	75.131	10.657	29.378	57.587	95.611
10	18.779	22.703	48.751	75.089	10.649	29.366	57.569	95.653
5	18.727	32.625	59.743	74.988	10.621	29.332	57.522	95.731
2	18.434	37.935	74.551	99.985	10.445	29.145	57.291	95.882
1	17.529	38.288	73.410	110.82	9.8882	28.562	56.616	95.789

TABLE 9.7 Curvature Effects on the Frequency Parameters $\Omega = \omega a^2 \sqrt{12\rho / E_{11} h^2}$ for Simply Supported [0°, 90°] Laminated Curved Thin Beam, $\ell / h = 100$; $E_{11}/E_{22} = 15$

unsymmetric lamination and coupling between axial and transverse deformation. This was *not* observed for isotropic curved beams (Chap. 4). In addition, a slight increase in curvature in S2 boundaries results in significant change in the frequency parameters. The table also shows results for the completely clamped case (i.e., C2 boundary condition), where a similar observation is made. More results can be found in Ref. [1].

9.6 Laminated Thick Curved Beams

A shear deformation beam theory (SDBT) will be developed here to treat thick curved beams. Shear deformation and rotary inertia will be included in the derivation of a thick beam theory. The middle surface strain and curvature change are [1]:

$$\epsilon_0 = \frac{\partial u_0}{\partial \alpha} + \frac{w_0}{R}, \quad \kappa = \frac{\partial \psi}{\partial \alpha}$$

$$\gamma = \frac{\partial w_0}{\partial \alpha} + \psi - \frac{u_0}{R} \tag{9.53}$$

where γ is the shear strain at the neutral axis, and ψ is rotation of a line element originally perpendicular to the longitudinal direction, about the out-of-plane direction.

Normal strain at an arbitrary point can be found from

$$\epsilon = \frac{1}{(1 + z/R)}(\epsilon_0 + z\kappa) \tag{9.54}$$

The presence of the $(1 + z/R)$ term in the denominator of the above equations should be handled with care. The term is *not* negligible when the beam is thick and is shown to have a significant impact [1]. Thus, it will be retained here for laminated thick beams. The force and moment resultants are the integrals of the stresses over the beam thickness (h):

$$[N, M, Q] = b\int_{-h/2}^{h/2} [\sigma, \sigma z, \tau] dz \tag{9.55}$$

where b is the width of the beam. For a laminated beam, with n number of layers, the above equations may be rewritten as:

$$[N, M, Q] = b\sum_{k=1}^{n} \int_{h_{k-1}}^{h_k} [\sigma, \sigma z, \tau] dz \tag{9.56}$$

Substituting (9.53) and (9.54) into (9.55), and carrying out the integration over the thickness yields:

$$\begin{bmatrix} N \\ M \\ Q \end{bmatrix} = \begin{bmatrix} (A_{11})_t & (B_{11})_t & 0 \\ (B_{11})_t & (D_{11})_t & 0 \\ 0 & 0 & A_{55} \end{bmatrix} \begin{bmatrix} \epsilon_0 \\ \kappa \\ \gamma \end{bmatrix} \tag{9.57}$$

where the $(A_{11})_t$, $(B_{11})_t$, $(D_{11})_t$ and $(A_{55})_t$ are the stiffness coefficients arising from the integration. The integration is carried exactly [1], with the z/R term, and yields

$$(A_{11})_t = R\sum_{k=1}^{N} b\bar{Q}_{11}^{(k)} \ln\left(\frac{R+h_k}{R+h_{k-1}}\right)$$

$$(B_{11})_t = R\sum_{k=1}^{N} b\bar{Q}_{11}^{(k)} \left[(h_k - h_{k-1}) - R\ln\left(\frac{R+h_k}{R+h_{k-1}}\right)\right]$$

$$(D_{11})_t = R\sum_{K=1}^{N} b\bar{Q}_{11}^{(k)} \left[\frac{1}{2}[(R+h_k)^2 - (R+h_{k-1})]^2)\right.$$

$$\left. -2R(h_k - h_{k-1}) + R^2 \ln\left(\frac{R+h_k}{R+h_{k-1}}\right)\right]$$

$$A_{55} = \frac{5}{4}\sum_{k=1}^{N} b\bar{Q}_{11}^{(k)} \left[(h_k - h_{k-1}) - \frac{4}{3h^2}(h_k^3 - h_{k-1}^3)\right] \tag{9.58}$$

Table 9.8 shows a comparison between the above terms (9.58) with those obtained earlier for thin beams (9.17). Equations (9.58) are accurate, and the comparison should give a hint on the limitation of the relations between the force and moment resultants and midsurface strains and curvature changes used in the thin curved beam theory. The difference becomes large for laminated composites with a high orthotropy ratio. This difference reaches or exceeds 7 percent for beams with orthotropy ratios of 15 and 40 and thickness ratio h/R of 0.2, which is taken as the limit of thick beam theories. In general, the z/R term should be included to obtain accurate equations for thick curved beams. The table shows that the term is indeed negligible for laminated thin curved beams as was done earlier in this chapter and for thin isotropic curved beams as was done in Sec. 4.13.

h/R	Approximate equations (9.17)			Accurate equations (9.58)		
	A_{11}/E_2bR	B_{11}/E_2bR^2	$D_{11}/E_2\,bR^3$	$(A_{11})_t/E_2bR$	$(B_{11})_t\,E_2bR^2$	$(D_{11})_t/E_2\,bR^3$
E_1/E_2 = 1 (i.e., single layer)						
0.01	0.01	0	0.00000008	0.01000008	−0.00000008	0.00000008
0.05	0.05	0	0.00001042	0.05001004	−0.00001042	0.00001042
0.10	0.10	0	0.00008333	0.10008346	−0.00008346	0.00008346
0.20	0.20	0	0.00066667	0.20067070	−0.00067070	0.00067070
Δ^*	−0.3%	–	−0.6%			
E_1/E_2 = 15						
0.01	0.08	0.000175	0.00000067	0.07982566	0.00017434	0.00000066
0.05	0.40	0.004375	0.00008333	0.39570700	0.00429300	0.00008200
0.10	0.80	0.017500	0.00066667	0.78314576	0.01685424	0.00064576
0.20	1.60	0.070000	0.00533333	1.53501321	0.06498679	0.00501321
Δ^*	4.2%	7.7%	6.4%			
E_1/E_2 = 40						
0.01	0.205	0.0004875	0.00000171	0.20451420	0.00048579	0.00000170
0.05	1.025	0.0121875	0.00021354	1.01302231	0.01197769	0.00020981
0.10	2.050	0.0487500	0.00170833	2.00289986	0.04710014	0.00164986
0.20	4.100	0.1950000	0.01366667	3.91776771	0.18223229	0.01276771
Δ^*	4.7%	7.0%	7.0%			

*The percentage difference between Eqs. (9.17) and (9.58) for $h/R = 0.2$

TABLE 9.8 A Comparison between Approximate and Accurate Equations for the Coefficients A_{11}, B_{11}, and D_{11} of [0°, 90°] Curved Beams

Considering rotary inertia terms, the equations of motion become [1]:

$$\frac{\partial N}{\partial \alpha}+\frac{Q}{R}=\bar{I}_1\frac{\partial^2 u_0}{\partial t^2}+\bar{I}_2\frac{\partial^2 \psi}{\partial t^2}-p_\alpha$$

$$-\frac{N}{R}+\frac{\partial Q}{\partial \alpha}=\bar{I}_1\frac{\partial^2 w_0}{\partial t^2}-p_n$$

$$\frac{\partial M}{\partial \alpha}-Q=\bar{I}_2\frac{\partial^2 u_0}{\partial t^2}+\bar{I}_3\frac{\partial^2 \psi}{\partial t^2} \tag{9.59}$$

where

$$\left(\bar{I}_1,\bar{I}_2,\bar{I}_3\right)=\left(I_1+2\frac{I_2}{R}+\frac{I_3}{R^2},I_2+\frac{I_3}{R},I_3\right)$$

and I_1, I_2, and I_3 are defined in (9.38). Note here that the above inertia terms ($\bar{I}_1,\bar{I}_2$, and $\bar{I}_3$)were obtained from similar considerations of the (1+z/R) in the numerator of the density integration equations [1]. Substituting previous constitutive and strain displacement equations into (9.59), the equations of motion (9.59) may be expressed in terms of the midsurface displacements and slope in matrix form, for a free vibration analysis, as:

$$\begin{bmatrix} L_{11} & L_{12} & L_{13} \\ L_{21} & L_{22} & L_{23} \\ L_{31} & L_{32} & L_{33} \end{bmatrix}\begin{bmatrix} u_0 \\ w_0 \\ \psi \end{bmatrix}-\begin{bmatrix} \bar{I}_1 & 0 & \bar{I}_2 \\ 0 & -\bar{I}_1 & 0 \\ \bar{I}_2 & 0 & \bar{I}_3 \end{bmatrix}\frac{\partial^2}{\partial t^2}\begin{bmatrix} u_0 \\ w_0 \\ \psi \end{bmatrix}=\begin{bmatrix} 0 \\ 0 \\ 0 \end{bmatrix} \tag{9.60}$$

where

$$L_{11}=A_{11}\frac{\partial^2}{\partial \alpha^2}-\frac{A_{55}}{R^2}$$

$$L_{12}=L_{21}=\left(\frac{A_{11}}{R}+\frac{A_{55}}{R}\right)\frac{\partial}{\partial \alpha}$$

$$L_{13}=L_{31}=B_{11}\frac{\partial^2}{\partial \alpha^2}+\frac{A_{55}}{R}$$

$$L_{22} = \frac{A_{11}}{R^2} - A_{55}\frac{\partial^2}{\partial\alpha^2}$$

$$L_{23} = L_{32} = \frac{B_{11}}{R}\frac{\partial}{\partial\alpha} - A_{55}\frac{\partial}{\partial\alpha}$$

$$L_{33} = D_{11}\frac{\partial^2}{\partial\alpha^2} - A_{55}$$

Note here that the subscript (t) in (9.58) is dropped for convenience. For symmetrically laminated beams, all terms containing B_{11} vanish but, unlike thick straight beams, coupling between extensional and bending deformation remains.

The strain energy stored in a beam during elastic deformation is

$$PE = \frac{1}{2}\int\left(N\epsilon_0 + M\frac{\partial\psi}{\partial\alpha} + Q\gamma\right)d\alpha \tag{9.61}$$

Writing the strain energy functional for the *kth* lamina, and summing for n number of layers yields

$$PE = \frac{1}{2}\int_L\left[A_{11}(\epsilon_0)^2 + B_{11}\epsilon_0\kappa + D_{11}\kappa^2 + A_{55}\gamma^2\right]d\alpha \tag{9.62}$$

Substituting the strain– and curvature–displacement relations into the above energy expression yields the strain energy functional in terms of the slopes and displacements

$$PE = \frac{1}{2}\int_L\left[A_{11}\left(\frac{\partial u_0}{\partial\alpha} + \frac{w_0}{R}\right)^2 + D_{11}\left(\frac{\partial\psi}{\partial\alpha}\right)^2\right.$$
$$\left. +2B_{11}\left(\frac{\partial u_0}{\partial\alpha} + \frac{w_0}{R}\right)\frac{\partial\psi}{\partial\alpha} + A_{55}\left(\frac{\partial w_0}{\partial\alpha} + \psi - \frac{u_0}{R}\right)^2\right]d\alpha \tag{9.63}$$

The work done by the external force components [p_α in the tangential (polar) direction, and p_n in the normal direction] can be done in a manner similar to that used earlier. The kinetic energy for the entire beam (including rotary inertia) is

$$KE = \frac{b}{2}\int_{l_k}\int\rho\left[\left(\left(1 + \frac{z}{R}\right)\frac{\partial u_0}{\partial t} + z\frac{\partial\psi}{\partial t}\right)^2\right]dz\,d\alpha$$

$$= \frac{b}{2}\int_L \int \rho \left[\left(\left(1+\frac{z}{R}\right)\frac{\partial u_0}{\partial t} \right)^2 + 2\left(1+\frac{z}{R}\right) z \frac{\partial u_0}{\partial t}\frac{\partial \psi}{\partial t} + \left(z\frac{\partial \psi}{\partial t} \right)^2 + \left(\frac{\partial w_0}{\partial t} \right)^2 \right] dz\, d\alpha \tag{9.64}$$

Integrating over z and summing for N number of layers yields

$$KE = \sum_{k=1}^{N} KE_k = \frac{b}{2}\int_L \left[\bar{I}_1 \left(\frac{\partial u_0}{\partial t} \right)^2 + \bar{I}_1 \left(\frac{\partial w_0}{\partial t} \right)^2 + 2\bar{I}_2 \left(\frac{\partial \psi}{\partial t}\frac{\partial u_0}{\partial t} \right) + \bar{I}_3 \left(\frac{\partial \psi}{\partial t} \right)^2 \right] d\alpha \tag{9.65}$$

One can get the consistent boundary conditions through a variational analysis [1]. On each boundary, one must specify three conditions:

$$u = 0 \quad \text{or} \quad N = 0$$

$$w = 0 \quad \text{or} \quad Q = 0$$

$$\psi = 0 \quad \text{or} \quad M = 0 \tag{9.66}$$

It is worth mentioning that for moderately thick curved beams, the term M/R in the first of the above equations drops. This is hard to find without the variational derivation. For elastically constrained boundaries, with the constraints being represented as translational and rotational springs at the beam edges, the boundary conditions are generalized to

$$\begin{aligned} k_u u_0 + N &= 0 \\ k_w w_0 + Q &= 0 \\ k_\psi \psi - M &= 0 \end{aligned} \tag{9.67}$$

at the edge $\alpha = -a/2$. The signs of N, M, and Q in the above equations change at the edge $\alpha = a/2$.

Similar to thin curved beams, the simple support S1 boundary conditions for thick beams at $\alpha = -a/2$ and $\alpha = a/2$ are $w_0 = N_\alpha = \partial\psi/\partial\alpha = 0$. The solution of the equations of motion with these boundary condition follow the same pattern for straight beams (9.41) by replacing x in (9.41) with α. Substituting these solutions into the equations of motion written in terms of displacement for moderately thick curved beams (9.60) yields

$$\begin{bmatrix} C_{11} & C_{12} & C_{13} \\ C_{21} & C_{22} & C_{23} \\ C_{31} & C_{32} & C_{33} \end{bmatrix} \begin{bmatrix} A_m \\ C_m \\ B_m \end{bmatrix} + \omega^2 \begin{bmatrix} \bar{I}_1 & 0 & \bar{I}_2 \\ 0 & -\bar{I}_1 & 0 \\ \bar{I}_2 & 0 & \bar{I}_3 \end{bmatrix} \begin{bmatrix} A_m \\ C_m \\ B_m \end{bmatrix} = \begin{bmatrix} 0 \\ 0 \\ 0 \end{bmatrix} \tag{9.68}$$

where

$$\begin{aligned} C_{11} &= -\alpha_m{}^2 A_{11} - \frac{A_{55}}{R^2} \\ C_{22} &= \frac{A_{11}}{R^2} + A_{55}\alpha_m{}^2 \\ C_{33} &= -D_{11}\alpha_m{}^2 - A_{55} \\ C_{21} &= -C_{12} = \alpha_m \left[\frac{A_{11}}{R} + \frac{A_{55}}{R} \right] \\ C_{31} &= C_{13} = -B_{11}\alpha_m{}^2 + \frac{A_{55}}{R} \end{aligned} \tag{9.69}$$

and

$$C_{23} = -C_{32} = \alpha_m \left[\frac{B_{11}}{R} - A_{55} \right]$$

Again, (9.68) is actually valid for problems of forced vibrations. The static problem results when the frequency is set to zero, and the free vibration problem arises when the pressure terms are zeros. Similar to thin beams, if one assumes that $u_{0m} = \cos(\alpha_m \alpha)$, $w_{0m} = \sin(\alpha_m \alpha)$, and $\psi_m = \cos(\alpha_m \alpha)$, then the problem of laminated curved beam having vertical hinge supports (i.e., $u_0 = \psi = Q = 0$) is solved. This has little, if any, practical significance.

The analysis of closed rings can be performed by assuming $u_0 = \sin(m\theta)$, $w_0 = \cos(m\theta)$ in the proposed exact solutions. Similar treatment can be made for moderately thick rings by further assuming $\psi_m = \sin(m\theta)$, which will yield a third-order determinant.

Table 9.9 shows a comparison between the results obtained using classical and shear deformation curved beam equations. The same frequency parameter is used. The thickness ratio h/R is varied from 0.01 to 0.2, which is taken as the limit of the moderately thick beam equations (to be proven when and if these equations are compared with those obtained by 3D theory of elasticity). The results show criteria for establishing the limit of the thin beam theory. The effect of shear deformation and rotary inertia increases for materials with higher orthotropy ratios. While the maximum difference (in the

	CBT				Shear deformation beam theory			
	m				*m*			
h/R	**1**	**2**	**3**	**4**	**1**	**2**	**3**	**4**
E_1/E_2 = 15								
0.01	4.0116	18.031	41.432	74.190	4.0094	18.000	41.286	73.738
0.02	3.9960	17.949	41.227	73.786	3.9885	17.839	40.681	72.095
0.05	3.9471	17.690	40.526	72.278	3.9109	17.089	37.667	64.041
0.10	3.8656	17.212	39.083	68.854	3.7419	15.329	31.300	49.452
0.20	3.7015	16.118	35.403	59.676	3.3312	11.808	21.481	31.295
E_1/E_2 = 40								
0.01	3.4145	15.349	35.270	63.157	3.4108	15.298	35.034	62.438
0.02	3.4014	15.274	35.081	62.778	3.3871	15.093	35.220	60.178
0.05	3.3559	15.034	34.424	61.355	3.2932	14.111	30.283	50.017
0.10	3.2809	14.591	33.077	58.157	3.0814	11.964	23.180	35.118
0.20	3.1315	13.592	29.743	49.981	2.5935	8.3979	14.446	20.425

Table 9.9 Exact Frequency Parameters $\Omega = \omega a^2 \sqrt{12\rho / E_{11} h^2}$ for simply supported [0°, 90°] laminated curved thin and moderately thick beams, $a/R = 1.0$, $G_{12}/E_2 = G_{13}/E_3 = 0.5$

fourth frequency) between the frequencies obtained by thin beam equations and moderately thick beam equations is 5.2 percent for a thickness ratio of 0.05 and single layer beams, this difference is 12.9 percent for [0°, 90°] cross-ply beams with an orthotropy ratio of 15 and it is 22.7 percent for an orthotropy ratio of 40. This shows clearly that even for what has been considered thin beams where CBT is applicable for isotropic beams, a SDBT is needed when composites are treated. As a result, we will pay close attention to shear deformation and rotary inertia in the upcoming sections and include such treatment. The difference between both theories becomes large, even for the fundamental frequencies, when the thickness ratio exceed 0.05, which is taken as the limit of the thin beam theory for fundamental frequencies. The difference between CBT and SDBT increases with higher thickness ratio and higher modes.

9.7 Laminated Thin Plates

An anisotropic material is one for which the material properties are different in different directions. For an anisotropic plate the 2D stress–strain equations of (6.10) are generalized, in matrix form, to

$$\begin{bmatrix} \sigma_x \\ \sigma_y \\ \tau_{xy} \end{bmatrix} = \begin{bmatrix} c_{11} & c_{12} & c_{13} \\ c_{12} & c_{22} & c_{23} \\ c_{13} & c_{23} & c_{33} \end{bmatrix} \begin{bmatrix} \epsilon_x \\ \epsilon_y \\ \gamma_{xy} \end{bmatrix} \tag{9.70}$$

thus, having six independent elastic coefficients (c_{ij}), instead of two (E and ν). This generalizes the $D\nabla^4 w$ term in (6.16) and (6.24) to the form

$$D_1 \frac{\partial^4 w}{\partial x^4} + D_2 \frac{\partial^4 w}{\partial x^3 \partial y} + D_3 \frac{\partial^4 w}{\partial x^2 \partial y^2} + D_4 \frac{\partial^4 w}{\partial x \partial y^3} + D_5 \frac{\partial^4 w}{\partial y^4} \tag{9.71}$$

Exact solutions of this generalization of (6.24) are not possible. However, the Ritz method may be applied straightforwardly.

An orthotropic material is a special case for which the elastic coefficients have principal directions, yielding $c_{13} = c_{23} = 0$ in (9.70), thereby eliminating the coupling between in-plane normal and shear stresses and strains. The resulting orthotropic plate occurs often in practical applications (e.g., an ordinary, isotropic plate having many parallel stiffening beam elements attached to it). The resulting differential equation of motion has $D_2 = D_4 = 0$ in (9.71). With the odd orders of derivatives vanishing, exact solutions of the type found in Sec. 6.2 are straightforwardly achievable for rectangular plates having two opposite edges simply supported.

If the stacking sequence of the layers is symmetric about the plate midplane, the resulting equations are those of anisotropic (or orthotropic) plate theory, as described above. However, if the stacking sequence produces a laminate which is unsymmetric about the plate midplane, then there is coupling between bending and midplane stretching during the vibratory motion [1,10]. The governing theory reflects this coupling, resulting in an eighth-order system of differential equations of motion, instead of the fourth-order equation one has for a symmetric laminate. As was found earlier for beams, the bending-stretching coupling may also reduce the natural frequencies of the plate considerably [11].

For *laminated* thin plates, where shear deformation and strains in the vertical direction are neglected, inplane strains include membrane components, and curvature changes which are the same as those (6.13). Strains at any point are

$$\begin{aligned} \epsilon_x &= \epsilon_{0x} + z\kappa_x \\ \epsilon_y &= \epsilon_{0y} + z\kappa_y \\ \gamma_{xy} &= \gamma_{0xy} + 2z\kappa_{xy} \end{aligned} \tag{9.72}$$

where the subscript (0) refers to the middle surface. Midsurface strains are

$$\epsilon_{0x} = \frac{\partial u_0}{\partial x}, \quad \epsilon_{0y} = \frac{\partial v_0}{\partial y}, \quad \gamma_{0\alpha\beta} = \frac{\partial v_0}{\partial x} + \frac{\partial u_0}{\partial y} \tag{9.73}$$

The midsurface curvature and twist changes are

$$\kappa_x = -\frac{\partial^2 w_0}{\partial x^2}, \quad \kappa_y = -\frac{\partial^2 w_0}{\partial y^2}, \quad \kappa_{xy} = -\frac{\partial^2 w_0}{\partial x \partial y} \tag{9.74}$$

where u_0 and v_0 are displacement components tangential to the middle surface in the x and y directions respectively; and w_0 is the normal displacement. For thin plates, the stress–strain equations for an element of material in the *kth lamina* may be written in the conventional form used for laminated plates [1] as in (9.11).

Although the layers are orthotropic in their material coordinates, (9.11) are the same as those of material with general anisotropy (9.70). This is a result of the lamination at an angle other than 0° or 90° with respect to the global (x,y) coordinates. In other words, the $\bar{Q}_{16}$ and $\bar{Q}_{26}$ coefficients result from a coordinate transformation and arise only if the plate coordinates (x,y) are neither parallel nor perpendicular to the fibers. Correspondingly, if the plate coordinates are parallel or perpendicular to the fibers, then the terms $\bar{Q}_{16}$ and $\bar{Q}_{26}$ are zero.

The stresses over the plate thickness (h) are integrated to obtain the force and moment resultants. Doing so results in equal inplane shear force resultants (i.e., $N_{xy} = N_{yx}$), and twisting moments (i.e., $M_{xy} = M_{yx}$).

Substituting (9.73) and (9.11) into Eqs. (6.11) and integrating stresses over the cross-section for force resultants, yields (9.12) and (9.13).

The equations of motion for laminated plates will include both inplane and out-of-plane forces. This can be achieved by combining the derivations made in Chap. 5 for membranes and those made in Chap. 6 for plates. The equations of motion for membranes (5.5), with inertia terms and inplane loading added, and those derived earlier for plates (6.15) may be written as:

$$\frac{\partial N_x}{\partial x}+\frac{\partial N_{xy}}{\partial y}+p_x=\bar{\rho}\frac{\partial^2 u}{\partial t^2}$$

$$\frac{\partial N_y}{\partial y}+\frac{\partial N_{xy}}{\partial x}+p_y=\bar{\rho}\frac{\partial^2 v}{\partial t^2}$$

$$\frac{\partial^2 M_x}{\partial x^2}+2\frac{\partial^2 M_{xy}}{\partial y\partial x}+\frac{\partial^2 M_y}{\partial y^2}+p_n=\bar{\rho}\frac{\partial^2 w}{\partial t^2} \tag{9.75}$$

where $\bar{\rho}$ is the average mass density of the plate per unit area of the midsurface, as given by

$$\bar{\rho}=\sum_{k=1}^{N}\rho^{(k)}\left(z_k-z_{k-1}\right) \tag{9.76}$$

$\rho^{(k)}$ is the density of the lamina (k) per unit volume. Substituting Eqs. (9.73) and (9.74) into (9.12) and then into (9.75), and multiplying the last equation by −1, yields

$$\begin{bmatrix} L_{11} & L_{12} & L_{13} \\ L_{21} & L_{22} & L_{23} \\ L_{31} & L_{32} & L_{33} \end{bmatrix}\begin{bmatrix} u_0 \\ v_0 \\ w_0 \end{bmatrix}+\begin{bmatrix} p_x \\ p_y \\ -p_n \end{bmatrix}=\begin{bmatrix} \bar{\rho} & 0 & 0 \\ 0 & \bar{\rho} & 0 \\ 0 & 0 & -\bar{\rho} \end{bmatrix}\frac{\partial^2}{\partial t^2}\begin{bmatrix} u_0 \\ v_0 \\ w_0 \end{bmatrix} \tag{9.77}$$

The L_{ij} coefficients of the above equations are

$$L_{11}=A_{11}\frac{\partial^2}{\partial x^2}+2A_{16}\frac{\partial^2}{\partial x\partial y}+A_{66}\frac{\partial^2}{\partial y^2}$$

$$L_{12}=L_{21}=A_{16}\frac{\partial^2}{\partial x^2}+\left\{A_{12}+A_{66}\right\}\frac{\partial^2}{\partial x\partial y}+A_{26}\frac{\partial^2}{\partial y^2}$$

$$L_{22}=A_{66}\frac{\partial^2}{\partial x^2}+2A_{26}\frac{\partial^2}{\partial x\partial y}+A_{22}\frac{\partial^2}{\partial y^2}$$

$$L_{13} = L_{31} = -B_{11}\frac{\partial^3}{\partial x^3} - B_{26}\frac{\partial^3}{\partial y^3} - 3B_{16}\frac{\partial^3}{\partial x^2 \partial y} - (B_{12} + 2B_{66})\frac{\partial^3}{\partial x \partial y^2}$$

$$L_{23} = L_{32} = -B_{16}\frac{\partial^3}{\partial x^3} - B_{22}\frac{\partial^3}{\partial y^3} - 3B_{26}\frac{\partial^3}{\partial x \partial y^2} - (B_{12} + 2B_{66})\frac{\partial^3}{\partial x^2 \partial y}$$

$$L_{33} = D_{11}\frac{\partial^4}{\partial x^4} + 4D_{16}\frac{\partial^4}{\partial x^3 \partial y} + 2(D_{12} + 2D_{66})\frac{\partial^4}{\partial x^2 \partial y^2} + 4D_{26}\frac{\partial^4}{\partial x \partial y^3} + D_{22}\frac{\partial^4}{\partial y^4} \tag{9.78}$$

Note that the L_{13} and L_{23} terms are functions of the stiffness parameters B_{ij}. These parameters have zero values if the plate is symmetrically laminated. This leads to decoupling the inplane vibration from the out-of-plane vibration. The out-of-plane vibration for a symmetrically laminated plate will then degenerate to $L_{33}\, w_0 + \bar{\rho}\, \ddot{w}_0 = 0$. In addition, if the plate is made of cross-ply lamination, the terms D_{16} and D_{26} become zero, which will reduce the equation to that of an orthotropic plate.

Strain energy functionals that are consistent with the above equations of motion can be written as [1]:

$$\begin{aligned} PE &= \frac{1}{2}\int_A \left\{ N_x \epsilon_{0x} + N_y \epsilon_{0y} + N_{xy}\gamma_{0xy} + M_x \kappa_x + M_y \kappa_y + M_{xy}\tau \right\} dA \\ &= \frac{1}{2}\int_A \Big\{ \left[A_{11}\left(\epsilon_{0x}\right)^2 + 2B_{11}\epsilon_{0x}\kappa_x + D_{11}\kappa_x^2 \right] \\ &\quad + \left[A_{22}\left(\epsilon_{0y}\right)^2 + 2B_{22}\epsilon_{0y}\kappa_y + D_{22}\kappa_y^2 \right] \\ &\quad + \left[A_{66}\left(\gamma_{0xy}\right)^2 + 2B_{66}\gamma_{0xy}\tau + D_{66}\tau^2 \right] \\ &\quad + 2\left[A_{12}\epsilon_{0x}\epsilon_{0y} + B_{12}\left(\epsilon_{0x}\kappa_y + \epsilon_{0y}\kappa_x\right) + D_{12}\kappa_x\kappa_y \right] \\ &\quad + 2\left[A_{16}\epsilon_{0x}\gamma_{0xy} + B_{16}\left(\epsilon_{0x}\tau + \gamma_{0xy}\kappa_x\right) + D_{16}\kappa_x\tau \right] \\ &\quad + 2\left[A_{26}\epsilon_{0y}\gamma_{0xy} + B_{26}\left(\epsilon_{0y}\tau + \gamma_{0xy}\kappa_y\right) + D_{26}\kappa_y\tau \right] \Big\} dA \end{aligned} \tag{9.79}$$

Substituting the strain– and curvature–displacement equations into the above equation yields the strain energy in terms of the displacements. This may be expressed as the sum of three parts:

$$PE = PE_s + PE_b + PE_{bs} \tag{9.80}$$

where PE_s is the part due to stretching alone, PE_b is the part due to bending alone, and PE_{bs} is the part due to bending–stretching coupling,

The PE_s is the part due to stretching alone:

$$U_s = \frac{1}{2}\int_A \left\{A_{11}\left(\frac{\partial u_0}{\partial x}\right)^2 + A_{22}\left(\frac{\partial v_0}{\partial y}\right)^2 + A_{66}\left(\frac{\partial v_0}{\partial x}+\frac{\partial u_0}{\partial y}\right)^2 + 2A_{12}\left(\frac{\partial u_0}{\partial x}\right)\left(\frac{\partial v_0}{\partial y}\right) + 2A_{16}\left(\frac{\partial u_0}{\partial x}\right)\left(\frac{\partial v_0}{\partial x}+\frac{\partial u_0}{\partial y}+\frac{2w_0}{R_{xy}}\right) + 2A_{26}\left(\frac{\partial v_0}{\partial y}\right)\left(\frac{\partial v_0}{\partial x}+\frac{\partial u_0}{\partial y}\right)\right\}dA \tag{9.81a}$$

PE_b is the part due to bending alone,

$$U_b = \frac{1}{2}\int_A \left\{D_{11}\left(\frac{\partial^2 w_0}{\partial x^2}\right)^2 + D_{22}\left(\frac{\partial^2 w_0}{\partial y^2}\right)^2 + 4D_{66}\left(\frac{\partial^2 w_0}{\partial x \partial y}\right)^2 + 2D_{12}\left(\frac{\partial^2 w_0}{\partial x^2}\right)\left(\frac{\partial^2 w_0}{\partial y^2}\right) + 4D_{16}\left(\frac{\partial^2 w_0}{\partial x^2}\right)\left(\frac{\partial^2 w_0}{\partial x \partial y}\right) + 4D_{26}\left(\frac{\partial^2 w_0}{\partial y^2}\right)\left(\frac{\partial^2 w_0}{\partial x \partial y}\right)\right\}dA \tag{9.81b}$$

and PE_{bs} is the part due to bending-stretching coupling,

$$U_{bs} = -\int_A \left\{B_{11}\left(\frac{\partial u_0}{\partial x}\right)\left(\frac{\partial^2 w_0}{\partial x^2}\right) + B_{22}\left(\frac{\partial v_0}{\partial y}\right)\left(\frac{\partial^2 w_0}{\partial y^2}\right) + 2B_{66}\left(\frac{\partial v_0}{\partial x}+\frac{\partial u_0}{\partial y}\right)\left(\frac{\partial^2 w_0}{\partial x \partial y}\right) + B_{12}\left[\left(\frac{\partial u_0}{\partial x}\right)\left(\frac{\partial^2 w_0}{\partial y^2}\right) + \left(\frac{\partial v_0}{\partial y}\right)\left(\frac{\partial^2 w_0}{\partial x^2}\right)\right] + B_{16}\left[2\left(\frac{\partial u_0}{\partial x}\right)\left(\frac{\partial^2 w_0}{\partial x \partial y}\right) + \left(\frac{\partial v_0}{\partial x}+\frac{\partial u_0}{\partial y}\right)\left(\frac{\partial^2 w_0}{\partial x^2}\right)\right] + B_{26}\left[2\left(\frac{\partial v_0}{\partial y}\right)\left(\frac{\partial^2 w_0}{\partial x \partial y}\right) + \left(\frac{\partial v_0}{\partial x}+\frac{\partial u_0}{\partial y}\right)\left(\frac{\partial^2 w_0}{\partial y^2}\right)\right]\right\}dA \tag{9.81c}$$

For symmetrically laminated plates, $B_{ij} = 0$ and, hence, $PE_{bs} = 0$. This will result in the inplane deformation being decoupled from the out-of-plane one. If one is interested in out-of-plane (or transverse) vibration only, the term PE_b will be the only term needed. These functionals can be used for approximate energy methods like the Ritz and finite element methods. Using the distributed external force components p_x and p_y in the tangential directions and p_z in the normal direction, the work done by the external forces (W) can be found [1]. Therefore, the total potential energy of the plate in its deformed shape is $\Pi = PE - W$. The kinetic energy of the entire plate is then found to be (neglecting rotary inertia terms):

$$KE = \sum_{k=1}^{N} KE_k = \frac{\bar{\rho}}{2}\int_A \left[\left(\frac{\partial u_0}{\partial t}\right)^2 + \left(\frac{\partial v_0}{\partial t}\right)^2 + \left(\frac{\partial w_0}{\partial t}\right)^2 \right] dA \tag{9.82}$$

Consider a thin rectangular plates with dimensions a and b, and thickness h. Plates with circular orthotropy are beyond the scope of this book, but may be found in the literature [1]. Rectangular cross-ply plates have the lamination angle for each layer making either a 0° or 90° with each of the boundaries. In other words, the fibers in each layer will be either parallel or perpendicular to each boundary. These plates may be symmetrically or unsymmetrically (i.e., asymmetrically) laminated about the middle surface. They are of particular interest as they do permit exact solutions for some boundary conditions. Such plates have

$$A_{16} = A_{26} = B_{16} = B_{26} = D_{16} = D_{26} = 0 \tag{9.83}$$

Thin rectangular plates with *symmetric* lamination may have four possible boundary conditions at each edge (similar to isotropic plates). Three of these boundary conditions are classical. These are simply supported, clamped, or free. The total possible combination of classical boundary conditions for such plates when they are isotropic is 21 as described in Chap. 6. Symmetrically laminated plates, although they have three possible classical boundaries per edge, will yield even higher numbers of combinations of boundaries for all edges. This is because of the reduced symmetry in laminated plates as compared with isotropic ones. Plates that are not symmetrically laminated will have 16 possible boundary conditions for each edge, with 12 of these being classical. This is because of the stretching-bending coupling that has to be considered for such plates and the inclusion of boundaries for inplane displacements.

The free boundaries, at x = constant, are

F1 $\quad Q_x = M_x = u_0 = v_0 = 0$

F2 $\quad Q_x = M_x = N_x = v_0 = 0$

$$\text{F3} \quad Q_x = M_x = u_0 = N_{xy} = 0$$

$$\text{F4} \quad Q_x = M_x = N_x = N_{xy} = 0 \tag{9.84a}$$

The simple support boundaries are

$$\text{S1} \quad w = M_x = u_0 = v_0 = 0$$

$$\text{S2} \quad w = M_x = N_x = v_0 = 0$$

$$\text{S3} \quad w = M_x = u_0 = N_{xy} = 0$$

$$\text{S4} \quad w = M_x = N_x = N_{xy} = 0 \tag{9.84b}$$

The clamped boundaries are

$$\text{C1} \quad w = \psi_x = u_0 = v_0 = 0$$

$$\text{C2} \quad w = \psi_x = N_x = v_0 = 0$$

$$\text{C3} \quad w = \psi_x = u_0 = N_{xy} = 0$$

$$\text{C4} \quad w = \psi_x = N_x = N_{xy} = 0 \tag{9.84c}$$

Among all the boundary combinations, only those that have two opposite edges simply supported permit having exact solutions (for homogeneous plates). When the solution procedure is extended to laminated composite plates, it is found that such exact solution can only be applied to two types of lamination sequences and boundary conditions:

1. Cross-ply plates with S2 ($w_0 = M_x = N_x = v_0 = 0$; at $x = 0, a$; and $w_0 = M_y = u_0 = N_y = 0$ at $y = 0$ and b) opposite boundaries. These are plates where the lamination angle for each layer makes either a 0° or a 90° with each of the boundaries. In other words, the fibers in each layer can be either parallel or perpendicular to each boundary. These plates could be symmetrically laminated or unsymmetrically (i.e., asymmetrically) laminated about the middle surface. Figure 9.6 shows examples of such lamination sequence. Note that the lamination sequences made of layers having 45° and –45° angles (e.g., [45°, –45°]) or 30° and –60° or similar fiber angles for each lamina are *not* considered here as cross-ply because, although these plies make right angles with each other, they do not make right angle with the edges of the plate. Cross-ply plates can have exact solutions when the two simply supported opposite boundaries are of the S2 type (i.e., shear diaphragms).

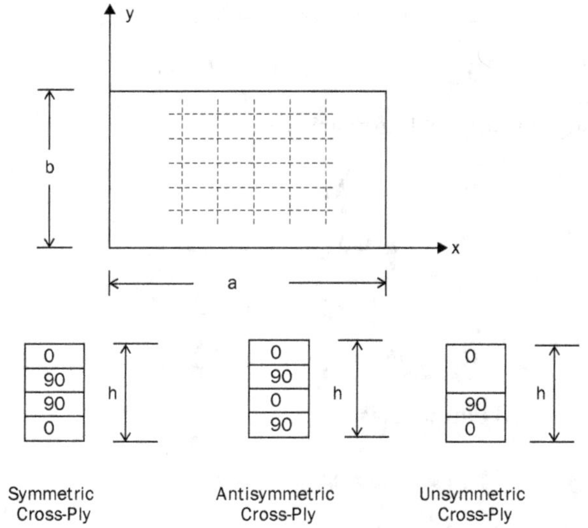

FIGURE 9.6 Plates with cross-ply lamination sequence.

2. Antisymmetrically laminated plates. These plates are constructed with laminates in which the fibers in each lamina above the middle surface making an angle θ with a coordinate is mirror-imaged with another one (of equal thickness) at exactly the same distance in the opposite z direction from the middle surface with fibers making negative the angle (i.e., $-\theta$) with the same coordinate. Figure 9.7 shows examples of such lamination sequence. Antisymmetrically laminated plates can have exact solutions when the two opposite boundaries are of the S3 type ($w_0 = M_x = u_0 = N_{xy} = 0$ at $x = 0$ and a, and $w_0 = M_y = u_0 = N_{xy} = 0$ at $y = 0$ and b).

Consider a plate with shear diaphragm (S2) boundaries on all four edges (described in more detail earlier in Chap. 7). The following solution satisfies the boundary conditions and the equations of motion:

$$u_o(x,y,t) = \sum_{m=0}^{M} \sum_{n=0}^{N} U_{mn} \cos(\alpha_m x)\sin(\beta_n y)\sin(\omega_{mn} t)$$

$$v_o(x,y,t) = \sum_{m=0}^{M} \sum_{n=0}^{N} V_{mn} \sin(\alpha_m x)\cos(\beta_n y)\sin(\omega_{mn} t)$$

$$w_o(x,y,t) = \sum_{m=0}^{M} \sum_{n=0}^{N} W_{mn} \sin(\alpha_m x)\sin(\beta_n y)\sin(\omega_{mn} t) \tag{9.85}$$

where $\alpha_m = \dfrac{m\pi}{a}$, $\beta_n = \dfrac{n\pi}{b}$, ω_{mn} is the natural frequency.

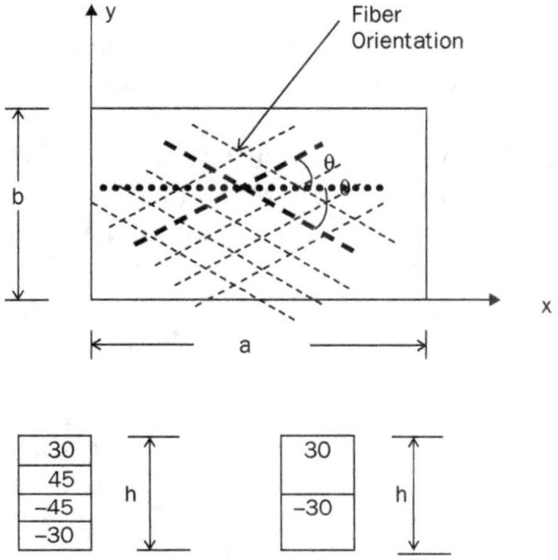

FIGURE 9.7 Plates with antisymmetric lamination sequence.

Also, the external forces can be expanded in a Fourier series [1]. Substituting (9.85) into (9.77), yields the following

$$\begin{bmatrix} C_{11} & C_{12} & C_{13} \\ C_{21} & C_{22} & C_{23} \\ C_{31} & C_{32} & C_{33} \end{bmatrix} \begin{bmatrix} U_{mn} \\ V_{mn} \\ W_{mn} \end{bmatrix} + \omega_{mn}^2 \begin{bmatrix} I_1 & 0 & 0 \\ 0 & I_1 & 0 \\ 0 & 0 & I_1 \end{bmatrix} \begin{bmatrix} U_{mn} \\ V_{mn} \\ W_{mn} \end{bmatrix} = \begin{bmatrix} -p_{\alpha mn} \\ -p_{\beta mn} \\ -p_{zmn} \end{bmatrix} \tag{9.86}$$

where

$$\begin{aligned}
C_{11} &= -A_{11}\alpha_m^2 - A_{66}\beta_n^2 \\
C_{12} &= C_{21} = -(A_{12} + A_{66})\alpha_m\beta_n \\
C_{22} &= -A_{66}\alpha_m^2 - A_{22}\beta_n^2 \\
C_{13} &= C_{31} = B_{11}\alpha_m^3 + (B_{12} + 2B_{66})\alpha_m\beta_n^2 \\
C_{23} &= C_{32} = B_{22}\beta_n^3 + (B_{12} + 2B_{66})\alpha_m^2\beta_n \\
C_{33} &= -\left[D_{11}\alpha_m^4 + 2(D_{12} + 2D_{66})\alpha_m^2\beta_n^2 + D_{22}\beta_n^4\right]
\end{aligned}$$

and where the p_{mn} are the Fourier coefficients of the external forces. The above equations are reduced to static problems if the natural frequency is set to zero and to free vibrations if the external forces are set to zero.

The plate with antisymmetric angle-ply laminate ($A_{16} = A_{26} = B_{11} = B_{12} = B_{22} = B_{66} = D_{16} = D_{26} = 0$) having another type of simply supported boundary conditions (S3) on all four edges is known to have an exact solution as well [1].

Boundary conditions that are different from those considered above *or* lamination sequences different than those considered (cross-ply and antisymmetric angle-ply) have no exact solutions. Approximate methods must be used in order to obtain results. The widely used finite element methods have been used by many engineers and researchers in the field to solve practical problems and find the dynamic behavior of composite plates. When parametric studies and/or optimization methods are used to maximize frequencies, finite elements becomes an expensive tool. Instead, for simple geometries, like the flat plates considered here, the less computationally demanding Ritz analysis can be employed.

The Ritz method with algebraic polynomial displacement functions is used here to solve the free vibration problem for laminated composite plates having various boundary conditions. Natural frequencies and mode shapes for plates having two adjacent free edges and the remaining edges simply supported, clamped, or free are presented. Convergence studies are made which demonstrate that accurate results (natural frequencies and mode shapes) can be obtained with this analysis. The effects of various parameters (material, fiber orientation, and boundary conditions) on the natural frequencies and mode shapes are studied.

For free vibrations of a plate having the rectangular planform, displacements are assumed as:

$$u(x,y,t) = U(x,y) \sin \omega t$$

$$v(x,y,t) = V(x,y) \sin \omega t$$

$$w(x,y,t) = W(x,y) \sin \omega t \tag{9.87}$$

Algebraic functions may be used as trial functions. The displacement trial functions, in terms of the nondimensional coordinates ξ and η, are taken as:

$$U(\xi,\eta) = \sum_{i=i_0}^{I} \sum_{j=j_0}^{J} \alpha_{ij} \xi^i \eta^j$$

$$V(\xi,\eta) = \sum_{k=k_0}^{K} \sum_{\ell=\ell_0}^{L} \beta_{k\ell} \xi^k \eta^\ell$$

$$W(\xi,\eta) = \sum_{m=m_0}^{M} \sum_{n=n_0}^{N} \gamma_{mn} \xi^m \eta^n \tag{9.88}$$

where $\xi = x/a$ and $\eta = y/b$.

The Ritz method requires satisfaction of geometric (forced) boundary conditions only. One can solve for many boundary conditions with the same analytical procedure by using a suitable selection of the value i_0, j_0, k_0, l_0, m_0, and n_0. Vibration problems for laminated plates having the boundary conditions XXFF, where X can be simply supported (S), clamped (C), or free (F) can be solved. One should keep in mind that for generally laminated plates there are four types of configurations for each of the simply supported, free, and clamped edge conditions.

For solving the free vibration problem, the displacement functions are substituted into the energy functional Eqs. (9.79) in order to get an expression for the maximum strain energy (PE_{max}) and into (9.81) in order to get an expression for the maximum kinetic energy (KE_{max}). The Ritz method requires minimization of the functional (KE_{max} − PE_{max}) with respect to the coefficient α_{ij}, $\beta_{k\ell}$, and γ_{mn}, which can be accomplished by setting:

$$\frac{\partial(KE_{\max}-PE_{\max})}{\partial\alpha_{ij}}=0,\ i=i_0,i_0+1,\ldots,I\ j=j_0,j_0+1,\ldots,J$$

$$\frac{\partial(KE_{\max}-PE_{\max})}{\partial\beta_{k\ell}}=0,\ k=k_0,k_0+1,\ldots,K\ \ell=\ell_0,\ell_0+1,\ldots,L$$

$$\frac{\partial(KE_{\max}-PE_{\max})}{\partial\gamma_{mn}}=0,\ m=m_0,m_0+1,\ldots,M\ n=n_0,n_0+1,\ldots,N \quad (9.89)$$

which yields a total of $(I - i_0 + 1) \times (J - j_0 + 1) + (K - k_0 + 1) \times (L - \ell_0 + 1) + (M - m_0 + 1) \times (N - n_0 + 1)$ simultaneous, linear, homogenous equations in an equal number of unknowns α_{ij}, β_{kl}, and γ_{mn}. The determinant of the coefficient matrix is set equal to zero which will yield a set of eigenvalues. Substituting each eigenvalue back into (9.89) yields the corresponding eigenvector. The mode shape corresponding to each frequency can be determined by substituting the eigenvector back into the displacement functions.

For symmetrically laminated plates all of the stretching–bending coupling terms vanish (i.e., $B_{ij} = 0$). This leads to decoupling the inplane displacements from the out-of-plane displacement. The possible combination of classical boundary conditions at $x = 0$ and at $y = 0$, which can be solved by the present method, reduces to 6 as was mentioned earlier. When the strain energy functionals are used to simulate boundary conditions at $x = a$ and at $y = b$, then all possible boundary conditions can be treated. Only the last equations of the sets of Eqs. (9.85) and (9.86) are needed for the transverse vibrations of symmetrically laminated plates. This leads to a system of linear, homogenous equations of the order $(M - m_0 + 1) \times (N - n_0 + 1)$. Comparisons among results from the present Ritz analysis and other analytical and experimental ones are described in Ref. [1].

Convergence studies are made for composite plates representative of those to be analyzed subsequently. These include symmetric laminates of three layers with stacking sequence $[\theta/-\theta/\theta]$. The angle θ lies between the fibers and the projection of the x-axis on the plate. Filamentary composite materials of graphite/epoxy (G/E) are considered,

$$\text{Graphite/epoxy }(G/E)\text{: } E_1 = 138.\text{ GPa}, E_2 = 8.96\text{ GPa}, G_{12} = 7.1\text{ GPa}, \nu_{12} = 0.30 \quad (9.90)$$

A typical plate of square planform ($a/b = 1$) is used. Convergence studies of the lowest eight frequency parameters $\Omega = \omega a^2\sqrt{\rho/E_1h^2}$ for graphite/epoxy plates having the six boundary conditions which will subsequently be analyzed can be found in Table 9.10. For each of the six boundary conditions, three solutions are presented. These solutions are obtained by using 36, 49, and 64 terms for the first,

B.C.	Det. Size	Ω					
		1	2	3	4	5	6
FFFF	36	1.6262	2.0910	3.7748	5.1285	5.1924	7.4462
	49	1.6203	2.0789	3.7177	5.0570	5.1460	7.2263
	64	1.6202	2.0784	3.7115	5.0517	5.0707	7.0800
SFFF	36	0.9171	2.5554	3.2899	4.5599	5.8347	7.7369
	49	0.9166	2.5371	3.2786	4.5284	5.7501	7.6644
	64	0.9165	2.5363	3.2754	4.5181	5.6929	7.5070
CFFF	36	0.6519	1.4392	3.1581	4.1996	5.6674	6.7157
	49	0.6513	1.4377	3.1253	4.1881	5.6427	6.6344
	64	0.6507	1.4372	3.1228	4.1839	5.6111	6.5257
SSFF	36	0.4671	1.8437	3.9371	4.6707	6.8341	8.2414
	49	0.4656	1.8248	3.9276	4.6559	6.7644	8.1051
	64	0.4644	1.8424	3.9263	4.6541	6.7514	8.0780
CSFF	36	1.0615	2.4227	5.0125	5.6484	7.9491	9.3699
	49	1.0606	2.4211	4.9812	5.6388	7.9011	9.1566
	64	1.0600	2.4205	4.9791	5.6243	7.8767	9.0771
CCFF	36	1.2913	3.0535	5.5559	6.2780	8.5898	10.367
	49	1.2912	3.0518	5.5534	6.2730	8.5715	10.241
	64	1.2907	3.0495	5.5463	6.2691	8.5507	10.236

TABLE 9.10 Convergence of the Frequency Parameter $\Omega = \omega a^2\sqrt{\rho/E_1h^2}$ for a Graphite/Epoxy Square ($a/b = 1$) Plate with [30°, −30°, 30°] Lamination Using Algebraic Polynomial Functions

second, and third solution, respectively. Equal number of terms are taken in each of the inplane directions. Zero frequencies corresponding to rigid body modes are not reported.

Convergence is observed to be reasonably good for engineering applications. The maximum difference between the 49- and the 64-term solutions is less than 3 percent for all the cases. The completely free boundary condition delivered the slowest convergence. As geometric constraints are imposed on the boundaries, convergence seems to improve, and the fastest convergence is observed for the plate with two adjacent clamped edges (i.e., CCFF plates). The maximum difference in the natural frequencies between the 49- and 64-term solutions for CCFF boundary conditions is 0.14 percent for G/E materials. From these studies, it may be considered that the 64-term solution for plates presents satisfactory convergence for many engineering applications. Therefore, a 64-term solution will be used in the subsequent analyses.

Natural frequencies and contour plots of the mode shapes for different boundary conditions are given by Qatu [12] and reported here for two boundary conditions in Figs. 9.8 and 9.9. The aspect ratio (a/b) is chosen to be one indicating square plates. Three-layer [θ, $-\theta$, θ] laminates are used. The lamination angle is varied from 0° to 90° with an increment of 15°. For FFFF, SSFF, and CCFF square plates, geometric symmetry about the line $\xi = \eta$ exists. This results in the frequencies for plates with the fiber angles θ = 60°, 75°, and 90° being the same as those with the fiber angles θ = 30°, 15°, and 0°, respectively.

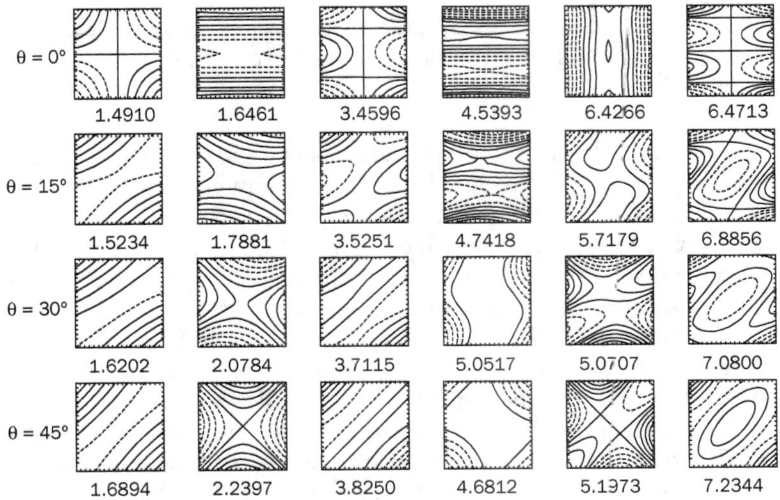

FIGURE 9.8 Mode shapes and frequency parameters $\omega a^2\sqrt{\rho/E_1h^2}$ for completely free (FFFF), G/E [θ,$-\theta$,θ] square plates.

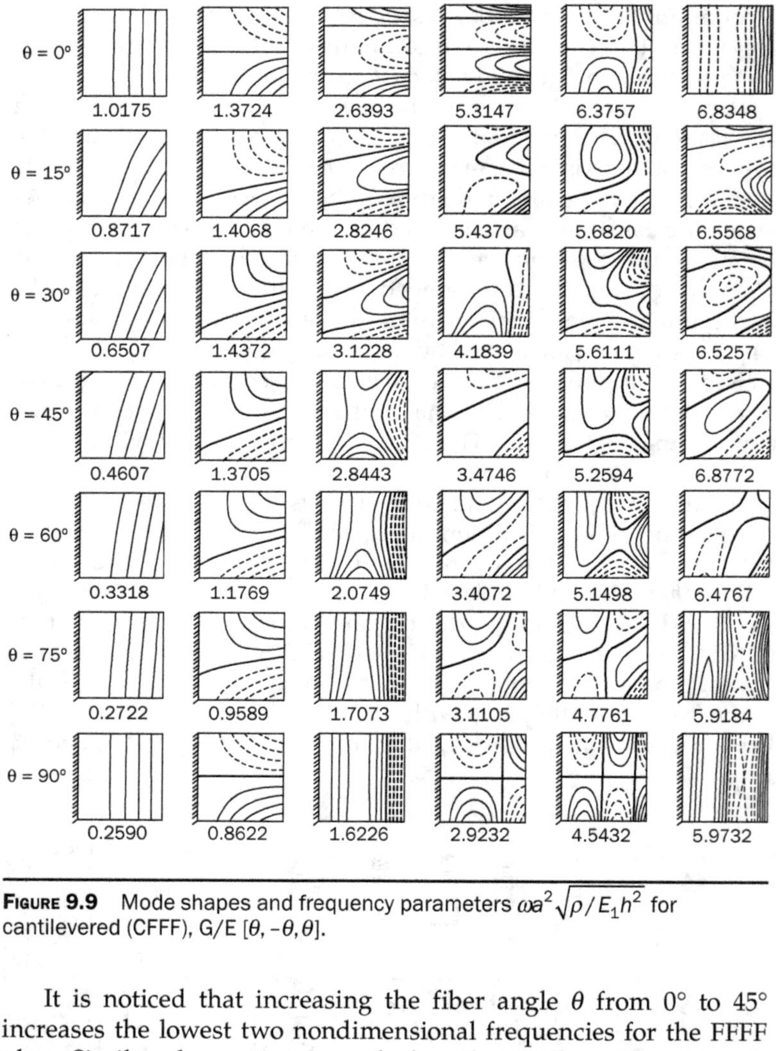

Figure 9.9 Mode shapes and frequency parameters $\omega a^2\sqrt{\rho/E_1h^2}$ for cantilevered (CFFF), G/E $[\theta, -\theta, \theta]$.

It is noticed that increasing the fiber angle θ from 0° to 45° increases the lowest two nondimensional frequencies for the FFFF plate. Similar observation is made for other plates with geometric symmetry about the $\xi = \eta$ line (i.e., SSFF and CCFF plates) [12]. From the symmetry of the problems, increasing the fiber angle from 45° to 90° decreases these frequencies. This shows that the maximum fundamental frequencies are obtained with a fiber angle of 45° for these boundary conditions. For that angle, the fibers are parallel (or perpendicular) to the line of geometric diagonal symmetry. Increasing the fiber angle from 0° to 45° increases the second lowest frequency by 37 percent for G/E material when the boundary conditions are completely free. Similar observations are true for the other two boundary conditions (i.e., SSFF and CCFF) as reported in Ref. [1].

For plates when the line of geometric symmetry is $\eta = 0.5$ (e.g., CFFF plates, Fig. 9.9) the behavior is different. For cantilever plates, it is observed that increasing the fiber orientation angle from 0° to 90° decreases the fundamental frequency, which corresponds to the first bending mode. Increasing the fiber angle from 0° to 90° decreases the lowest natural frequency by 75 percent. For the second lowest frequency, which corresponds to the first twisting mode, the maximum natural frequency is obtained when the fiber angle is 30°.

For the completely free boundary condition, one should note that four possible symmetry classes exist in the displacement functions chosen. For example, one can chose the coordinates at the center and then chose $m = n = 0, 2, 4, \ldots$ for the doubly symmetric (i.e., symmetric about both the ξ and η axes) modes. Similar choices can be made for the other three symmetry classes. These symmetry classes about the ξ and η axes (once chosen at the center) exist for the isotropic, orthotropic, and cross-ply plates only. This symmetry is lost for plates with angle-ply lamination, and one should keep all the terms in the polynomials. For the special case of diagonally orthotropic angle-ply laminates which are made of 45° angle layers, the symmetry classes exist about the diagonals (Fig. 9.8). The gradual change in contour lines with increasing θ is evident.

For plates with SFFF and CFFF (i.e., cantilever) boundaries, only two classes of symmetry are possible in the displacement functions (i.e., the displacement functions can be either symmetric or antisymmetric about the ξ-axis, once located to pass through the center). Figure 9.9 gives the first six mode shapes for the cantilever case. For example, to obtain the symmetric modes one could choose $n = 0, 2, 4, \ldots$. This symmetry/antisymmetry in the mode shapes can be seen for isotropic and cross-ply laminated plates. For angle-ply plates, the symmetry classes are lost, and one should keep all the terms in the analysis. One should keep in mind that one zero frequency exists for SFFF plates which corresponds to the rigid body mode (rotation about the η axis) [12].

More numerical results on laminated plates will be discussed in the section of laminated thin shallow shells. It is placed in that section to draw comparisons between such plates when they are flat and when they have curvature (i.e., become a shallow shell).

Plates with other geometries are discussed in the literature [1]. Natural frequencies for cantilevered laminated composite right triangular and trapezoidal plates and completely free triangular and trapezoidal plates were presented in detail [13,14].

9.8 Thick Plates

As discussed earlier in Sec. 6.8, thick plates are ones with a thickness smaller by approximately one order of magnitude when compared with other plate parameters, particularly, its vibration mode shape wavelength (thickness is smaller than 1/10th of the smallest of the wavelengths). Thick plate theories (also referred to as shear deformation plate theories, or SDPT) require the inclusion of shear deformation and rotary inertia factors when compared with thin or classical plate theory (CPT). This section will present the set of equations that can be used for thick composite plates. The treatment that follows is an extension of Mindlin's theory to laminated plates.

In thick plate theories, the midplane plate displacements are expanded in terms of its thickness. Such an expansion can be of a first or a higher order. In the case of first-order expansion, the theories are referred to as first-order shear deformation theories. Thick plate theories still use the assumption of negligible stretching in the z-direction (i.e., $\epsilon_z = 0$). This assumption is generally valid except within the vicinity of a highly concentrated force. The displacements are assumed as

$$
\begin{aligned}
u(x,y,z) &= u_0(x,y) + z\psi_x(x,y) \\
v(x,y,z) &= v_0(x,y) + z\psi_y(x,y) \\
w(x,y,z) &= w_0(x,y)
\end{aligned}
\tag{9.91}
$$

where u_0, v_0, and w_0 are midsurface displacements of the plate and ψ_α and ψ_β are midsurface rotations. The above equations are the basis of a typical first-order SDPT and will constitute the only assumption made in this development when compared with the 3D theory of elasticity. The strains at any point in the plate can be found using:

$$
\begin{aligned}
\epsilon_x &= (\epsilon_{0x} + z\kappa_x) \\
\epsilon_y &= (\epsilon_{0y} + z\kappa_y) \\
\epsilon_{xy} &= (\epsilon_{0xy} + z\kappa_{xy}) \\
\epsilon_{yx} &= (\epsilon_{0yx} + z\kappa_{yx}) \\
\text{or} \quad \gamma_{xy} &= \gamma_{0xy} + z\tau; \quad \gamma_{xy} = \epsilon_{xy} + \epsilon_{yx}; \\
\text{and} \quad \tau &= \kappa_{xy} + \kappa_{yx}
\end{aligned}
\tag{9.92}
$$

and for vertical shear strains

$$
\begin{aligned}
\gamma_{xz} &= \gamma_{0xz} \\
\gamma_{yz} &= \gamma_{0yz}
\end{aligned}
\tag{9.93}
$$

Note that when the shear strains ϵ_{xy} and ϵ_{yx} are equal, the engineering shear strain γ_{xy} ($= \epsilon_{xy} + \epsilon_{yx}$) will be used. The midsurface strains are:

$$
\epsilon_{0x} = \frac{\partial u_0}{\partial x}, \qquad \epsilon_{0y} = \frac{\partial v_0}{\partial y}
$$

$$
\gamma_{0xy} = \frac{\partial v_0}{\partial x} + \frac{\partial u_0}{\partial y}
$$

$$\gamma_{0xz} = \frac{\partial w_0}{\partial x} + \psi_x, \qquad \gamma_{0yz} = \frac{\partial w_0}{\partial y} + \psi_y \tag{9.94}$$

and the curvature and twist changes are:

$$\kappa_x = \frac{\partial \psi_x}{\partial x}, \quad \kappa_y = \frac{\partial \psi_y}{\partial y}, \quad \tau = \frac{\partial \psi_y}{\partial x} + \frac{\partial \psi_x}{\partial y} \tag{9.95}$$

The subscript (0) will refer to the middle surface in the subsequent equations.

The force and moment resultants are obtained by integrating the stresses over the plate thickness. The normal and shear force resultants are found by integrating the stresses over the thickness. The bending and twisting moment resultants are found by integrating the stress couples over the thickness. Carrying out the integration over the thickness, from layer to layer, yields (9.12) and

$$\begin{bmatrix} Q_x \\ Q_y \end{bmatrix} = \begin{bmatrix} A_{55} & A_{45} \\ A_{45} & A_{44} \end{bmatrix} \begin{bmatrix} \gamma_{0xz} \\ \gamma_{0yz} \end{bmatrix} \tag{9.96a}$$

where

$$\left. A_{ij} = \sum_{k=1}^{N} K_i K_j \bar{Q}_{ij}^{(k)} (h_k - h_{k-1}) \right\} \qquad i, j = 4, 5 \tag{9.96b}$$

and K_i and K_j are shear correction factors to compensate for the nonconstant shear distribution over the cross-section.

The equations for motion can be derived for plates using Newton's second law and a differential element of the plate or by using Hamilton's principle from energy expressions [1]. The equations of motion are:

$$\begin{aligned}
\frac{\partial N_x}{\partial x} + \frac{\partial N_{yx}}{\partial y} + q_x &= \left(I_1 \ddot{u}_0 + I_2 \ddot{\psi}_x\right) \\
\frac{\partial N_y}{\partial y} + \frac{\partial N_{xy}}{\partial x} + q_y &= \left(I_1 \ddot{v}_0 + I_2 \ddot{\psi}_y\right) \\
\frac{\partial Q_x}{\partial x} + \frac{\partial Q_y}{\partial y} + q_n &= \left(I_1 \ddot{w}_0\right) \\
\frac{\partial M_x}{\partial x} + \frac{\partial M_{yx}}{\partial y} - Q_x + m_x &= \left(I_2 \ddot{u}_0 + I_3 \ddot{\psi}_x\right) \\
\frac{\partial M_y}{\partial y} + \frac{\partial M_{xy}}{\partial x} - Q_y + m_y &= \left(I_2 \ddot{v}_0 + I_3 \ddot{\psi}_y\right)
\end{aligned} \tag{9.97}$$

where the two dots represent the second derivative of these terms with respect to time, m_x and m_y are body couples, and where the inertia terms are defined as

$$I_i = \sum_{k=1}^{N} \int_{h_{k-1}}^{h_k} \rho^{(k)} z^{i-1} dz \tag{9.98}$$

The boundary terms for the boundaries with x = constant are

$$\begin{aligned} N_{0x} - N_x &= 0 & \text{or} \quad u_0 &= 0 \\ N_{0xy} - N_{xy} &= 0 & \text{or} \quad v_0 &= 0 \\ Q_{0x} - Q_x &= 0 & \text{or} \quad w_0 &= 0 \\ M_{0x} - M_x &= 0 & \text{or} \quad \psi_x &= 0 \\ M_{0xy} - M_{xy} &= 0 & \text{or} \quad \psi_y &= 0 \end{aligned} \tag{9.99}$$

Possible combination of boundary conditions at each edge is given in Table 9.11.

Boundary type	Conditions
Free Boundaries	
F1	$Q_x = M_x = u = v = \psi_y = 0$
F2	$Q_x = M_x = N_x = v = \psi_y = 0$
F3	$Q_x = M_x = u = N_{xy} = \psi_y = 0$
F4	$Q_x = M_x = N_x = N_{xy} = \psi_y = 0$
F5	$Q_x = M_x = u = v = M_{xy} = 0$
F6	$Q_x = M_x = N_x = v = M_{xy} = 0$
F7	$Q_x = M_x = u = N_{xy} = M_{xy} = 0$
F8	$Q_x = M_x = N_x = N_{xy} = M_{xy} = 0$
Simple Support Boundaries	
S1	$w = M_x = u = v = \psi_y = 0$
S2	$w = M_x = N_x = v = \psi_y = 0$
S3	$w = M_x = u = N_{xy} = \psi_y = 0$
S4	$w = M_x = N_x = N_{xy} = \psi_y = 0$
S5	$w = M_x = u = v = M_{xy} = 0$
S6	$w = M_x = N_x = v = M_{xy} = 0$
S7	$w = M_x = u = N_{xy} = M_{xy} = 0$
S8	$w = M_x = N_x = N_{xy} = M_{xy} = 0$

TABLE 9.11 Combinations of Boundary Conditions for Thick Shells at Each Edge x = Constant [1]

Boundary type	Conditions
Clamped Boundaries	
C1	$w = \psi_x = u = v = \psi_y = 0$
C2	$w = \psi_x = N_x = v = \psi_y = 0$
C3	$w = \psi_x = u = N_{xy} = \psi_y = 0$
C4	$w = \psi_x = N_x = N_{xy} = \psi_y = 0$
C5	$w = \psi_x = u = v = M_{xy} = 0$
C6	$w = \psi_x = N_x = v = M_{xy} = 0$
C7	$w = \psi_x = u = N_{xy} = M_{xy} = 0$
C8	$w = \psi_x = N_x = N_{xy} = M_{xy} = 0$

Table 9.11 Combinations of Boundary Conditions for Thick Shells at Each Edge x = Constant [1] *(Continued)*

The equilibrium equations can be written in terms of displacements. These equations are proven useful when exact solutions are desired. The equations can be written as:

$$L_{ij}u_i + M_{ij}\ddot{u}_i = q \tag{9.100}$$

The stiffness parameters L_{ij} in (9.100) are

$$L_{11} = A_{11}\frac{\partial^2}{\partial x^2} + 2A_{16}\frac{\partial^2}{\partial x \partial y} + A_{66}\frac{\partial^2}{\partial y^2}$$

$$L_{12} = A_{16}\frac{\partial^2}{\partial x^2} + (A_{12} + A_{66})\frac{\partial^2}{\partial x \partial y} + A_{26}\frac{\partial^2}{\partial y^2}$$

$$L_{13} = 0$$

$$L_{14} = B_{11}\frac{\partial^2}{\partial x^2} + 2B_{16}\frac{\partial^2}{\partial x \partial y} + B_{66}\frac{\partial^2}{\partial y^2}$$

$$L_{15} = B_{16}\frac{\partial^2}{\partial x^2} + (B_{12} + B_{66})\frac{\partial^2}{\partial x \partial y} + B_{26}\frac{\partial^2}{\partial y^2}$$

$$L_{22} = A_{66}\frac{\partial^2}{\partial x^2} + 2A_{26}\frac{\partial^2}{\partial x \partial y} + A_{22}\frac{\partial^2}{\partial y^2}$$

$$L_{23} = 0$$

$$L_{24} = B_{16}\frac{\partial^2}{\partial x^2} + (B_{12} + B_{66})\frac{\partial^2}{\partial x \partial y} + B_{26}\frac{\partial^2}{\partial y^2}$$

$$L_{25} = B_{66}\frac{\partial^2}{\partial x^2} + 2B_{26}\frac{\partial^2}{\partial x \partial y} + B_{22}\frac{\partial^2}{\partial y^2}$$

$$L_{33} = -A_{55}\frac{\partial^2}{\partial x^2} - 2A_{45}\frac{\partial^2}{\partial x \partial y} - A_{44}\frac{\partial^2}{\partial y^2}$$

$$L_{34} = -A_{55}\frac{\partial}{\partial x} - A_{45}\frac{\partial}{\partial y}$$

$$L_{35} = -A_{45}\frac{\partial}{\partial x} - A_{44}\frac{\partial}{\partial y}$$

$$L_{44} = -A_{55} + D_{11}\frac{\partial^2}{\partial x^2} + 2D_{16}\frac{\partial^2}{\partial x \partial y} + D_{66}\frac{\partial^2}{\partial y^2}$$

$$L_{45} = -A_{45} + D_{16}\frac{\partial^2}{\partial x^2} + (D_{12} + D_{66})\frac{\partial^2}{\partial x \partial y} + D_{26}\frac{\partial^2}{\partial y^2}$$

$$L_{55} = -A_{44} + D_{66}\frac{\partial^2}{\partial x^2} + 2D_{26}\frac{\partial^2}{\partial x \partial y} + D_{22}\frac{\partial^2}{\partial y^2} \tag{9.101}$$

The mass parameters in Eq. (9.100) are

$$\begin{aligned} M_{ji} &= M_{ij} \\ M_{11} &= M_{22} = -M_{33} = -I_1 \\ M_{14} &= M_{25} = -I_2 \\ M_{44} &= M_{55} = -I_3 \\ \text{all other } M_{ij} &= 0 \end{aligned} \tag{9.102}$$

The displacement and loading vectors are

$$u_i = \left[u_o, v_o, w_o, \psi_x, \psi_y\right]^T, \quad \text{and}$$
$$q_i = \left[-q_x, -q_y, +q_n, -m_x, -m_y\right]^T \tag{9.103}$$

The above equations describe a thick plate with general lamination sequence. If the plate is symmetrically laminated, then all the B_{ij} terms vanish. This will make $L_{14} = L_{15} = L_{24} = L_{25} = 0$. The inplane displacements (u_0 and v_0) will then be de-coupled from the out-of-plane displacement and shear deformation. This will reduce the equations and subsequent treatment significantly. Equations (9.100) will include a 3 × 3 differential operator, for a symmetrically laminated thick plate, instead of the above 5 × 5. The order of the whole system of differential equations reduces from 10 to 6.

Energy functional can be derived easily from the above equations in a fashion similar to that done for thin plates.

Plates that are not symmetrically laminated will have 16 possible boundary conditions for each edge, with 12 of these being classical. This is because of the stretching–bending coupling that has to be considered for such plates. Symmetrically laminated thick plates can have six possible classical boundary conditions at each edge. Unsymmetrically laminated plates can have up to 24 classical possible boundary conditions at each edge (Table 9.11). The combinations of boundary conditions are higher for thick rectangular plates when compared with thin plates.

Of particular importance and ease of use is the problem with all edges being simply supported.

Consider a plate that is made of a cross-ply laminate. The differential operators L_{ij} in the equations of motion (9.100) and (9.101) become

$$
\begin{aligned}
L_{11} &= A_{11}\frac{\partial^2}{\partial x^2} + A_{66}\frac{\partial^2}{\partial y^2} \\
L_{12} &= (A_{12} + A_{66})\frac{\partial^2}{\partial x \partial y} \\
L_{13} &= 0 \\
L_{14} &= B_{11}\frac{\partial^2}{\partial x^2} + B_{66}\frac{\partial^2}{\partial y^2} \\
L_{15} &= (B_{12} + B_{66})\frac{\partial^2}{\partial x \partial y} \\
L_{22} &= A_{66}\frac{\partial^2}{\partial x^2} + A_{22}\frac{\partial^2}{\partial y^2} \\
L_{23} &= 0 \\
L_{24} &= (B_{12} + B_{66})\frac{\partial^2}{\partial x \partial y} \\
L_{25} &= B_{66}\frac{\partial^2}{\partial x^2} + B_{22}\frac{\partial^2}{\partial y^2} \\
L_{33} &= -A_{55}\frac{\partial^2}{\partial x^2} - A_{44}\frac{\partial^2}{\partial y^2} \\
L_{34} &= -A_{55}\frac{\partial}{\partial x} \\
L_{35} &= -A_{44}\frac{\partial}{\partial y} \\
L_{44} &= -A_{55} + D_{11}\frac{\partial^2}{\partial x^2} + D_{66}\frac{\partial^2}{\partial y^2} \\
L_{45} &= (D_{12} + D_{66})\frac{\partial^2}{\partial x \partial y} \\
L_{55} &= -A_{44} + D_{66}\frac{\partial^2}{\partial x^2} + D_{22}\frac{\partial^2}{\partial y^2}
\end{aligned}
\tag{9.104}
$$

Consider a plate, with shear diaphragm (S2) boundaries on all four edges. That is, the following boundary conditions apply:

$$N_x = w_0 = v_0 = M_x = \psi_y = 0 \quad \text{for the edges} \quad x = 0, a$$

$$N_y = w_0 = u_0 = M_y = \psi_x = 0 \quad \text{for the edges} \quad y = 0, b \tag{9.105}$$

The following displacement and slope functions satisfy both the equations of motion and boundary conditions

$$u_0(x,y,t) = \sum_{m=1}^{\infty}\sum_{n=1}^{\infty} U_{mn}\cos(\alpha_m x)\sin(\beta_n y)\sin(\omega_{mn}t)$$

$$v_0(x,y,t) = \sum_{m=1}^{\infty}\sum_{n=1}^{\infty} V_{mn}\sin(\alpha_m x)\cos(\beta_n y)\sin(\omega_{mn}t)$$

$$w_0(x,y,t) = \sum_{m=1}^{\infty}\sum_{n=1}^{\infty} W_{mn}\sin(\alpha_m x)\sin(\beta_n y)\sin(\omega_{mn}t)$$

$$\psi_x(x,y,t) = \sum_{m=1}^{\infty}\sum_{n=1}^{\infty} \psi_{xmn}\cos(\alpha_m x)\sin(\beta_n y)\sin(\omega_{mn}t)$$

$$\psi_y(x,y,t) = \sum_{m=1}^{\infty}\sum_{n=1}^{\infty} \psi_{ymn}\sin(\alpha_m x)\cos(\beta_n \beta y)\sin(\omega_{mn}t) \tag{9.106}$$

where $\alpha_m = m\pi/a$, $\beta_n = n\pi/b$, and U_{mn}, V_{mn}, W_{mn}, $\psi_{\alpha mn}$, and $\psi_{\beta mn}$ are arbitrary coefficients.

Substituting the above equations into equations of motion and using a Fourier expansion for the loading functions yields $[K]\{\Delta\} + (\omega_{mn})^2\,[M]\,\{\Delta\} = -\{F\}$, where $[K]$, and $[M]$ are the stiffness and mass symmetric 5 × 5 matrices, respectively; ω_{mn} is the frequency with m number of half sine waves in the x direction and n number of half sine waves in the y direction; $\{F\}$ is the forcing function, which only has sinusoidal terms; and $\{\Delta\} = \{U_{mn}, V_{mn}, W_{mn}, \psi_{xmn}, \psi_{ymn}\}^T$. The K_{ij} coefficients are

$$K_{11} = -A_{11}\alpha_m^2 - A_{66}\beta_n^2, \qquad K_{12} = -(A_{12} + A_{66})\alpha_m\beta_n, \qquad K_{13} = 0$$
$$K_{14} = -B_{11}\alpha_m^2 - B_{66}\beta_n^2, \qquad K_{15} = -(B_{12} + B_{66})\alpha_m\beta_n$$

$$K_{22} = -A_{66}\alpha_m^2 - A_{22}\beta_n^2, \qquad K_{23} = 0, \qquad K_{24} = -(B_{12} + B_{66})\alpha_m\beta_n$$
$$K_{25} = -B_{66}\alpha_m^2 - B_{22}\beta_n^2, \qquad K_{33} = -A_{55}\alpha_m^2 - A_{44}\beta_n^2$$

$$K_{34} = -A_{55}\alpha_m, \qquad K_{35} = -A_{44}\beta_n, \qquad K_{44} = -A_{55} - D_{11}\alpha_m^2 - D_{66}\beta_n^2$$
$$K_{45} = -(D_{12} + D_{66})\alpha_m\beta_n, \qquad K_{55} = -A_{44} - D_{66}\alpha_m^2 - D_{22}\beta_n^2$$

$$\tag{9.107}$$

If the forcing function $\{F\}$ is assumed to be zero, the above equations can be used directly to find the natural frequencies of free vibration. Unless stated otherwise, the following natural frequency parameter is used in the subsequent analysis:

$$\Omega = \omega a^2 \sqrt{\rho / E_2 h^2} \tag{9.108}$$

Table 9.12 shows the frequency parameter for cross-ply square plates with relatively high orthotropy ratio ($E_1/E_2 = 15$) using the classical plate theory (CPT) as well as a shear deformation plate theory (SDPT) presented earlier. Two lamination sequence values are considered [0°, 90°], and [0°, 90°, 90°, 0°]. Four thickness ratios (a/h) of 100, 20, 10, and 5 are used. The differences between the predictions of SDPT and those of CPT are less than 1 percent for very thin plates with a thickness ratio (a/h) of 100. The difference reaches 3.2 percent for plates with thickness ratio of 20, 11.8 percent for plates with thickness ratio of 10, and 40 percent for plates with thickness ratios of 5 (where the shear deformation plate theory is not proven to apply and must be compared with results obtained by the theory of elasticity). It does seem reasonable to assume that classical plate theory (CPT) applies only for plates having a thickness ratio of 20 or higher. This theory overpredicts the natural frequency as expected. It is also important to note that the nondimensional frequency parameter chosen here does not change with thickness for the predictions made using classical plate theory (CPT) for symmetrically laminated cross-ply plates. It does change slightly for asymmetrically laminated plates. Interestingly, the accuracy of CPT in predicting frequencies is higher for asymmetrically laminated plates than it is for symmetrically laminated ones.

Table 9.13 shows the frequency parameter for cross-ply square plates with relatively high orthotropy ratio ($E_1/E_2 = 25$) using shear deformation plate theory (SDPT) and classical plate theory (CPT).

Lamination	**[0°, 90°]**		**[0°, 90°, 90°, 0°]**	
Theory	**SDPT**	**CPT**	**SDPT**	**CPT**
$a/h = 100$	8.56394	8.56847	12.26147	12.27733
$a/h = 20$	8.44807	8.55811	11.90100	12.27733
$a/h = 10$	8.11956	8.52569	10.97163	12.27733
$a/h = 5$	7.14661	8.39526	8.77840	12.27733

TABLE 9.12 Frequency Parameters $\Omega = \omega a^2 \sqrt{\rho / E_2 h}$ for Shear Diaphragm Supported, Rectangular, Cross-Ply Plates ($a/b = 1$, $E_1/E_2 = 15$, $G_{12}/E_2 = 0.5$, $G_{13}/E_2 = 0.5$, $G_{23}/E_2 = 0.5$, $\nu_{12} = 0.25$, $K^2 = 5/6$) for Shear Deformation Plate Theory (SDPT) and Classical Plate Theory (CPT)

Lamination Theory	[0°, 90°]		[0°, 90°, 0°]		[0°, 90°, 90°, 0°]	
	SDPT	CPT	SDPT	CPT	SDPT	CPT
a/h = 100	9.6873	9.6960	15.183	15.228	15.184	15.228
a/h = 10	8.9001	9.6436	12.163	15.228	12.226	15.228

TABLE 9.13 Frequency Parameter $\Omega = \omega a^2 \sqrt{\rho / E_2 h^2}$ for Cross-Ply Plates ($a/b = 1$, $E_1/E_2 = 25$, $G_{12}/E_2 = 0.5$, $G_{13}/E_2 = 0.5$, $G_{23}/E_2 = 0.2$, $\nu_{12} = 0.25$, $k^2 = 5/6$) for Shear Deformation Plate Theory (SDPT) and Classical Plate Theory (CPT)

Three lamination sequence values are considered [0°, 90°], [0°, 90°, 0°], and [0°, 90°, 90°, 0°]. Two thickness ratios (a/h) of 100 and 10 are used. The results show that for the thin plate with $a/h = 100$, the difference between the two theories is minimal. Thus, the classical plate theory is certainly accurate. For a thick plate with $a/h = 10$, the difference between both theories reaches more than 20 percent, showing the lack of accuracy that the classical plate theory (CPT) has in predicting vibrations of thick plates.

9.9 Laminated Shallow Shells

For very thin shells, the assumptions which are used in the previously described thin shell theory (Chap. 7) will be used here. Mainly, it is assumed that the shell is thin enough such that the ratio of the thickness compared to any of the shell's radii of curvature or its width or length is negligible compared to unity, the normals to the middle surface remain straight and normal to the midsurface when the shell undergoes deformation, and no normal strain exists in the direction normal to the middle surface.

The first of the above assumptions assures that certain parameters in the shell equations (including the z/R term) as described for curved beams earlier can be neglected. The second of the above assumptions allows for neglecting shear deformation and rotary inertia. It also allows for making the inplane displacement to be linearly varying through the shell's thickness. Consider thin laminated shallow shells with the projections of the fibers in each layer to the planform being straight. It simplifies the equations and the subsequent treatment of the problem considerably to represent the equations of such shells using the rectangular coordinates (i.e., planform coordinates). The inplane displacement can then be described in the same manner as plates. This is true even if the boundaries of such shells are not rectangular in shape.

Consider now a shallow shell (as discussed in Chap. 7) with the radii of curvature R_x, R_y, and R_{xy} being constant (Fig. 7.3). Midsurface strains can then be written as in (7.5); noting that the displacement being midsurface displacement. It should be noted that a term that appears in the denominator of strains of shells [e.g., $\epsilon_x = (\epsilon_{0x} + z\kappa_x)/(1+z/R_x)$] is negligible for shallow shells, thus yielding strain expressions similar to those for plates [1].

Curvature changes are described in (9.74). Employing shallow shell assumptions, the first two equations of motion for shallow shells are the same as those for plates (9.75). The third equation of motion is:

$$-\left(\frac{2N_{xy}}{R_{xy}} + \frac{N_x}{R_x} + \frac{N_y}{R_y}\right) + \frac{\partial^2 M_x}{\partial x^2} + 2\frac{\partial^2 M_{xy}}{\partial y \partial x} + \frac{\partial^2 M_y}{\partial y^2} + p_n = \bar{\rho}\frac{\partial^2 w_o}{\partial t^2} \qquad (9.109)$$

where $\bar{\rho}$ is the average mass density (9.76). Note that the above equation is similar to that of (7.2).

The midsurface stresses are determined using equations similar to 6.11 for plates, except that for shells we have an expression in the denominator [e.g., $N_\alpha = \int_{-h/2}^{h/2} \sigma_\alpha(1+z/R_\beta)dz$] that becomes negligible for shallow shells [1] yielding the same expressions used for plates (9.12) and (9.13).

Further manipulation of the equations of motion, they can be written in terms of displacements as in (9.77), where the L_{ij} coefficients for laminated thin shallow shells are

$$L_{11} = A_{11}\frac{\partial^2}{\partial x^2} + 2A_{16}\frac{\partial^2}{\partial x \partial y} + A_{66}\frac{\partial^2}{\partial y^2}$$

$$L_{12} = L_{21} = A_{16}\frac{\partial^2}{\partial x^2} + (A_{12} + A_{66})\frac{\partial^2}{\partial x \partial y} + A_{26}\frac{\partial^2}{\partial y^2}$$

$$L_{22} = A_{66}\frac{\partial^2}{\partial x^2} + 2A_{26}\frac{\partial^2}{\partial x \partial y} + A_{22}\frac{\partial^2}{\partial y^2}$$

$$L_{13} = L_{31} = -\left\{B_{11}\frac{\partial^3}{\partial x^3} + B_{26}\frac{\partial^3}{\partial y^3} + 3B_{16}\frac{\partial^3}{\partial x^2 \partial y} + \{B_{12} + 2B_{66}\}\frac{\partial^3}{\partial x \partial y^2}\right\}$$
$$+\left\{\frac{A_{11}}{R_x} + \frac{A_{12}}{R_y} + 2\frac{A_{16}}{R_{xy}}\right\}\frac{\partial}{\partial x} + \left\{\frac{A_{16}}{R_x} + \frac{A_{26}}{R_y} + 2\frac{A_{66}}{R_{xy}}\right\}\frac{\partial}{\partial y}$$

$$L_{23} = L_{32} = -\left\{B_{16}\frac{\partial^3}{\partial x^3} + B_{22}\frac{\partial^3}{\partial y^3} + 3B_{26}\frac{\partial^3}{\partial x \partial y^2} + \{B_{12} + 2B_{66}\}\frac{\partial^3}{\partial x^2 \partial y}\right\}$$
$$+\left\{\frac{A_{16}}{R_x} + \frac{A_{26}}{R_y} + 2\frac{A_{66}}{R_{xy}}\right\}\frac{\partial}{\partial x} + \left\{\frac{A_{12}}{R_x} + \frac{A_{22}}{R_y} + \frac{A_{26}}{R_{xy}}\right\}\frac{\partial}{\partial y}$$

$$L_{33} = \left\{ D_{11}\frac{\partial^4}{\partial x^4} + 4D_{16}\frac{\partial^4}{\partial x^3 \partial y} + 2(D_{12}+2D_{66})\frac{\partial^4}{\partial x^2 \partial y^2} \right.$$
$$\left. + 4D_{26}\frac{\partial^4}{\partial x \partial y^3} + D_{22}\frac{\partial^4}{\partial y^4} \right\}$$
$$-2\left\{ \left(\frac{B_{11}}{R_x} + \frac{B_{12}}{R_y} + 2\frac{B_{16}}{R_{xy}} \right)\frac{\partial^2}{\partial x^2} + 2\left(\frac{B_{16}}{R_x} + \frac{A_{26}}{R_y} + 2\frac{B_{66}}{R_{xy}} \right)\frac{\partial^2}{\partial x \partial y} \right.$$
$$\left. + \left(\frac{B_{12}}{R_x} + \frac{B_{22}}{R_y} + 2\frac{B_{26}}{R_{xy}} \right)\frac{\partial^2}{\partial y^2} \right\}$$
$$+ \left\{ \frac{A_{11}}{R_x^2} + 2\frac{A_{12}}{R_x R_y} + \frac{A_{22}}{R_y^2} \right.$$
$$\left. + 4\left(\frac{A_{16}}{R_{xy} R_x} + \frac{A_{26}}{R_{xy} R_x} + \frac{A_{66}}{R_{xy}^2} \right) \right\} \tag{9.110}$$

For symmetrically laminated shells, all terms containing B_{ij} vanish.

The strain energy functional for a shallow shell can be written as in Eqs. (9.79). Substituting the strain– and curvature–displacement equations into (9.79) yields the strain energy in terms of the displacements. This may be expressed as the sum of three parts as described for plates in (9.80).

The PE_s is the part due to stretching alone,

$$PE_s = \frac{1}{2}\int_A \left\{ A_{11}\left(\frac{\partial u_0}{\partial x} + \frac{w_0}{R_x} \right)^2 + A_{22}\left(\frac{\partial v_0}{\partial y} + \frac{w_0}{R_y} \right)^2 \right.$$
$$+ A_{66}\left(\frac{\partial v_0}{\partial x} + \frac{\partial u_0}{\partial y} + \frac{2w_0}{R_{xy}} \right)^2$$
$$+ 2A_{12}\left(\frac{\partial u_0}{\partial x} + \frac{w_0}{R_x} \right)\left(\frac{\partial v_0}{\partial y} + \frac{w_0}{R_y} \right)$$
$$+ 2A_{16}\left(\frac{\partial u_0}{\partial x} + \frac{w_0}{R_\alpha} \right)\left(\frac{\partial v_0}{\partial x} + \frac{\partial u_0}{\partial y} + \frac{2w_0}{R_{xy}} \right)$$
$$\left. + 2A_{26}\left(\frac{\partial v_0}{\partial y} + \frac{w_0}{R_y} \right)\left(\frac{\partial v_0}{\partial x} + \frac{\partial u_0}{\partial y} + \frac{2w_0}{R_{xy}} \right) \right\} dA \tag{9.111}$$

PE_b is the part due to bending alone,

$$PE_b = \frac{1}{2}\int_A \left\{ D_{11}\left(\frac{\partial^2 w_0}{\partial x^2}\right)^2 + D_{22}\left(\frac{\partial^2 w_0}{\partial y^2}\right)^2 + 4D_{66}\left(\frac{\partial^2 w_0}{\partial x \partial y}\right)^2 + 2D_{12}\left(\frac{\partial^2 w_0}{\partial x^2}\right)\left(\frac{\partial^2 w_0}{\partial y^2}\right) + 4D_{16}\left(\frac{\partial^2 w_0}{\partial x^2}\right)\left(\frac{\partial^2 w_0}{\partial x \partial y}\right) + 4D_{26}\left(\frac{\partial^2 w_0}{\partial y^2}\right)\left(\frac{\partial^2 w_0}{\partial x \partial y}\right)\right\} dA \tag{9.112}$$

and PE_{bs} is the part due to bending–stretching coupling,

$$\begin{aligned} PE_{bs} = -\int_A \Bigg\{ & B_{11}\left(\frac{\partial u_0}{\partial x} + \frac{w_0}{R_y}\right)\left(\frac{\partial^2 w_0}{\partial x^2}\right) + B_{22}\left(\frac{\partial v_0}{\partial y} + \frac{w_0}{R_y}\right)\left(\frac{\partial^2 w_0}{\partial y^2}\right) \\ & + 2B_{66}\left(\frac{\partial v_0}{\partial x} + \frac{\partial u_0}{\partial y} + \frac{2w_0}{R_{xy}}\right)\left(\frac{\partial^2 w_0}{\partial x \partial y}\right) \\ & + B_{12}\left[\left(\frac{\partial u_0}{\partial x} + \frac{w_0}{R_x}\right)\left(\frac{\partial^2 w_0}{\partial y^2}\right) + \left(\frac{\partial v_0}{\partial y} + \frac{w_0}{R_y}\right)\left(\frac{\partial^2 w_0}{\partial x^2}\right)\right] \\ & + B_{16}\left[2\left(\frac{\partial u_0}{\partial x} + \frac{w_0}{R_x}\right)\left(\frac{\partial^2 w_0}{\partial x \partial y}\right) + \left(\frac{\partial v_0}{\partial x} + \frac{\partial u_0}{\partial y} + \frac{2w_0}{R_{xy}}\right)\left(\frac{\partial^2 w_0}{\partial x^2}\right)\right] \\ & + B_{26}\left[2\left(\frac{\partial v_0}{\partial y} + \frac{w_0}{R_y}\right)\left(\frac{\partial^2 w_0}{\partial x \partial y}\right) + \left(\frac{\partial v_0}{\partial x} + \frac{\partial u_0}{\partial y} + \frac{2w_0}{R_{xy}}\right)\left(\frac{\partial^2 w_0}{\partial y^2}\right)\right]\Bigg\} dA \end{aligned} \tag{9.113}$$

Note again that for symmetrically laminated shells, $B_{ij} = 0$, and hence, $U_{bs} = 0$.

Consider a shallow shell that is made of a cross-ply laminate, thus $A_{16} = A_{26} = B_{16} = B_{26} = D_{16} = D_{26} = 0$. Assuming the radius of twist to be infinity, (i.e., $R\alpha_\beta = \infty$), Eqs. (9.77) or ($L_{ij}u_i + M_{ij}\ddot{u}_i = q$) for plates can be used for shallow shells with L_{ij} parameters defined in (9.110).

Consider an open shell with shear diaphragm (S2) boundaries on all four edges. Equations (9.85) give a solution that satisfies the boundary conditions and the equations of motion exactly. Also, the external forces can be expanded in a Fourier series. Substitute the solution of Eqs. (9.85) in the reduced equations of motion [Eq. (9.77) after being specialized further for cross-ply laminates]; these equations of motion can then be written as (9.86). The stiffness coefficients for laminated thin shallow shells are

$$
\begin{aligned}
C_{11} &= -A_{11}\alpha_m^2 - A_{66}\beta_n^2 \\
C_{12} &= C_{21} = -(A_{12} + A_{66})\alpha_m\beta_n \\
C_{22} &= -A_{66}\alpha_m^2 - A_{22}\beta_n^2 \\
C_{13} &= C_{31} = B_{11}\alpha_m^3 + (B_{12} + 2B_{66})\alpha_m\beta_n^2 + \left\{\frac{A_{11}}{R_x} + \frac{A_{12}}{R_y}\right\}\alpha_m \\
C_{23} &= C_{32} = B_{22}\beta_n^3 + (B_{12} + 2B_{66})\alpha_m^2\beta_n + \left\{\frac{A_{12}}{R_x} + \frac{A_{22}}{R_y}\right\}\beta_n \\
C_{33} &= -\Big[D_{11}\alpha_m^4 + 2(D_{12} + 2D_{66})\alpha_m^2\beta_n^2 + D_{22}\beta_n^4 \\
&\quad + 2\left\{\left(\frac{B_{11}}{R_x} + \frac{B_{12}}{R_y}\right)\alpha_m^2 + \left(\frac{B_{12}}{R_x} + \frac{B_{22}}{R_y}\right)\beta_n^2\right\} \\
&\quad + \left\{\frac{A_{11}}{R_x^2} + 2\frac{A_{12}}{R_xR_y} + \frac{A_{22}}{R_y^2}\right\}\Big]
\end{aligned}
\tag{9.114}
$$

Again, free vibrations equation results when the loads are set to zero. Results for cross-ply laminated shallow shells will be described in the next section when thick shallow shells are treated.

The Ritz method with algebraic polynomials can be used for laminated thin shallow shells in a manner similar to that used for plates as described in (9.88) and (9.89) with the energy functionals (9.111)–(9.113).

Convergence studies are made for a series of composite plates and shallow shells representative of those to be analyzed subsequently. These include four-layer symmetric laminates with stacking sequence $[\theta, -\theta, -\theta, \theta]$. The angle θ lies between the fibers and the projection of the x-axis on the shell. Filamentary composite materials of graphite/epoxy are considered. Other stacking sequence and materials are considered in Ref. [15]. Three types of shell curvatures were considered; these were spherical ($R_y/R_x = +1$), cylindrical ($R_y/R_x = 0$), and hyperbolic paraboloidal ($R_y/R_x = -1$). Figure 7.6 shows these types of curvatures. A moderate shallowness ($a/R = 0.5$) ratio is taken.

Table 9.14 shows a convergence study of the frequency parameter $\Omega = \omega a^2\sqrt{\rho/E_1h^2}$ for completely free composite plates and shallow shells. Other convergence studies can be found in Ref. [15] for different materials. It is observed that the material with the smaller orthotropy (i.e., E-glass/epoxy) yields slightly faster convergence

R_y/R_x	Determinant Size	$\Omega = \omega a^2 \sqrt{\rho / E_1 h^2}$									
		1	**2**	**3**	**4**	**5**	**6**	**7**	**8**	**9**	**10**
Plate	64	1.914	2.255	4.664	5.098	5.470	7.929	8.548	10.22	13.05	13.69
	100	1.914	2.255	4.664	5.097	5.468	7.923	8.518	10.12	12.97	13.03
	144	1.914	2.255	4.664	5.097	5.468	7.923	8.517	10.12	12.94	13.03
	108	2.292	3.333	6.020	6.886	8.756	11.90	13.87	17.32	19.29	24.60
+1	147	2.284	3.325	5.886	6.796	8.421	11.34	13.21	15.71	18.05	21.32
	192	2.283	3.323	5.871	6.781	8.394	11.11	12.81	15.32	17.12	20.70
	108	2.221	5.204	5.641	7.863	8.238	13.42	13.80	14.60	16.12	18.64
0	147	2.217	5.134	5.575	7.717	8.018	12.91	13.32	14.28	15.38	17.63
	192	2.216	5.124	5.568	7.682	7.991	12.55	12.98	14.00	14.93	17.19
	108	2.122	4.961	4.975	8.940	9.655	14.82	15.68	16.98	20.19	20.50
−1	147	2.117	4.886	4.898	8.853	9.202	13.85	15.24	16.81	18.98	19.83
	192	2.115	4.882	4.887	8.809	9.106	13.36	14.94	16.77	18.28	19.46

TABLE 9.14 Convergence of the Frequency Parameter for Graphite/Epoxy [30°, −30°, −30°, 30°] Plates and Shallow Shells on Square Planform, $a/h = 100$, $a/R_y = 0.5$

than does that with the larger orthotropy ratio (i.e., graphite/epoxy) on the average [15]. The maximum difference between the 64-term and the 144-term solutions is about 5 percent for graphite/epoxy. For the remaining frequencies, convergence occurred for three significant figures only. From these studies, it was decided to choose 64 terms (i.e., $M = N = 7$) for subsequent solutions for plates. The maximum difference between the 147-term solution and the 192-term solution for spherical shells is 5.3 percent (for the tenth frequency); and that for cylindrical shells is 2.4 percent, while it is only 1.8 percent for hyperbolic paraboloidal shells. Unlike plates, convergence occurred for the first three significant figures for most of the fundamental frequencies with 192 terms. For the rest of the frequencies, convergence occurred for the first two significant figures only. From these studies, it was decided to use 192-terms for laminated shallow shells (i.e., $I = J = K = L = M = N = 7$). This is also consistent with using $192/3 = 64$ terms for laminated plates (where only w is needed).

Table 9.15 gives the first eight nondimensional frequencies for four-layer $[\theta, -\theta, -\theta, \theta]$ angle-ply plates and shallow shells. The fiber angle varies between 0° and 45°. Because of the symmetry of the problem, frequencies for $\theta = 60°$, 75°, and 90° are the same as those for $\theta = 30°$, 15°, and 0°, respectively, when $a/b = 1$. It is found that the fundamental frequencies of shallow shells are virtually the same as these of the plate when $\theta = 0°$. This is due to the fact that the fundamental mode shape in such cases has node lines along the two symmetry axes, requiring no significant stretching of the middle surface. However, the higher frequencies are all significantly affected by curvature. Increasing θ causes the fundamental frequency to increase for all the shell geometries used here. The maximum fundamental frequency is obtained at $\theta = 45°$ and for higher frequencies it can exist at any angle θ between 0° and 45°.

Maximum fundamental frequencies are obtained for spherical shells and minimum frequencies are found for plates and/or hyperbolic paraboloidal shells. For spherical shells, changing the angle from 0° to 45° increases the fundamental frequency by almost 50 percent.

Studies are made for symmetrically laminated cross-ply plates and shallow shells of square planform. [0°], [0°, 90°, 0°], and [90°, 0°, 90°] laminates are considered with graphite/epoxy material in Ref. [1]. Two shallowness ratios (a/R) were used for each type of shell. Table 9.16 gives the fundamental nondimensional frequency parameters. Because of the symmetry of the problem, both laminates [0°, 90°, 0°] and [90°, 0°, 90°] give the same results, except for cylindrical shells. Increasing the curvature increases the fundamental frequencies for all shells except [90°, 0°, 90°] G/E circular cylindrical shells. For higher frequencies, increasing the curvature increases the frequencies.

Symmetrically laminated plates and shallow shells having $a/b = 1$ are also studied. A single-layer [45°] and an angle-ply laminate

R_y/R_x	θ (degrees)	$\Omega = \omega a^2 \sqrt{\rho / E_1 h^2}$							
		1	**2**	**3**	**4**	**5**	**6**	**7**	**8**
Plate	0	1.491	1.646	3.459	4.538	6.424	6.471	7.132	8.854
	15	1.653	1.843	3.904	4.773	5.902	7.217	7.237	9.185
	30	1.914	2.255	4.664	5.097	5.468	7.923	8.517	10.135
	45	2.233	2.275	4.655	5.043	5.934	9.183	9.423	10.204
	0	1.509	3.961	4.233	6.781	7.652	8.553	9.974	14.683
+1	15	1.836	3.829	4.834	6.768	7.715	9.774	10.992	15.638
	30	2.283	3.323	5.871	6.781	8.394	11.119	12.806	15.316
	45	2.296	3.093	6.459	6.712	9.001	11.172	13.484	14.841
	0	1.514	4.514	6.455	6.755	7.092	9.198	10.647	12.175
0	15	1.805	4.959	6.070	7.126	7.258	10.748	11.517	13.224
	30	2.216	5.124	5.568	7.682	7.991	12.551	12.982	13.997
	45	2.243	4.007	5.586	8.129	8.578	10.409	11.706	15.364
	0	1.507	4.235	4.541	7.056	8.956	9.062	11.758	13.358
−1	15	1.762	4.457	4.678	7.657	8.441	11.671	12.197	14.615
	30	2.115	4.882	4.887	8.809	9.106	13.363	14.940	16.774
	45	2.240	4.968	5.102	9.450	9.481	13.873	16.450	18.223

TABLE 9.15 Effect of Varying θ on the Frequency Parameters Ω of Graphite/Epoxy Four Layer $[\theta, -\theta, -\theta, \theta]$ Shallow Shells; $a/b = 1$, $a/h = 100$, $a/R_y = 0.5$

Laminate	R_α/R_β	a/R_α	1	2	3	4	5	6	7	8
[0°]	Plate	0	1.491	1.646	3.460	4.539	6.4276	6.471	7.132	9.009
	+1	0.1	1.493	2.319	3.511	5.534	6.577	6.706	7.340	9.345
		0.5	1.509	3.961	4.232	6.781	7.652	8.553	9.974	14.68
	0	0.1	1.493	2.420	3.509	5.694	6.472	6.572	7.145	9.099
		0.5	1.514	4.514	6.445	6.755	7.092	9.198	10.65	12.17
	−1	0.1	1.493	2.398	3.505	5.732	6.565	6.753	7.208	9.038
		0.5	1.507	4.235	4.541	7.056	8.956	9.062	11.76	13.36
[0°, 90°, 0°]	Plate	0	1.504	2.040	3.682	5.622	6.361	7.042	7.245	9.070
	+1	0.1	1.507	3.071	3.840	6.506	6.868	7.504	7.518	9.890
		0.5	1.521	4.373	5.032	6.954	7.912	10.21	11.31	16.73
	0	0.1	1.507	3.394	3.840	6.361	6.830	7.142	7.512	9.255
		0.5	1.525	6.344	6.413	7.015	7.910	11.72	12.28	16.77
	−1	0.1	1.506	3.245	3.823	6.826	7.042	7.275	7.501	9.086
		0.5	1.518	4.814	5.140	7.042	9.217	11.041	12.97	18.05
[45°]	Plate	0	1.378	2.215	3.120	4.565	4.777	6.086	7.938	7.960
	+1	0.1	1.881	2.219	3.557	4.753	4.911	7.278	8.047	8.135
		0.5	2.247	2.824	5.484	5.932	6.868	9.397	10.15	12.13

	0	0.1	1.486	2.615	3.235	4.621	4.961	6.412	8.048	8.088
		0.5	1.582	3.903	4.043	5.836	6.724	9.089	9.287	11.96
	−1	0.1	1.378	3.121	3.541	4.580	5.352	6.097	8.096	8.388
		0.5	1.376	3.119	4.754	6.216	8.701	9.354	12.65	12.68
[45°, −45°, 45°]	Plate	0	1.689	2.240	3.825	4.681	5.197	7.234	8.697	8.786
	+1	0.1	2.245	2.288	4.301	5.242	5.376	8.488	8.903	8.932
		0.5	2.276	2.963	5.973	6.336	8.202	10.21	12.03	13.68
	0	0.1	1.785	2.887	3.946	4.873	5.533	7.555	8.866	8.969
		0.5	1.814	4.094	4.606	7.117	7.534	10.03	10.75	13.76
	−1	0.1	1.690	3.827	4.270	4.708	5.916	7.242	9.129	9.345
		0.5	1.691	3.843	5.001	7.314	9.591	10.67	14.85	17.97

Table 9.16 Frequency Parameters $\Omega = \omega a^2 \sqrt{\rho / E_1 h^2}$ of Completely Free G/E Shallow Shells; $a/b = 1$, $a/h = 100$

[45°, −45°, 45°] are used. Two curvature ratios and three types of shells, as well as plates, are studied. It is important to notice that, because of the symmetry of the problem [45°, −45°, 45°] and [−45°, 45°, −45°] laminates give the same results for all the cases. Table 9.16 gives the fundamental nondimensional frequencies. Even more strongly than what was found for cross-ply laminates, the three-layer angle-ply laminates give higher frequencies than the single-layer configurations in every case.

For the *G/E* material used, increasing the curvature increases the fundamental frequencies for spherical and cylindrical shells and decreases them slightly for hyperbolic paraboloidal shells when the single-layer configurations are used. Increasing the curvature ratio is found to have larger effects on the single-layer configuration than the multi-layer one.

It is also observed that a slight curvature ($a/R = 0.1$, corresponding to an arc length of 5.7°) causes a large increase in the fundamental frequency for the spherical shell. It is observed that the 45° angle-ply laminates give higher frequencies than cross-ply laminates for most of the results when the aspect ratio is 1.

Figure 9.10 gives the first six mode shapes with various lamination angles for spherical shells. The aspect ratio (a/b) is taken as 1, and the thickness ratio (a/h) is taken as 100, the curvature ratio (a/R) is taken as 0.5, a three-layer laminate (θ, $-\theta$, θ) is considered. For these shells, mode shapes for $\theta = 60°$, 75°, and 90° are similar to those given for the $\theta = 30°$, 15°, and 0°, respectively, and can be obtained easily by changing the coordinates.

For the completely free boundary condition, symmetry exists about the inplane axes for the isotropic and orthotropic (e.g., cross-ply) shells. This symmetry is lost for shells with angle-ply lamination. For

FIGURE 9.10 Mode shapes and frequency parameters $\omega a^2\sqrt{\rho/E_1h^2}$ for completely free [θ,-θ,θ], G/E spherical shells, a/b = 1, a/h = 100, a/R = 0.5.

the special case of diagonally orthotropic angle-ply laminates which are made of 45° angle layers, symmetry exists about the diagonals for spherical and hyperbolic paraboloidal shells only. The gradual change in contour lines with increasing θ is evident in the figures.

As the lamination angle increases, the order of the mode shapes may vary. There is a tendency for the nodal lines to go in the direction of the fibers (i.e., lamination angle). For spherical and hyperbolic paraboloidal shell [1] with diagonal orthotropy (i.e., $\theta = 45°$), almost all the nodal lines are diagonal. Furthermore, it is observed that the crossing of the nodal lines for orthotropic materials (i.e., $\theta = 0°$ or 90°) is replaced by curve veering for angle-ply laminates for most of the mode shapes.

The mode shapes can also be used to explain the natural frequencies. For spherical shells, changing the angle from 0° to 45° gradually changes the first mode shape to one similar to the second mode shape which has a considerably larger natural frequency. That is why the first frequency parameter increases considerably as the lamination angle increases. Changing the lamination angle from 0° to 45° does the opposite for the second mode shape which becomes similar to the first mode shape. This explains why the second natural frequency decreases as the lamination angle increases.

A cantilevered doubly curved shallow shell is shown in Fig. 9.11. Convergence studies are made for a series of cantilevered plates and shallow shells representative of those to be analyzed subsequently [1]. Comparisons are made with results obtained experimentally and using finite element methods [1].

Studies are made for symmetrically laminated plates and shallow shells of square planform ($a/b = 1$) having either a single layer or four layers in an angle-ply stacking sequence. Results are presented for the strongly orthotropic graphite/epoxy (G/E) composites. Table 9.17

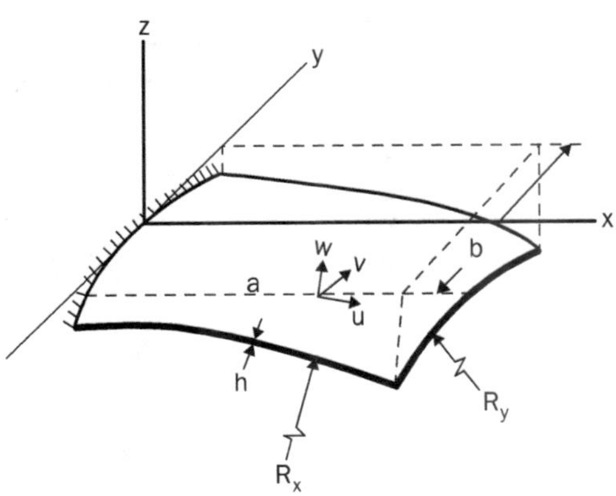

FIGURE 9.11 A doubly-curved cantilevered shallow shell.

θ (degrees)	1	2	3	4	5	6	7	8
Plates								
0	1.018	1.372	2.639	5.315	6.376	6.835	8.246	10.22
15	0.906	1.470	2.946	5.652	5.817	6.697	8.695	10.71
30	0.708	1.613	3.464	4.556	5.986	7.294	9.535	12.47
45	0.501	1.599	3.012	4.173	5.665	8.447	9.485	9.944
60	0.346	1.368	2.134	4.057	5.369	6.835	7.893	9.638
75	0.274	1.038	1.714	3.361	4.798	6.230	6.795	8.694
90	0.259	0.862	1.623	2.923	4.543	5.973	6.670	8.197
Spherical shells								
0	1.547	1.888	4.420	6.650	8.223	9.257	15.18	18.24
15	1.664	1.967	4.800	6.637	8.414	10.42	16.30	17.03
30	1.606	2.044	5.189	6.352	8.646	11.86	15.14	16.60
45	1.349	1.986	4.723	6.112	8.654	11.54	13.95	16.16
60	1.161	1.673	4.349	5.456	7.820	10.71	12.48	15.52
75	1.052	1.273	3.989	4.659	7.254	8.952	10.52	14.60
90	0.991	1.076	3.517	4.369	7.197	7.769	9.534	13.16
Cylindrical shells								
0	1.962	2.334	4.558	5.518	8.193	8.210	10.26	11.80
15	2.081	2.628	5.180	6.026	7.854	8.875	11.20	13.13

30	2.066	2.931	5.084	7.091	7.890	9.508	13.74	14.89
45	1.866	2.606	4.847	7.334	7.761	10.95	14.28	14.75
60	1.552	2.014	5.041	6.395	7.416	12.30	12.53	13.53
75	1.208	1.674	4.442	6.193	6.838	9.914	10.59	12.03
90	1.027	1.555	3.899	5.973	6.825	8.482	9.701	10.70
Hyperbolic paraboloidal shells								
0	1.473	1.512	5.693	6.956	10.09	10.83	15.93	16.50
15	1.273	1.881	5.348	7.697	10.86	11.95	16.84	17.63
30	1.209	2.146	5.179	8.248	11.35	13.87	17.64	18.41
45	1.150	2.092	4.983	7.988	11.01	15.15	16.63	18.17
60	1.038	1.747	4.497	7.019	9.893	13.73	14.24	15.34
75	0.947	1.311	3.914	5.931	8.561	10.68	12.26	13.23
90	0.973	1.023	3.578	5.340	8.141	8.845	11.28	12.41

TABLE 9.17 Effect of Varying θ on the Frequency Parameters $\Omega = \omega a^2 \sqrt{\rho / E_1 h^2}$ of Graphite/Epoxy Four Layer $[\theta, -\theta, -\theta, \theta]$ Cantilevered Shallow Shells; $a/b = 1$, $a/h = 100$, $b/R_y = 0.5$

presents the fundamental nondimensional frequency parameters for four-layer angle-ply plates and shallow shells. Other materials and lamination sequences are considered [16]. The fiber orientation angle varies between 0° and 90° by an angle increment of 15°.

It is observed that increasing θ causes the fundamental frequency to decrease monotonically for the plates and for most of the shells considered. However, for spherical and cylindrical shells, the maximum fundamental frequency occurs in the vicinity of $\theta = 15°$.

It is interesting to notice that the percent change in the fundamental frequency between the 0° angle and the 90° angle is considerably more for G/E than that for E/E materials [16]. It is also observed that changing the fiber angle has a much greater effect on plates than for shells. The least change in the frequency parameter with increasing θ occurs for hyperbolic paraboloidal shells. The effect of changing the angle on higher frequencies is less than that for the fundamental frequency.

Further studies showed that increasing the curvature increases the frequencies considerably [16]. Its effect is maximum for cylindrical shells and minimum for hyperbolic paraboloidal shells.

Figure 9.12 gives the first six mode shapes with various lamination angles for cylindrical shells. Mode shapes for spherical and hyperbolic paraboloidal can be found in Ref. [1]. Unlike the completely free case, there is no diagonal symmetry here. The mode shapes for all the lamination angles of the three types of shells are considered.

For cantilevered cross-ply shells, two classes of symmetry are possible in the displacement functions. The displacements can be either symmetric or antisymmetric about the x-axis. For the flat plate and when $\theta = 0°$, the first six mode shapes were, in order, the first spanwise bending mode (1B), the first torsional mode (1T), the first chordwise bending mode (1C), the second chordwise bending mode (2C), the second torsional mode (2T), and finally the second bending mode (2B) [1]. Unlike flat plates, for all the shells considered here, the first mode is the first torsional mode when $\theta = 0°$ or 90°. The shape of the mode for shells is not as clear as that for plates. For the lamination angle $\theta = 0°$, the order of the mode shapes is

a. For spherical shells 1T, 1B, 1C, 2T, 2B, 2C
b. For cylindrical shells 1T, 1C, 1B, 2C, 2T, 3C
c. For hyperbolic paraboloidal shells 1T, 1B, 1C, 2T, 2B, 2C

The order of the mode shapes changes for the 90° lamination angle, but the first mode shape remains the torsional one. More chordwise bending modes tend to exist for the 0° lamination angle than the 90° one. For lamination angles between 0° and 90°, the fundamental modes are combinations of the above modes. A gradual change in the mode shapes is observed as the lamination angle increases. In some cases, this gradual change in the mode shapes

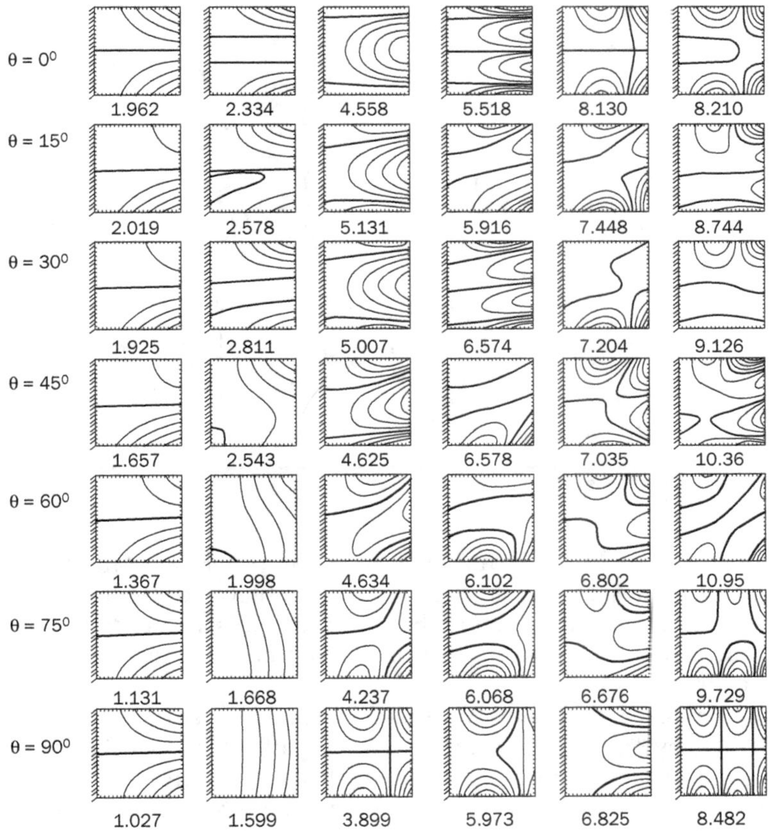

FIGURE 9.12 Mode shapes and frequency parameters $\omega a^2\sqrt{\rho/E_1 h^2}$ for cantilevered G/E cylindrical shells. a/b = 1, a/h = 100, a/R = 0.5.

results in a completely different one when the lamination angle becomes 90°.

Looking at these mode shapes more closely, one observes that for most of the mode shapes large displacements are observed in limited regions of the shell, such as at a corner, whereas most of the shell displaces very little. This is characteristic of thin shells, where one observes much more activity near the edges or corners of the shell.

An interesting phenomenon is observed for spherical shells with $\theta = 45°$ [17]. For these shells, the mode shapes can be approximately divided into two categories of symmetric and antisymmetric about the x-axis.

A twisted plate (Fig. 9.13) is actually a shell. In one set of coordinates it is only twisted. But in other coordinates it has curvatures as well. The cantilevered twisted plate is a mechanical element of considerable technical significance. It has many

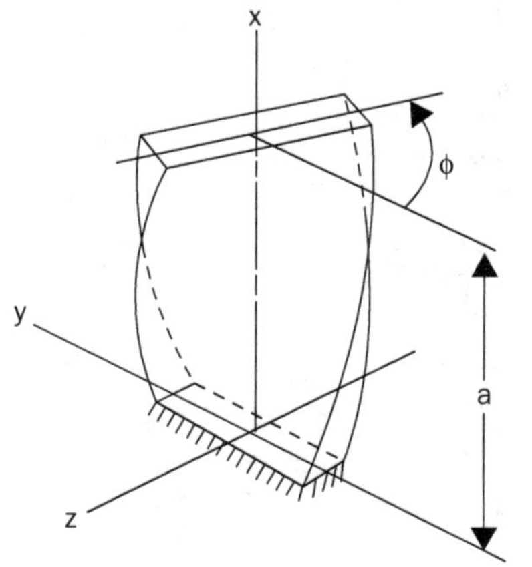

Figure 9.13 A twisted plate.

applications such as in turbomachinery, impeller, and fan blades. The vast majority treated blades as cantilevered beams. Such representation is inaccurate if higher frequencies are needed or the blade is short. A more accurate model can be made using shallow shell theory.

A twisted plate is characterized by its middle surface, which is defined by $z = xy/R_{xy}$. The constant R_{xy} identify the radius of twist. A typical twisted plate is shown in Fig 9.13. Natural frequencies and mode shapes of laminated cantilever plates having pretwist is presented in Ref. [18]. Laminated shallow shell theory and the Ritz method are used here. The effects of many parameters like twist angle, lamination angle, stacking sequence, and thickness and orthotropy ratios on the natural frequencies and mode shapes of twisted cantilevered composite plates is also studied [18].

A typical plate of moderate thickness (b/h = 100) and twist (ϕ = 30°) and square planform (a/b = 1) was used in Ref. [18]. Convergence studies were made of the lowest frequency parameters $\Omega = \omega a^2\sqrt{\rho/E_1h^2}$ for graphite/epoxy. The maximum number of terms used is 192 by choosing 8 terms in both x and y directions for all the three components of displacements.

The maximum fundamental frequencies are observed when the fibers are perpendicular to the clamped edge (i.e., θ = 0°) in all cases. Frequencies and mode shapes were given in Fig. 9.9 for untwisted cantilevered plates. For the fundamental frequency, which corresponds to the first spanwise bending mode, increasing the fiber angle has a very large effect. The natural frequencies when θ = 90° are about one quarter of those when θ = 0° when G/E material is used for all twist angles. This difference is much less for E/E materials. For higher frequencies, the maximum frequency parameters can be found at a fiber angle between 0° and 30° and occasionally at a fiber angle of 45°, with some exceptions for *E/E* materials. Although thin untwisted plate theory predicts the same transverse vibration frequency parameters $\Omega = \omega a^2\sqrt{\rho/E_1h^2}$ when the thickness ratio is changed, new mode shapes can be introduced for some modes of the thicker plates, which correspond to inplane vibration modes and not transverse ones. This could not be seen if the inplane terms (i.e., *U* and *V*) are removed, which is normally done in vibration studies made for untwisted symmetrically laminated thin plates.

Figures 9.14 shows the fundamental six frequency parameters and corresponding mode shapes for laminated twisted graphite/epoxy square plates for an angle of twist of 45°. Other angles of twist are given in the literature [18].

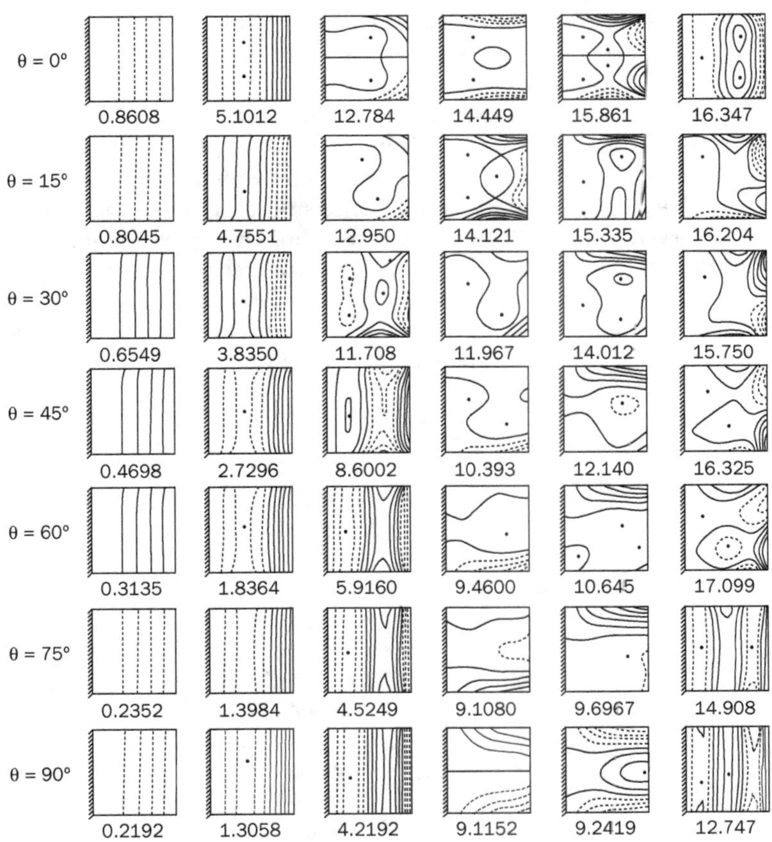

FIGURE 9.14 Mode shapes and frequency parameters $\omega a^2\sqrt{\rho/E_1h^2}$ for graphite/epoxy cantilevered plates with a 45° twist angle.

For twisted plates, unlike flat plates, the second mode is not the torsional mode. In fact, for $\phi = 30°$ and $45°$, the second mode is the second bending mode; the first torsional mode is the third mode when $\theta = 0°$ and the fourth one when $\theta = 90°$, being preceded by three bending modes. The number of symmetric modes such as 1B, 2B, 3B, 1C and 2C when $\theta = 90°$ increases as the angle of twist increases. For $\phi = 0°$, three of the first six mode shapes are symmetric, whereas for $\phi = 45°$, five are symmetric.

For fiber angles between 0° and 90°, coupling between the modes exists, and the strength of the coupling increases both with increasing fiber angle and increasing twist angle. With large twist angle (Fig. 9.14), one observes many modes where large displacements occur only in limited regions of the plate, such as in a corner, whereas most of the plate displaces very little. This is characteristic of thin shells, and the twisted plate is in actuality a shell.

A doubly cantilevered shallow shell (i.e., two adjacent edges clamped and the others free) is shown in Fig. 9.15. A convergence study for moderately thin ($b/h = 100$) plates and shallow shells having square planform ($a/b = 1$) was performed [19]. The depth of the shell is taken at the limits of shallow shell theory ($b/R_y = 0.5$), which is found to yield slower convergence than for shallower shells. Results for the flat plate are shown for comparison. Three layer $[\theta, -\theta, \theta]$ laminates are considered in the analysis. A representative lamination angle of 30° has been adopted in the convergence studies. Converging results are observed with a 49 terms used for laminated plates. Table 9.18 presents the first eight nondimensional frequency parameters for three-layer plates and shallow shells of moderate curvature (b/R_y = 0.5) for graphite/epoxy materials. The fiber orientation angle varies between 0° and 90° by an angle increment of 15°.

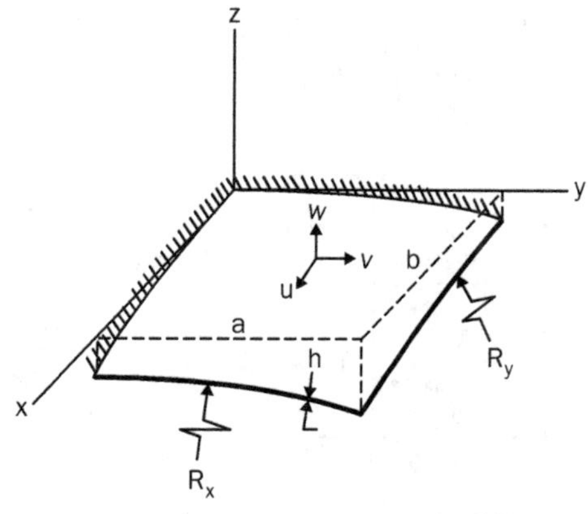

Figure 9.15 A doubly cantilevered shallow shell.

θ (degrees)	1	2	3	4	5	6	7	8
Spherical shells								
0	3.015	4.875	8.933	9.932	14.75	18.90	20.41	23.00
15	3.344	4.782	9.806	10.52	16.18	19.19	21.41	25.65
30	3.652	4.726	10.78	11.40	17.44	19.77	23.88	27.44
45	3.754	4.692	11.41	11.45	17.85	20.53	25.70	28.04
Cylindrical shells								
0	2.162	5.091	7.247	8.171	9.938	11.72	12.84	15.07
15	2.490	5.603	8.002	8.894	11.00	12.93	13.85	15.72
30	2.709	6.209	8.472	10.02	13.31	14.37	16.28	17.94
45	2.787	6.358	8.738	11.54	13.59	16.24	19.39	20.77
60	2.774	5.870	8.698	11.93	14.32	16.08	21.74	21.74
75	2.564	5.519	8.116	10.23	15.64	15.76	18.13	22.15
90	2.248	5.098	7.735	9.229	14.68	15.22	18.14	21.36
Hyperbolic paraboloidal shells								
0	1.551	5.970	10.38	11.90	15.23	15.81	19.41	20.73
15	1.976	7.273	11.49	13.94	15.57	17.92	18.95	21.51
30	2.229	8.050	13.20	15.15	16.56	18.56	20.86	22.85
45	2.306	8.316	14.87	16.26	17.12	18.70	21.54	21.66

Table 9.18 Effect of Varying θ on the Frequency Parameters $\Omega = \omega a^2 \sqrt{\rho / E_1 h^2}$ of Graphite/Epoxy Three Layer [θ, $-\theta$, θ] Doubly Cantilevered Shallow Shells; $a/b = 1$, $b/h = 100$, $b/R_y = 0.5$

Natural frequencies for spherical and hyperbolic paraboloidal shells as well as those for plates are the same when the fiber angle θ is 60°, 75°, and 90° as those when it is 30°, 15°, and 0° respectively. This is due to the symmetry of the problem about the diagonal for the shallow shells of square planform studied here.

It is observed that increasing θ has its largest effect on the fundamental frequencies of hyperbolic paraboloidal shells for all the results. The effect of increasing the lamination angle from 0° to 45° is a 36 percent increase in fundamental frequencies for hyperbolic paraboloidal, 22 percent for cylindrical, 24 percent for spherical shells, compared with only 9.3 percent for plates. For deeper shallow shells ($b/R_y = 0.5$), it is observed that increasing the angle from 0° to 45° causes the fundamental frequencies to increase by 24 percent, 29 percent, and 48 percent for spherical, cylindrical and hyperbolic paraboloidal shells respectively. Vibration of cantilevered and completely free composite triangular and trapezoidal shallow shells has also been studies [20, 21].

9.10 Laminated Thick Shallow Shells

Thick shallow shells are shells with a thickness smaller by approximately one order of magnitude when compared with other shell parameters, like its vibration mode shape wavelength and/or radii of curvature. A thick shallow shell theory requires the inclusion of shear deformation and rotary inertia factors to be accurate.

The midsurface strains for a shallow shell are:

$$\epsilon_{0x} = \frac{\partial u_0}{\partial x} + \frac{w_0}{R_x}, \quad \epsilon_{0y} = \frac{\partial v_0}{\partial y} + \frac{w_0}{R_y}$$

$$\epsilon_{0xy} = \frac{\partial u_0}{\partial y} + \frac{w_0}{R_{xy}}, \quad \epsilon_{0yx} = \frac{\partial v_0}{\partial x} + \frac{w_0}{R_{xy}}$$

$$\text{or} \quad \gamma_{0xy} = \frac{\partial v_0}{\partial x} + \frac{\partial u_0}{\partial y} + \frac{2w_0}{R_{xy}} \tag{9.115}$$

Transverse shear strains and curvature changes are similar to those for plates (9.94) and (9.95).

The equations of motion for a laminated shallow shell are similar to those for a plate (9.97) with the exception for the third equation. For shallow shells, this equation is

$$-\left(\frac{N_x}{R_x} + \frac{N_y}{R_y} + \frac{N_{xy} + N_{yx}}{R_{xy}}\right) + \frac{\partial}{\partial x}(Q_x) + \frac{\partial}{\partial y}(Q_y) + q_n = (I_1 \ddot{w}_0) \tag{9.116}$$

The equilibrium equations can be written in terms of displacements ($L_{ij}u_i + M_{ij}\ddot{u}_i = q$) same as (9.101) where the L_{ij} are listed below, and the M_{ij} coefficients are given in Eqs. (9.102).

$$L_{11} = A_{11}\frac{\partial^2}{\partial x^2} + 2A_{16}\frac{\partial^2}{\partial x \partial y} + A_{66}\frac{\partial^2}{\partial y^2}$$

$$L_{12} = A_{16}\frac{\partial^2}{\partial x^2} + (A_{12} + A_{66})\frac{\partial^2}{\partial x \partial y} + A_{26}\frac{\partial^2}{\partial y^2}$$

$$L_{13} = \left[\frac{A_{11}}{R_x} + \frac{A_{12}}{R_y} + \frac{2A_{16}}{R_{xy}}\right]\frac{\partial}{\partial \alpha} + \left[\frac{A_{16}}{R_x} + \frac{A_{26}}{R_y} + \frac{2A_{66}}{R_{xy}}\right]\frac{\partial}{\partial y}$$

$$L_{14} = B_{11}\frac{\partial^2}{\partial x^2} + 2B_{16}\frac{\partial^2}{\partial x \partial y} + B_{66}\frac{\partial^2}{\partial y^2}$$

$$L_{15} = B_{16}\frac{\partial^2}{\partial x^2} + (B_{12} + B_{66})\frac{\partial^2}{\partial x \partial y} + B_{26}\frac{\partial^2}{\partial y^2}$$

$$L_{22} = A_{66}\frac{\partial^2}{\partial x^2} + 2A_{26}\frac{\partial^2}{\partial x \partial y} + A_{22}\frac{\partial^2}{\partial y^2}$$

$$L_{23} = \left[\frac{A_{26}}{R_y} + \frac{A_{16}}{R_x} + \frac{2A_{66}}{R_{xy}}\right]\frac{\partial}{\partial x} + \left[\frac{A_{22}}{R_y} + \frac{A_{12}}{R_x} + \frac{2A_{26}}{R_{xy}}\right]\frac{\partial}{\partial y}$$

$$L_{24} = B_{16}\frac{\partial^2}{\partial x^2} + (B_{12} + B_{66})\frac{\partial^2}{\partial x \partial y} + B_{26}\frac{\partial^2}{\partial y^2}$$

$$L_{25} = B_{66}\frac{\partial^2}{\partial x^2} + 2B_{26}\frac{\partial^2}{\partial x \partial y} + B_{22}\frac{\partial^2}{\partial y^2}$$

$$L_{33} = -A_{55}\frac{\partial^2}{\partial x^2} - 2A_{45}\frac{\partial^2}{\partial x \partial y} - A_{44}\frac{\partial^2}{\partial y^2} + \frac{A_{11}}{R_x^2} + \frac{2A_{12}}{R_x R_y} + \frac{A_{22}}{R_y^2}$$
$$+ \frac{4}{R_{xy}}\left[\frac{A_{16}}{R_x} + \frac{A_{26}}{R_y} + \frac{A_{66}}{R_{xy}}\right]$$

$$L_{34} = \left[-A_{55} + \frac{B_{11}}{R_x} + \frac{B_{12}}{R_y} + \frac{2B_{16}}{R_{xy}}\right]\frac{\partial}{\partial x} + \left[-A_{45} + \frac{B_{16}}{R_x} + \frac{B_{26}}{R_y} + \frac{2B_{66}}{R_{xy}}\right]\frac{\partial}{\partial y}$$

$$L_{35} = \left[-A_{45} + \frac{B_{16}}{R_x} + \frac{B_{26}}{R_y} + \frac{2B_{66}}{R_{xy}}\right]\frac{\partial}{\partial x} + \left[-A_{44} + \frac{B_{12}}{R_x} + \frac{B_{22}}{R_y} + \frac{2B_{26}}{R_{xy}}\right]\frac{\partial}{\partial y}$$

$$L_{44} = -A_{55} + D_{11}\frac{\partial^2}{\partial x^2} + 2D_{16}\frac{\partial^2}{\partial x \partial y} + D_{66}\frac{\partial^2}{\partial y^2}$$

$$L_{45} = -A_{45} + D_{16}\frac{\partial^2}{\partial x^2} + (D_{12} + D_{66})\frac{\partial^2}{\partial x \partial y} + D_{26}\frac{\partial^2}{\partial y^2}$$

$$L_{55} = -A_{44} + D_{66}\frac{\partial^2}{\partial x^2} + 2D_{26}\frac{\partial^2}{\partial x \partial y} + D_{22}\frac{\partial^2}{\partial y^2} \tag{9.117}$$

Both symmetric and unsymmetric lamination sequences have 24 possible classical boundary conditions at each edge, 8 for each of the classical boundary conditions (i.e., free, simply supported and clamped).

Like cross-ply laminated plates, shallow shells that have two opposite edges simply supported can permit exact solutions. Similar to plates, the problem with all edges being simply supported S2 have a direct, relatively straightforward solution.

Consider again cross-ply laminates, with $A_{16} = A_{26} = B_{16} = B_{26} = D_{16} = D_{26} = 0$. An exact solution similar to (9.106) can be employed for thick shallow shells with shear diaphragm supports at $x = 0$ and a, and $y = 0$ and b (similar to plates). Substituting these into the equations of motion, using a Fourier expansion for the loading functions yields the equations $[K]\{\Delta\}+(\omega_{mn})^2\,[M]\,\{\Delta\} = -\{F\}$ similar to that of plates. For shallow shells, the K_{ij} coefficients are

$$K_{11} = -A_{11}\alpha_m^2 - A_{66}\beta_n^2$$

$$K_{12} = -(A_{12} + A_{66})\alpha_m\beta_n$$

$$K_{13} = \left[\frac{A_{11}}{R_x} + \frac{A_{12}}{R_y}\right]\alpha_m$$

$$K_{14} = -B_{11}\alpha_m^2 - B_{66}\beta_n^2$$

$$K_{15} = -(B_{12} + B_{66})\alpha_m\beta_n$$

$$K_{22} = -A_{66}\alpha_m^2 - A_{22}\beta_n^2$$

$$K_{23} = \left[\frac{A_{22}}{R_y} + \frac{A_{12}}{R_x}\right]\beta_n$$

$$K_{24} = -(B_{12} + B_{66})\alpha_m\beta_n$$

$$K_{25} = -B_{66}\alpha_m^2 - B_{22}\beta_n^2$$

$$K_{33} = -A_{55}\alpha_m^2 - A_{44}\beta_n^2 - \left[\frac{A_{11}}{R_x^2} + \frac{2A_{12}}{R_xR_y} + \frac{A_{22}}{R_y^2}\right]$$

$$K_{34} = \left[-A_{55} + \frac{B_{11}}{R_x} + \frac{B_{12}}{R_y}\right]\alpha_m$$

$$K_{35} = \left[-A_{44} + \frac{B_{12}}{R_x} + \frac{B_{22}}{R_y}\right]\beta_n$$
$$K_{44} = -A_{55} - D_{11}\alpha_m^2 - D_{66}\beta_n^2$$
$$K_{45} = -(D_{12} + D_{66})\alpha_m\beta_n$$
$$K_{55} = -A_{44} - D_{66}\alpha_m^2 - D_{22}\beta_n^2 \tag{9.118}$$

The mass matrix is the same as that defined for plates. Table 9.19 shows the frequency parameter $\Omega = \omega a^2\sqrt{\rho/E_2h^2}$ for cross-ply hyperbolic paraboloidal shallow shells on square planform ($a/b = 1$) and with material properties representative of graphite/epoxy ($E_1/E_2 = 15$, $G_{12}/E_2 = 0.5$, $G_{13}/E_2 = 0.5$, $G_{23}/E_2 = 0.5$, $\nu_{12} = 0.25$). A typical shear correction factor is used ($K^2 = 5/6$). Results are presented by using a shear deformation shallow shell theory (SDSST) and classical shallow shell theory (CSST). Cross-ply laminates with both symmetric [0°, 90°, 90°, 0°] and unsymmetric [0°, 90°] and [90°, 0°] lamination sequences are used in the analysis. Curvature ratios are varied from a flat plate to that of the limit of shallow shell theory ($a/R = 0.5$). Tables 9.20 and 9.21 show similar results for both cylindrical and spherical shells, respectively.

The first observation made here is the difference between the shear deformation shell theory (SDSST) and the classical shell theory (CSST). Both theories gave results that agree to the third significant figure for a thickness ratio of 100. The difference between the two theories reaches 3 percent for hyperbolic paraboloidal shells, 2.5 percent for cylindrical shells and 1.5 percent for spherical shells when a thickness ratio of 20 is used, indicating the validity of the classical theory for shells with this thickness ratio. When the thickness ratio of 10 is used, the difference between the two theories exceeds 10 percent for the most part, showing the lack of accuracy of the classical theory in predicting frequencies for such moderately thick shallow shells.

The second important observation to be made is on the impact of unsymmetric lamination. Lower frequency parameters were obtained for unsymmetrically laminated cross-ply plates when compared with symmetrically laminated ones. The unsymmetric lamination decreases the frequencies significantly for almost all results presented here except the thin, moderately deep, spherical shells, where its impact is noticed to be minimal. This can be due to the high influence of membrane forces of these shells on the frequency parameter. Furthermore, the [0°, 90°] lamination gave slightly higher frequencies than the [90°, 0°] lamination. This is particularly true for thicker and deeper shells and more so for hyperbolic paraboloidal shells than cylindrical shells. The two lamination sequences showed no difference for spherical shells.

Lamination Theory		[0°, 90°, 90°, 0°]		[0°, 90°]		[90°, 0°]	
	a/R	SDSST	CSST	SDSST	CSST	SDSST	CSST
$a/h = 100$	0	12.2615	12.3773	8.56394	8.56847	8.56394	8.56847
	0.1	12.2491	12.2649	8.55914	8.56847	8.55143	8.55594
	0.2	12.2121	12.2279	8.53710	8.54161	8.52183	8.52632
	0.5	11.9622	11.9776	8.37296	8.37737	8.33708	8.34143
$a/h = 20$	0	11.9010	12.2773	8.44807	8.55811	8.44807	8.55811
	0.1	11.8890	12.2649	8.45815	8.56879	8.42109	8.53026
	0.2	11.8530	12.2276	8.45115	8.56208	8.37769	8.48572
	0.5	11.6104	11.9763	8.33038	8.44045	8.15772	8.26102
$a/h = 10$	0	10.9716	12.2773	8.11956	8.52569	8.11956	8.52569
	0.1	10.9605	12.2647	8.14483	8.55570	8.07848	8.47898
	0.2	10.9273	12.2269	8.15383	8.56844	8.02233	8.41638
	0.5	10.7031	11.9722	8.08313	8.50237	7.77388	8.14496

TABLE 9.19 Frequency Parameter $\Omega = \omega a^2 \sqrt{\rho / E_2 h^2}$ for Cross-Ply Hyperbolic Paraboloidal Shallow Shells ($a/b = 1$, $E_1/E_2 = 15$, $G_{12}/E_2 = 0.5$, $G_{13}/E_2 = 0.5$, $G_{23}/E_2 = 0.5$, $\nu_{12} = 0.25$, $K^2 = 5/6$) for Shear Deformation Shallow Shell Theory (SDSST) and Classical Shallow Shell Theory (CSST)

Lamination Theory	a/R	[0°, 90°, 90°, 0°]		[0°, 90°]		[90°, 0°]	
		SDSST	CSST	SDSST	CSST	SDSST	CSST
$a/h = 100$	0	12.2615	12.3773	8.56394	8.56847	8.56394	8.56847
	0.1	13.9561	13.9703	10.8575	10.8616	10.8526	10.8567
	0.2	18.0992	18.1107	15.8445	15.8484	15.8303	15.8342
	0.5	35.1759	35.1838	28.2155	28.2471	27.7925	27.8270
$a/h = 20$	0	11.9010	12.2773	8.44807	8.55811	8.44807	8.55811
	0.1	11.9700	12.3442	8.55769	8.66733	8.53889	8.64780
	0.2	12.1740	12.5418	8.86048	8.96840	8.82171	8.92813
	0.5	13.4883	13.8195	10.6898	10.7893	10.5752	10.6707
$a/h = 10$	0	10.9716	12.2773	8.11956	8.52569	8.11956	8.52569
	0.1	10.9867	12.2897	8.15924	8.56701	8.12590	8.52847
	0.2	11.0316	12.3270	8.24451	8.65203	8.17744	8.57455
	0.5	11.3342	12.5784	8.75225	9.14960	8.57836	8.94972

TABLE 9.20 Frequency Parameter $\Omega = \omega a^2 \sqrt{\rho / E_2 h^2}$ for Cross-Ply Cylindrical Shallow Shells ($a/b = 1$, $E_1/E_2 = 15$, $G_{12}/E_2 = 0.5$, $G_{13}/E_2 = 0.5$, $G_{23}/E_2 = 0.5$, $\nu_{12} = 0.25$, $K^2 = 5/6$) for Shear Deformation Shallow Shell Theory (SDSST) and Classical Shallow Shell Theory (CSST)

Lamination	a/R	[0°, 90°, 90°, 0°] SDSST	[0°, 90°, 90°, 0°] CSST	[0°, 90°] SDSST	[0°, 90°] CSST
a/h = 100	0	12.2615	12.3773	8.56394	8.56847
	0.1	18.1175	18.1290	15.8534	15.8573
	0.2	29.3005	29.3090	27.9620	27.9666
	0.5	66.5695	66.5774	66.0054	66.0139
a/h = 20	0	11.9010	12.2773	8.44807	8.55811
	0.1	12.1863	12.5545	8.84994	8.95725
	0.2	13.0000	13.3466	9.95202	10.0534
	0.5	17.5209	17.7854	15.4499	15.5392
a/h = 10	0	10.9716	12.2773	8.11956	8.52569
	0.1	11.0428	12.3397	8.21896	8.62173
	0.2	11.2522	12.5236	8.50843	8.90183
	0.5	12.5718	13.6975	10.2492	10.5975

TABLE 9.21 Frequency Parameter $\Omega = \omega a^2\sqrt{\rho/E_2h^2}$ for Cross-Ply Spherical Shallow Shells ($a/b = 1$, $E_1/E_2 = 15$, $G_{12}/E_2 = 0.5$, $G_{13}/E_2 = 0.5$, $G_{23}/E_2 = 0.5$, $\nu_{12} = 0.25$, $K^2 = 5/6$) for Shear Deformation Shallow Shell Theory (SDSST) and Classical Shallow Shell Theory (CSST)

The third observation made here is on the influence of curvature. The influence of curvature is noticed to be the strongest for spherical shells and the weakest for hyperbolic paraboloidal shells. For the later shells, increasing the curvature is observed to reduce the frequency parameter slightly. For spherical shells, a slight curvature increase to a curvature ratio of $a/R = 0.1$ results in 50 percent increase in the frequency parameters associated with thin shells. Its impact is less for thicker shells. Additional detailed results can be found in Ref. [1].

9.11 Laminated Cylindrical Shells

Consider Fig. 7.10, the following equations represent the midsurface strains and curvature changes for a thin cylindrical shell:

$$\epsilon_{0\alpha} = \frac{\partial u_o}{\partial \alpha}, \quad \epsilon_{0\beta} = \frac{\partial v_o}{\partial \beta} + \frac{w_o}{R}, \quad \gamma_{0\alpha\beta} = \frac{\partial v_o}{\partial \alpha} + \frac{\partial u_o}{\partial \beta}$$

$$k_\alpha = -\frac{\partial^2 w_o}{\partial \alpha^2}, \quad k_\beta = \frac{\partial}{\partial \beta}\left(\frac{v_o}{R}\right) - \frac{\partial^2 w_o}{\partial \beta^2}, \quad \tau = \frac{\partial}{\partial \alpha}\left(\frac{v_o}{R}\right) - 2\frac{\partial^2 w_o}{\partial \beta \partial \alpha} \tag{9.119}$$

The β subscript is dropped because there is only one radius of curvature for cylindrical shells ($R = R_\beta$).

Consider a laminated composite thin shell. Applying the Kirchhoff hypothesis of neglecting shear deformation, assuming that ϵ_z is negligible, as is customarily done in thin shell theory (or classical shell theory, CST), and integrating the stresses over the thickness of the shell yields Eqs. (9.12) and (9.13). Note that these equations are the same as those for laminated plates, which are valid for *thin* shells. Note also that the orthotropy of each layer is assumed to follow the coordinate system.

By taking a cylindrical differential element and applying equilibrium conditions on it (or through variational principles), the equations of motion for a cylindrical shells are:

$$\frac{\partial N_\alpha}{\partial \alpha} + \frac{\partial N_{\alpha\beta}}{\partial \beta} + p_\alpha = \bar{I}_1 \frac{\partial^2 u_o}{\partial t^2}$$

$$\frac{\partial N_\beta}{\partial \beta} + \frac{\partial N_{\alpha\beta}}{\partial \alpha} + \frac{1}{R}\left[\frac{\partial M_\beta}{\partial \beta} + \frac{\partial M_{\alpha\beta}}{\partial \alpha}\right] + p_\beta = \bar{I}_1 \frac{\partial^2 v_o}{\partial t^2}$$

$$-\frac{N_\beta}{R} + \frac{\partial^2 M_\alpha}{\partial \alpha^2} + 2\frac{\partial^2 M_{\alpha\beta}}{\partial \beta \partial \alpha} + \frac{\partial^2 M_\beta}{\partial \beta^2} + p_n = \bar{I}_1 \frac{\partial^2 w_o}{\partial t^2} \tag{9.120}$$

where p_α, p_β and p_n are pressure components. The above equations can be written in term of the coordinates (x, θ) as was done in Chap. 7 for isotropic shells by substituting the following expression into them:

$$\alpha = x, \quad \beta = R\theta$$

$$\frac{\partial}{\partial \beta} = \frac{1}{R}\frac{\partial}{\partial \theta} \tag{9.121}$$

Substituting (9.119) and (9.12) into Eqs. (9.120) yields the equations of motion in terms of displacement similar to (9.77). The L_{ij} coefficients of these equations for a laminated thin cylindrical shells are:

$$L_{11} = A_{11}\frac{\partial^2}{\partial \alpha^2} + 2A_{16}\frac{\partial^2}{\partial \alpha \partial \beta} + A_{66}\frac{\partial^2}{\partial \beta^2}$$

$$L_{12} = L_{21} = \left\{A_{16} + \frac{B_{16}}{R}\right\}\frac{\partial^2}{\partial \alpha^2} + \left\{(A_{12} + A_{66}) + \frac{(B_{12} + B_{66})}{R}\right\}\frac{\partial^2}{\partial \alpha \partial \beta}$$
$$+ \left\{A_{26} + \frac{B_{26}}{R}\right\}\frac{\partial^2}{\partial \beta^2}$$

$$L_{22} = \left\{A_{66} + 2\frac{B_{66}}{R} + \frac{D_{66}}{R^2}\right\}\frac{\partial^2}{\partial\alpha^2} + \left\{2A_{26} + 4\frac{B_{26}}{R} + 2\frac{D_{26}}{R^2}\right\}\frac{\partial^2}{\partial\alpha\partial\beta}$$
$$+ \left\{A_{22} + 2\frac{B_{22}}{R} + \frac{D_{22}}{R^2}\right\}\frac{\partial^2}{\partial\beta^2}$$

$$L_{13} = L_{31} = -B_{11}\frac{\partial^3}{\partial\alpha^3} - B_{26}\frac{\partial^3}{\partial\beta^3} - 3B_{16}\frac{\partial^3}{\partial\alpha^2\partial\beta}$$
$$-(B_{12} + 2B_{66})\frac{\partial^3}{\partial\alpha\partial\beta^2} + \left\{\frac{A_{12}}{R}\right\}\frac{\partial}{\partial\alpha} + \left\{\frac{A_{26}}{R}\right\}\frac{\partial}{\partial\beta}$$

$$L_{23} = L_{32} = \left\{-B_{16} - \frac{D_{16}}{R}\right\}\frac{\partial^3}{\partial\alpha^3} + \left\{-B_{22} - \frac{D_{22}}{R}\right\}\frac{\partial^3}{\partial\beta^3}$$
$$+ \left\{-3B_{26} - 3\frac{D_{26}}{R}\right\}\frac{\partial^3}{\partial\alpha\partial\beta^2}$$
$$+ \left\{-(B_{12} + 2B_{66}) - \frac{(D_{12} + 2D_{66})}{R}\right\}\frac{\partial^3}{\partial\alpha^2\partial\beta}$$
$$+ \left\{\frac{A_{26}}{R} + \frac{B_{26}}{R^2}\right\}\frac{\partial}{\partial\alpha} + \left\{\frac{A_{22}}{R} + \frac{B_{22}}{R^2}\right\}\frac{\partial}{\partial\beta}$$

$$L_{33} = \left\{D_{11}\frac{\partial^4}{\partial\alpha^4} + 4D_{16}\frac{\partial^4}{\partial\alpha^3\partial\beta} + 2(D_{12} + 2D_{66})\frac{\partial^4}{\partial\alpha^2\partial\beta^2}\right.$$
$$\left.+4D_{26}\frac{\partial^4}{\partial\alpha\partial\beta^3} + D_{22}\frac{\partial^4}{\partial\beta^4}\right\}$$
$$-2\left\{\left(\frac{B_{12}}{R}\right)\frac{\partial^2}{\partial\alpha^2} + 2\left(\frac{B_{26}}{R}\right)\frac{\partial^2}{\partial\alpha\partial\beta}\right.$$
$$\left.+\left(\frac{B_{22}}{R}\right)\frac{\partial^2}{\partial\beta^2}\right\} + \left\{\frac{A_{22}}{R^2}\right\} \qquad (9.122)$$

The strain energy functional for a cylindrical shell made of laminated composite can be written as in (9.81) where parts of the potential energy for a cylindrical shell are

$$PE_s = \frac{1}{2}\int_A \left\{ A_{11}\left(\frac{\partial u_o}{\partial \alpha}\right)^2 + A_{22}\left(\frac{\partial v_o}{\partial \beta} + \frac{w_o}{R}\right)^2 + A_{66}\left(\frac{\partial v_o}{\partial \alpha} + \frac{\partial u_o}{\partial \beta}\right)^2 \right.$$

$$+ 2A_{12}\left(\frac{\partial u_o}{\partial \alpha}\right)\left(\frac{\partial v_o}{\partial \beta} + \frac{w_o}{R}\right)$$

$$+ 2A_{16}\left(\frac{\partial u_o}{\partial \alpha}\right)\left(\frac{\partial v_o}{\partial \alpha} + \frac{\partial u_o}{\partial \beta}\right)$$

$$\left. +2A_{26}\left(\frac{\partial v_o}{\partial \beta} + \frac{w_o}{R}\right)\left(\frac{\partial v_o}{\partial \alpha} + \frac{\partial u_o}{\partial \beta}\right)\right\} dA \tag{9.123a}$$

$$PE_b = \frac{1}{2}\int_A \left\{ D_{11}\left(\frac{\partial^2 w_o}{\partial \alpha^2}\right)^2 + D_{22}\left(\frac{1}{R}\frac{\partial v_o}{\partial \beta} - \frac{\partial^2 w_o}{\partial \beta^2}\right)^2 \right.$$

$$+ D_{66}\left(\frac{1}{R}\frac{\partial v_o}{\partial \alpha} - 2\frac{\partial^2 w_o}{\partial \alpha \partial \beta}\right)^2$$

$$+ 2D_{12}\left(-\frac{\partial^2 w_o}{\partial \alpha^2}\right)\left(\frac{1}{R}\frac{\partial v_o}{\partial \beta} - \frac{\partial^2 w_o}{\partial \beta^2}\right)$$

$$+ 2D_{16}\left(-\frac{\partial^2 w_o}{\partial \alpha^2}\right)\left(\frac{1}{R}\frac{\partial v_o}{\partial \alpha} - 2\frac{\partial^2 w_o}{\partial \alpha \partial \beta}\right)$$

$$\left. +2D_{26}\left(\frac{1}{R}\frac{\partial v_o}{\partial \beta} - \frac{\partial^2 w_o}{\partial \beta^2}\right)\left(\frac{1}{R}\frac{\partial v_o}{\partial \alpha} - 2\frac{\partial^2 w_o}{\partial \alpha \partial \beta}\right)\right\} dA \tag{9.123b}$$

$$PE_{bs} = \int_A \left\{ -B_{11}\left(\frac{\partial u_o}{\partial \alpha}\right)\left(\frac{\partial^2 w_o}{\partial \alpha^2}\right) + B_{22}\left(\frac{\partial v_o}{\partial \beta} + \frac{w_o}{R}\right)\left(\frac{1}{R}\frac{\partial v_0}{\partial \beta} - \frac{\partial^2 w_o}{\partial \beta^2}\right) \right.$$

$$+ B_{66}\left(\frac{\partial v_o}{\partial \alpha} + \frac{\partial u_o}{\partial \beta}\right)\left(\frac{1}{R}\frac{\partial v_0}{\partial \alpha} - 2\frac{\partial^2 w_o}{\partial \alpha \partial \beta}\right)$$

$$+ B_{12}\left[\left(\frac{\partial u_o}{\partial \alpha}\right)\left(\frac{1}{R}\frac{\partial v_0}{\partial \beta} - \frac{\partial^2 w_o}{\partial \beta^2}\right) - \left(\frac{\partial v_o}{\partial \beta} + \frac{w_o}{R}\right)\left(\frac{\partial^2 w_o}{\partial \alpha^2}\right)\right]$$

$$+ B_{16}\left[\left(\frac{\partial u_o}{\partial \alpha}\right)\left(\frac{1}{R}\frac{\partial v_0}{\partial \alpha} - 2\frac{\partial^2 w_o}{\partial \alpha \partial \beta}\right) - \left(\frac{\partial v_o}{\partial \alpha} + \frac{\partial u_o}{\partial \beta}\right)\left(\frac{\partial^2 w_o}{\partial \alpha^2}\right)\right]$$

$$+ B_{26}\left\{\left[\left(\frac{\partial v_o}{\partial \beta} + \frac{w_o}{R}\right)\left(\frac{1}{R}\frac{\partial v_0}{\partial \alpha} - 2\frac{\partial^2 w_o}{\partial \alpha \partial \beta}\right)\right.\right.$$

$$\left.\left.\left. +\left(\frac{\partial v_o}{\partial \alpha} + \frac{\partial u_o}{\partial \beta}\right)\left(\frac{1}{R}\frac{\partial v_0}{\partial \beta} - \frac{\partial^2 w_o}{\partial \beta^2}\right)\right]\right]\right\} dA \tag{9.123c}$$

Note that for symmetrically laminated shells, $B_{ij} = 0$ and, hence, $U_{bs} = 0$. The kinetic energy of the entire cylindrical shell is expressed with the same equations used for plates (9.82).

Thin cylindrical shells can have up to 16 boundary conditions at each edge. Twelve of these are classical boundary conditions (i.e., free, simply supported, and clamped). This leads to numerous combinations of boundary conditions, particularly when the shells are open. Like shallow shells, only those open cylindrical shells that have two opposite edges simply supported can permit exact solutions (for cross-ply lamination). Interestingly, for closed shells, this constitutes all possible boundary conditions the shell may have. The problem with all edges being simply supported with shear diaphragm boundaries has a direct, relatively straightforward solution. Interestingly enough, closed shells with shear diaphragm boundaries at each of the opposing two edges permit the same exact solution. The reason is that conditions at the nodal lines of these shells simulate a shear diaphragm boundary. This can be mathematically proven with the obtained exact solution.

Consider a cylindrical shell that is made of a cross-ply laminate, thus $A_{16} = A_{26} = B_{16} = B_{26} = D_{16} = D_{26} = 0$. The differential parameters in the equations of motion L_{ij} become:

$$L_{11} = A_{11}\frac{\partial^2}{\partial\alpha^2} + A_{66}\frac{\partial^2}{\partial\beta^2}$$

$$L_{12} = L_{21} = \left\{(A_{12} + A_{66}) + \frac{(B_{12} + B_{66})}{R}\right\}\frac{\partial^2}{\partial\alpha\partial\beta}$$

$$L_{22} = \left\{A_{66} + 2\frac{B_{66}}{R} + \frac{D_{66}}{R^2}\right\}\frac{\partial^2}{\partial\alpha^2} + \left\{A_{22} + 2\frac{B_{22}}{R} + \frac{D_{22}}{R^2}\right\}\frac{\partial^2}{\partial\beta^2}$$

$$L_{13} = L_{31} = -B_{11}\frac{\partial^3}{\partial\alpha^3} - (B_{12} + 2B_{66})\frac{\partial^3}{\partial\alpha\partial\beta^2} + \frac{A_{12}}{R}\frac{\partial}{\partial\alpha}$$

$$\begin{aligned} L_{23} = L_{32} = &-\left\{B_{22} + \frac{D_{22}}{R}\right\}\frac{\partial^3}{\partial\beta^3} \\ &-\left\{(B_{12} + 2B_{66}) + \frac{(D_{12} + 2D_{66})}{R}\right\}\frac{\partial^3}{\partial\alpha^2\partial\beta} \\ &+\left\{\frac{A_{22}}{R} + \frac{B_{22}}{R^2}\right\}\frac{\partial}{\partial\beta} \end{aligned}$$

$$L_{33} = D_{11}\frac{\partial^4}{\partial\alpha^4} + 2(D_{12} + 2D_{66})\frac{\partial^4}{\partial\alpha^2\partial\beta^2} + D_{22}\frac{\partial^4}{\partial\beta^4}$$
$$-2\left\{\left(\frac{B_{12}}{R}\right)\frac{\partial^2}{\partial\alpha^2} + \left(\frac{B_{22}}{R}\right)\frac{\partial^2}{\partial\beta^2}\right\} + \frac{A_{22}}{R^2} \tag{9.124}$$

Consider an open cylindrical shell with shear diaphragm (S2) boundaries on all four edges. The solution in (9.85) with x replaced by α and y by β satisfies the boundary conditions and the equations of motion exactly for open cylindrical shells. These equations can be used in analyzing closed shells by substituting b with πR. With $\beta = R\theta$; the above solution can be written in terms of θ for closed cylindrical shells as

$$u_0(\alpha,\beta,t) = \sum_{m=0}^{M}\sum_{n=0}^{N} U_{mn}\cos(\alpha_m\alpha)\sin(n\theta)\sin(\omega_{mn}t)$$
$$v_0(\alpha,\beta,t) = \sum_{m=0}^{M}\sum_{n=0}^{N} V_{mn}\sin(\alpha_m\alpha)\cos(n\theta)\sin(\omega_{mn}t)$$
$$w_0(\alpha,\beta,t) = \sum_{m=0}^{M}\sum_{n=0}^{N} W_{mn}\sin(\alpha_m\alpha)\sin(n\theta)\sin(\omega_{mn}t)$$
$$\text{where } \alpha_m = \frac{m\pi}{a},\ \omega_{mn} \text{ is the natural frequency.} \tag{9.125}$$

Note also that for such shells, at the nodal points, where $\theta = 0$ and π/n (for each n), this solution yields the inplane displacement u_0 (which is tangential to the boundaries $\theta = 0$ and π/n, i.e., $\beta = 0, R\pi/n$) and transverse displacement w_0 to be zero. The other inplane displacement v_0 normal to the boundaries $\theta = 0$ and π/n (i.e., $\beta = 0$, $R\pi/n$) is not zero. These are the geometric constraints for shear diaphragm boundaries. The forces and moments at these nodal lines can be determined [1] showing that for higher modes, each segment of the shell containing double half sine waves can actually be treated as an independent shell with shear diaphragm boundary conditions vibrating at its fundamental frequency. As will be seen later, this is an important observation for treating closed cylindrical shells having arbitrary boundaries at $\alpha = 0$ and a.

Substitute the general solution of open shells into the equations of motion in terms of displacement yields (9.86) with the following coefficients:

$$C_{11} = -A_{11}\alpha_m^2 - A_{66}\beta_n^2$$
$$C_{12} = C_{21} = -\left\{(A_{12} + A_{66}) + \frac{(B_{12} + B_{66})}{R}\right\}\alpha_m\beta_n$$

$$C_{22} = -\left\{A_{66} + 2\frac{B_{66}}{R} + \frac{D_{66}}{R^2}\right\}\alpha_m^2 - \left\{A_{22} + 2\frac{B_{22}}{R} + \frac{D_{22}}{R^2}\right\}\beta_n^2$$

$$C_{13} = C_{31} = B_{11}\alpha_m^3 + (B_{12} + 2B_{66})\alpha_m\beta_n^2 + \left\{\frac{A_{12}}{R}\right\}\alpha_m$$

$$C_{23} = C_{32} = \left\{B_{22} + \frac{D_{22}}{R}\right\}\beta_n^3 + \left\{(B_{12} + 2B_{66}) + \frac{(D_{12} + 2D_{66})}{R}\right\}\alpha_m^2\beta_n + \left\{\frac{A_{22}}{R} + \frac{B_{22}}{R^2}\right\}\beta_n$$

$$C_{33} = -\left\{D_{11}\alpha_m^4 + 2(D_{12} + 2D_{66})\alpha_m^2\beta_n^2 + D_{22}\beta_n^4\right\} + 2\left\{\left(\frac{B_{12}}{R}\right)\alpha_m^2 + \left(\frac{B_{22}}{R}\right)\beta_n^2\right\} + \left\{\frac{A_{22}}{R^2}\right\} \tag{9.126}$$

Many comparisons are made in the literature to show that the above equations are indeed accurate for thin laminated shells [1]. More complicated equations that can be obtained using the assumptions of Flügge or Sanders in shell theory. Figure 7.12 shows the sine waves and possible mode shapes along the circumferential direction of the shell.

The nondimensional frequency parameter $\Omega = \omega R\sqrt{\rho/E_1}$ is used in all subsequent results. In the results obtained for laminated composite shells, the [0°, 90°] and [90°, 0°] lamination sequences are used in most of these results because they generally represent a shell with the maximum stretching–bending coupling effects (i.e., B_{ij} terms are the highest).

Table 9.22 presents results for graphite/epoxy closed and open [0°, 90°] laminated shells. The material properties for graphite/epoxy are $E_1 = 20.02 \times 10^6$ psi, $E_2 = 1.3\times10^6$ psi, $G_{12} = 1.03 \times 10^6$ psi, $\nu_{12} = 0.3$. These are presented in (9.90) in SI units. Other studies are in Table 9.23 which lists the natural frequencies for [90°, 0°], [0°, 90°]$_s$, and [90°, 0°]$_s$ graphite/epoxy closed cylindrical shells. The thickness ratio of (R/h = 100) is used. Results obtained previously for [0°, 90°] shells are not repeated here but can be used from Table 9.22 for comparisons.

The first important observation is that unsymmetric lamination does not always yield lower frequencies. For example, when $a/mR >$ 2.0, the unsymmetric lamination sequence [90°, 0°] yielded higher fundamental frequency than the symmetric lamination [0°, 90°]$_s$. This is generally not the case for plates.

Shells with the outer fibers in the circumferential direction of the shell ([90°, 0°] and [90°, 0°]$_s$) give higher frequencies when compared with shells having the outer fibers in the longitudinal

a/mR_β	*n*										
	0	**1**	**2**	**3**	**4**	**5**	**6**	**7**	**8**	**9**	**10**
Closed cylindrical shells											
8	0.08907	0.04903	0.02140	0.01482	0.02075	0.03235	0.04723	0.06495	0.08543	0.10865	0.13460
4	0.17815	0.11600	0.06212	0.03740	0.03013	0.03561	0.04846	0.06552	0.08578	0.10891	0.13483
2	0.35629	0.23975	0.14556	0.09650	0.07025	0.05915	0.06070	0.07187	0.08936	0.11121	0.13651
1	0.71259	0.44205	0.29163	0.20869	0.15935	0.12928	0.11295	0.10837	0.11412	0.12826	0.14882
0.5	0.73565	0.63822	0.49616	0.38973	0.31622	0.26575	0.23171	0.21047	0.20015	0.19960	0.20777
Open cylindrical shells (*b*/*R* = 2)											
8	0.08907	0.02984	0.01508	0.02864	0.05197	0.08227	0.11934	0.16314	0.21367	0.27093	0.33490
4	0.17815	0.08060	0.03544	0.03305	0.05293	0.08263	0.11958	0.16337	0.21390	0.27117	0.33515
2	0.35629	0.17846	0.09168	0.06095	0.06305	0.08650	0.12156	0.16474	0.21507	0.27227	0.33623
1	0.71259	0.34488	0.20014	0.13638	0.11052	0.11270	0.13631	0.17420	0.22209	0.27813	0.34152
0.5	0.73565	0.55382	0.37757	0.27836	0.22450	0.20102	0.20207	0.22342	0.26061	0.31001	0.36929

TABLE 9.22 Frequency Parameters for [0°, 90°] Graphite/Epoxy Closed Cylindrical Shells, $R/h = 100$

a/mR	n										
	0	1	2	3	4	5	6	7	8	9	10
[90°, 0°]											
8	0.08907	0.04925	0.02182	0.01553	0.02143	0.03298	0.04789	0.06570	0.08632	0.10971	0.13587
4	0.17815	0.11653	0.06276	0.03836	0.03147	0.03699	0.04975	0.06678	0.08707	0.11030	0.13637
2	0.35629	0.24079	0.14668	0.09770	0.07184	0.06126	0.06312	0.07433	0.09177	0.11359	0.13892
1	0.71259	0.44339	0.29322	0.21025	0.16105	0.13136	0.11560	0.11160	0.11777	0.13216	0.15284
0.5	0.73090	0.63573	0.49559	0.38999	0.31694	0.26689	0.23335	0.21273	0.20311	0.20327	0.21208
$[0°, 90°]_s$											
8	0.08907	0.04914	0.02155	0.01449	0.01923	0.02949	0.04285	0.05883	0.07732	0.09829	0.12174
4	0.17815	0.11627	0.06244	0.03765	0.02961	0.03354	0.04462	0.05980	0.07798	0.09882	0.12220
2	0.35629	0.24031	0.14621	0.09717	0.07079	0.05892	0.05880	0.06778	0.08289	0.10222	0.12483
1	0.71259	0.44316	0.29322	0.21064	0.16165	0.13173	0.11501	0.10915	0.11271	0.12405	0.14149
0.5	0.73919	0.64338	0.50396	0.40026	0.32947	0.28158	0.24969	0.22981	0.21966	0.21786	0.22343
$[90°, 0°]_s$											
8	0.08907	0.04914	0.02251	0.02341	0.04017	0.06422	0.09403	0.12935	0.17014	0.21638	0.26807
4	0.17815	0.11627	0.06276	0.04187	0.04602	0.06615	0.09483	0.12978	0.17043	0.21661	0.26827
2	0.35629	0.24027	0.14626	0.09875	0.07888	0.08180	0.10211	0.13351	0.17262	0.21809	0.26940
1	0.71259	0.44270	0.29247	0.21022	0.16385	0.14166	0.14031	0.15694	0.18738	0.22800	0.27656
0.5	0.73296	0.63664	0.49550	0.38978	0.31790	0.27151	0.24589	0.23906	0.24950	0.27500	0.31295

TABLE 9.23 Lamination Effects on the Frequency Parameter for Graphite/Epoxy [0°, 90°] Closed Cylindrical Shells

direction (i.e., [0°, 90°] and $[0°, 90°]_s$). The effect of lamination sequence is observed to be less for higher frequencies and/or lower length ratio (a/mR). Also, lamination sequence effects are insignificant for frequencies with $n < 2$.

The orthotropy ratio has a considerable effect on all the frequencies [1]. Its effects on the fundamental frequency parameters are found to be higher for short shells ($a/mR = 0.5$) than longer ones. This is reversed when $n = 0$, where the orthotropy effects are found to be higher for longer shells ($a/mR = 8$) than shorter ones. The fundamental frequencies are observed to occur at a lower n value for lower orthotropy ratios. Similar studies of the effect of the orthotropy ratio on [90°, 0°] laminates yielded similar results.

9.12 Vibrations of Other Laminated Shells

As was discussed in Chap. 7, shells can have several geometrical shapes or curvature. Among the mostly used ones are cylindrical (which may have circular or noncircular cross-sections), spherical, and conical shells. All of these (except the cylindrical shell with a noncircular cross-section) can actually be characterized as shells of revolution.

Laminated shells offer the same of the differential equations as other isotropic shells, but are far more complex. Reference [1] offers the list of the set of equations (strain–displacement, stress resultant, equilibrium, and boundary conditions) for many of the above shells including cylindrical, barrel, spherical, and conical shells. It offers these equations for *both* thin and thick shells where shear deformation and rotary inertia is included.

Interestingly, exact solutions for cross-ply shells with shear diaphragm boundaries at the ends are possible. These exact solutions were used by many researchers for both thin shells [22] and thick shells [23].

It should be noted here that Eqs. (9.13) have been used in the literature for shells. This is proven inaccurate for laminated, deep thick shells. Alternative accurate equations are first developed by Qatu [24] and are available in Refs. [1,25]. These equations should be used for thick laminated shells.

Table 9.24 shows comparisons of the natural frequency parameters obtained using the accurate stress resultants as shown in Refs. [1,24,25] and those of plates (9.13) used for cylindrical open shells having shear diaphragm boundaries. Ye and Soldatos [26] used the 3D theory of elasticity and obtained exact solutions for cylindrical shells without making the assumption of shallowness made by Bhimaraddi [27]. First-order shear deformation shell theories (SDST) were used by Librescu et al. [28] and Bhimaraddi [27], who also used higher order shear deformation shell theories (HSDT).

Despite their added complexity, the results shown indicate that HSDST theories do not always yield better results than first-order

	Qatu [25]	Ye and Soldatos [26]	Bhimaraddi [27]			Librescu [28]	
R/a	SDST	3D	3D	SDST	HSDST	SDST	HSDST
1	10.643	10.6973	10.409	10.748	10.919		
2	9.4428	9.4951	9.3627	9.3653	9.5664		
3	9.1755	9.1155	9.1442	9.0563	9.2642		
4	9.0731		9.0613	8.9403	9.1506		
5	9.0221	9.0616	9.0200	8.8840	9.0953	8.931	8.959
10	8.9446	8.9778	8.9564	8.8026	9.0150	8.897	8.933
20	8.9194	8.9477	8.9341	8.7779	8.9904	8.894	8.934
∞	8.9001	8.9248	8.9179	8.7640	8.9761	8.900	8.944

TABLE 9.24 Nondimensional Frequency Parameters $\Omega = \omega a^2\sqrt{\rho/E_2h^2}$ for [0°, 90°] Cylindrical Shells ($E_1/E_2 = 25$, $G_{12}/E_2 = 0.5$, $G_{13}/E_2 = 0.5$, $G_{23}/E_2 = 0.2$, $\nu_{12} = 0.3$, $k^2 = 5/6$, $a/b = 1$, $a/h = 10$)

SDST theories (when compared with the 3D theory of elasticity). In all SDST and HSDST theories of both Librescu et al. and Bhimaraddi, the results deviate from the 3D theory of elasticity as the shell becomes deeper. This indicates that the error in these theories may very well be due to the fact that they used (9.13) in the stiffness coefficients which are not accurate for deep thick shells. When accurate stiffness parameters were used [25] closer approximation to the 3D results when compared with the SDST theories of Librescu et al. [28] and Bhimaraddi [27].

Table 9.25 lists the natural frequencies for [0°, 90°] graphite/epoxy closed shell. The results show clearly that gross error can occur when using CST especially for thicker and shorter shells. One important observation is that thin shell theory predicted a fundamental frequency at $n = 4$ for the short shells with a thickness ratio of 20. Thick shell theory predicted the fundamental frequency to be associated with $n = 5$ for such shells. For shorter shells, the fundamental frequency tends to occur at higher n values. It is worth noting that the fundamental mode occurs at $n = 2$ for longer shells, at $n = 3$ for intermediate shells, and at $n = 3$ or 4 for shorter shells.

In addition, Ref. [1] offers equations for energy functionals for various shell geometries (barrel, spherical, cylindrical, and conical) that can be readily used in a numerical approach like the Ritz or finite element methods.

Several survey articles exist in the literature that focus on recent developments on shells. Shallow shells were reviewed in Ref. [29] and laminated shells were reviewed in Ref. [30] and most recently in Ref. [31].

	n								
a/mR	**0**	**1**	**2**	**3**	**4**	**5**	**6**	**7**	**8**
Classical shell theory (CST)									
8	0.08911	0.04810	0.04115	0.10098	0.19178	0.30881	0.45132	0.61884	0.81087
4	0.17822	0.11386	0.07120	0.10845	0.19525	0.31151	0.45383	0.62128	0.81324
2	0.35644	0.23644	0.15148	0.14888	0.21639	0.32633	0.46632	0.63268	0.82399
1	0.71288	0.44995	0.31809	0.28021	0.31421	0.40272	0.53077	0.69013	0.87686
0.5	0.92024	0.82358	0.72155	0.68285	0.69753	0.75633	0.85381	0.98615	1.15019
Shear deformation shell theory (SDST)									
8	0.08907	0.04787	0.04057	0.09878	0.18492	0.29238	0.41839	0.56051	0.71644
4	0.17815	0.11364	0.07039	0.10592	0.18807	0.29474	0.42053	0.56254	0.71838
2	0.35629	0.23605	0.15033	0.14556	0.20804	0.30813	0.43137	0.57211	0.72715
1	0.71258	0.44696	0.31433	0.27329	0.30084	0.37807	0.48809	0.62072	0.77029
0.5	0.87845	0.78699	0.68362	0.63829	0.64250	0.68613	0.76206	0.86455	0.98873

TABLE 9.25 Comparisons among Shell Theories for [0°, 90°] Closed Cylindrical *G/E* Shells, *R/h* = 10

References

1. M. S. Qatu, *Vibrations of Laminated Shells and Plates*, Elsevier, 2004, 409 pp.
2. J. R. Vinson, and R. L. Sierakowski, *The Behavior of Structures Composed of Composite Materials*, 2nd ed., Kluwer Academic Publishers, 2002.
3. A. K. Kaw, *Mechanics of Composite Materials*, 2nd ed., CRC Press, 2005.
4. M. S. Qatu, "Theories and analyses of thin and moderately thick laminated composite curved beams," *Int. J. Solids Struct.*, 30 (20)(1993): 2743–2756.
5. H. Zinberg and M. F. Symonds. "The development of an advanced composite tail rotor driveshaft," Presented at the 26th annual forum of the American Helicopter Society, Washington, DC, 1970.
6. C. D. Kim and C. W. Bert, "Critical speed analysis of laminated composite hollow drive shafts," *Composite Eng.*, 3 (1993): 633–43.
7. C. W. Bert, and C. D. Kim, "Whirling of composite-material driveshafts including bending–twisting coupling and transverse shear deformation," *J. Vib. Acoust.*, 117 (1995): 17–21.
8. M. Y. Chang, J. K. Chen, and C. Y. Chang, "A simple spinning laminated composite shaft model," *Int. J. Solids Struct.*, 41 (2004): 637–62.
9. J. Iqbal and M. S. Qatu, "Transverse vibration of a two-segment cross-ply composite shafts with a lumped mass," *Composite Struct.*, 92 (2010): 1126–31.
10. J. M. Whitney, *Structural Analysis of Laminated Anisotropic Plates*, Technomic Publishing, Inc., 1987.
11. J. M. Whitney and A. W. Leissa, "Analysis of heterogeneous anisotropic plates," *J. Appl. Mech.*, 36 (1969): 261–66.
12. M. S. Qatu, "Free vibration of laminated composite rectangular plates," *Int. J. Solids Struct.*, 28 (8)(Aug. 1991): 941–54.
13. M. S. Qatu, "Natural frequencies for cantilevered laminated composite right triangular and trapezoidal plates," *Composite Sci. Tech.*, 51 (1994): 441–49.
14. M. S. Qatu, "Vibrations of laminated composite completely free triangular and trapezoidal plates," *Int. J. Mech. Sci.*, 36 (9)(1994): 797–809.
15. M. S. Qatu, and A. W. Leissa, "Free vibrations of completely free doubly-curved laminated composite shallow shells," *J. Sound. Vib.*, 151 (1991): 9–29.
16. M. S. Qatu, and A. W. Leissa. "Natural frequencies for cantilevered doubly-curved laminated composite shallow shells," *Composite Struct.*, 17 (1991): 227–56.
17. M. S. Qatu, "Mode shape analysis of laminated composite shallow shells," *J Acoust. Soc. Amer.*, 92 (1992): 1509–20.
18. M. S. Qatu, and A. W. Leissa, "Vibration studies for laminated composite twisted cantilever plates," *Int. J. Mech. Sci.*, 33 (1991): 927–40.
19. M. S. Qatu, "Vibrations of doubly-cantilevered laminated composite thin shallow shells," *Thin-Walled Struct.*, Vol. 15, pp. 235-248, 1993.
20. M. S. Qatu, "Vibration of Cantilevered Composite Triangular and Trapezoidal Doubly-Curved Shallow Shells," *Acta Mechanica.*, 108 (1995): 63–75, 1995.
21. M. S. Qatu, "Natural Vibration of Free Laminated Composite Triangular and Trapezoidal Shallow Shells," *Composite Struct.* 31 (1)(1995): 9–19.
22. M. S. Qatu, "Theory and vibration analysis of laminated barrel thin shells," *J. Vib. Control.*, 5 (1999): 851–89.
23. M. S. Qatu, "Theory and vibration analysis of laminated barrel thick shells," *J. Vib. Control*, 10 (2004): 319–41.
24. M. S. Qatu, "Accurate Stress Resultant Equations for Laminated Composite Deep, Thick Shells," *Composites for the Pressure Vessel Industry*, ASME-PVP, 302 (1995): 39–46.
25. M. S. Qatu, "Accurate Theory for Laminated Composite Deep Thick Shells," *Int. J. Solids Struct.*, 36 (19)(1999): 2917–41.
26. J. Q. Ye and K. P. Soldatos, "Three-dimensional vibrations of laminated cylinders and cylindrical panels with a symmetric or an antisymmetric cross-ply lay-up," *Composite Eng.*, 4 (1994): 429–44.

27. A. Bhimaraddi, "Free vibration analysis of doubly curved shallow shells on rectangular planform using 3D elasticity theory," *Int. J. Solids Struct.*, 27 (7) (1991): 897–913.
28. L. Librescu, A. A. Khdeir, and D. Frederick, "A shear-deformable theory for laminated composite shallow shell-type panels and their response analysis i: free vibration and buckling," *Acta Mechanica.*, 76 (1989): 1–33.
29. M. S. Qatu, "Review of shallow shell vibration research," *Shock Vib. Dig.*, 24 (1992): 3-15.
30. M. S. Qatu, "Recent research advances in the dynamic behavior of shells: 1989-2000, Part 2: homogeneous shells," *Appl. Mech. Rev.*, 55 (5)(2002): 415–35.
31. M. S. Qatu, R. W. Sullivan, and W. Wang, "Recent research advances in the dynamic behavior of composite shells: 2000-2009," *Composite Struct.*, 93 (2010): 14–31.

Problems

1 Use (9.4) for the transformation matrix and perform the matrix multiplication in (9.7) to prove the coefficients in (9.8).

2 Consider a laminate made of graphite epoxy with material properties given in (9.90) with a total thickness h. Determine all laminate parameters A_{ij}, B_{ij}, and D_{ij} for

A. A single layer laminate of [0]°,

B. A cross-ply laminate of [0, 90, 0, 0]°

C. An angle-ply laminate of [30, 45, −45, −30]°

D. A symmetrical laminate [30, 45, 45, 30]°

E. An asymmetric laminate [30, 30, 60, 60]°

3 For the previous problem, comment on the type of couplings that exist in each of the laminates that makes it different from those of isotropic materials. Compare your findings with those of Example 9.1.

4 Consider (9.22). Develop a closed form solution for the frequency parameters for laminated thin beams.

5 Consider a simply supported laminated straight beam with length ℓ and width b ($= \ell/20$). Consider two thickness ratios ($h = b$ and $h = 2b$) and a cross-ply [0, 90] and angle ply [30, 60] laminate. For each of the four possible combinations of lamination and thickness ratio, find the following:

A. the natural frequency parameters using a 3D finite element model with any of the available commercial packages

B. the frequency parameters using expression (9.13) and (9.27) for the stiffness parameters using thin beam theory

C. the frequency parameters using expression (9.13) and (9.27) for the stiffness parameters using shear deformation beam theory

D. Comment on the accuracy of the frequency parameters obtained by using expressions (9.13) and (9.27) for each of the four cases.

6 A. Prove the relations presented in (9.58) for curved thick beams. Use (9.54) in your proof.

B. Expand (9.54) in a Taylor series and develop relations similar to (9.58) by considering the first two terms. Compare the results obtained in (9.58) with those obtained using the new formula.

7 Consider a laminated thin curved beam with simple support boundaries. The exact solution is presented in (9.52). Formulate a closed form solution for such beams using this formula.

8 Exact solutions are possible for laminated curved beams. For Arbitrary boundary conditions, the displacements in the equations of motion can be assumed as

$$u_0(x,t) = Ce^{s\alpha} \sin \omega t, \quad w_0(x,t) = De^{s\alpha} \sin \omega t$$

Substitute the above equations into the equations of motion.

A. Determine the determinant that can yield the natural frequencies.

B. Discuss the possible roots for the characteristic equations. Are they real? Are they positive?

C. Generate a possible solution based on these roots.

9 A. Expand the denominator of (9.54) using a geometric series (or a Taylor series expansion). Keep two terms in the expansion series and neglect the rest of the terms. Substitute the stress–strain relations into (9.55) and develop approximate expressions for the stiffness parameters.

B. Compare the results with (9.58) and (9.17).

C. Substitute the parameters found in Table 9.8 and obtain results for h/R of 0.05, 0.1, and 0.2. Compare these with those found in Table 9.8. Are the terms developed using Taylor series accurate?

10 Consider a symmetric cross-ply laminated thin plate with a rectangular planform of dimensions a and b having all edges with shear diaphragm boundary conditions.

A. Develop a closed-form expression for the natural frequency parameters for each of the possible half sine combinations (m and n).

B. Find the frequency parameters for such plates for aspect ratios of 1 and 2.

11 A rectangular plate with dimensions a and b is made of antisymmetric angle ply ($A_{16} = A_{26} = B_{11} = B_{12} = B_{22} = B_{66} = D_{16} = D_{26} = 0$) composites and has S3 type of boundary condition:

$$w = u = N_{xy} = M_x = 0 \quad \text{for the edges} \quad x = 0, a$$

$$w = v = N_{xy} = M_y = 0 \quad \text{for the edges} \quad y = 0, b$$

(Note the difference between the S3 boundaries and the shear diaphragm boundaries. The shear diaphragm boundaries allow motion in the inplane direction that is perpendicular to the boundary and prevents it in the inplane direction tangential to the boundary. The S3 has opposite treatment of the inplane displacement supports.)

A. Rewrite the equations of motion in terms of displacement for these composites.

B. Search for an exact solution (similar to (6.85)) and show that it satisfies the equations of motion and boundary conditions.

C. Develop the exact solutions similar to (6.86).

D. Find nondimensional frequency parameters for G/E materials for $a/h = 100$, 20, and 10.

12 Repeat the previous problem for a shallow cylindrical shell $R = 2a$. Find if an exact solution is possible. If so, compare the results obtained with those found in Problem 7.

13 Derive relations in (9.126) for cross-ply laminated cylindrical shells having shear diaphragm boundary conditions.

APPENDIX A

Summary of One Degree-of-Freedom Vibrations (with Viscous Damping)

Equation of Motion

$$m\ddot{x} + c\dot{x} + kx = F_0 \sin \Omega t$$

Free Undamped Vibration

$$F_0 = c = 0$$

$$x = A \sin \omega t + B \cos \omega t, \quad \omega = \sqrt{\frac{k}{m}}$$

Free Damped Vibrations

$$F_0 = 0$$

Overdamped: $\zeta > 1$, $x = e^{-\zeta \omega t}\left(Ae^{\sqrt{(\zeta^2 - 1)}\,\omega t} + Be^{-\sqrt{(\zeta^2 - 1)}\,\omega t}\right)$

Critically damped: $\zeta = 1, \quad x = e^{-\omega t}(A + Bt)$

Underdamped: $\zeta < 1, \quad x = e^{-\zeta\omega t}\left[A\sin\sqrt{1-\zeta^2}\,\omega t + B\cos\sqrt{1-\zeta^2}\,\omega t\right]$

$$\zeta = \frac{c}{c_c}, \quad c_c = 2\sqrt{km} = 2m\omega$$

Forced Damped Vibration (Steady-State Part)

$$x = A\sin\Omega t - B\cos\Omega t = C\sin(\Omega t - \phi), \quad \tan\phi = \frac{B}{A}$$

$$\frac{A}{\delta_{st}} = \frac{1-(\Omega/\omega)^2}{\Delta}, \quad \frac{B}{\delta_{st}} = \frac{2\zeta(\Omega/\omega)}{\Delta}$$

$$\frac{C}{\delta_{st}} = \frac{1}{\sqrt{\Delta}}, \quad \delta_{st} = \frac{F_0}{k}, \quad \Delta = \left[1-\left(\frac{\Omega}{\omega}\right)^2\right]^2 + \left[2\zeta\left(\frac{\Omega}{\omega}\right)\right]^2$$

Excitation by Nonsinusoidal, But Periodic, Forcing Function

$$F(t) = \sum_{n=1}^{\infty} F_n \sin n\Omega t, \quad F_n = \frac{z}{\tau}\int_0^{\tau} F(t)\sin n\Omega t\,dt$$

$$\tau = \frac{2\pi}{\Omega} = \text{period of forcing function}$$

$$x = \sum_n (A_n \sin n\Omega t - B_n \cos n\Omega t) = \sum_n C_n \sin(n\Omega t - \phi_n), \quad \tan\phi_n = \frac{B_n}{A_n}$$

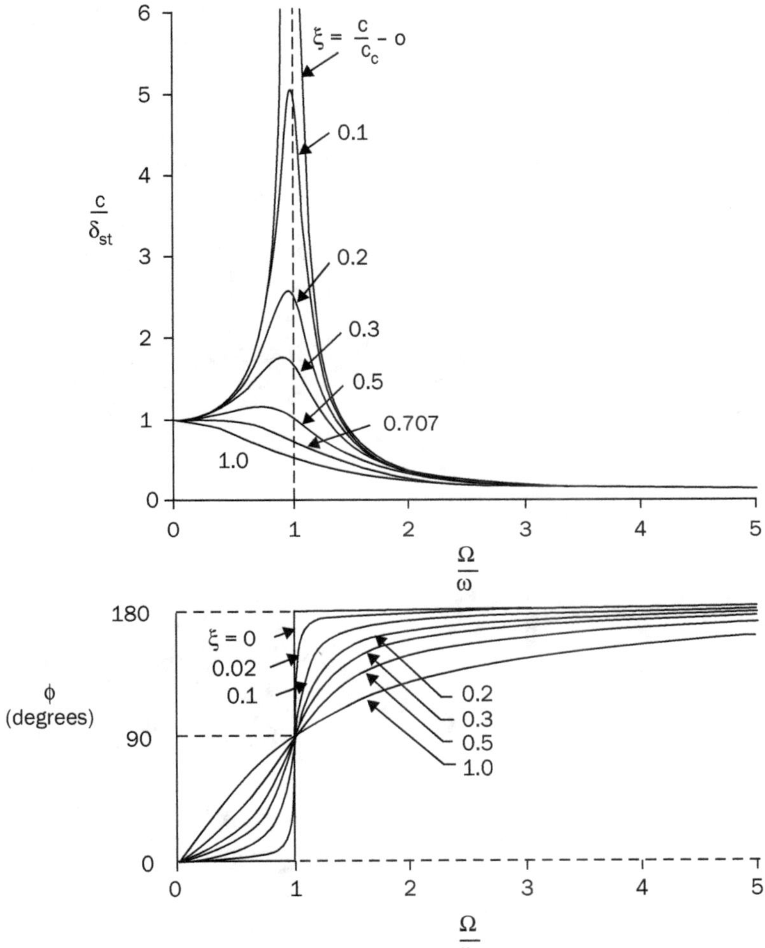

FIGURE A.1 Amplitude ratio and phase angle for the motion of single degree-of-freedom, forced vibration (viscous damping).

APPENDIX B

Bessel Functions: Some Useful Information

Differential Equations and Their Solutions

$$z^2 \frac{d^2y}{dz^2} + z\frac{dy}{dz} + \left(k^2z^2 - \nu^2\right)y = 0 \quad \begin{cases} y = C_1 J_\nu(kz) + C_2 Y_\nu(kz), \text{ always} \\ \\ y = C_1 J_\nu(kz) + C_2 J_{-\nu}(kz), \ \nu \text{ not an integer} \end{cases}$$

$$z^2 \frac{d^2y}{dz^2} + z\frac{dy}{dz} + \left(-k^2z^2 - \nu^2\right)y = 0 \quad \begin{cases} y = C_1 I_\nu(kz) + C_2 K_\nu(kz), \text{ always} \\ \\ y = C_1 I_\nu(kz) + C_2 I_{-\nu}(kz), \ \nu \text{ not an integer} \end{cases}$$

Series Expansions

$$J_0(z) = 1 - \frac{z^2}{2^2} + \frac{z^4}{2^2 4^2} - \frac{z^6}{2^2 4^2 6^2} + \cdots = \sum_{i=0}^{\infty} (-1)^i \frac{(z/2)^{2i}}{(i!)^2}$$

$$J_n(z) = \left(\frac{z}{2}\right)^n \left[\frac{1}{n!} - \frac{(z/2)^2}{1!(n+1)!} + \frac{(z/2)^4}{2!(n+2)!} + \cdots\right]$$

$$= \left(\frac{z}{2}\right)^n \sum_{i=0}^{\infty} (-1)^i \frac{(z/2)^{2i}}{i!(n+i)!} \qquad \text{(integer } n\text{)}$$

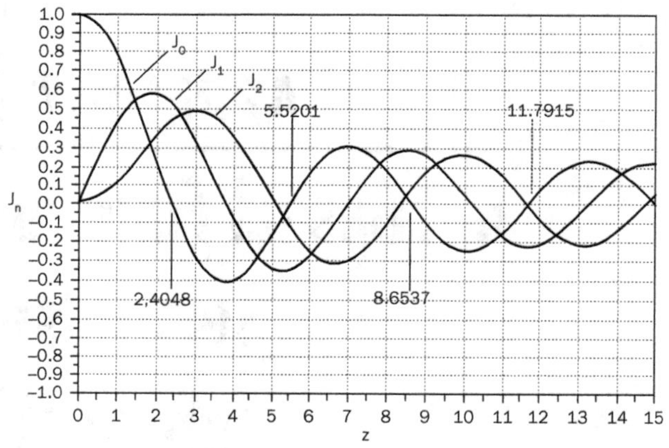

FIGURE B.1 $J_n(z)$ for $n = 0, 1, 2$ and zeros of $J_0(z)$.

$$J_\nu(z)=\frac{(z/2)^\nu}{\Gamma(\nu+1)}\left[1-\frac{(z/2)^2}{(\nu+1)}+\frac{(z/2)^4}{2!(\nu+1)(\nu+2)}-\frac{(z/2)^6}{3!(\nu+1)(\nu+2)(\nu+3)}\cdots\right]$$

$$=\left(\frac{z}{2}\right)^\nu\sum_{i=0}^{\infty}(-1)^i\frac{(z/2)^i}{i!\Gamma(\nu+i+1)} \qquad \text{(arbitrary } \nu\text{)}$$

where Γ is the gamma function (cf. [1], Chap. 6, and [2], Appendix I)

$$Y_0(z)=\frac{2}{\pi}\left[\gamma+\ln\left(\frac{z}{2}\right)\right]J_o(z)-\frac{2}{\pi}\sum_{i=0}^{\infty}\frac{(-1)^i(z/2)^{2i}}{(i!)^2}\left[1+\frac{1}{2}+\frac{1}{3}+\cdots+\frac{1}{i}\right]$$

$$\gamma=0.57721\ 56649\ldots \quad \text{(Euler's constant)}$$

$$I_0(z)=1+\frac{z^2}{2^2}+\frac{z^4}{2^2\,4^2}+\frac{z^6}{2^2\,4^2\,6^2}+\cdots=\sum_{i=0}^{\infty}\frac{(z/2)^{2i}}{(i!)^2}$$

$$I_\nu(z)=\left(\frac{z}{2}\right)^\nu\sum_{i=0}^{\infty}\frac{(z/2)^i}{i!\Gamma(\nu+i+1)},\quad I_n(z)=i^{-n}J_n(iz)$$

$$J_n(-z)=(-1)^n J_n(z)\,,\ I_n(-z)=(-1)^n I_n(z)$$

For the complicated series expansions of $Y_\nu(z)$ and $K_\nu(z)$, see Refs. [1] and [2].

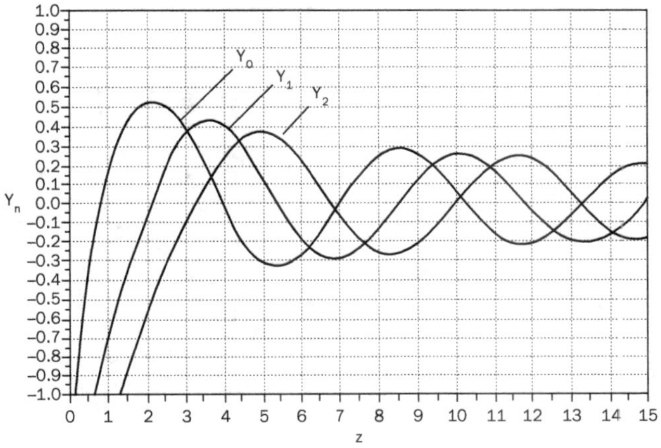

FIGURE B.2 $Y_n(z)$ for n = 0, 1, 2.

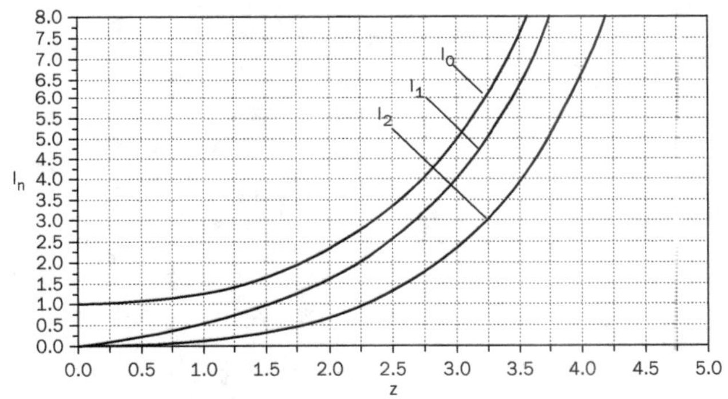

FIGURE B.3 $I_n(z)$ for n = 0, 1, 2.

Recursion Formulas for Generating Higher Order Functions

$$J_{\nu+1}(x) = \frac{2\nu}{x} J_\nu(x) - J_{\nu-1}(x) \qquad \text{(same form for } Y_\nu\text{)}$$

$$I_{\nu+1}(x) = -\frac{2\nu}{x} I_\nu(x) + I_{\nu-1}(x)$$

$$K_{\nu+1}(x) = \frac{2\nu}{x} K_\nu(x) + K_{\nu-1}(x)$$

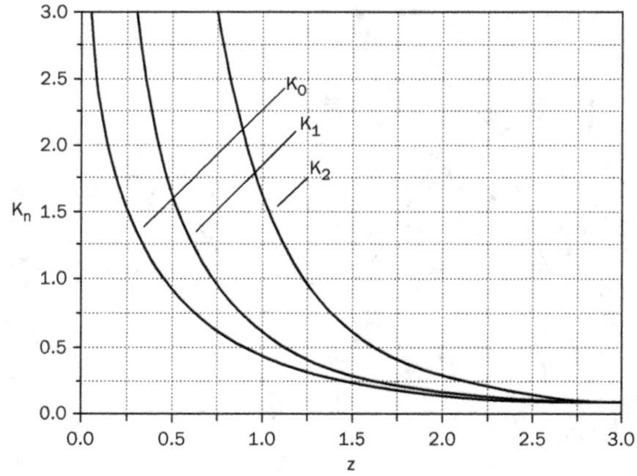

FIGURE B.4 $K_n(z)$ for n = 0, 1, 2.

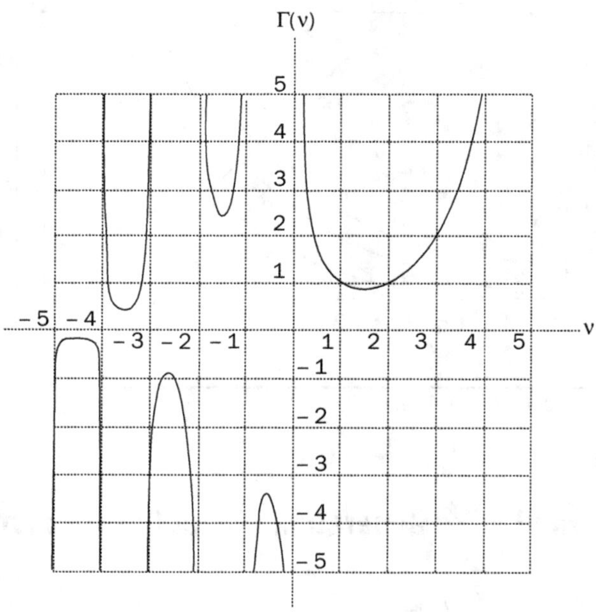

FIGURE B.5 Gamma function.

Differentiation Formulas (see [4])

$$J_v'(z) = \frac{v}{z}J_v(z) - J_{v+1}(z) = -\frac{v}{z}J_v(z) + J_{v-1}(z) \qquad \text{same form for } Y_v'$$

$$I_v'(z) = \frac{v}{z}I_v(z) + I_{v+1}(z) = -\frac{v}{z}I_v(z) + I_{v-1}(z)$$

$$K_v'(z) = \frac{v}{z}K_v(z) - K_{v+1}(z) = -\frac{v}{z}K_v(z) - K_{v-1}(z)$$

$$J_v'(kz) = \frac{dJ_v(kz)}{d(kz)}, \quad \frac{d}{dz}J_v(kz) = k\,J_v'(kz)$$

References and Bibliography

1. M. Abramowitz and I. A Stegun, *Handbook of Mathematical Functions*, U.S. Government Printing Office, 1964.
2. N. W. McLachlan, *Bessel Functions for Engineers*, 2nd ed., Oxford University Press, 1955.
3. W. G. Bickley, *Bessel Functions and Formulae*, Cambridge University Press, 1953.
4. A. Gray, G. B. Mathews, and T. M. MacRobert, *A Treatise on the Theory of Bessel Functions*, 2nd ed., Macmillan and Co., 1931.
5. G. N. Watson, *A Treatise on the Theory of Bessel Functions*, Cambridge University Press, 1958.
6. H. W. Dwight, *Tables of Integrals and Other Mathematical Data*, The MacMillan Co., 1957.
7. E. Jahnke, F. Emde, and F. Lösch, *Tables of Higher Functions*, 6th ed., McGraw-Hill Book Co., 1960.

APPENDIX C

Hyperbolic Functions: Some Useful Relations

$$\cosh x = \frac{e^x + e^{-x}}{2}, \quad \sinh x = \frac{e^x - e^{-x}}{2}, \quad \tanh x = \frac{\sinh x}{\cosh x}$$

$$\text{sech}\, x = \frac{1}{\cosh x}, \quad \text{csch}\, x = \frac{1}{\sinh x}, \quad \text{ctnh}\, x = \frac{1}{\tanh x}$$

$$\cosh^2 x - \sinh^2 x = 1$$

$$\sinh(x \pm y) = \sinh x \bullet \cosh y \pm \cosh x \bullet \sinh y$$

$$\cosh(x \pm y) = \cosh x \bullet \cosh y \pm \sinh x \bullet \sinh y$$

$$\tanh(x \pm y) = \frac{\tanh x \pm \tanh y}{1 \pm \tanh x \tanh y}$$

$$\sinh x = x + \frac{x^3}{3!} + \frac{x^5}{5!} + \frac{x^7}{7!} + \cdots \qquad (\text{all } x)$$

$$\cosh x = 1 + \frac{x^2}{2!} + \frac{x^4}{4!} + \frac{x^6}{6!} + \cdots \qquad (\text{all } x)$$

$$\tanh x = x - \frac{x^3}{3} + \frac{2}{15}x^5 - \frac{17}{315}x^7 - \frac{62}{2835}x^9 - \cdots \quad (x^2 < \pi^2/4)$$

$$\sinh(0) = 0, \quad \cosh(0) = 1, \quad \tanh(0) = 0$$

$$\sinh(\infty) = \infty, \quad \cosh(\infty) = \infty, \quad \tanh(\infty) = 1$$

$$\frac{d}{dx}(\sinh \alpha x) = \alpha \cosh \alpha x, \quad \frac{d}{dx}(\cosh \alpha x) = \alpha \sinh \alpha x$$

$$\frac{d}{dx}(\tanh \alpha x) = \alpha \operatorname{sech}^2 \alpha x$$

Index

Note: Page numbers followed by "f" indicate material in figures; by "t", in tables; by "p", in problems.

A

B

C

G

H

N

O

P

Q

R

T

V

W

Y

Z